# An Introduction to
# Gravity Currents
# and Intrusions

T0203649

# An Introduction to
# Gravity Currents
# and Intrusions

## Marius Ungarish

Department of Computer Science
Technion – Israel Institute of Technology
Haifa, Israel

**CRC Press**
Taylor & Francis Group
Boca Raton London New York

CRC Press is an imprint of the
Taylor & Francis Group, an **informa** business

A CHAPMAN & HALL BOOK

CRC Press
Taylor & Francis Group
6000 Broken Sound Parkway NW, Suite 300
Boca Raton, FL 33487-2742

First issued in paperback 2019

© 2009 by Taylor & Francis Group, LLC
CRC Press is an imprint of Taylor & Francis Group, an Informa business

No claim to original U.S. Government works

ISBN-13: 978-1-58488-903-8 (hbk)
ISBN-13: 978-0-367-38568-2 (pbk)

This book contains information obtained from authentic and highly regarded sources. Reasonable efforts have been made to publish reliable data and information, but the author and publisher cannot assume responsibility for the validity of all materials or the consequences of their use. The authors and publishers have attempted to trace the copyright holders of all material reproduced in this publication and apologize to copyright holders if permission to publish in this form has not been obtained. If any copyright material has not been acknowledged please write and let us know so we may rectify in any future reprint.

Except as permitted under U.S. Copyright Law, no part of this book may be reprinted, reproduced, transmitted, or utilized in any form by any electronic, mechanical, or other means, now known or hereafter invented, including photocopying, microfilming, and recording, or in any information storage or retrieval system, without written permission from the publishers.

For permission to photocopy or use material electronically from this work, please access www.copyright.com (http://www.copyright.com/) or contact the Copyright Clearance Center, Inc. (CCC), 222 Rosewood Drive, Danvers, MA 01923, 978-750-8400. CCC is a not-for-profit organization that provides licenses and registration for a variety of users. For organizations that have been granted a photocopy license by the CCC, a separate system of payment has been arranged.

**Trademark Notice:** Product or corporate names may be trademarks or registered trademarks, and are used only for identification and explanation without intent to infringe.

### Library of Congress Cataloging-in-Publication Data

Ungarish, M. (Marius), 1951-
    An introduction to gravity currents and intrusions / Marius Ungarish.
      p. cm.
    Includes bibliographical references and index.
    ISBN 978-1-58488-903-8 (hard back : alk. paper)
    1. Gravity. 2. Gravitational fields. 3. Density currents. 4. Fluid dynamics--Mathematical models. I. Title.

QB334.U54 2009
532'.05--dc22                                       2009009355

**Visit the Taylor & Francis Web site at**
**http://www.taylorandfrancis.com**

**and the CRC Press Web site at**
**http://www.crcpress.com**

To the memory of Moshe Israeli, mentor and friend.

# Contents

# Preface

The phenomena of gravity currents and intrusions are at the core of a wide and active area of academic research and engineering applications. The time is ripe for a new book which, I hope, will increase the chance that future generations of researchers and engineers will share a more up-to-date systematic theoretical background. In private conversations, seminars, conferences, and as a referee, it is necessary to explain again and again to young authors that Benjamin's speed formula is one equation for two unknowns, to clarify again and again what the difference is between a box-model approach and a solution of the equations of motion, and to emphasize again and again that the behavior of a current released from a reservoir of the same depth as that of the ambient fluid into which it propagates is not representative of the currents released from a submerged shallow reservoir. The need for a book which clarifies these and related fundamental matters is evident.

The double problem is that the topic considered in this book is wide and has received considerable attention. This is a real obstacle for the young researcher who is eager to leap to the frontier. A systematic and critical reading of all the available literature is impossible (the long lists of references which are now fashionable in most papers are symptomatic). This is also a problem for a prospective author of an introductory text. It is impossible, and rather useless, to present a comprehensive summary of all, or even most, of the published contributions. Some works can be included, others must be left out. The decision is sometimes arbitrary, sometimes a matter of ignorance or distraction (one of the few privileges of a professor that were not abolished).

The objective is to provide a solid introduction on which the researcher or engineer can build additional results. This means the presentation, with sufficient detail but yet within a simple mathematical framework, of a more-or-less unified theoretical approach for the interpretation and prediction of the gravity current flow. The attempt is to develop this theory from first principles, and to focus mostly on up-to-date methods and results with a wide range of applicability or with apparent potential for extensions. For confidence building and completeness, some experimental and Navier-Stokes simulation results must also be discussed. To achieve this objective it is necessary, in my opinion, to study mostly motion in quite ordinary circumstances: currents which propagate on a horizontal surface, in systems of incompressible simple fluids, with small or moderate density differences, without special interfacial effects and with negligible interference with the "far end" boundary of the container. Complications introduced by inclined boundaries, interaction with obstacles, genuine three-dimensional geometries, underflow, interfacial diffusion or chemical reactions, surface tension, non-Newtonian behavior, compressibility, and porous media are not considered, or only briefly discussed. Also not considered are exotic setups like a current in a converging wedge, or extreme cases like avalanches or a dam-break of a water reservoir in the atmosphere.

The simple systems which are the objective of this book still display numerous ramifications (some of them with fascinating effects) which are followed in the text: two-dimensional and axisymmetric (cylindrical) geometries, rotating and non-rotating frames, inertial and viscous reaction to the driving forces, stratified and non-stratified ambient fluids, particle-driven flow, and more.

The gravity current problem is like Ravel's "Bolero." It has a simple and modest beginning: a sound here, an echo there. Then more instruments join the music. The pace increases, the instruments combine in unexpected yet beautifully controlled results. The lead of the melody passes from one form to the other, the drums vibrate, and the sound reemerges again and again with new vigor and confidence. It becomes difficult to keep track of the instruments and the music, it is all just one big excitement. Similarly, the study of gravity currents starts modestly, with a bit of saltwater here and a container of freshwater there. Then additional effects, balances, and equations appear. The same equations produce steady-state and self-similar results, a bit of rotation generates lenses, inertial-terms dominance gives way to a shear-dominated regime, discontinuities appear and vanish. The investigation passes back and forth from simple analysis to experiments and numerical simulation, and reemerges again and again with improved confidence. All in all, it can be a source of great professional satisfaction.

Because of the various branchings of the subject a linear sequential reading, although preferable, may be inadequate for some readers. To facilitate jumps between topics I divided the book into two parts: I. Non-stratified ambient currents; and II. Stratified ambient currents and intrusions. I also attempted to make the different topics self-contained. This requires some repetition of definitions, figures, scalings and conclusions.

In writing this book I have benefited from direct and indirect help and support of colleagues and family. It is in particular a pleasure to mention the useful and inspiring collaboration with H. E. Huppert at the Institute of Theoretical Geophysics, Department of Applied Mathematics and Theoretical Physics, University of Cambridge. I thank Michal Ungarish for her efficient assistance with the preparation of the figures and cover, and Tamar Zemach for carefully reading the draft of the manuscript and making very useful comments. Helpful clarifications and discussions concerning various papers mentioned in the book were contributed by R. T. Bonnecaze, M. R. Flynn, M. A. Hallworth, K. R. Helfrich, J. R. Lister, T. Maxworthy, E. Meiburg, and M. G. Worster.

P. A. Davidson, in the preface of his book on magnetohydrodynamics, has expressed doubts about if people read prefaces. What about the book itself? I recall a discussion with H. P. Greenspan concerning the impact of his book

*The theory of rotating fluids.* The book was cited widely, and translated into other languages. Yet Greenspan mentioned, with disappointment, that there were misprints in the book, and only very few were pointed out by readers; about 90 % he found by himself. According to this statistic, all the readers together have really read only 10 % of the book (approximately 32 pages); the other 90 % of the book was read mostly by the author. On the other hand, perhaps the majority of the readers considered it safer to keep the discovery of the misprint to themselves. Therefore, to both improve the statistics and liberate the readers, I wish to say that I will be grateful to readers who will inform me about misprints in my book. I did not plant any deliberately, but I know they must be there.

Given these doubts, why write a book? Perhaps this is not a good time to ask this question. The Preface appears first, but is actually written last, when the book is ready and the author is rushed by a submission deadline. However, there are some good reasons. First (and I believe this), a book is useful to the field even if people do not read it fully or very carefully. One gets some ideas, inspiration, and references by browsing, or selective reading of one of the sections, or by misprint-hunting as suggested above. Moreover, a colleague may decide that the book is not good (it is not necessary to read a book to reach such a fair conclusion), and will thus immediately write a better book himself - an indirect, but clearly real benefit for the field. Second (so I was told, and this agrees with observations), the author gets some credit even if people do not read his (her) book. Third (just an inspired guess), the average modest prospective author is bound to theorize that his book will be the rare exception, thoroughly read from cover to cover including preface and all, by a large and appreciative audience; this bold conjecture calls for experimental verification. On the other hand, at this point it rather makes sense to recall the sound words of an experienced author, J. Swift's Gulliver: "but, I have now done with all such visionary Schemes for ever."

So here is the end of the author's "Schemes." The next pages are the territory of the reader - and I am confident he (or she) will make the best of it.

M. U.

# COPYRIGHT ACKNOWLEDGMENT

Some figures and tables were reproduced from journals with permission as follows.

*Acta Mechanica*, Copyright Springer-Verlag.

*European Journal of Mechanics B/Fluids*, Copyright Elsevier.

*Journal of Fluid Mechanics*, Copyright Cambridge University Press.

*Physics of Fluids*, Copyright American Institute of Physics.

The full details of the publication are given in the citation in the caption of the figure or table, and the corresponding copyright is as given above.

# Chapter 1

## *Introduction*

A gravity current appears when fluid of one density, $\rho_c$, propagates into another fluid of a different density, $\rho_a$, and the motion is mainly in the horizontal direction. A gravity current is formed when we open the door of a heated house and cold air from outside flows over the floor into the less dense warm air inside. A gravity current is formed when we pour honey on a pancake and we let it spread out on its own. Gravity currents originate in many natural and industrial circumstances and are present in the atmosphere, lakes and oceans as winds, cold or warm streams or currents, polluted discharges, etc.

The study of gravity currents has generated a quite large literature, including some partial review papers (e.g., Griffiths 1986; Felix 2002; Huppert 2006) and the well known book of Simpson (1997). The start of quantitative study (or modeling) of the gravity current is attributed to von Kármán in 1940 (see Huppert 2006 and § 5.5 below), but the related dam-break problem (see § 2.5) was solved by Saint-Venant about a century earlier. It is of course impossible to cite and summarize in detail all these topics, contributions, discussions, and suggestions.

The objective here is to present a more-or-less unified theoretical approach for the interpretation and prediction of the gravity current flow. The attempt is to develop this theory from first principles and focus mostly on methods and results with a wide range of applicability or with apparent potential for extensions. On the other hand, we try to maintain the mathematical simplicity of the presentation. The analytical (or semi-analytical) results are of course only approximations. Their acceptance requires a confidence-building process. To this end comparisons with experimental and Navier-Stokes simulation data will be discussed.

The basic configuration for the understanding of these flows is sketched in Fig. 1.1. The typical gravity current is thin and long. By this we mean that the ratio of the vertical length scale of the current, $L_z$, to the horizontal length scale, $L_x$, is small. The $L_z/L_x \ll 1$ geometric property has very significant dynamical implications. In particular, the small $L_z/L_x$ implies that the accelerations in the normal-vertical direction are small, and hence the corresponding hydrostatic $z$-momentum balance prevails during propagation in the $x$ direction. This provides justification to the simplification which produce the powerful shallow-water (SW) and lubrication theory formulations used for the modeling of the "inviscid current" and "viscous current," respectively. These theories are the backbone of the analytical investigation.

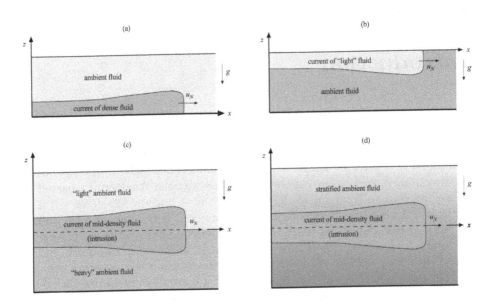

**FIGURE 1.1**: Schematic description of typical gravity current configurations (a) bottom current of more dense (heavy) fluid, $\rho_c > \rho_a$; (b) top (surface) current of less dense (light) fluid, $\rho_c < \rho_a$; (c) intrusion of "mixed" fluid in a sharply stratified ambient; (d) intrusion of "mixed" fluid in a linearly-stratified ambient, $\rho_c = \rho_a(z = 0)$.

The gravity current is driven by the gravity body forces, but in a rather subtle way. The gravitational acceleration acts in the vertical $(-z)$ direction, while the gravity current is driven to propagate mainly horizontally in direction $x$ say. This is actually a very common occurrence: a patch of water on the floor, or syrup on the pancake, spreads out in the horizontal direction because they are more dense ("heavier") than the embedding air. To be more specific, the fluids under consideration are in almost hydrostatic balance in the $z$ direction, $\partial p/\partial z \approx -\rho g$, where $p$ and $\rho$ are the pressure and the density, respectively. The difference in $\rho$ between the fluids gives rise to pressure difference in the $x$ direction, which is then balanced by the dynamic reaction (i.e., a non-trivial velocity field with a major horizontal component which we denote by $u$). The gravity effect associated with the density difference is referred to as the reduced gravity, $g'$. In a two-fluid system, where $\rho_c$ and $\rho_a$ are the constant densities of the current and of the ambient fluid, we define

$$\Delta\rho = \rho_c - \rho_a; \quad \epsilon_a = \frac{\Delta\rho}{\rho_a}; \quad \epsilon_c = \frac{\Delta\rho}{\rho_c}; \tag{1.1}$$

$$g'_a = |\epsilon_a|g; \quad g'_c = |\epsilon_c|g; \tag{1.2a}$$

and

$$g' = \frac{|\Delta\rho|}{\max(\rho_a, \rho_c)} \quad g = \min(g'_a, g'_c). \tag{1.2b}$$

This definition of the reduced gravity seems awkward, and it will be soon simplified. The advantage of the definition (1.2a) is that the resulting $g'$ covers appropriately the large range of $\rho_c/\rho_a$ attainable by various fluid combination. In all these circumstances $g' < g$, which reflects the physical understanding that the reduced gravity cannot exceed the natural force. On the other hand, we are clearly not interested in $g' = 0$ cases.

The velocity of propagation of the current can be estimated from simple energy conversion arguments. Suppose that the current is released from rest in a lock of height (thickness) $h_0$. The mean potential energy of a particle in the lock is $(1/2)g'h_0$ per unit mass. If this is converted to the kinetic energy $(1/2)U^2$ we obtain

$$U = (g'h_0)^{1/2}. \tag{1.3}$$

Although obtained by a very simplistic argument and open to numerous objections, this result turns out to be a good starting value for the speed of propagation of the typical gravity current.

We can now proceed to a rough classification of the gravity current flow fields and systems.

---

## 1.1   Classification

*Constant (fixed)/non-constant volume.* Most of our discussion is concerned with currents of a constant volume, $\mathcal{V}$. We perceive the current as a body enclosed in a clear interface. In the laboratory the current is usually marked by a small amount of dye which enhances the contrast with the ambient fluid. In nature the visual differentiation between current and ambient may be difficult when liquid-liquid and gas-gas systems are concerned. Currents in the atmosphere are made observable by dust and smog, and currents in lakes and seas may carry noticeable pollutants or sand particles (see Simpson 1997 and Huppert 2006 for illustrations). For clearly defined density differences the volume $\mathcal{V}$ may decrease due to drainage into a porous horizontal boundary, or may increase due to a source (typically at the origin). Volume variations of the current may also occur due to entrainment or mixing, and due to particle settling in particle-driven currents. In the last two cases it may even be difficult to define a clear-cut interface between the current and the ambient.

*Inviscid/viscous.* As mentioned above, the gravity current is driven by the $\partial p/\partial x$ effect. The major velocity-field reaction which balances this forcing can be inertial (inviscid) or viscous. The Reynolds number

$$Re = \frac{UL}{\nu} \tag{1.4}$$

gives an indication of the ratio of these two effects. Here $U$, $L$, and $\nu$ are the typical velocity, length, and kinematic viscosity of the current. A current is inviscid or inertial when $Re \gg 1$, and viscous when $Re$ is not large. To start with, we use the velocity estimate (1.3) and take $L = h_0$ to obtain

$$Re = \frac{\sqrt{g'h_0}h_0}{\nu}. \tag{1.5}$$

Further justification and refinements will be presented later. The value of this formal $Re$ provides a feedback to our previous assumption: if large, it justifies the use of (1.3); and if small, it means that the current is viscous and the velocity is significantly smaller than the estimated (1.3). A current released from rest is typically in the inviscid regime, unless thicknesses of less than 1 cm and high-viscosity fluids are concerned. Consider for example a gravity current created in the laboratory in a rectangular tank filled with freshwater. We mix some salt in a sub-volume of 10 cm height and length, to obtain a density difference of about 1%. Using $g' = 10 \,\mathrm{cm\,s^{-2}}$, $h_0 = 10 \,\mathrm{cm}$, and $\nu = 10^{-2} \,\mathrm{cm^2\,s^{-1}}$ we estimate $Re = 10^4$. A larger height or more salt will increase $Re$. Now let us switch to a sweet current concerning honey or syrup whose viscosity is about three thousand times larger than that of water. Consider a pool of this viscous fluid with typical dimensions of 10 cm released on a flat surface surrounded by air (say an idealization of a jar containing this material which is accidentally broken on the floor). This "dam-break" releases a current with, approximately, $h_0 = 10 \,\mathrm{cm}$, $\nu = 30 \,\mathrm{cm^2\,s^{-1}}$, and $g' \approx g = 980 \,\mathrm{cm\,s^{-2}}$. The estimated $Re = 33$ indicates that even this current is not truly viscous. However, this syrup will quite fast spread to a thickness of below 1 cm, for which we obtain $Re < 1$. In this case the velocity estimate (1.3) cannot be trusted; the more relevant estimate is developed in Chapter 11.

We indeed emphasize that the gravity current is a time-dependent phenomenon, which implies that the ratio of inertial to viscous effects may (and in many cases will) change with $t$. Typically, as the current spreads out both its speed and thickness decay. An initially inviscid gravity current may become viscous, with a remarkable drop of the subsequent rate of spread. Thus, for a reliable differentiation between inviscid and viscous control of the current, the indication provided by the formal (or initial) $Re$, given by (1.5), is insufficient. The more stringent question is for how long does an inviscid current remain inviscid. For a current which propagates over a significant distance it will be necessary to monitor the ratio of inertia to viscous forces as a function of time or distance.

Surprisingly perhaps, there are circumstances in which the current starts in the viscous regime, and after some propagation becomes inertial. This unexpected behavior concerns currents produced by inflow, not by the prototype lock-release mechanism. The details are discussed in § 4.2.3.

*Boussinesq/non-Boussinesq.* A current system is of Boussinesq type when the density differences between the current and the ambient are relatively small. In a two-fluid case this can be expressed as $\rho_a \approx \rho_c$, or $\rho_c/\rho_a \approx 1$, or $g'/g \ll 1$, or $|\epsilon_a| \approx |\epsilon_c| \ll 1$. We can arbitrary decide that density differences of up to 10% (or 20%) are within this domain, because the transition to the non-Boussinesq cases is not dramatic. The variations of kinematic viscosity are also small in a Boussinesq system. Many flows in the atmosphere and oceans are of Boussinesq type. The addition of salt (NaCl) to water yields a density increase of about 20% close to saturation at room temperature ($\nu$ increases by about 60%). On the other hand, a gravity current of liquid in gas (and in particular the classical dam-break problem concerning a reservoir of water in the atmosphere) can be regarded as the limit $\rho_c/\rho_a \to \infty$. We shall consider mostly Boussinesq systems. Their mathematical analysis, Navier-Stokes simulation, and experimental realization are simpler, and their interpretation covers a broader range of boundary conditions. For example, a bottom and a surface (ceiling, top) inviscid gravity current with same $|\rho_c - \rho_a|$ and initial geometry are mirror images in the Boussinesq approximations. Non-Boussinesq systems are still awaiting a comprehensive investigation. Roughly, the Boussinesq approximation means that the difference between $\rho_a$ and $\rho_c$ is used in the buoyancy term of the momentum equation only; in all other places $\rho_c = \rho_a$. This implies that the driving force is well reproduced, but the reaction of the fluids to this forcing is distorted by the relative amount $|\rho_c/\rho_a - 1|$, approximately. Since the driving force is correct, upon a careful scaling the Boussinesq results can provide some useful guiding lines for non-Boussinesq systems.

*Homogeneous/stratified ambient.* This important distinction is concerned with the density $\rho_a$ of the ambient fluid as a function of the vertical coordinate $z$. The case of homogeneous constant $\rho_a$ in the entire ambient fluid is obviously the simpler and the more investigated. The stratification can be either sharp, i.e., a jump between two relatively thick layers of homogeneous fluid, or a continuous (typically linear) smooth transition.

The last case is the most interesting, difficult, and as expected, the less investigated one. Essentially, any result derived for the non-stratified case depends now on an additional parameter $S$ which reflects the importance of the stratification, with $S = 0$ as the non-stratified limit. When $S > 0$ the driving force is reduced. In general, gravity currents and intrusions in a continuously-stratified ambient propagate more slowly than in a homogeneous ambient for similar geometry and density pertinent differences, but the details need elaboration. The analysis is complicated by the fact that the propagating current displaces the isopycnals and thus modifies the prescribed density of the ambient. Moreover, gravity currents and intrusions in a stratified ambient create, interact, and ultimately combine with internal gravity waves.

The first part of this book is concerned with currents in a non-stratified ambient, and the second part with currents and intrusions in a stratified ambient.

*Gravity current/intrusion.* A gravity current propagates on a well defined lower or upper geometric boundary of the system, such as the bottom or top of a channel, including the free surface (when $\rho_c < \rho_a$). An intrusion propagates horizontally inside a stratified fluid, typically like an isolated wedge which does not touch the horizontal boundaries. The guiding surface, $z = 0$ say, is the plane of neutral buoyancy for the intruding fluid, i.e., $\rho_c \approx \frac{1}{2}[\rho_a(z = 0^+) + \rho_a(z = 0^-)]$. Many results developed for the gravity current can be applied to intrusions. Actually, for an inviscid Boussinesq system subject to some simple symmetries about the plane of neutral buoyancy, the intrusion can be reduced to a superposition of a bottom and top (surface) gravity currents flowing with the same speed on the neutral buoyancy plane. We reiterate that the motion of the gravity current and intrusion is mostly horizontal, in contrast to the buoyant "plume."

*Two-dimensional (2D) rectangular geometry/axisymmetric (cylindrical) geometry.* In the cylindrical case considered here the axis of symmetry is vertical, and the current propagates in the radial direction ($r$). The curvature terms in the axisymmetric system introduce some non-trivial differences to the 2D current. In particular, we note that a cylindrical current may be divergent (spread out in the positive $r$ direction) or convergent (propagate from the periphery of the container to the center), and this has no counterpart in the 2D case.

*Rotating/non-rotating frames.* Suppose that the system in which the current propagates is rotating with $\Omega$ about the vertical axis $z$. This introduces new effects. Although the centrifugal acceleration of the system $\Omega^2 L$ (say) is usually small as compared to the gravitational acceleration $g$, the Coriolis acceleration competes well with the inertial terms on the privilege to counterbalance $g'$, with remarkable consequences. It is convenient to distinguish between two typical cases. For an axisymmetric gravity current, the Coriolis acceleration is in the direction of spread, hinders the propagation, and eventually renders the current into a steady-state lens. For a current in a $x$-long box bounded by vertical side-walls, the major Coriolis acceleration is in the direction $y$, say, perpendicular to the side-walls. To achieve this acceleration, the current agglomerates near the side-wall, $y = 0$ say, that can provide the necessary "push." This renders a boundary current, i.e., one that is thicker near the "pushing" wall and thinner (or even detached) at the opposite wall of the channel. (An observer attached to the channel may say that the current is deflected toward a side-wall by a Coriolis force.)

*Compositional/particle driven.* The density difference of $\rho_c$ to $\rho_a$ is typically a result of different concentrations of a dissolved material like salt, or of different temperatures. This is referred to as a compositional driving effect. The different $\rho_c$ may also be a result of a suspension of non-neutrally buoyant (i.e., settling) particles in the fluid whose original density is $\rho_a$. The difference between the domain of the suspension and the original clear fluid generates a particle-driven current. Combinations of the two effects may also occur. The density of the compositional current is uniform in space and constant in time (typically, when entrainment is unimportant). This facilitates the distinction between the current and the ambient and the tracking of the appropriate interface. On the other hand, the density of the particle-driven current behaves like the volume fraction of the settling particles in the suspension, which is in general a space-dependent, diminishing with time, function. The coupling between the motion of the current to that of the particles evidently complicates the problem. The settling and dilution of particles blur the interface between the particle-driven current and the ambient, and eventually the current disappears (runs out of particles).

This short classification indicates that the gravity current is a very complex, multi-faced, and parameter-rich physical manifestation. Nevertheless, the gravity current also turns out to be a modeling-friendly phenomena. Indeed, visualizations of the real flow field reveal an extremely complicated three-dimensional motion, with an irregular interface, billows, mixing, and instabilities. On the other hand, there are mathematical models based on a long line of assumptions such as hydrostatic pressure, sharp interface, Boussinesq system, thin layer, idealized release conditions. This enables us to determine the behavior of the averaged variables entirely from theoretical considerations. There are no stringent theoretical estimates of the errors expected when this theory is applied to real fluids and containers. Consequently, the practical use of this theory is rather based on order-of-magnitude arguments and a confidence-building interaction of comparisons between predictions and observations. A superficial comparison of the thickness at a certain point, or the crude addition of the possible errors contributed by the various simplification, may produce the pessimistic view that these models are useless. The more optimistic expectation is that in the real systems the errors are smaller than the estimated bounds, and there is some cancellation of deviations from the idealized conditions.

The facts support the optimistic approach. Qualitatively, the theory is able to provide the governing dimensionless parameters and the salient features of the various flow regimes. Quantitatively, the simple models predict velocities of propagation which agree with experiments within a few percent, sometimes within the range of the experimental errors. This is nicely illustrated by the high-Reynolds-number axisymmetric current. Experiments with various saltwater systems were performed and the current was visualized by a tiny amount of dye mixed in the heavier fluid behind the lock. Hallworth, Huppert, and Ungarish (2001) used a full circular container of about 13 m diameter and Patterson et al. (2006) used a wedge sector of angle 10° and length 2.35 m; the height of the fluid was about 80 cm in the first configuration and 30 cm in the second. The values of $g'$ varied from 4.9 to 43.8 $cm\,s^{-2}$. The observations (from above and from the side) revealed complex velocity fields and shapes of the interface, which, in details, differ significantly from the simplified flow which is amenable to analytical solution (the shallow-water, one-dimensional, smooth, stagnant ambient, inviscid idealization). A comparison between the relevant experimental data, Navier-Stokes (NS) simulations, and SW modeling was attempted by Ungarish (2007a). It shows that the dimensionless reduction of the propagation data according to the SW simplification shows that in all these different flow fields the major governing dimensionless parameter is the depth ratio of ambient to lock, $H$, as predicted by the simple SW model. Moreover, the SW scaling collapses the experimental data to the theoretical curve of $r_N$ (radius of propagation) as a function of $t$ (time) within about 5%, which is also the estimated experimental error. Axisymmetric NS simulations provide richer and more accurate details of the flow field, but the results become incoherent after a limited propagation. Surprisingly, the SW model remains in good agreement with the measurements for a significantly longer time. (Three-dimensional NS simulations of this flow field are richer in detail and in better agreement with experiments than the axisymmetric simulations, but they may require a continuous run of several weeks on powerful computers, as reported by Cantero, Balachandar, and Garcia 2007.) This example will be discussed in detail in § 6.7.

This example, and other similar agreements between the manageable theoretical models and the realistic data, indicates that the main motion of the gravity current is actually governed by some simple and robust underlying balances which can be isolated from the complex appearances. On the other hand, this does not mean that it is easy to develop a reliable model, or that any simple approximation is valuable. There are surprises and failures. Finding the proper compromise between physical complexity and mathematical simplicity in the solution of a specific problem is a matter of art as well as of science, and here experience and insight are essential. On account of the body of knowledge developed during the last five decades, we attempt to present, develop, and evaluate the tools and results which are the backbone of typical gravity current problems and the platform for further research on this topic.

## 1.2    The Navier-Stokes equations

For the record, we write down the balance equations which will be used, mostly in simplified forms, in our discussion. The gravitational acceleration $g$ acts in the $-z$ direction. We use an Eulerian description of the flow field as a function of position $\mathbf{r}$ and time $t$. The density is $\rho(\mathbf{r}, t)$, and the thermodynamic pressure is $\mathcal{P}(\mathbf{r}, t)$.

In an *inertial frame* the velocity field is $\mathbf{v}(\mathbf{r}, t)$. We employ the following dimensional balance equations:

1. Continuity of volume

$$\nabla \cdot \mathbf{v} = 0; \tag{1.6}$$

2. Momentum balance

$$\rho \frac{D\mathbf{v}}{Dt} = -\nabla \mathcal{P} - \rho g \hat{z} + \mu \nabla^2 \mathbf{v}; \tag{1.7}$$

3. Density transport

$$\frac{\partial \rho}{\partial t} + \mathbf{v} \cdot \nabla \rho = \kappa \nabla^2 \rho. \tag{1.8}$$

Here $D/Dt$ is the "substantial" derivative; explicit forms are presented in Appendix C.2. The coefficients of dynamic viscosity, $\mu$, and of component diffusion, $\kappa$, will be specified later. Usually, the effects associated with $\kappa$ are very small. The inviscid form of the foregoing set of equations, obtained by dropping completely the viscous terms, is usually referred to as the Euler equations.

A significant part of the pressure distribution $\mathcal{P}$ is determined by a static component, associated with the body-force potential, $\mathcal{B}$, and a constant representative density, $\bar{\rho}$. This component can be canceled out from the momentum balance by the use of the reduced (or excess) pressure, $p = \mathcal{P} + \bar{\rho}\mathcal{B}$. To be specific, let $\bar{\rho}$ be a constant reference density of the ambient fluid (for the non-stratified cases, $\bar{\rho} = \rho_a$ is a good choice). Define

$$p = \mathcal{P} + \bar{\rho}gz + C, \tag{1.9}$$

where $C$ is a constant. Equation (1.7) can be rewritten as

$$\rho \frac{D\mathbf{v}}{Dt} = -\nabla p - (\rho - \bar{\rho})g\hat{z} + \mu \nabla^2 \mathbf{v}. \tag{1.10}$$

The difference between $p$ and $\mathcal{P}$ is unimportant in the differential momentum equations, but may be relevant on a free boundary (where $\mathcal{P} = \text{const}$), in pressure force evaluations and in other integrals of the momentum equations. Some care is required to sort this out when encountered.

In some cases it is convenient to analyze the current in a *rotating frame*. Suppose that the coordinate system introduced above is also rotating with constant $\Omega \hat{z}$. In particular, $\mathbf{v}$ is now the velocity of a fluid particle measured in this rotating system. Equations (1.6) and (1.8) are unchanged. The momentum equation (1.7) is modified: the centrifugal and Coriolis acceleration terms (multiplied by $\rho$) must be added on the LHS. This addition reads $\rho[2\Omega\hat{z} \times \mathbf{v} + \Omega^2 \hat{z} \times (\hat{z} \times \mathbf{r})]$. In a cylindrical coordinate system $\{r, \theta, z\}$ the centrifugal acceleration of the system is given by the more common $-\Omega^2 r \hat{r}$. The reduced pressure can be defined as

$$p = \mathcal{P} + \bar{\rho}[gz - \frac{1}{2}\Omega^2 r^2] + C. \qquad (1.11)$$

In the rotating system the counterpart of (1.7) is

$$\rho[\frac{D\mathbf{v}}{Dt} + 2\Omega\hat{z} \times \mathbf{v}] = -\nabla\mathcal{P} - \rho[g\hat{z} - \Omega^2 r\hat{r}] + \mu\nabla^2\mathbf{v}; \qquad (1.12)$$

and the counterpart of (1.10) is

$$\rho[\frac{D\mathbf{v}}{Dt} + 2\Omega\hat{z} \times \mathbf{v}] = -\nabla p - (\rho - \bar{\rho})[g\hat{z} - \Omega^2 r\hat{r}] + \mu\nabla^2\mathbf{v}. \qquad (1.13)$$

Again, in the RHS of (1.12)-(1.13) $r$ is the radial coordinate in a cylindrical system of coordinates. The centrifugal acceleration contributed by the rotation of the frame in which we measure the motion behaves like a body-force and can be regarded as an addition (usually small) to the gravitational acceleration $g$. The Coriolis term is a different story because it is proportional to the speed and may effectively compete with the inertial terms in shaping the motion.

In both the rotating and non-rotating formulations given above it was assumed that the system coordinates (to be specific, the origin $O$) does not accelerate. For a *system with an acceleration* $\mathbf{a}_O$ of the origin $O$, it is necessary to add $\rho\mathbf{a}_O$ on the LHS of the momentum equation.

These apparently simple equations, with appropriate quite standard initial and boundary conditions, and subject to some plausible assumptions, cover most of the phenomena discussed in this book. Some additional balance (or closure) equations are necessary for particle-driven currents, porous media effects, mutual diffusion between the current and the ambient, and explicit turbulence. The accurate and detailed solution of these equations can be achieved only on a computer. We shall postpone this approach for a while, and in any case, this is not our main objective. Our major task is to reduce these equations to simplified comprehensible forms from which useful analytical information and insights can be extracted.

# Part I

# Non-stratified ambient currents

# Chapter 2

## Shallow-water (SW) formulation for high-Re flows

We consider the simple, yet very relevant, problem of a two-dimensional current which propagates at the bottom of a large container filled with a relatively deep ambient fluid; see Fig. 2.1.

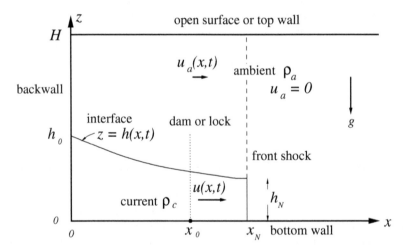

**FIGURE 2.1:** Schematic description of the gravity current configurations for a bottom current of more dense (heavy) fluid released from behind a lock.

The volume of the current is fixed and we neglect viscosity (both internal and friction at boundaries) effects. The densities of the current and ambient are the constant, $\rho_c$ and $\rho_a$, respectively, and $\rho_c > \rho_a$. When necessary, we use the subscripts $a$ and $c$ to denote the fields pertinent to the ambient and current, respectively. We use dimensional variables (unless specified otherwise). Let $x, z$ be the horizontal and vertical axes, with gravity $g$ acting in the $-z$ direction, and $z = 0$ is the bottom wall. The position of the upper boundary is $z = H$; the constant value $H$ is also the height of the upstream ambient fluid into which the current propagates. We assume that the current propagates in the positive $x$ direction. The motion to the left is restricted by the solid backwall at $x = 0$, but the opposite "far end wall" of the container is at a sufficiently large distance to make its effect unimportant.

The objective is to develop a simplified set of equations which provides an insightful description of the system. The variables of major interest are the shape and the horizontal motion of the current (the volume of dense fluid). The sharper task is to determine these variables entirely from theoretical considerations. Then, of course, we would like to know how reliable and accurate this prediction is. The shape will be represented by the spread $x_N(t)$ and thickness $h(x, t)$, and the motion by the $z$-averaged $\bar{u}(x, t)$. To this end, we proceed as follows. We develop the equation of motion of the interface under the assumption that the velocity field is known; the result can also be viewed as a $z$-integrated continuity equation. Then we consider the pressure fields in the ambient and in the current. We introduce the shallow-water (SW) simplification and, by manipulating the vertical momentum equation, obtain a simple relationship between the pressure in the current and $h(x, t)$. This is next used to eliminate the pressure from the $x$-momentum balance and to manipulate the remaining terms into a satisfactory form. The result can also be viewed as a $z$-integrated $x$-momentum equation. The set of this "continuity" and "momentum" equations must be closed by a front condition which is also introduced.

---

## 2.1　Motion of the interface and the continuity equation

Observations of real flows and numerical simulations indicate that the change of density at the interface of the gravity current is sharp. It is both practically and mathematically convenient to consider this interface as a discontinuity of the density, a "kinematic shock." Let us denote by $\Sigma(\mathbf{r}, t) = 0$ the locus of this discontinuity, and formulate the equation of motion. The full differential of $\Sigma(\mathbf{r}, t) = 0$ yields

$$\frac{\partial \Sigma}{\partial t} dt + (\nabla \Sigma) \cdot d\mathbf{r} = 0, \tag{2.1}$$

therefore

$$\frac{\partial \Sigma}{\partial t} + (\nabla \Sigma) \cdot \mathbf{V}_\Sigma = 0, \tag{2.2}$$

where $\mathbf{V}_\Sigma \equiv d\mathbf{r}/dt$ is the rate of displacement of the interface, which in the present problem is also the velocity of the fluid at the interface. Since $\nabla \Sigma$ is in the normal direction, it is clear that the displacement of the interface is governed only by the normal velocity component of the fluid at the interface. The immediate implication is that at the transition between the current and the ambient the normal velocity is continuous on the interface, but the tangential component may be discontinuous (this is an accepted simplification in inviscid approximations).

To be specific, the interface under consideration in the present configuration has two components; see Fig. 2.1. The first is a vertical front (also referred to as nose, and denoted by the subscript $N$), which is given by $\Sigma_N = x - x_N(t) = 0$ $(0 \le z \le h_N(t))$. The second, and major, part of the interface is the inclined, sometimes piecewise inclined-horizontal, surface which is given by $\Sigma_h = z - h(x, t) = 0$ $(0 \le x \le x_N(t))$. Here $x_N$ and $h_N$ are the position and height of the nose. The velocity of the fluid is given by $\mathbf{v} = u\hat{x} + w\hat{z}$. Using (2.2) ($\mathbf{V}_\Sigma$ is the value of $\mathbf{v}$ at the interface) we obtain the motion of the interface as

$$\frac{dx_N}{dt} = u_N; \tag{2.3a}$$

$$\frac{\partial h}{\partial t} + u_h \frac{\partial h}{\partial x} - w_h = 0, \tag{2.3b}$$

where the subscript $N$ denotes the nose, and the subscript $h$ denotes the inclined part of the interface. These equations are usually supplemented by given initial conditions for $x_N$ and $h(x)$ at $t = 0$. As expected, when the velocity field of the current is known we can calculate the position of the interface from purely kinematic considerations.

Equation (2.3b) can be rewritten as follows. First, we draw the attention of the reader to a very useful formula in the context of $z$-averaged simplifications: Leibniz's rule for differentiation of an integral; see Appendix C. Using this formula, the no-penetration condition $w = 0$ on $z = 0$, and the incompressible continuity equation (1.6), $\nabla \cdot \mathbf{v} = 0$, we write

$$\frac{\partial}{\partial x} \int_0^h u\,dz = u_h \frac{\partial h}{\partial x} + \int_0^h \frac{\partial u}{\partial x} dz = u_h \frac{\partial h}{\partial x} - w_h. \tag{2.4}$$

Consequently, for a 2D flow over a solid boundary, (2.3b) can be rewritten as

$$\frac{\partial h}{\partial t} + \frac{\partial}{\partial x} \int_0^h u\,dz = 0, \tag{2.5}$$

which is also referred to as the continuity equation of the 2D current.

We keep in mind that in the derivation of (2.5) we assumed that $z = 0$ is an impermeable boundary. Otherwise, a non-zero $w(x, z = 0)$ appears in the RHS of (2.5). This is consistent with the expectation that when the current gains (losses) fluid during propagation on a porous boundary its thickness increases (decreases) faster than usual.

The use of the equation of motion for the infinitesimally thin interface between domains of prescribed density relaxes the need to solve the density field. Consequently, in the approximate models for the gravity current discussed below we shall be concerned with the volume continuity and momentum balance, but not with the density transport equation (1.8). The density fields for the current and ambient domains, separated by the interface, are supposed to satisfy this equation. In particular, the widely used constant $\rho_c$ and $\rho_a$ are exact solutions of (1.8).

## 2.2   One-layer model

We assume that the motion in the ambient is negligible as compared to that of the dense fluid. A more stringent criterion is that the acceleration of the ambient fluid is small as compared to that of the current. This is readily justified when the thickness of the ambient layer above the current, $H - h$, is large as compared with the thickness of the current, $h$, and $(\rho_a/\rho_c)$ is not large. For a Boussinesq system, this turns out to be a relevant hypothesis for the analysis of the motion in the current even for a moderate thickness ratio, say 2. This will be discussed in Chapter 5.

We use the subscripts $a$ and $c$ to denote the ambient and current layers when necessary to avoid ambiguity. When no subscript is used we refer to the current (unless stated otherwise). We employ the reduced pressure, $p = \mathcal{P} + \rho_a gz$, where $\mathcal{P}$ is the hydrodynamic pressure. We define

$$\Delta\rho = \rho_c - \rho_a. \tag{2.6}$$

In the ambient fluid, where $\mathbf{v}_a = \mathbf{0}$, the solution of the Navier-Stokes equations is simple. The continuity equation (1.6) is trivially satisfied, while the $x$ and $z$ components of the momentum equation (1.7) yield the hydrostatic balance

$$0 = -\frac{\partial p_a}{\partial x}; \quad 0 = -\frac{\partial p_a}{\partial z}. \tag{2.7}$$

The solution of (2.7) is

$$p_a = C, \tag{2.8}$$

where $C$ is some constant.

Consider the lower layer of dense fluid below the ambient. Suppose that in the horizontal motion the scales for length, velocity, and time are $L_x$, $U$, and $L_x/U$; while the vertical counterparts are $L_z$, $W$, and $L_z/W$. Here we introduce the shallow-water (SW) assumption that $(L_z/L_x) = \delta \ll 1$. The continuity equation shows that $W = (L_z/L_x)U = \delta U$. Let us evaluate the orders of magnitude of the inertial terms in the momentum equations which balance the pressure gradient. For the $x$ direction we obtain

$$\rho_c[u_t + uu_x + wu_z] = \rho_c O(\frac{U^2}{L_x}) = -\frac{\partial p_c}{\partial x}. \tag{2.9}$$

For the $z$ direction we obtain

$$\rho_c[w_t + uw_x + ww_z] = \rho_c O(\frac{W^2}{L_z}) = \rho_c \delta O(\frac{U^2}{L_x}) = -\frac{\partial p_c}{\partial z} - \Delta\rho g. \tag{2.10}$$

Here the $O(\ldots)$ means the order of magnitude. Let us integrate the last equation with respect to $z$, keeping in mind that the length scale of this variable is $L_z = \delta L_x$. We obtain $p_c(x,z,t) = -\Delta\rho g z + f(x,t) + \rho_c \delta^2 O(U^2)$. Here $f(x,t)$ is some unspecified function, but a substitution into (2.9) proves that this must be of magnitude $\rho_c O(U^2)$. Thus, within a relative error $\delta^2$, we can discard the contribution of the inertial terms in the vertical momentum equation. In other words, in the $z$ direction the hydrostatic balance

$$\frac{\partial p_c}{\partial z} = -\Delta\rho g \qquad (2.11)$$

remains valid in spite of the significant motion in the horizontal direction.

We conclude that, in the framework of SW inviscid assumptions, it is justified to approximate the pressure in the current by

$$p_c(x,z,t) = -\Delta\rho g z + f(x,t) \quad (0 \le x \le x_N(t), 0 \le z \le h(x,t)). \qquad (2.12)$$

Matching of $p_c$ with $p_a$ in the ambient produces an important result. We argue that the pressure can be discontinuous about vertical fronts (shocks), but it must be continuous on inclined and horizontal portions of the interface. As shown later in Chapter 3, a pressure discontinuity produces a quite fast motion of the interface (with a typical velocity of the order of $U$). Since the main pressure field is $z$-dependent, a rapid and short displacement in the $z$ direction is bound to eliminate (smooth out) a possible pressure difference about the interface $z = h(x,t)$. On the other hand, the nose propagates in the horizontal direction and encounters the same pressure all the way. This preserves the pressure discontinuity at $x = x_N(t)$. This behavior of the pressure field supports the basic feature of the gravity current which is a dominant horizontal propagation accompanied by slow variations of thickness.

We therefore introduce, as a major component of the SW description, the *pressure continuity* condition on the inclined and horizontal portions of the interface

$$p_a = p_c \quad \text{at } z = h(x,t) \quad (0 \le x \le x_N(t)). \qquad (2.13)$$

However, we keep in mind that in general $h(x,t)$ may have jumps which form the vertical portions of the interface. A pressure discontinuity (shock) is present across these vertical portions of the interface. The common case is the shock at $x = x_N(t)$.

Applying the pressure continuity principle to (2.8) and (2.12) renders the previously unspecified $f(x,t) = \Delta\rho g h(x,t) + C$. Substitution into (2.12) yields

$$p_c(x,z,t) = -\Delta\rho g z + \Delta\rho g h(x,t) + C \quad (0 \le x \le x_N(t), 0 \le z \le h(x,t)). \qquad (2.14)$$

Finally, the longitudinal pressure gradient in the current can be written as

$$\frac{\partial p_c}{\partial x} = \Delta\rho g \frac{\partial h(x,t)}{\partial x}. \qquad (2.15)$$

This is an important *fundamental* outcome which concerns the main internal driving force of the current. We confirmed that this effect is provided by the buoyancy (or reduced gravity) acceleration, expressed by the density difference times $g$, which justifies the term gravity (or density) current. This force decelerates the current where $(\partial h/\partial x)$ is positive, and accelerates it where $(\partial h/\partial x)$ is negative. We note that this effect is independent of $z$. This indicates that, in the present approximation, the speed $u$ of the current which is initially $z$-independent remains so during propagation. Moreover, this property also facilitates the development of the governing momentum equation to which we proceed now.

We start with the inviscid $x$-momentum equation of the current in conservation form

$$\frac{\partial u}{\partial t} + \frac{\partial u^2}{\partial x} + \frac{\partial}{\partial z}(uw) = -\frac{1}{\rho_c}\frac{\partial p_c}{\partial x}. \tag{2.16}$$

(This follows from the combination of (1.6) and (1.10) subject to $\bar{\rho} = \rho_a$ and $\mu = 0$.) We integrate (2.16) across the current to obtain

$$\int_0^h \left[\frac{\partial u}{\partial t} + \frac{\partial u^2}{\partial x}\right] dz + u_h w_h = -\frac{1}{\rho_c}\int_0^h \frac{\partial p_c}{\partial x}dz, \tag{2.17}$$

where we used the boundary condition $uw = 0$ on $z = 0$, and the subscript $h$ denotes values at the interface $z = h$. On the other hand, it can be verified that

$$\frac{\partial}{\partial t}\int_0^h u\,dz + \frac{\partial}{\partial x}\int_0^h u^2 dz - u_h\left(\frac{\partial h}{\partial t} + u_h\frac{\partial h}{\partial x} - w_h\right) = \text{LHS of (2.17)} . \tag{2.18}$$

The term in the brackets on the LHS of (2.18) vanishes on account of the kinematic conditions of the interface expressed by equation (2.3b). The combination of (2.17) and (2.18) yields

$$\frac{\partial}{\partial t}\int_0^h u\,dz + \frac{\partial}{\partial x}\int_0^h u^2 dz = -\frac{1}{\rho_c}\int_0^h \frac{\partial p_c}{\partial x}. \tag{2.19}$$

We introduce the straightforward definition of the $z$-averaged velocity

$$\bar{u}(x,t) = \frac{1}{h}\int_0^h u\,dz, \tag{2.20}$$

and of the velocity shape factor

$$\varsigma = \frac{1}{h\bar{u}^2}\int_0^h u^2 dz = 1 + \frac{1}{h}\int_0^h \left(1 - \frac{u}{\bar{u}}\right)^2 dz. \tag{2.21}$$

Equation (2.19), which expresses the integrated form of the inviscid momentum balance, can be rewritten as

$$\frac{\partial \bar{u}h}{\partial t} + \frac{\partial \varsigma\bar{u}^2 h}{\partial x} = -\frac{1}{\rho_c}\int_0^h \frac{\partial p_c}{\partial x}dz. \tag{2.22}$$

Now we apply the following approximations to (2.22). First, we use the SW result (2.15) to eliminate the pressure. Second, we argue that the deviations of $\varsigma$ from 1 can be neglected in many cases of interest. For example, when the initial $u$ is $z$-independent, $u$ and $\bar{u}$ remain identical and $\varsigma = 1$. When the typical difference between $u$ and $\bar{u}$ is 20%, the value of $\varsigma$ is expected to increase by about 4%; see (2.21). We therefore use $\varsigma = 1$ in our subsequent analysis. (For a discussion of cases with larger shape factors see, for example, Hogg and Pritchard 2004.) The result is the simplified SW inviscid momentum equation

$$\frac{\partial \bar{u}h}{\partial t} + \frac{\partial \bar{u}^2 h}{\partial x} = -\frac{\Delta\rho}{\rho_c} g \frac{1}{2} \frac{\partial h^2}{\partial x}. \tag{2.23}$$

For completeness we also rewrite here the continuity equation (2.5) after the substitution of (2.20) as

$$\frac{\partial h}{\partial t} + \frac{\partial h\bar{u}}{\partial x} = 0. \tag{2.24}$$

This is the required simplified system of two equations for the two variables $h(x, t)$ and $\bar{u}(x, t)$. We have the necessary initial conditions at $t = 0$, and the obvious boundary condition $u = 0$ at $x = 0$, but the conditions at the front $x = x_N(t)$ need further consideration.

A detailed discussion with pertinent references can be found in Klemp, Rotunno, and Skamarock (1994). The essentials are as follows. When $\rho_c/\rho_a \to \infty$ the resistance of the ambient fluid is unimportant, and the formulation reduces to the classical Saint-Venant dam-break solution in which the height of the current can drop to zero at the nose, while the speed $u_N$ is provided by the maximum speed available on the characteristics coming from the reservoir (see Stansby, Chegini, and Barnes 1998; Hogg and Pritchard 2004, and also Chapter 10 below). We are not interested in this case. On the other hand, for a finite density ratio $\rho_c/\rho_a$, and in particular for the Boussinesq fluid systems with $\rho_c/\rho_a \approx 1$, the front of the current propagates like a "wall" of finite height, $h_N$, and the speed is smaller than the maximum value available on the characteristic from the reservoir. This is supported by the theoretical results presented later in Chapter 3 and comparisons with experimental observations. The consensus is that the required boundary condition is

$$u_N = \left(\frac{\Delta\rho}{\rho_a} g\right)^{1/2} h_N^{1/2} Fr(a), \qquad \text{and} \quad a \le a_{\max} \approx 0.5, \tag{2.25}$$

where $N$ denotes the nose (front), $a = h_N/H$ is the height ratio of nose to ambient, and $Fr(a)$ is a Froude number (actually a function) of the order of 1. The restriction on $a$ comes from energy considerations. Roughly, it means that the increase of energy in the ambient activated by the current (the return flow above the front) cannot exceed the decrease of energy in the current. We assume that there are no sources of energy in the system.

The $Fr(a)$ function expresses the volume and momentum balances of a moving jump of pressure and velocity between the stationary fluid in front of

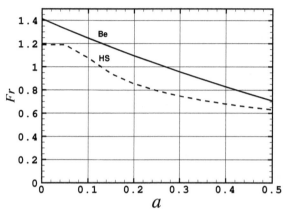

**FIGURE 2.2**:   Froude functions of Benjamin (Be) and Huppert-Simpson (HS), given by equations (2.26) and (2.27), respectively.

the current, and the moving fluid just behind $x_N(t)$. The theoretical result is Benjamin's Froude number function,

$$Fr(a) = \left[ \frac{(2-a)(1-a)}{1+a} \right]^{1/2}. \qquad (2.26)$$

A more practical suggestion is the Froude number function of Huppert and Simpson (1980), referred to as the HS $Fr$,

$$Fr(a) = \begin{cases} 1.19 & (0 \leq a < 0.075) \\ \frac{1}{2}a^{-1/3} & (0.075 \leq a \leq 1). \end{cases} \qquad (2.27)$$

This has been obtained as an experimental curve-fit counterpart to Benjamin's formula. It predicts a smaller value than Benjamin's for a given $a$, which means a lower speed for a given nose height $h_N$. The justification is that in real fluid systems there are friction and mixing losses which reduce the speed of propagation below the idealization of Benjamin. Various applications indicate that this is a more convenient formula than that of Benjamin, because (1) The SW predictions of $u_N$ are closer to observations, in general; (2) The case of a deep current with constant $Fr$ is attained for the practical $h_N/H < 0.075 = 1/13.3$, while Benjamin's $Fr$ is a constant only in the abstract limit $H/h_N \to \infty$. This feature is relevant to the self-similar propagation and box-model approximations which are discussed later. We shall use the HS $Fr$ as the prototype in this book. However, we must keep in mind that only Benjamin's $Fr$ has been derived with some theoretical rigor. We shall return to this topic in Chapter 3, § 5.5, and Chapter 14. The $Fr$ functions of Benjamin and HS are displayed in Fig. 2.2.

The formulation is now complete. This is a good point for asking the following question: what will change in the foregoing equations if we consider a current of light (less dense than the ambient, $\rho_c/\rho_a < 1$) fluid which propagates on the top boundary (or ceiling), instead of the current of heavy (more dense than the ambient, $\rho_c/\rho_a > 1$) which propagates on the bottom boundary? The quite intuitive answer is that we should replace $\Delta\rho = \rho_c - \rho_a$ with $-\Delta\rho$ and measure $h(x,t)$ downward from the position $z = H$. Indeed, if we consider $h(x,t)$ as being the thickness of the current (a positive quantity by definition), and replace $\Delta\rho$ by $|\Delta\rho|$, then the continuity, momentum, and front condition equations become valid for both top (light) and bottom (heavy) currents which propagate on the appropriate boundary. This, however, does not mean that in general these currents are mirror images; but they just obey the same set of equations. This apparently surprising observation will be elaborated in Chapter 10. Here we introduce a useful assumption which, among other simplifications, preserves the mirror-image symmetry between the heavy and the light currents.

## 2.2.1  Boussinesq simplification

We introduce here the Boussinesq assumption that density difference is small, i.e., $\rho_c/\rho_a \approx 1$. We define

$$\Delta\rho = \rho_c - \rho_a, \quad \epsilon = \epsilon_a = \frac{\Delta\rho}{\rho_a}. \tag{2.28}$$

The use of Boussinesq assumption means that we restrict our analysis to $|\epsilon| \ll 1$, unless stated otherwise. Formally, we consider an expansion of the variables in powers of $\epsilon$. We substitute this expansion into the SW equations (including the front condition), and keep only the leading terms. The relative errors of $u$ and $h$ appear to be about $0.5|\epsilon|$, but it is difficult to obtain a sharp estimate. (We keep in mind that there are also models for non-Boussinesq systems which can be employed if this error seems consequential in our application; this will be discussed later in Chapter 10.) The physical meaning of the Boussinesq approximation is that the density difference enters the system only in the context of the gravitational (and later also in the centrifugal) buoyancy force; otherwise, the system is a homogeneous incompressible fluid. This is, however, an appropriate simplification. As mentioned in the Introduction (Chapter 1), many geophysical currents are driven by small density variations. From the point of view of our study, this approximation eliminates the dependency on the $\rho_c/\rho_a$ parameter. The discussion and the insights are simpler and sharper.

To be specific, in the present one-layer SW model the Boussinesq approximation is implemented as follows. We chose $\rho_a$ to be our reference density. Consequently,

$$\rho_c = \rho_a(1 + \epsilon) \tag{2.29}$$

and

$$g' = \frac{|\Delta\rho|}{\rho_a}g = g'_a = |\epsilon|g. \tag{2.30}$$

The continuity equation (2.24) does not contain density terms and hence is not affected by the Boussinesq assumption. We substitute (2.29)-(2.30) into (2.23) and (2.25) to obtain

$$\frac{\partial \bar{u}h}{\partial t} + \frac{\partial \bar{u}^2 h}{\partial x} = -\frac{|\Delta\rho|}{\rho_a(1+\epsilon)}g\frac{1}{2}\frac{\partial h^2}{\partial x} = -g'\frac{1}{2}\frac{\partial h^2}{\partial x}\cdot\frac{1}{1+\epsilon} =$$

$$-g'\frac{1}{2}\frac{\partial h^2}{\partial x} \quad [+O(\epsilon)]. \tag{2.31}$$

$$u_N = g'^{1/2}h_N^{1/2}\,Fr(a), \quad \text{and} \quad a \le a_{\max} \approx 0.5. \tag{2.32}$$

The Boussinesq approximation means that the last $O(\epsilon)$ term on the RHS of the momentum equation (2.31) is discarded. This is expected to introduce a similar relative error, $O(\epsilon)$, in the $u^2$ terms which appear on the LHS of this equation. The continuity equation and the front condition are unaffected by this approximation. The fact that the approximation is made in only one of the three equations which govern the model seems to mitigate the loss of accuracy. We also recall that this approximation is made in addition to serious previous simplifications (inviscid, shallow-water, averaged flow). We must hope that the errors do not accumulate.

Since $\rho_c \approx \rho_a$ we could choose $\rho_c$ as the reference density, and use the reduced gravity $g'_c = (|\Delta\rho|/\rho_c)g$ instead of $g'_a$. In this case the RHS of the momentum equation contains exactly $g'_c$ and is unaffected by the Boussinesq approximation. The front condition contains $g'_c(1+\epsilon)$, which is approximated by $g'_c$ within an $O(\epsilon)$ error. The details are left for an exercise. Evidently, the subsequent solution and accuracy is the same as for the choice of $\rho_a$ as the reference density. The use of $\rho_a$ as reference seems, however, more convenient. It facilitates the discussion of different currents in the same ambient. Also, in experimental sets it is easier to vary the densities of the released currents with respect to $\rho_a$ than vice versa.

In what follows we adopt $\rho_a$ and $g' = g'_a$ as the representative density and reduced gravity for the Boussinesq systems, unless specified otherwise.

## 2.2.2   Scaling

For the lock-release problem it is natural to scale the vertical lengths with $x_0$ and the heights with $h_0$. The reference density is $\rho_a$, and $g' = g'_a$ defined by (2.30). The use of the front condition (2.32) indicates that the proper velocity and time scales are

$$U = (g'h_0)^{1/2}; \quad T = \frac{x_0}{U}. \tag{2.33}$$

We now switch all dimensional variables, denoted below by an asterisk, to dimensionless variables as follows

$$\{x^*, z^*, h^*, H^*, t^*, u^*, p^*\} = \{x_0 x, h_0 z, h_0 h, h_0 H, Tt, Uu, \rho_a U^2 p\}. \quad (2.34)$$

An inspection of the continuity, momentum, and front condition equations (2.24), (2.31), and (2.32), and of the pressure equation (2.15), confirms the consistency of this scaling for the variables of interest.

The equations of motion can be written in conservation form by expressing the $\partial/\partial x$ terms as the divergence of a flux. The conserved variables are $h$ and $uh$. Another form is the direct, primitive-variable formulation, for $h$ and $u$. This is convenient for the calculation of the characteristics of the (hyperbolic) system and is referred to as the characteristic form. We substitute the scaling presented above into the continuity, momentum, and front condition equations (2.24), (2.31), and (2.32). For simplicity, we also drop the overbar which denotes the $z$-averaged velocity. We keep in mind that in the SW framework we solve for the averaged $u(x, t)$. The results are given below.

## 2.2.3   Summary 1

In conservation form the dimensionless equations of motion can be written as

$$\frac{\partial h}{\partial t} + \frac{\partial}{\partial x}(uh) = 0, \quad (2.35)$$

and

$$\frac{\partial}{\partial t}(uh) + \frac{\partial}{\partial x}\left(u^2 h + \frac{1}{2}h^2\right) = 0. \quad (2.36)$$

In characteristic form, this becomes

$$\begin{bmatrix} h_t \\ u_t \end{bmatrix} + \begin{bmatrix} u & h \\ 1 & u \end{bmatrix} \begin{bmatrix} h_x \\ u_x \end{bmatrix} = \begin{bmatrix} 0 \\ 0 \end{bmatrix}. \quad (2.37)$$

The characteristic curves and relationships provide useful information for the solution of the system. The methodology for deriving the characteristics is summarized in Appendix A.1. Comparing (2.37) with (A.1) we readily identify that in the present case $\mathcal{H} = h$, $\mathcal{U} = u$, $A = 1$, $B = u$, and $S_1 = S_2 = 0$. Following the procedure outlined in Appendix A.1, and with an obvious change of notation, we obtain the characteristic balances

$$h^{-1/2}dh \pm du = 0 \quad \text{on} \quad \frac{dx}{dt} = c_\pm = u \pm h^{1/2}. \quad (2.38)$$

The nose (front) condition needed at $x = x_N(t)$ is

$$u_N = h_N^{1/2} \, Fr(h_N/H), \quad \text{and} \quad \frac{h_N}{H} \le a_{max} \approx 0.5, \quad (2.39)$$

where

$$Fr(h_N/H) = \begin{cases} 1.19 & (0 \le h_N/H < 0.075) \\ 0.5H^{1/3}h_N^{-1/3} & (0.075 \le h_N/H \le 1). \end{cases} \tag{2.40}$$

The initial and boundary conditions are provided by the physical situation under consideration. Here the fluid is released from rest from a lock of unit length and height (in scaled form). At $x = 0$ (the left-hand-side end of the sketched system) the fluids are bounded by a solid vertical wall. This is expressed as

$$u(x, t = 0) = 0; \quad h(x, t = 0) = 1; \quad (0 \le x \le 1); \tag{2.41}$$

$$u(x = 0, t) = 0 \quad (t \ge 0). \tag{2.42}$$

To calculate the propagation, the system of equations for $h(x, t)$ and $u(x, t)$ is supplemented by the kinematic condition

$$\frac{dx_N}{dt} = \dot{x}_N = u_N; \quad x_N(t = 0) = 1. \tag{2.43}$$

The $x$ domain of solution is $[0, x_N(t)]$ which increases with time.

Analytical results of the foregoing SW formulation may be obtained for restricted circumstances. The analytical behavior of $h(x, t)$ and $u(x, t)$ in the initial "dam-break" stage and in the large-time self-similar stage will be presented later. In general the resulting SW system requires numerical solution. Both numerical and analytical procedures may be facilitated by the following transformation.

---

## 2.3 A useful transformation

A gravity current may spread out over a significant distance as compared to the initial $x_N(0) = 1$. It is convenient to keep track of it in a standard domain, $[0, 1]$ say. This is achieved by introducing a stretched horizontal coordinate,

$$y = y(x, t) = \frac{x}{x_N(t)} \quad (0 \le y \le 1). \tag{2.44}$$

Note that

$$\frac{\partial y}{\partial t} = -x\frac{\dot{x}_N}{x_N^2} = -y\frac{\dot{x}_N}{x_N}; \quad \frac{\partial y}{\partial x} = \frac{1}{x_N}; \tag{2.45}$$

where the overdot denotes differentiation in time. Now the description of the current is changed from $h = h(x, t)$ and $u = u(x, t)$ to $h = h(y, t)$ and $u = u(y, t)$. For obtaining the details we follow the systematic approach of

Bonnecaze, Huppert, and Lister (1993). In the previously derived governing equations the partial derivatives of the original dependent variables must be replaced according to the chain rule. Using (2.45) this yields

$$\left(\frac{\partial}{\partial t}\right)_x = \left(\frac{\partial}{\partial t}\right)_y + \frac{\partial y}{\partial t}\left(\frac{\partial}{\partial y}\right)_t = \left(\frac{\partial}{\partial t}\right)_y - y\frac{\dot{x}_N}{x_N}\left(\frac{\partial}{\partial y}\right)_t; \qquad (2.46)$$

$$\left(\frac{\partial}{\partial x}\right)_t = \frac{\partial y}{\partial x}\left(\frac{\partial}{\partial y}\right)_t = \frac{1}{x_N}\left(\frac{\partial}{\partial y}\right)_t; \qquad (2.47)$$

where the subscripts denote the fixed variable in the differentiation.

Consider the velocity in the original system $dx/dt = \lambda$. Following the displacement of the same particle along the $y$ coordinate during $dt$ we find that $dx = x_N dy + y\dot{x}_N dt$. Consequently, the transformation of the velocity (rate of displacement) along a trajectory is given by

$$\frac{dy}{dt} = \frac{1}{x_N}(\lambda - y\dot{x}_N) = \frac{1}{x_N}\left(\frac{dx}{dt} - y\dot{x}_N\right). \qquad (2.48)$$

The last result is useful for the transformation of the characteristic lines (the balances between the dependent variables on the characteristics are not influenced, except for the direct substitution of (2.44) where necessary).

### 2.3.1  Summary 2

We apply the transformation (2.44)-(2.47) to the system of SW governing equations (2.37). The result is

$$y = y(x,t) = \frac{x}{x_N(t)} \qquad (0 \le y \le 1). \qquad (2.49)$$

$$\begin{bmatrix} h_t \\ u_t \end{bmatrix} + \frac{1}{x_N(t)}\begin{bmatrix} u - y\dot{x}_N & h \\ 1 & u - y\dot{x}_N \end{bmatrix}\begin{bmatrix} h_y \\ u_y \end{bmatrix} = \begin{bmatrix} 0 \\ 0 \end{bmatrix}. \qquad (2.50)$$

The characteristic balances for this system can be obtained directly, or by using (2.38) combined with (2.48). The derivation is left for an exercise. We obtain

$$h^{-1/2}dh \pm du = 0 \quad \text{on} \quad \frac{dy}{dt} = \frac{1}{x_N}\left(u - y\dot{x}_N \pm h^{1/2}\right). \qquad (2.51)$$

The initial and boundary conditions (2.41)-(2.43) are not affected by the transformation, with the understanding that $y = 0$ is the left boundary, and that the nose condition (2.39) is applied at $y = 1$. Indeed, the $y$ domain of solution is now $[0,1]$ for all times.

## 2.4   The full behavior by numerical solution

The knowledge accumulated to this point allows us to proceed to a global solution of the typical gravity current. This is achieved by a numerical finite-difference solution. Here we use a Lax-Wendroff scheme and a grid with 400 intervals in the domain $0 \leq y \leq 1$ (i.e., $0 \leq x \leq x_N(t)$). The details are presented in Appendix A.2.

To be specific, we solved here the dimensionless system for a current released from a rectangular lock. Initially $h = 1$ and $u = 0$ in the domain $0 \leq x \leq x_N(0) = 1$. The thickness ratio of the ambient to lock is $H = 3$. The current is subject to the HS $Fr$ condition, and this is the only equation which uses the value of $H$. We recall that the aspect ratio $h_0/x_0$ of the lock has been implicitly incorporated in the scaling. The results represent, in reduced form, a wide range of physical settings.

The SW gravity current results display three stages of propagation. The first is called "slumping," from the release to $t = 3$ at least. The shape of the interface and the distribution of the velocity change dramatically in this stage, but *the nose maintains a constant speed and height*. This stage can be sub-divided in two: First, a dam-break motion propagates from the gate into the reservoir (during $t \leq 1$). This stage of motion will be analyzed in detail later; here we just give a brief description (see Fig. 2.5). During this motion the interface in the reservoir has a horizontal portion beneath which there still is stationary fluid. The next part of the interface has a negative slope, corresponding to a rarefaction (expansion) pressure wave, under which fluid is accelerated. The leading part of the current is a rectangle of constant height $h_N$, which spreads like a slug in front of the lock. Second, after all the fluid is set into motion, the height of the interface in the reservoir domain descends quickly, a positive slope of $h$ appears, and a deceleration expands toward the front. The slumping stage in the one-layer model prevails for about $t = 3$; this is the typical time necessary for the characteristic from the opened lock to sweep the fluid in the reservoir, and then, after reflection from the backwall, to propagate to the nose.

The second stage is a transition without special features. The fluid is under the influence of both the initial conditions and the nose boundary conditions. The initial conditions are eventually forgotten. This second stage is less interesting than the first and third stages, which we consider as major stages in the propagation of an inviscid current of fixed volume.

The third stage is of self-similar motion ($t \geq 8$, approximatively, in Fig. 2.3). The variables $h$ and $u$ maintain their $x$-dependent shape up to a $t$-depending scaling. The speed $u_N$ decays like $t^{-1/3}$. This type of propagation will be discussed in more detail later.

According to the present results the current will propagate to infinity in the self-similar mode. This apparently absurd outcome is because the hindering

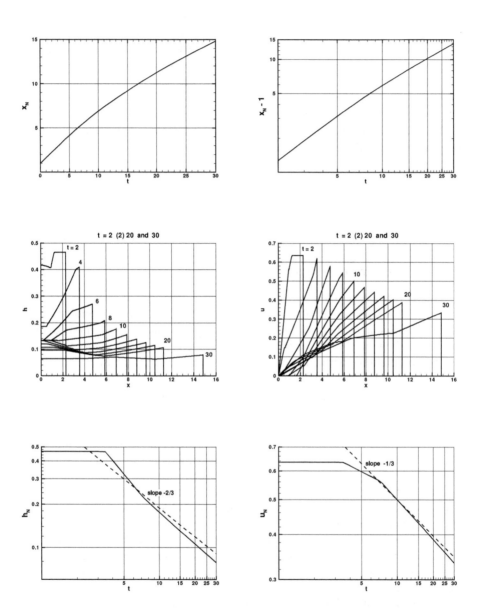

**FIGURE 2.3:** SW results for $H = 3$ ($Fr$ given by HS formula). Top frames: the propagation $x_N$ vs. $t$ and $x_N - 1$ vs. $t$ on log-log axes. Middle frames: the $h$ and $u$ profiles as a function of $x$ for various $t$. Bottom frames: $h_N$ and $u_N$ vs. $t$, log-log axes. Lines (dashed) with slopes $-2/3$ and $-1/3$ are also show to emphasize the self-similar behavior at large $t$. Remark: $t = t_1(\Delta)t_2$ denotes the times from $t_1$ to $t_2$ at interval $\Delta$.

viscous effects were fully discarded. Since both the thickness and velocity of the current decrease with time, the importance of the viscous friction relative to the inertial terms increases monotonically during the self-similar propagation. After some (typically quite long) distance of propagation the present inviscid model becomes invalid. Afterward, the viscous terms control the propagation. The length of the current when transition occurs, $x_V$, depends on the Reynolds number. This will be solved in § 2.7.

The features revealed by the finite-difference solution of the SW equations are supported and further clarified by strong and useful analytical results, in particular for the dam-break and self-similar periods. The derivations of these solutions are our next tasks.

---

## 2.5  Dam-break stage

The dam-break problem provides elegant and insightful analytical solutions to the SW equations. It illustrates the importance of the characteristics in the development of the flow field. This problem is therefore essential to the understanding of the initial motion of a gravity current, a good test-case for the prediction power of the SW models, and an efficient tool for testing numerical solvers of the SW equations.

The closely related dam-break flow of a water reservoir open to the atmosphere is a well known and extensively researched problem in hydraulics, starting with Saint-Venant's solution in 1843; see for example Billingham and King (2000). In the original problem the presence of the embedding ambient fluid is unimportant because the density ratio of ambient to current, $\rho_a/\rho_c$, is very small. The generation of a gravity current by the release of a stationary volume of fluid of given density from behind a lock into an ambient fluid of a (slightly) different density is a dam-break type problem in which the ambient fluid has significant influence on the major motion of the released fluid. First, the "reduced gravity" (based on the relative density difference) replaces the full gravitational acceleration of the classical water-air configuration; and second, the gravity current is subject to an additional condition, namely, that the front of the propagating fluid (i.e., the nose of the current) advances like a "wall" in the embedding fluid. This approach is attributed to Abbott (1961).

The system under consideration is sketched in Fig. 2.4. A given volume of homogeneous fluid of density $\rho_c$ is initially at rest in a rectangular reservoir of height $h_0$ and length $x_0$, bounded by a "dam" (or gate) at $x = 0$ and a solid wall at $x = -x_0$. This is embedded by an ambient stationary fluid of a smaller density, $\rho_a$. We assume that the height of the dense fluid does not exceed that of the ambient, $H \geq h_0$.

This system is kept in equilibrium by the dam (whose thickness is assumed

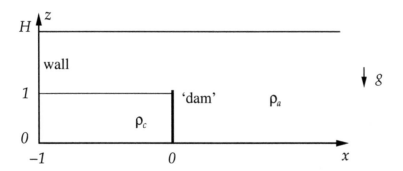

**FIGURE 2.4**: Schematic description of the dam-break system at $t = 0$. Note that the dam is at $x = 0$. $x$ and $z$ are scaled with $x_0$ and $h_0$, respectively.

negligibly small as compared to the length of the reservoir). There is a significant pressure difference between the sides of the dam. The corresponding force is $0.5\Delta\rho g h_0^2$ (dimensional) per unit width. We can say that this is the buoyancy force, $F_B$, of the fluid. This force is balanced by the dam in the static state. When the dam is suddenly removed, this force sets the fluid in motion and a dynamic balance appears. The behavior of the fluid in this new balance is derived here, under the assumption that the viscous effects are unimportant. We consider a Boussinesq system.

At time $t = 0$ the dam is instantaneously removed, and a two-dimensional current commences to spread in the direction $x$. It is convenient to keep in mind that $\partial p_c / \partial x = 0$ in the reservoir when the dam is removed. This implied that the motion of the dense fluid develops quickly at $x \approx 0$, but the fluid inside the reservoir is not instantaneously affected. Actually, this fluid remains in equilibrium until a backward-propagating rarefaction (or expansion) wave brings the information that a decrease of pressure occurred in the fluid at a larger $x$, and hence motion gradually develops. We say that the end of the dam-break stage occurs when the motion that has started at the dam reaches the backwall $x = -x_0$ of the reservoir. (This is a quite arbitrary definition; here it means that all the fluid in the reservoir has been set in motion, and a significant change in the mathematical description must be introduced.)

We use dimensionless variables, scaled according to (2.33)-(2.34). *In this section the origin $x = 0$ is set at the position of the dam.* Initially, the height of the released current is 1, its length is 1, and the speed is zero. The height of the ambient fluid is $H$, assumed $\geq 1$.

The continuity and momentum equations are given by (2.37). For the present solution it is important that this system is hyperbolic. We can apply the powerful method of characteristics which reduces the PDE equations for $h(x,t)$ and $u(x,t)$ into ODE equations. Using the standard procedure, we obtained in § 2.2.3 the characteristic paths

$$\frac{dx}{dt} = c_{\pm} = u \pm h^{1/2}, \tag{2.52}$$

and the relationships between the variables on these trajectories

$$h^{-1/2}dh \pm du = 0. \tag{2.53}$$

The initial conditions for the dam-break problem are: zero velocity and unit dimensionless height and length at $t = 0$. The boundary conditions are: at $x = -1$ the velocity is zero, and an additional condition for the velocity $u$ at the nose $x = x_N(t)$ should be considered. In the classical hydraulic dam-break problem the nose height of the water (that propagates against the negligible resistance of the embedding air) decreases smoothly to zero at $x_N(t)$, and hence the velocity of the tip of the current follows straightforwardly from the internal solution. Actually, there is no sharp front or nose in this case (see also § 10.3.1). For the Boussinesq gravity current the nose, that encounters the resistance of the quite dense embedding fluid, is a discontinuity. We can view it as a moving piston, or interface jump, of height $h_N$ and velocity $u_N$. The justification is provided by the theoretical analysis of Chapter 3 and by the fact that this condition produces good agreements with experimental and numerical results concerning the speed of propagation. The appropriate boundary condition is (2.39), with a given $Fr$ as a function of $h_N/H$. To be specific, we use the HS $Fr$ function (2.40). We shall see that the features of the dam-break solution are only slightly influenced by the exact details of this function.

The dam-break problem is of particular interest to the understanding of the inviscid gravity current because the existence of a relatively simple analytical solution (which is derived by the method of characteristics). The typical expected behavior that guides the analysis is presented in Fig. 2.5, in the physical plane $xz$ and in the characteristic plane $xt$. Point $W$ represents the fixed position of the wall ($x = -1$). Points $L$ and $M$ represent the moving loci of the end of the unperturbed domain and start of nose-dominated domain, respectively. The interface $h(x,t)$ is inclined in the $LM$ domain. The letter $N$ represents the moving nose. Point $O$ is the origin $x = 0, t = 0$ in the characteristic plane. We remark that $M$ is not necessarily to the right of the origin as sketched in Fig. 2.5.

The equation (2.53) of the balances on the characteristics can be easily integrated. The result is

$$u + \Upsilon(h) = \Gamma_{+} \quad \text{on} \quad \frac{dx}{dt} = c_{+}; \tag{2.54}$$

$$u - \Upsilon(h) = \Gamma_{-} \quad \text{on} \quad \frac{dx}{dt} = c_{-}; \tag{2.55}$$

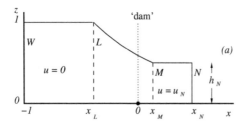

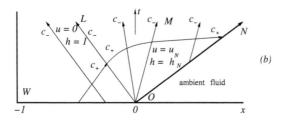

**FIGURE 2.5:** Schematic description of the current during the initial dam-break stage. (a) geometry in $xz$ plane; (b) characteristics in $xt$ plane (here point $O$ is the origin).

where $\Gamma_\pm$ are constants,

$$\Upsilon(h) = \int_0^h \sqrt{h'}dh' = 2\sqrt{h}, \tag{2.56}$$

and $c_\pm$ is given by (2.52).

The solution is now straightforward. At a point $(x, t)$ of interest we have two equations, (2.54) and (2.55), for the two unknowns $u$ and $h$. We must carefully determine the origin of the two characteristics (denoted by $+$ and $-$) which arrive at this point; this specifies the values of the appropriate $\Gamma_\pm$. We thus obtain $u$ and $h$. In the present case this procedure is successful because there are no shocks (i.e., contradictory, or insufficient, information at some points), the trajectories of the characteristics are easily determined, and the trajectory of the nose (the line $ON$) is straight.

Consider the $WOL$ characteristic region. This domain is covered by characteristics that start at $t = 0$ in the range $x < 0$ and carry the information $u = 0$ and $h = 1$, i.e., $\Gamma_+ = -\Gamma_- = \Upsilon(1) = 2$. Substitution of these conditions into (2.54)-(2.55) yields $u = 0$ and $h = 1$ in the entire $WOL$ domain. For these values we obtain $c_\pm = \pm 1$, i.e., the characteristics that carry these values are straight lines. The largest $x$ of the pertinent domain is given by the line $OL$:

$$x_L(t) = t \cdot c_- = -t. \tag{2.57}$$

The result (2.57) is valid for $t \leq 1$, until $L$ reaches the wall at $x = -1$. This occurrence, at $t = 1$, is considered here as the end of the dam-break stage. Subsequently, all the fluid is moving. The boundary condition at $x = -1$ becomes time dependent, and this complicates the analysis by the method of characteristics. Note, however, that the nose "does not know" yet that the expansion wave has reached the wall. The "last" $c_+$ characteristic which carries the $h = 1$ information to the nose is emanated at $t = 1^-$. The nose is supplied with this information until this "last" characteristic arrives at $x_N$, which is at about $t = 3$. During this time, which we call slumping interval, the nose moves with the constant $u_N$ established by the dam-break process.

Consider the solution of (2.54)-(2.55) in the *LOM* domain of characteristics. Assume that $c_+$ characteristics from the above-mentioned *WOL* domain enter the present domain, i.e., on $c_+$

$$u + \Upsilon(h) = \Upsilon(1) = 2. \tag{2.58}$$

We argue that the relevant $c_-$ characteristics are emanated from point $O$ as a fan, as follows. During a short initial time, $t_1 \ll 1$, the interface of the fluid about $x = 0$ must adjust to the new situation created by the removal of the gate. The lower part, of fast fluid particles, will propagate as a vertical front, $z \leq h_N$. The "last" fluid particle will remain at $z = 1$. In between, there is a thin $x$ layer of fluid in which $h$ decreases with $x$. On the scale of the process, this layer is seen as the point $x = 0, t = 0$ where the initial value of $h$ for the characteristics decreases continuously from 1 to $h_N$. This is expressed as

$$u - \Upsilon(h) = \Gamma_- \qquad (h_N \leq h \leq 1) \tag{2.59}$$

on the rays

$$\frac{dx}{dt} = c_- = u - h^{1/2}, \quad x(t = 0) = 0. \tag{2.60}$$

The intersection of (2.58) and (2.59), and use of (2.60) yield

$$u = \Upsilon(1) - \Upsilon(h) = 2(1 - \sqrt{h}) = C_1, \tag{2.61}$$

and $h = C_2$ on the ray

$$\frac{x}{t} = u - h^{1/2} = 2 - 3\sqrt{h}. \tag{2.62}$$

These rays "fan out" from the origin and do not intersect. The last ray in the fan corresponds to $h = 1$, and this recovers the trajectory $x_L(t)$ given by (2.57). The first ray in the fan corresponds to $h = h_N$ and yields the trajectory of point $M$ as follows

$$x_M(t) = (2 - 3\sqrt{h_N})t. \tag{2.63}$$

The combination of (2.61) and (2.62) can be reformulated as

$$h = \frac{1}{9}\left(2 - \frac{x}{t}\right)^2; \quad u = \frac{2}{3}\left(1 + \frac{x}{t}\right); \quad (-1 \leq \frac{x}{t} \leq 2 - 3\sqrt{h_N}). \tag{2.64}$$

This gives the typical parabolic with $x$ shape of the interface in the expansion domain; see line $LM$ in Fig. 2.5 (a). The dependency of $h$ and $u$ in the expansion domain $LOM$ is on $(x/t)$, which can be defined as the similarity variable of this domain. Thus, the scaling length $x_0$ can be chosen arbitrary to describe the motion until $L$ encounters the backwall. This makes sense: the dam-break motion does not depend on the length of the reservoir until all the fluid is set into motion.

Suppose that $h_N \leq 4/9$; we shall see below that this corresponds to $Fr \geq 1$. In this case $x_M \geq 0$; see (2.63). In other words, the expansion fan covers the position of the lock, $x = 0$. This is the typical situation for a deep ambient. In this case, an interesting result of (2.64) is that at the position of the lock, $x = 0$, the height of the current is $4/9$ and the speed is $2/3$, independent of $t$ and on the speed of the nose.

It is easy to verify that $c_+ > c_-$ in the $LOM$ domain. Consequently, the "+" characteristics from the reservoir, which carry the value $\Gamma_+ = 2$, cover the expansion fan domain and then proceed beyond the $OM$ line, into the $MON$ domain.

Finally, consider the characteristic region $MON$ where $u = u_N$ and $h = h_N$. The intersection of (2.61) with (2.39),

$$u_N = \Upsilon(1) - \Upsilon(h_N) = 2 - 2\sqrt{h_N} = Fr(h_N)\sqrt{h_N}, \qquad (2.65)$$

provides (uniquely) the value of $h_N$, and then of $u_N$. The $c_-$ characteristics in the $MON$ domain are lines parallel to $OM$ and carry the consistent information $u = u_N$, $h = h_N$. This closes the dam-break solution: we obtained $u$ and $h$ for all the current domain $-1 \leq x \leq x_N(t)$, for $0 < t \leq 1$.

In general, (2.65) must be solved numerically, and results for the $Fr$ functions of Benjamin and Huppert-Simpson are shown in Fig. 2.6. Formally, we can rewrite (2.65) in the more insightful form as

$$h_N = \left(\frac{1}{1 + Fr/2}\right)^2; \qquad (2.66a)$$

$$u_N = \frac{Fr}{1 + Fr/2}; \qquad (2.66b)$$

but we keep in mind that $Fr$ on the RHS must correspond to the value of $h_N$ on the LHS of (2.66)(a). When $Fr$ *is a constant*, as expected for a deep ambient, its direct substitution into (2.66) provides the solution for the initial stage of propagation with constant speed. For $Fr = 1.19$ this yields $h_N = 0.393$ and $u_N = 0.746$; for $Fr = \sqrt{2}$ the values are $h_N = 0.343$ and $u_N = 0.828$. The $Fr$ functions of interest in the Boussinesq domain decrease to about 0.5 when $h_N/H$ increases to 0.5. This explains the dependency on $H$ observed in Fig. 2.6.

The result (2.66) indicates that when the thickness of the ambient, $H$, decreases then $u_N$ decreases. This is consistent with observations (to be discussed later, but see Fig. 5.11). On the other hand, the value of $h_N$ increases

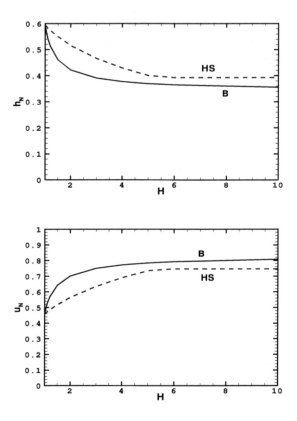

**FIGURE 2.6:** Nose height and speed in dam-break analytical solution as functions of $H$, for $Fr$ functions of Benjamin and HS. One-layer SW model.

when $H$ decreases. These trends of $u_N$ and $h_N$ appear because the gravity current encounters a larger resistance from the ambient fluid when $H$ decreases, and therefore the speed decreases while the pressure force at the nose is increased to counterbalance the resistance. For $H < 1.2$ (approximately) we obtain $h_N/H > 0.5$, which violates the energy restriction. This is an indication that the one-layer SW model needs correction in shallow ambient cases. An *ad hoc* remedy is to impose the choking restriction $h_N = 0.5H$ and the corresponding speed $u_N = Fr(0.5)(0.5H)^{1/2}$. The rigorous approach to this difficulty is the two-layer model discussed in Chapter 5; we note that a similar dam-break behavior appears, but the expansion domain is complicated by a backward-propagating shock.

It is important to emphasize that the speed of the nose is time-independent for both constant and $h_N$ dependent $Fr$. The prediction that the gravity current propagates initially with constant speed has wide experimental support.

This will be elaborated in Chapter 5. The fact that the SW model provides a closed prediction and interpretation of this experimental robust property of the gravity current lends strong support to this theoretical approach. The values of the predicted $u_N$ are also in fair agreement with measurements. This will be discussed in Chapter 5.

The instantaneous appearance of a finite $u_N$ in a fluid released from rest is physically problematic. The instantaneous drop of height from $h = 1$ to $h_N$ at $x = 0$ poses a similar difficulty. It is therefore evident that the present solution of the SW equations implies a short adjustment stage of time $t_1$ in which large accelerations (including in the vertical direction), and perhaps viscous effects, are significant. The solution is therefore valid only for $t > t_1$. We can use the previous results, and in particular (2.64), to estimate the magnitude of the initial accelerations. It is left for an exercise to show that consistency with the assumptions of the SW model requires $t_1 = O(h_0/x_0)$. A typical comparison between the analytical and numerical SW dam-break results is presented in Fig. 2.7. The agreement is, as expected, excellent, which provides a straight-forward verification of both solutions. The backward-moving expansion wave reaches the wall at $t = 1$, which explains the discrepancy between the present analytical and finite-difference solutions at later times.

A real laboratory saltwater Boussinesq current is shown in Fig. 2.8 in three frames for the dimensionless $t = 0.79, 1.4, 2.1$ for comparison. The current propagates with the constant $u_N = 0.66$, approximately, which is 12% below the analytical prediction 0.75 (for the HS $Fr$).

Here we also use some Navier-Stokes simulations (which can be regarded as "numerical experiments") to gain more insight into the initial behavior of the current. Results of the Navier-Stokes (NS) solution are presented in Figs. 2.9-2.10, following Ungarish (2005a). The details of the numerical approach are given in Appendix B. Briefly, the simulation was performed via a finite-difference solution of the Navier-Stokes equations introduced in § 1.2. The specific formulation is in a two dimensional Cartesian geometry, $x$ and $z$, for the variables $(u, w, p, \phi)$. Here $\phi$ is the reduced density function $[\rho(x, z, t)/\rho_a - 1]/\epsilon$, whose initial value is 1 in the current and 0 in the ambi-ent. No-slip conditions are applied on the bottom, back, and front boundaries of the container (tank), while the top boundary of the container is treated as a frictionless lid. This makes the simulation compatible with real rectangular open-tank laboratory systems. Forward-time and central-space discretizations are used in the continuity and momentum equations, and the MacCormack forward-backward explicit method is used for the density function transport equation (derived from (1.8)) to achieve a sharp numerical interface.

We consider currents with $H = 3$. In the simulations $\epsilon = 0.05$ and $Re = Uh_0/\nu = 8333$, the initial aspect ratio is $x_0/h_0 = 4$, the horizontal length of the channel is 3.5, and the grid is $200 \times 200$. This configuration is compatible with typical laboratory saltwater experiments in an open tank of about $15\,\text{cm} \times 200\,\text{cm}$ (height × length) and at least $10\,\text{cm}$ width.

Density contour plots at $t = 1$ and 1.5 are displayed in Fig. 2.9. The NS

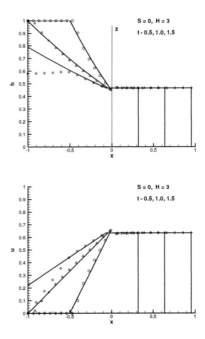

**FIGURE 2.7:** Analytical (lines) and SW numerical (symbols) results, of $h(x)$ and $u(x)$ at $t = 0.5, 1.0, 1.5$, for $H = 3$ and HS $Fr$ function. ($S = 0$ means no stratification; as discussed later.) (Ungarish 2005a, with permission.)

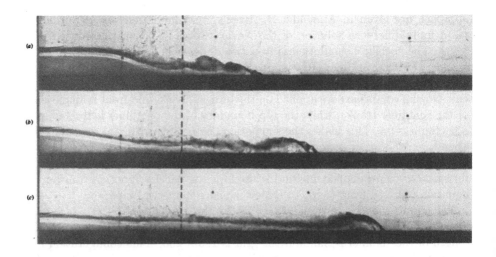

**FIGURE 2.8:** Shadowgraph of a gravity current with $h_0 = 7$ cm, $x_0 = 50$ cm, $g' = 47$ cm s$^{-2}$ at $t = 2.2, 3.8$, and $6.0$ s (in dimensionless units, $t = 0.79, 1.4$, and $2.1$) after release. The current is deep: $H = 7$ (dimensionless). The dotted vertical line indicates the position of the lock gate, and the thin lines are at $10$ cm intervals. (Rottman and Simpson 1983, with permission.) Here the reference speed and time are $U = 18$ cm s$^{-1}$ and $T = 2.8$ s.

simulations display a fair global agreement with the SW predictions. The dense fluid displays, after a quick adjustment motion, flow-field domains that can be identified with the regions of Fig. 2.5: $MN$ following the head, $LM$ with inclined interface, and unperturbed $LW$. The nose is a prominent discontinuity and propagates with velocity close to the predicted $u_N$. Indeed, the SW prediction (with the HS $Fr$) is that $u_N = 0.635$, and hence at $t = 1.5$ the nose would be at $x_N = 0.95$. This is in very good agreement with the position of the nose obtained in the NS simulations, Fig. 2.9. The rarefaction wave of the simulated interface reaches the wall $x = -1$ at (approximately) the time predicted by the SW theory. However, the shape of the interface in domain $LM$ differs from the SW predictions, perhaps as a result of the strong shear about the interface in the real flow.

Contour lines of the simulated $u(x, z, t)$ fields at $t = 1$ and 2 are shown in Fig. 2.10. The comparison with the SW results indicates fair agreement in the $WM$ (practically, $x < 0$) domain; however, in the domain of fluid that trails the nose $u$ displays some fluctuations about the constant (average) value predicted by the SW theory. This is not surprising, since the nose of a real gravity current is known to be a very complicated flow field (Simpson and Britter 1979, Härtel et al. 2000), and the SW theory lumps this behavior in an idealized jump condition.

In summary: the analytical solution of the SW equations predicts an initial motion with constant $h_N$ and $u_N$ (typical values are 0.4 and 0.7, respectively), which is clearly dominated by the forward and backward propagation of information on characteristics. During the dam-break interval of $t = 1$ a rarefaction wave propagates from the dam to the backwall of the reservoir, and the interface in this domain is parabolic with a negative slope. At $t = 1$ this wave is reflected, and a more complex motion develops in the rear part of the current and starts to expand toward the nose. However, the nose continues to propagate with the constant $h_N$ and $u_N$ for a while, until the reflected wave reaches the nose (at $t \approx 3$). This is the end of the slumping stage. The dam-break solution provides a good explanation to the behavior of the current provided in § 2.4 by the numerical solution of the SW equations for initial times, and is in fair agreement with laboratory experiments and NS simulations.

When $H < 2$, some deviations from this one-layer model solution appear. The most pronounced discrepancy during the dam-break stage is the formation of a backward-moving discontinuity (jump) of $h$ at position $L$ (which is subsequently reflected at $x = -1$ as a forward-moving bore). In addition, when $H$ approaches 1, the slumping distance of propagation with constant $u_N$ is significantly prolonged, and in some circumstances the maximum velocity of the $c_+$ characteristic imposes some limitations on the propagation of the nose. To obtain the details a two-layer model must be used, as discussed later in Chapter 5. The one-layer model misses these effects for $H < 2$, but predicts fairly well the initial velocity of propagation. For large $H$ ($\geq 3$, say) the differences between the two and one-layer models are rather academic.

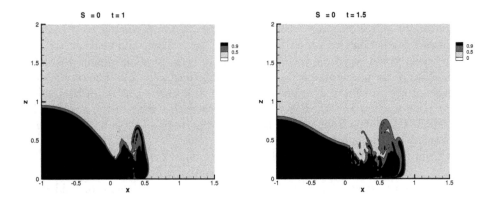

**FIGURE 2.9**:   NS simulation results: contours of the density function $\phi$ at two times, $H = 3$. (Ungarish 2005a, with permission.)

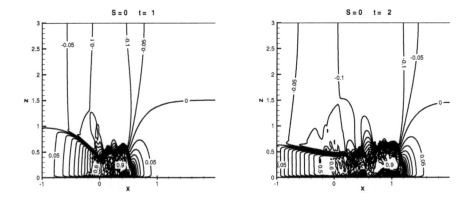

**FIGURE 2.10**:   NS results: contours of the horizontal velocity $u(x, z)$ at two times, $H = 3$. (Ungarish 2005a, with permission.)

We also note that the dam-break behavior discussed here is concerned with a straight vertical dam. The geometry of the reservoir is important to the initial motion of the characteristics and acceleration of the fluid. This type of effects will be discussed later. However, these initial conditions are eventually "forgotten" by the current. At large times the solution displays self-similar profiles which scale with the volume of the released fluid, irrespective of the initial shape. We now proceed to this topic.

---

## 2.6   Similarity solution

Similarity (or self-similar) solutions are the result of some transformations which reduce the system of the PDE equations dependent on $x$ and $t$ to a system of ODE of a single variable, which we denote $y$ (not to be confused with the Cartesian coordinate). The new similarity independent variable is a combination of $x$ and $t$, typically $y = x/t^\beta$, where $\beta$ is some positive constant. In the present context the implication is that, after some significant propagation, the current "forgets" the initial conditions. Afterward, (a) $u$ and $h$ attain and maintain a generic constant shape, modified by some time-dependent scale factors; and (b) the propagation is with $t$ at some power (to be specific, $x_N = Kt^\beta$). Some hints in this directions are provided by the numerical solution of the SW equation for $t > 10$ displayed in Fig. 2.3. The formal analysis is quite complicated; see Grundy and Rottman (1985a) (following Fanneløp and Waldman 1972 and Hoult 1986). Here we present a short derivation of the essential results.

We use the transformation $y = x/x_N(t)$ and the corresponding equations of motion (2.50), subject to the boundary condition

$$u(y = 1, t) = \dot{x}_N = Fr \, [h(y = 1, t)]^{1/2}. \tag{2.67}$$

In addition, the similarity solution is subject to global volume conservation.

The similarity solution considered here is relevant for large times after release. The current is thin and we therefore consider $Fr$ to be a constant. We assume

$$x_N(t) = Kt^\beta; \quad h(y,t) = \varphi(t)\mathcal{H}(y); \quad u(y,t) = \dot{x}_N\mathcal{U}(y); \tag{2.68}$$

where $K$ and $\beta$ are some positive constants. The similarity coordinate is $y = x/(Kt^\beta)$ in the range $[0, 1]$. The obvious boundary condition is $\mathcal{U}(1) = 1$.

The form of $h$ is inspired by continuity consideration; we expect $\varphi = C\mathcal{V}/x_N(t)$, where $\mathcal{V}$ is the volume of the current, in general a time dependent variable, but a constant in the present case.

The objective is to determine $\varphi(t)$, $\beta$, $\mathcal{U}(y)$, $\mathcal{H}(y)$, and $K$.

Consider the boundary condition at the nose, (2.67). To satisfy it we use

$$\varphi(t) = \beta^2 K^2 t^{2\beta-2} = \dot{x}_N^2; \tag{2.69}$$

and

$$\mathcal{H}(1) = Fr^{-2}. \tag{2.70}$$

(This convenient choice of $\mathcal{H}(1)$ is not unique, but the final results are unaffected.)

Consider the integral volume continuity (global volume conservation) of the current, and use (2.68) and (2.69) to write

$$\mathcal{V} = \int_0^{x_N} h(x,t)dx = \int_0^1 h(y,t)x_N dy = x_N\varphi(t)\int_0^1 \mathcal{H}(y)dy =$$

$$\beta^2 K^3 t^{3\beta-2} \int_0^1 \mathcal{H}(y)dy. \tag{2.71}$$

When $\mathcal{V}$ is a constant, this equation is satisfied by

$$\beta = \frac{2}{3} \tag{2.72}$$

and

$$K = \left[\frac{4}{9}\int_0^1 \mathcal{H}(y)dy\right]^{-1/3} \mathcal{V}^{1/3} = K_1\mathcal{V}^{1/3}. \tag{2.73}$$

We note that the value of $\beta$ in the self-similar motion is determined by the nose condition (the dependency of $u_N$ on $h_N$) and by the global volume conservation. Therefore, $\beta$ is expected to be a robust property of the gravity current motion, quite insensitive to the details of the $\mathcal{H}$ and $\mathcal{U}$ profiles.

Next we use the equations of motion (2.50). Substitution of (2.68) and (2.69) yields, after some arrangement, the continuity equation as

$$(\mathcal{U} - y)\frac{\mathcal{H}'}{\mathcal{H}} + \mathcal{U}' = 2\left(\frac{1}{\beta} - 1\right); \tag{2.74}$$

and the momentum equation as

$$\mathcal{H}' + (\mathcal{U} - y)\mathcal{U}' + \left(1 - \frac{1}{\beta}\right)\mathcal{U} = 0; \tag{2.75}$$

where the prime denotes $y$ derivative. Recall that the boundary conditions at $y = 1$ are $\mathcal{U} = 1$ and $\mathcal{H} = Fr^{-2}$. In general, the system (2.74)-(2.75) must be solved numerically. However, for some values of $\beta$ the equations can be decoupled and simple solutions follow. This happens in the present case.

Indeed, for the present $\beta = 2/3$ the RHS of (2.74) is equal to 1. This equation admits the simple solution

$$\mathcal{U}(y) = y. \tag{2.76}$$

This simple $\dot{x}_N y$ behavior of $u$ derived above turns out to be a very versatile outcome. First, the boundary condition $u(y = 0, t) = 0$ is satisfied. Second, this provides a significant simplification to the equations of motion (2.50): the diagonal terms in the matrix vanish. For this reason it makes sense to start the search for a similarity solution with the assumption $\mathcal{U}(y) = y$; in currents of fixed volume this shortcut works, or at least indicates what modifications are necessary for obtaining the solution.

To calculate $\mathcal{H}(y)$, we substitute (2.76) and $\beta = 2/3$ into the momentum equation (2.75). We obtain

$$\mathcal{H}'(y) - \frac{1}{2}y = 0. \tag{2.77}$$

Integration, subject to the condition $\mathcal{H}(1) = Fr^{-2}$, gives

$$\mathcal{H}(y) = \frac{1}{Fr^2} - \frac{1}{4} + \frac{1}{4}y^2. \tag{2.78}$$

Finally, we calculate $K$. With the result (2.78) we go back to the outcome of volume conservation (2.73), and obtain

$$K_1 = \left[ \frac{27}{2} \frac{Fr^2}{6 - Fr^2} \right]^{1/3}; \quad K = K_1 \mathcal{V}^{1/3}. \tag{2.79}$$

To be specific $K_1 = 1.61$ for $Fr = 1.19$ and $1.890$ for $Fr = \sqrt{2}$.

We summarize:

$$x_N = Kt^{2/3}; \quad h = \dot{x}_N^2 \left[ \frac{1}{Fr^2} + \frac{1}{4}(y^2 - 1) \right]; \quad u = \dot{x}_N y. \tag{2.80}$$

Since $\dot{x}_N = (2/3)Kt^{-1/3}$, (2.80) can also be expressed as

$$x_N = Kt^{2/3}; \quad h = \frac{4}{9}K^3 x_N^{-1} \left[ \frac{1}{Fr^2} + \frac{1}{4}(y^2 - 1) \right]; \quad u = \frac{2}{3}K^{3/2} x_N^{-1/2} y. \tag{2.81}$$

It is now appropriate to check the physical consistency of the mathematical solution for $0 \leq y \leq 1$. An inspection of the results shows that the thickness $h$ may become negative for small $y$; see (2.78). The condition for $\mathcal{H} \geq 0$ in the entire domain is $Fr^2 \leq 2$. For Boussinesq systems this condition is satisfied, according to both theory and experimental evidence. The validity of the similarity solution in non-Boussinesq systems will be discussed later.

The similarity profiles are displayed in Fig. 2.11.

The advantage of the similarity solution is that this is an exact and simple analytical solution of the equations and boundary conditions. The major results concerning propagation with $t^{2/3}$, decay of thickness with $t^{-2/3}$, and linear dependency of $u$ on $x$ are a versatile description of the behavior at large $t$. This behavior was indicated by the numerical results displayed in Fig. 2.3, but (2.80) is a general and clear-cut deduction.

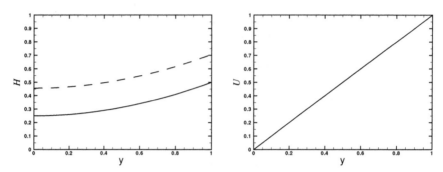

**FIGURE 2.11:** Similarity profiles for 2D SW current. $\mathcal{H}(y)$ is given for $Fr = 1.19$ (solid line) and $Fr = \sqrt{2}$ (dashed line). The similarity profile $\mathcal{U}$ does not depend on $Fr$.

However, this result is actually a weak prediction tool. The self-similar solution cannot satisfy simple initial conditions of $u$ and $h$ at $t = 0$. We imposed only the condition of a prescribed volume, expressed by $\mathcal{V}$ in (2.79). The incompatibility with initial conditions is even more severe. It can be easily proved that the substitution $t + \gamma$ for $t$, where $\gamma$ is a constant, does not affect the validity of the previous results. This is sometimes referred to as a "virtual origin" property of the similarity result. Practically, this means that the solution may be inaccurate unless we specify the additional constant $\gamma$. This can be determined by some matching with a known result (which satisfies the initial conditions) at some time (say $t_m = 10$) where some overlap is expected. For example, we can calculate $\gamma$ by the requirement that at $t = 10$ the value of $x_N$ predicted by the finite-difference solution of the SW equations (or provided by some accurate experiment) is equal to $K(10 + \gamma)^{2/3}$ corresponding to the similarity solution. For the case illustrated in Fig. 2.3 we have $x_N(10) = 6.90$ and $K = 1.61$, which yield $\gamma = -1.12$.

It is not clear *a priori* that a given current will attain the self-similar behavior, or when this will happen. Experiments and numerical solutions of the SW equations (Rottman and Simpson 1983 Fig. 12, and Grundy and Rottman 1985a) show that the self-similar propagation is a robust feature of the gravity current. Strictly speaking, the solution $h(y, t)$ and $u(y, t)$ of the SW equations for a current released from a rectangular lock will attain a close agreement with the idealized counterparts given by (2.80) only after a time of propagation of the order of 100. However, the leading portion of the gravity current (say $0.7 < y \leq 1$) seems to develop the essentials of the similarity behavior at a quite early stage, typically one lock-length after the end of the slumping stage (propagation with constant speed). We can argue that the self-similar decay of nose velocity cannot start before the backward-forward motion of the rarefaction (expansion) wave produced at $t = 0$ is accomplished. First, the released current is flattened to a rectangle of height $h_N \approx 0.5$, approximately, at about $t = 3$. Next, a characteristic from the wall at $x = 0$ propagates to the front with speed $u_N + h_N^{1/2}$, approximately, with the information that the thickness of the tail must decrease. This reaches the front at $t \approx 4$. This gives the estimate for the time when the current is able to "forget" the initial conditions (the end of the slumping phase) and start adjusting to the self-similar propagation. This description is valid for a non-shallow ambient $H > 2$. In a shallow ambient $H < 2$ the transition to self-similar motion is further delayed by the facts that: (a) the motion of the backward-moving rarefaction wave is constrained; and (b) $a = h_N/H$ is not small, and hence the value of $Fr(a)$ is still not sufficiently close to the constant value assumed in the self-similar solution. To summarize: the self-similar solution is a reasonable approximation after a propagation of the current to $x_N = 4$ when $H > 2$. The $H < 2$ cases require significantly longer distances of adjustment, to be discussed later.

We also note that some difficulties emerge in the comparison of the analytical similarity solution with the numerical solution of the SW equations with given initial conditions. Theoretically, the agreement is bound to improve as $t$ attains larger and larger values. However, for large $t$ the values of $u$ and $h$ are small. The numerical accuracy of time-marching schemes cannot be maintained when $h$ and $u$ become very small while the numerical errors accumulate. As shown by Grundy and Rottman (1985a), a transformation of variables resolves this deficiency; some details are presented later in § 16.3.3.1.

Clearly, the self-similar result is a quite weak prediction tool. However, it may provide a good test for a computer code for the solution of the SW equation as explained in Appendix A.2. Also, it provides a quick indicator for experimental results for high-Reynolds number currents: the decay of speed $u_N$ with $t^{-1/3}$ marks the self-similar inviscid phase, and a stronger decay (as discussed later) reveals the dominance of viscous effects. This is easily detected when the results are displayed on a log-log plot. Note, however, that the behavior of the speed $u_N$ provides a more reliable test than the behavior of $x_N$. This can be inferred from an inspection of Fig. 2.3: we see that on a log-log plot the changes of $u_N$ with time (last frame) are sharper than these of $x_N$ or $x_N - 1$ (first two frames). Indeed, upon the transition from slumping to similarity the slope of $u_N$ changes from 0 to $-1/3$, while that of $x_N - 1$ changes from $2/3$ to $1/3$.

Rescaling is necessary to finalize the similarity solution. We mentioned that in the self-similar propagation the current has forgotten the initial conditions. The careful reader may object, because the scalings of our results are still determined by $x_0$ and $h_0$ of the lock. To eliminate this difficulty, we introduce a new scaling length, $L^*$, for both vertical and horizontal lengths, and also new scales for time and speed, as follows,

$$L^* = \mathcal{V}^{*1/2}; \quad T^* = (L^*/g')^{1/2}; \quad U^* = L^*/T^*; \qquad (2.82)$$

where $\mathcal{V}^*$ is the dimensional volume of the current (per unit width of the 2D container).

Substituting this re-scaling into (2.80) we find that nothing changes, except for the fact that now $\mathcal{V} = 1$, in (2.79). The details are left for an exercise. This indeed proves that the similarity solution is independent of the shape of the lock. We also note the coincidence $\mathcal{V} = 1$ obtained when the original scaling (2.33)-(2.34) is used and the geometry of the lock is rectangular.

### 2.6.1 Extensions

For currents with inflow at the origin, when $\mathcal{V} = qt^\alpha$, similarity solutions of the form (2.68) can also be obtained. The similarity coordinate is, again, $y = x/x_N(t)$. The boundary conditions at $y = 1$ and the function $\varphi(t)$ are as in the foregoing solution. However, the global volume conservation (2.71) imposes $\beta = (\alpha + 2)/3$.

In general, however, the similarity profile of $u$ is not linear with $y$. The corresponding function $\mathcal{U}(y)$ and $\mathcal{H}(y)$ must be determined (numerically) from the system (2.74)-(2.75); moreover, jumps (hydraulic shocks) in the profiles may appear. The details are beyond the scope of this text, and can be found in Grundy and Rottman (1985b) and Gratton and Vigo (1994). (For the convenience of the reader, we note the following differences of notation: $\delta, \zeta, \beta$ in these papers corresponds to $\beta, y, Fr$ in the present text.) Also note that these papers do not specify the value of $Fr$.

An interesting exception is the case of inflow with $\alpha = 1$, for which $\beta = 1$. The resulting simplified system (2.74)-(2.75) shows that the constant plug flow profiles $u(y, t) = FrC$ and $h(y, t) = C^2$ for $0 \leq y \leq 1$ are a similarity solution of the SW formulation. It is left for an exercise to verify this and calculate the value of $C$.

The similarity propagation with $t^{(\alpha+2)/3}$ is, in general, also well reproduced by the box-model approximations which will be discussed later in Chapter 4.

---

## 2.7   The validity of the inviscid approximation

The importance of the viscous friction on the motion of the current increases with time and distance of propagation. Even for quite large values of $Re = (g'h_0)^{1/2}h_0/\nu$, the inviscid SW formulation may become invalid at moderate values of $x_N$. To monitor this effect, we use the previous results for $u$ and $h$ to estimate the time-dependent ratio of global inertial, $F_I$, to viscous, $F_V$, effects. The inertia per unit volume (dimensional) is well represented by $\rho_c u u_x$, and the viscous force per unit area (dimensional) is expected to be proportional to $\rho_c \nu u/h$. Switching to dimensionless variables with the aid of (2.34), we write the ratio of the inertial forces in the volume of the current to the viscous force on the contact area as

$$\frac{F_I}{F_V} = \left(\frac{Uh_0}{\nu}\right)\frac{h_0}{x_0}\frac{\int_0^{x_N(t)} uu_x h\,dx}{\int_0^{x_N(t)}(u/h)\,dx} = Re\frac{h_0}{x_0}\theta(t).$$  (2.83)

The function $\theta(t)$, or more conveniently $\theta(x_N)$, is expected to be of the order of unity at the beginning of the propagation, and to decay to quite small values eventually. This function can be easily calculated from the SW results for $u(x,t)$ and $h(x,t)$. The typical behavior of $\theta$ for a rectangular current configuration with $H = 3$ is displayed in Fig. 2.12. This function is indeed of the order 1 during the early stages of propagation and decreases monotonically to very small values for $t \approx 20$ and $x_N \approx 10$.

Using these results we can estimate the importance of the viscous terms, and also the limit of validity of the inviscid assumption, for a real gravity current with given $Re$ and $h_0/x_0$. The inviscid theory is expected to be relevant for, roughly, $F_I/F_V > 1$ (a more pessimistic researcher may prefer $> 3$). Let us denote by $x_V$ the position of propagation at which $F_I/F_V = 1$ (i.e., where $\theta$ is already so small that it exactly counteracts the large $Reh_0/x_0$). When the current spreads to $x_N > x_V$ it enters into the domain of $F_V > F_I$ where viscous influence is important. This provides the estimate of $x_N < x_V$ as the domain of validity of the inviscid SW model. The present $\theta$ results, Fig. 2.12, provide a sharp practical means for evaluating $x_V$ for a realistic current of given $Re(h_0/x_0)$. Roughly, for $Re(h_0/x_0) = 10^4$ the value of $x_V$ is 15.

The transition of the flow from inertial to viscous dominance is not necessarily sharp. There are, however, some experimental observations which indicate an abrupt change of the pattern of propagation as a function of $t$ when viscous effects become significant. The inertia-dominated slumping and self-similar propagation are so well defined by $u_N =$ Const and $u_N \propto t^{-1/3}$, respectively, that a different (actually stronger) deceleration is easily detected. Some details will be given later.

A useful analytical approximation for $\theta$ and $x_V$ can be developed. The typical current becomes viscous after a significant propagation, i.e., after reaching the stage of self-similar propagation. It is therefore both convenient and accurate to use the exact analytical results (2.81) to evaluate $\theta$. To be specific, the integral at the numerator of (2.83) yields

$$I_I = \left(\frac{4}{9}K^3\right)^2 \frac{1}{x_N^2}\int_0^1 (c + \frac{1}{4}y^2)y\,dy,$$  (2.84)

and the integral at the denominator of (2.83) is

$$I_V = \frac{3}{2}K^{-3/2}x_N^{3/2}\int_0^1 \frac{y\,dy}{c + \frac{1}{4}y^2},$$  (2.85)

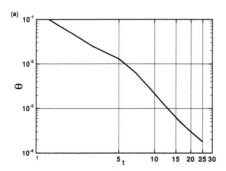 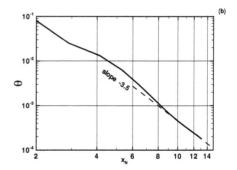

**FIGURE 2.12:** SW results. The coefficient $\theta$, see (2.83), for $H = 3$ (a) as a function of time; and (b) as a function of $x_N$, displaying also the slope of the similarity derivation (2.86) .

where $c = Fr^{-2} - (1/4)$. Consequently

$$\theta = \left(\frac{2}{3}\right)^5 K_1^{15/2} \int_0^1 (c + \frac{1}{4}y^2)y\,dy \left[\int_0^1 \frac{y\,dy}{c + \frac{1}{4}y^2}\right]^{-1} \mathcal{V}^{5/2} x_N^{-7/2} =$$

$$C_V \mathcal{V}^{5/2} x_N^{-7/2}. \quad (2.86)$$

The calculations give $C_V = 1.56$ for $Fr = 1.19$ and $C_V = 2.11$ for $Fr = \sqrt{2}$.

The position of the nose when $F_I/F_V$ attains the value 1 is calculated by combining (2.83) with (2.86). We obtain

$$x_V = \left(Re\frac{h_0}{x_0}C_V \mathcal{V}^{5/2}\right)^{2/7}. \quad (2.87)$$

This analysis provides some quite sharp insights. We found that the dominance of the inertial forces decays like $x_N^{-7/2}$, and that $C_V$ is about 1. The aspect ratio of the lock, $h_0/x_0$, does not enter the SW formulation, but we see that this parameter affects in a non-trivial manner the range of applicability of the inviscid results, $x_V$. For $Re < 10^4$, $h_0/x_0 < 0.5$ and a rectangular lock $\mathcal{V} = 1$ we obtain $x_V < 14$.

We mentioned that the box-model approximations are in fair agreement with the similarity solution for large times. Here we note that box-model estimates concerning the ratio of inertial to viscous effects (as used by Huppert 1982; Didden and Maxworthy 1982) are in good agreement with the results (2.86) and (2.87). The derivation starts with (2.83). Assuming that the current is a box of height $h_N(t)$, and accordingly $u_x = u_N/x_N$, we estimate $\theta \approx u_N h_N^2 x_N^{-1}$. Substituting $u_N = Fr h_N^{1/2}$, and using the volume conservation in the box $h_N = x_N^{-1}$, we recover the result $\theta \approx C x_N^{-7/2}$, from which (2.87) follows. A more detailed discussion of the box models is given in Chapter 4. The calculation based on the similarity solution is of course more rigorous.

Another advantage of the present calculation is that the similarity behavior, and hence (2.86), are not restricted to a rectangular lock. Using the rescaling (2.82), the position $x_V$ where viscous forces become as important as inertial forces is given by

$$x_V = \left[ \frac{(g'L^{*3})^{1/2}}{\nu} C_V \right]^{2/7} \tag{2.88}$$

for a current of general initial shape; here $x_V$ is scaled with $L^* = \mathcal{V}^{*1/2}$.

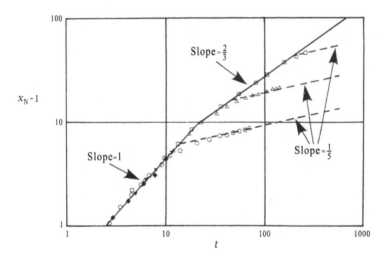

**FIGURE 2.13:** $x_N - 1$ as a function of $t$, experimental values for various currents released from a lock with $H = 1$. (Rottman and Simpson 1983, with permission.)

How good are these estimates? Comparisons with experiments indicate that they are very reliable, even sharp. The main argument in these comparisons is that when a lock-released real current changes suddenly its speed of propagation to a value significantly smaller than predicted by the SW theory, it must be due to a viscous influence. This is subject to the assumption that other obvious hindering effects (like obstacles, waves, loss of buoyancy) can be excluded. For a current in a simple rectangular geometry the change from self-similar propagation with $t^{2/3}$ to a slower one turns out to be sharp, and can be easily detected. The evidence of laboratory experiments is that in containers shorter than $x_V$ a high $Re$ current reaches the front wall without displaying significant deviations from the expected SW propagation. (Similar observations were made for the counterpart axisymmetric current subject to the cylindrical geometry modifications discussed in § 6.6.) For a sharper comparison we consider Fig. 2.13. This shows experimental measurements of propagation for Boussinesq gravity currents released from a rectangular lock with $H = 1$ for three values of $Reh_0/x_0 = 530, 6.0 \cdot 10^3$ and $7.8 \cdot 10^4$. The results are displayed as $x_N - 1$ vs. $t$ on log-log axes which facilitates the distinction between various stages of propagation. According to our estimate (2.87) (with $C_V = 1.56$) the expected values of $x_V$ are $6.8, 14$, and $28$, respectively. The experimental transition to the viscous regime in the first case was at $x_N \approx 6$, but we note that the current was still in the slumping stage at this point. The experimental transition values in the second and third cases were 15 and 43, respectively. Allowing for some reasonable experimental errors, the overall agreement is good. We recall that there is some uncertainty in our expectation that $F_I/F_V = 1$ is the criterion for transition between inviscid to viscous regimes (our definition of $x_V$), and that our modeling of the viscous effects is a quite crude approximation. With this in mind, we conclude that the SW theory provides a simple and satisfactorily accurate means for determining the limit of validity of the inviscid approximation and transition to the viscously-dominated regime.

Finally, we remark that for a current which propagates on a free surface we expect a smaller viscous influence than estimated above. However, there are experimental observations which indicate that surface currents mirror bottom currents (at least in the Boussinesq domain). This has been attributed to the effect of impurities. The details are presented in § 11.1.4. This topic requires further investigation, and meanwhile the present $x_V$ can be regarded as a lower bound of the length where viscous effects become important.

# Chapter 3

## The steady-state current and nose jump conditions

### 3.1 Benjamin's analysis

The front conditions (2.25) and (2.26) are an essential ingredient in the SW analysis of the gravity current. These (or closely related) closures are in general applied to the time-dependent SW equations. In this chapter we shall consider the derivation and interpretation of these front conditions.

For both historical and conceptual reasons we start with an apparently different problem: the steady-state propagation of a 2D inviscid gravity current in a channel. The pertinent discussion and results are in general referred to as Benjamin's analysis, following the classical paper by Benjamin (1968).

In this chapter we use *dimensional variables* unless stated otherwise.

Benjamin's analysis considers a steady-state half-infinite current in an infinite horizontal channel of height $H$; see Fig. 3.1(a). The current is assumed to move like a slug (except for a small zone behind the head which may be influenced by interfacial wave-breaking), and the tail is assumed of constant thickness $h$. The origin of the coordinate system is attached to the foremost point of the current, $O$. In this system the ambient fluid appears to be moving over an obstacle (the current) from right to left say; the foremost point of this obstacle is at the bottom and is also a point of stagnation for the ambient fluid. In the far upstream (right) and the far downstream (left) domains the flow is assumed to be uniform and parallel, i.e., horizontal. Benjamin argued that this does not contradict the presence of some complex and even turbulent motion about the head. The turbulent fluctuations are expected to vanish downstream. The important point in this configuration is that the parallel flow in the far upstream and downstream domains is an exact one-dimensional solution of the inviscid steady-state equations of motion (the Euler equations). The normal (vertical) pressure balances are hydrostatic in these domains.

However, this situation can be attained and sustained only when the flow fields in the two fluids are compatible. Following Benjamin, we consider the main matching conditions between the left and right domains which must be satisfied. We use a rectangular control volume which is bounded by the bottom and top walls of the channel, $z = 0, H$, and by two vertical planes in the upstream and downstream positions, where the streamlines are horizontal. On the right side the uniform speed is $u_r = U$.

### 3.1.1 Volume continuity and flow-force balance

The first matching condition is volume continuity for the ambient fluid

$$HU = (H - h)u_l, \quad \text{or} \quad u_l = \frac{UH}{H - h} = \frac{U}{1 - a} \qquad (3.1)$$

where $a = h/H$. The subscripts $r$ and $l$ denote the right and left boundaries of the control volume.

The second matching condition concerns the $x$-momentum balance. In steady-state the acceleration of the control volume vanishes. Since no viscous stresses are present on the boundaries of this volume, the $x$-momentum balance is reduced to the balance between the flow forces on the left and right, expressed as

$$\int_0^H (\rho u^2 + p)_l dz = \int_0^H (\rho u^2 + p)_r dz. \qquad (3.2)$$

On the right $u = U$ and $p = p_a = 0$. On the left side we have

$$u_l = \begin{cases} 0 & (0 \le z \le h) \\ \dfrac{U}{1 - a} & (h \le z \le H); \end{cases} \qquad (3.3)$$

$$p_l = \begin{cases} \frac{1}{2}\rho_a U^2 - \Delta\rho g z & (0 \le z \le h) \\ \frac{1}{2}\rho_a U^2 - \Delta\rho g h & (h \le z \le H); \end{cases} \qquad (3.4)$$

where $\Delta\rho = \rho_c - \rho_a$, and the $(1/2)\rho_a U^2$ term is the value of the pressure at the stagnation point.

Substituting into (3.2) we rewrite the momentum balance as

$$\rho_a \left(\frac{U}{1 - a}\right)^2 (H - h) + \frac{1}{2}\rho_a U^2 H - \int_0^h \Delta\rho g z\, dz - \int_h^H \Delta\rho g h\, dz = \rho_a U^2 H. \qquad (3.5)$$

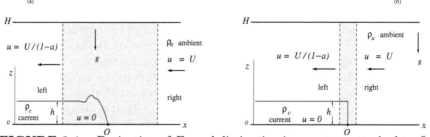

**FIGURE 3.1:** Derivation of $Fr$ and dissipation in a system attached to $O$. (a) Control volume for Benjamin's steady-state current; (b) control volume about the front shock in SW model.

Solving for $U$ we obtain the classical Benjamin result

$$U = \left[\frac{(2-a)(1-a)}{1+a}\right]^{1/2} \left(\frac{\Delta\rho}{\rho_a}g\right)^{1/2} h^{1/2}, \qquad (3.6)$$

which can be expressed as

$$U = Fr(a)\left(g_a' h\right)^{1/2}, \qquad (3.7)$$

where

$$Fr(a) = \left[\frac{(2-a)(1-a)}{1+a}\right]^{1/2} \quad \text{and} \quad g_a' = \frac{\Delta\rho}{\rho_a}g. \qquad (3.8)$$

This proves that a steady-state current of thickness $h$ (measured in the parallel-flow domain) moves with a well-defined speed. The $Fr(a)$ is the well-known Froude function (or number) of Benjamin, plotted in Fig. 2.2. The fundamental relationship (3.6) (or (3.7)-(3.8)) is the outcome of volume and momentum balances applied to the control volume illustrated in Fig. 3.1(a). The speed of the current is determined by the resistance of the ambient fluid. Roughly, the driving force is proportional to $\Delta\rho g h$, and the dynamic reaction is proportional to $\rho_a U^2$. This explains the presence of $g_a' h$ in (3.7). The coefficient $Fr$ (or rather $Fr^2$) lumps the details of the geometric engagement between the forces.

### 3.1.2 Energy considerations

The third condition concerns the mechanic energy. Here, however, the matching between the left and right domains is more flexible than before. Benjamin argued that the system can be in steady state without conserving energy. The only stringent requirement is that no external supply of energy should be necessary to sustain the steady flow of the current, but an energy loss (dissipation) can be vindicated by the non-uniform flow about the nose inside the control volume.

Consider the rate of energy flow, $u(p + \frac{1}{2}\rho u^2)$ integrated from $z = 0$ to $H$. The difference between the leaving (to the left) and arriving (from the right) terms is the rate of energy dissipation inside the control volume. Using (3.4) and subsequently (3.7) this can be expressed as

$$\dot{D} = UH\left[-\Delta\rho g h + \frac{1}{2}\rho_a\left(\frac{U}{1-a}\right)^2\right] = UH\Delta\rho g h\left[-1 + \frac{1}{2}\frac{Fr^2}{(1-a)^2}\right] =$$

$$= -\frac{1}{2}\rho_a^{-1/2}(\Delta\rho g)^{3/2}h^{5/2}Fr(a)\frac{1-2a}{1-a^2} = -\frac{1}{2}\rho_a(g_a')^{3/2}h^{5/2}Fr(a)\frac{1-2a}{1-a^2}. \qquad (3.9)$$

The only energy-conserving, i.e., non-dissipative ($\dot{D} = 0$), current is obtained for $a = 1/2$. For $a > 1/2$ we find $\dot{D} > 0$ which means that the current

which leaves the control volume carries more energy than the inflow counterpart. A steady-state increase of energy is physically unacceptable. This
renders the restriction $a \leq 1/2$ as a necessary condition for an attainable
flow field, even if the energetic details of the solution are not of interest. We
summarize that Benjamin's analysis predicts that a steady-state current will
occupy at most half of the thickness of the channel into which it propagates
(except for the head domain). The speed of propagation is determined by volume and momentum balances, and the thickness restriction is an associated
energy constraint.

The energy dissipation of the system can be correlated with the loss of total
head, $\Delta$. This variable expresses the deviation (drop) of the total pressure
from the ideal Bernoulli balance on a streamline of the ambient fluid between
the right and left domains, scaled with $\Delta \rho g$, as follows

$$\Delta = \left\{ \frac{1}{2}\rho_a U^2 + p_r - \left[ \frac{1}{2}\rho_a \left( \frac{U}{1-a} \right)^2 + p_l \right] \right\} \frac{1}{\Delta \rho g}. \tag{3.10}$$

Since we use the pressure reduced with $+\rho_a g z$ the Bernoulli equation for
the ambient fluid does not involve the usual potential energy terms. This
equation is understood to be on a streamline of the ambient. However, we
notice that $p_r = 0$ and, according to (3.4), $p_l = -\Delta \rho g h + (1/2)\rho_a U^2$ uniformly
in the domain of ambient fluid. Consequently, we obtain the same $\Delta$ for all
streamlines. Substituting these pressures and (3.6) into (3.10) we obtain

$$\Delta = h \frac{1}{2} a \frac{1-2a}{1-a^2}. \tag{3.11}$$

It can be verified that the same result is provided by

$$\Delta = \frac{-\dot{\mathcal{D}}}{\rho_a g_a' U H}. \tag{3.12}$$

Both the head loss and rate of dissipation are indicators of the energetic
behavior, but some care is required in the interpretation. The only zero of
(3.11) is at $a = 1/2$, the half-channel thickness current. Thus, zero dissipation
corresponds to a zero head loss, as expected from ideal flows. For $a > 1/2$
equation (3.11) yields a negative head loss which is physically unacceptable,
again in agreement with previous results. However, for $a < 1/2$ the dissipation
of the system and the head loss on a streamline may display different trends,
and some confusion may arise. The difficulty appears because $\Delta$ expresses a
distribution of the dissipation over the depth of the ambient; see (3.12). The
main effect under consideration is the current of thickness $h$, and the relevant
dissipation is thus proportional to $h$. The distribution of this finite quantity
over a large depth may give the wrong impression of diminution. This is in
particular notable for the deep ambient case, as follows.

The deep current, or deep ambient case, is expressed by the limit $a = h/H \to 0$, for a fixed $h$. In this case the previous analysis gives (a) $Fr = \sqrt{2}$;

(b) the rate of dissipation (in absolute value) tends to the maximum, see (3.9); and (c) $\Delta$ approaches 0, see (3.11). Thus, the behavior of $\Delta$ may give the wrong impression of energy conservation. Actually, the result that $\Delta \to 0$ when $a \to 0$ indicates that in this limit the Bernoulli equation is a good approximation for the flow of the ambient fluid, in spite of the fact that the standard conditions for the validity of this equation were not imposed. Let us use this observation. For the $a \to 0$ case we perform the Bernoulli balance on a streamline from $z = 0$ on the right (where $p_a = 0$) to $z = h$ on the left (where $p_a = (1/2)\rho_a U^2 - \Delta \rho g h$). Moreover, because the thickness of the current is small, the volume conservation of the ambient fluid produces the trivial results $u_l = u_r = U$. We obtain the Bernoulli equation

$$-\frac{1}{2}\rho_a U^2 + \left[ \frac{1}{2}\rho_a U^2 + (\frac{1}{2}\rho_a U^2 - \Delta \rho g h) \right] = 0. \qquad (3.13)$$

The solution for $U$ yields $Fr = \sqrt{2}$. We emphasize that here the flow-force balance was not used, yet the same $Fr$ is obtained as for the previously derived Benjamin result (3.8) for $a \to 0$.

These different derivations of the result $Fr = \sqrt{2}$ for the deep current reproduce the different approaches of Kármán (1940) and Benjamin. The latter emphasized that the derivation by Bernoulli's equation is not rigorous because the system is dissipative and there is no reason to assume *a priori* that this is a good approximation. However, Benjamin also pointed out that the more rigorous derivation based on the flow-force balance is open to objections in the $a \to 0$ limit. In any case, the conclusion was that for the deep ambient the best available theoretical result is $Fr = \sqrt{2}$, and a significant[1] dissipation is present.

### 3.1.3  Summary

Benjamin's analysis is apparently the most celebrated single theoretical contribution to the investigation of the gravity current. It is, however, not a complete prediction tool. This is the point to recall that the balances which produced $Fr$ considered a steady-state solution without specifying how (and if) this can be attained from realistic initial conditions. The price is that at the end of the analysis we have only the result (3.6), which is one equation for two variables, $U$ and $h$. The restriction $a \leq 1/2$ is of little use in obtaining a more explicit solution. The associated dissipation is also an intriguing issue. Where does this energy go to, and/or where does it come from? It is evident that some initial and boundary conditions are necessary for a particular application. Moreover, the use of Benjamin's results in realistic flow fields requires some reconciliation with the time dependency of the variables.

---

[1] "Compared to what?" will ask the careful reader. Compared to the work at the nose performed by the moving pressure force, for example. This will be clarified later.

## 3.2   Jump condition

The reconciliation between the steady-state solution and a real gravity current (in particular produced by lock-release) is achieved by using Benjamin's results as a jump condition at the front of the current calculated from the SW equations. We follow the presentation of Ungarish (2008). The front is now a shock, and the key factor is to perform the balances in a system attached to this shock. We consider a control volume about the front as illustrated in Fig. 3.1(b). As the thickness of this discontinuity becomes infinitesimally small, the volume of fluid inside and its inertia diminish. Consequently, the balances between the flow properties on the right and left of this control volume satisfy the same rules as the steady-state counterpart. In other words, the same volume-flux continuity, flow-force equilibrium, and energy-flux budget which were applied in the steady-state problem are valid for the discontinuity. The results of the previous analysis can be straightforwardly implemented as jump conditions with the obvious change of notation $U = u_N, h = h_N$. This is the justification for: (a) the use of the boundary condition (2.25) in general circumstances, such as for time-dependent currents, and for axisymmetric currents (discussed later); and (b) the comment that Benjamin's $Fr$ has been derived with rigor.

The combination of Benjamin's result with the time-dependent SW formulation closes a circle. We showed that in the slumping stage the lock-released SW current displays a leading region (following the front) of constant and uniform height and velocity. This is consistent with the horizontal steady-state tail assumed by Benjamin. We shall show later (in particular in § 5.5) that the pertinent velocity and energy results obtained by the two-layer SW model are in full agreement with Benjamin's steady-state solution. These SW solutions also throw more light on the energy dissipation revealed by Benjamin's analysis. In particular, we learn that it is insightful to compare the rate of dissipation to the rate of work preformed by the nose of the current, $\dot{W}_N$. The rate of work, pressure-force times velocity, is given by $(1/2)(\Delta\rho g)h_N^2 u_N$ which is $(1/2)(\Delta\rho g)h^2 U$ in the notation of Fig. 3.1(b). Combining with (3.9) and using the value of $Fr(a)$ given by (3.8) we obtain

$$\frac{-\dot{D}}{\dot{W}_N} = \frac{1 - 2a}{1 - a^2}. \qquad (3.14)$$

This gives a simple meaning to "large" or "small" dissipation. For a deep current, $a \to 0$, all the work of the nose is lost by the ambient fluid, and for a half-depth current $a = 1/2$ all the work of the nose is recovered as kinetic energy in the ambient. It is interesting that this result does not depend on the densities of the involved fluids, but only on the geometric ratio $a$.

We note that the analysis in this section does not involve the Boussinesq assumption. The reduced gravity which appears in the results is $g'_a$, and when these results are used for Boussinesq systems we just replace it with $g'$. Formally, a large $g'_a$ may produce a large $U$ by (3.7), but in practice the relevant thickness $h$ (at the nose) tends to be small for currents with large $g'_a$. A limiting example is the classical dam-break concerning a reservoir of water of height $h_0$ in the ambient air. Although $g'_a \gg g$, the nose of the resulting current becomes so thin that the speed of propagation is no more than $(2gh_0)^{1/2}$. This will be considered in Chapter 10. This illustrates again the limitations of (3.7): it is difficult to extract reliable information about a certain gravity current configuration from one equation with two unknowns.

An inspection of the previous results suggest the following simple interpretation

$$U = Fr(a) \left[ \frac{p_c(z=0) - p_a(z=0)}{\rho_a} \right]^{1/2}. \tag{3.15}$$

This relates the pressure head of the front of the current with the dynamic pressure of the current. The idea is that (3.15) provides a separation between the driving pressure effect, and the hindrance effects; the latter are lumped into $Fr(a)$. A smaller $Fr$ means that a larger part of the available pressure is spent on hindrance effects, such as the need to accelerate the ambient fluid in the upper layer, or friction. (The careful reader has probably noticed that $p_a(z) = 0$. This term is kept to emphasize that pressure difference is the driving effect.)

We mentioned that the HS $Fr$ formula (2.27) is an empirical adjustment of Benjamin's result. The theoretical $Fr$ is a function of $a = h/H$ only, which changes from about 0.7 to about 1.4 when $a$ decreases from 0.5 to 0. The empirical result is a function of the same variable, but the values are smaller by typically 15%. The jump-condition analysis can be implemented to support the attribution of this reduction to friction effects in the momentum budget. Suppose that the (small) fraction $\delta$ of the incoming momentum flux is lost on mixing and friction (with the current nose and the walls). The effect on $Fr$ is estimated by multiplying the RHS of (3.5) by $1 - \delta$, and solving again for $U$. We then divide by $(g'_a h)^{1/2}$ and obtain

$$Fr_f(a) = Fr_B(a) \left[ 1 + 2\delta \frac{1-a}{1+a} \right]^{-1/2} \approx Fr_B(a) \left[ 1 - \delta \frac{1-a}{1+a} \right], \tag{3.16}$$

where the subscript $f$ denotes the friction effect and the subscript $B$ denotes the ideal Benjamin result. We note that the friction reduces the value of $Fr$ for any $a$. In particular, when $\delta = 0.25$ the result (3.16) is in fair agreement with the HS $Fr$ predictions. The proper trend is thus explained, but we still do not know what determines the magnitude of $\delta$.

The extension of Benjamin's analysis to a stratified ambient will be presented in Chapter 14. An interesting outcome is that (3.15), with "of the shelf" $Fr(a)$ formulas derived for the homogeneous currents, is a good and useful approximation for stratified ambients.

# Chapter 4

## Box models for 2D geometry

For some quick estimates of the expected behavior of the current we can use a bold approximation which is referred to as the box-model solution. We start with the assumption that the current has a simple shape, a well defined "box." In the present 2D case the box is a rectangle of length $x_N(t)$ and height $h_N(t)$. The main objective is to determine the behavior of these two variables, which means that we need two equations.

The first governing equation of the box model is provided by the obvious volume continuity requirement

$$x_N(t)h_N(t) = \mathcal{V}. \tag{4.1}$$

The second equation must involve some dynamic considerations. In general, simplified forms for the velocity and reduced gravity fields are postulated, so that the integrals of the buoyancy, inertial and viscous effects can be obtained as closed and simple expressions. Then some governing momentum integral balance is applied, with the objective to derive an equation for $u_N(t)$. The integration of this equation provides $x_N(t)$. Surprisingly, for many cases of interest an explicit solution of the form $x_N = Kt^\beta$ (with constant $\beta$ and $K$) can be obtained. This is illustrated below.

This approximation is a loose combination of dimensional analysis and similarity assumptions. It lacks rigor and actually involves assumptions which are evidently incompatible with observations, more rigorous solutions, and boundary conditions. Nevertheless, this simplification provides some useful information concerning the values of $\beta$ and of the dependency of $K$ on the parameters of the problem. Confidence has been gained by comparisons with experiments and other results. Attention: in § 4.1 we use dimensionless variables, and in § 4.2 we use dimensional variables.

---

## 4.1 Fixed volume current with inertial-buoyancy balance

*In this section we use dimensionless variables.* The scaling is given by (2.33)-(2.34). This will facilitate the comparison of the box-model solution with the previous SW results.

In the particular case of inertial-buoyancy dominated motion we can skip the integral balance and go directly to the boundary condition (2.39)

$$u_N = \frac{dx_N}{dt} = Fr(a)[h_N(t)]^{1/2} \tag{4.2}$$

where $a = h_N(t)/H$. The justification is provided by the expectation that the box-model and SW propagation are governed by the same mechanism, so we can start with the best available result which is already in a sufficiently simple form for use in the box model. It is left for an exercise to show that if we start with the formal buoyancy-inertial balance $F_B = F_I$ (see below) we obtain a quite similar equation for $u_N$, but with a constant $Fr = \sqrt{2}$.

Combining (4.1) and (4.2) we obtain one equation for $x_N(t)$ (or for $h_N(t)$). Formally, for the case of a fixed volume $\mathcal{V}$, the solution is

$$t - t_I = \frac{1}{\mathcal{V}^{1/2}} \int_{x_I}^{x_N(t)} [Fr(a(x))]^{-1} x^{1/2} dx \tag{4.3}$$

where $a = \mathcal{V}/(Hx)$. Here $t_I$ and $x_I$ indicate that some initial condition $x_N = x_I$ at $t = t_I$ must be specified.

Consider the case of a deep current with a *constant* $Fr$. The result of (4.3) can be expressed as

$$x_N = \left[\frac{3}{2}Fr\mathcal{V}^{1/2}(t - t_I) + x_I^{3/2}\right]^{2/3} = \left(\frac{3}{2}Fr\right)^{2/3} \mathcal{V}^{1/3}(t + \gamma)^{2/3}. \tag{4.4}$$

Here $\gamma$ is a constant which combines the values of the initial $t_I$ and $x_I$.

The result (4.4) is remarkably close to the similarity solution (2.80). The propagation is with $t^{2/3}$, and the coefficient is proportional to $\mathcal{V}^{1/3}$ in both solutions (see (2.79)). The ratio $(1.5Fr)^{2/3}/K_1$ is 0.92 for $Fr = 1.19$ and 0.88 for $Fr = \sqrt{2}$. This means that the box model underpredicts the speed of propagation by about 10% (as compared to the similarity solution). This is explained by the fact that the self-similar current has a higher nose $h_N$ for a given $x_N$. Note that the box-model approximation $h(x,t) = h_N(t)$ implies a $x$-independent axial velocity and hence $u(x,t)$ is a linear function of $x$. This is, again, in agreement with the self-similar solution. Moreover, as in the self-similar result, there is a time-shift constant $\gamma$ which must be determined by some initial conditions that may depend on the previous stage of propagation, i.e., the slumping stage. This renders the box-model result also a quite weak prediction tool. We summarize that the box-model result (4.4) provides a plausible approximation for large times. However, if an analytical solution of the SW equations is available, there is actually no need for this model. It may still be convenient to use it for scaling arguments or for initial estimates of experimental data.

Equation (4.3) is amenable to analytical solutions also for non-constant $Fr$. In particular, consider the effect of the non-constant branch of HS $Fr$

formula (2.27) for a current with $\mathcal{V} = 1$. Substitution into (4.3) yields, after some algebra and use of the initial condition $x_I = 1$ at $t_I = 0$,

$$x_N = \left[ \frac{7}{12} H^{1/3} t + 1 \right]^{6/7} \qquad (h_N = x_N^{-1} \geq 0.075H). \qquad (4.5)$$

Contrary to the long-time result (4.4), the present outcome lacks a clear-cut mathematical similarity with the initial behavior of a SW current. One may expect (see Huppert and Simpson 1980) that the use of (4.5) for the initial motion, matched with (4.4) for larger times, reproduces the behavior of a real current, i.e., a dam-break slumping which then evolves into the self-similar stage. This expectation is not fulfilled in general. A real current displays a constant velocity slumping interval of several lock lengths, while (4.5) predicts a decreasing velocity from the start. The condition $a < 0.5$ cannot be imposed on (4.5). This makes the result especially unreliable for $H < 2$. For example, for the full-depth case $H = 1$, the box-model current spreads to $x_N = 2$ while $h_N/H$ is in the non-physical range. On the other hand, for an initially deep ambient $H^{-1} < 0.075$, the stage of propagation (4.5) does not appear, and the box-model motion starts with the similarity behavior (4.4). In both cases the initial motion is mispredicted.

The conclusion is clear: the box model is not a reliable substitute for the SW equations of motion and their solution. Actually, the box model changes the formulation from a PDE hyperbolic set with initial and boundary conditions to an ODE equation with one initial condition for $x_N$. This is bound to reduce the accuracy of the solution and the resemblance with the real physical system.

---

## 4.2 Inflow volume change $\mathcal{V} = qt^\alpha$

*In this section we use dimensional variables.* This provides a more direct insight into the balances and results which appear. For simplicity, we use the Boussinesq approximation, but this restriction can be relaxed with some care. Here we follow the discussions of Didden and Maxworthy (1982) and Huppert (1982).

We consider a gravity current of changing volume in time according to $qt^\alpha$ where $q$ ($> 0$) and $\alpha$ ($\geq 0$) are given constants. The $\alpha = 0$ case gives the simple fixed-volume current, and the $\alpha = 1$ case corresponds to a current supplied by a constant-rate influx (say, by an open tap). The $\alpha > 1$ cases may represent a leak or eruption which worsens with time.

The volume continuity requirement of the box model is

$$x_N(t)h_N(t) = \mathcal{V} = qt^\alpha. \qquad (4.6)$$

In the particular case $\alpha = 0$ we can replace $q = \mathcal{V}$, where $\mathcal{V}$ is the volume (per unit width) and the dimensions are $cm^2$. In general, the dimensions of $q$ are

$cm^2 s^{-\alpha}$. For definiteness, let the source be at $x = 0$. We assume that the inflow velocity is sufficiently small to prevent formation of a jet. This can be expressed as $u_i/(g'h_i)^{1/2} < 1$, where the subscript $i$ denotes the inlet position. This condition implies that characteristics from the interior can reach the inlet; see second equation of (2.38) (recall that this is in dimensionless form).

The following simplified behavior is assumed:

$$x_N = Kt^\beta; \quad h_N = \frac{\mathcal{V}}{x_N} = \frac{q}{K}t^{\alpha-\beta}; \quad (4.7a)$$

$$u_N = \beta K t^{\beta-1}; \quad u(x,t) = u_N \frac{x}{x_N} = \beta t^{-1}x, \quad (4.7b)$$

where $\beta$ and $K$ are constants. For later use we also rewrite the first dependency in (4.7a) as

$$t = \left(\frac{x_N}{K}\right)^{\frac{1}{\beta}}. \quad (4.8)$$

The main objective is to determine the values of $\beta$ and $K$. The other global properties of the flow follow.

The relevant forces (per unit width or the current) can now be estimated as follows. The driving "buoyancy" force is the integral of $\partial p_c/\partial x$ over the volume of the current. This effect is represented by the pressure force on the periphery "dam" which bounds the current in the box

$$F_B = \frac{1}{2}\rho_c g' h_N^2. \quad (4.9)$$

We note that some care may be required when using this term in the box-model context. The pressure forces in a lock (box) tend to push the fluid at the boundary $x_N$ in the positive $x$ direction. However, the solution of the SW equations clearly shows that in some cases the pressure forces decelerate the fluid. The deceleration-acceleration is determined by the positive-negative inclination of the interface, an effect not properly incorporated in the box model. The inertial forces are

$$F_I = \rho_c h_N \int_0^{x_N} u u_x dx = \frac{1}{2}\rho_c u_N^2 h_N, \quad (4.10)$$

and the viscous forces are

$$F_V = \rho_c \nu \int_0^{x_N} \frac{u}{h} dx = \frac{1}{2}\rho_c \nu \frac{u_N}{h_N}x_N. \quad (4.11)$$

The ratio of inertial to viscous forces can be expressed as

$$\frac{F_I}{F_V} = \frac{1}{\nu}u_N h_N^2 \frac{1}{x_N} = \frac{1}{\nu}\beta\left(\frac{q}{K}\right)^2 t^{2\alpha-2\beta-1}. \quad (4.12)$$

The thickness ratio $h_N/x_N$ behaves like $t^{\alpha-2\beta}$; see (4.7a). The assumption of a thin current restricts the validity of the analysis to

$$\alpha - 2\beta < 0. \quad (4.13)$$

## 4.2.1 Inertial-buoyancy balance

We start again with the shortcut that the propagation of the inertial current is governed by the jump conditions at the nose. Using (2.39) and (4.7) we write

$$u_N = Fr(g'h_N)^{1/2}, \quad \text{i.e.,} \quad \beta K t^{\beta-1} = Fr\, g'^{1/2} \left(\frac{q}{K}\right)^{1/2} t^{(\alpha-\beta)/2}. \quad (4.14)$$

Assuming that $Fr$ *is a constant*, we equate the powers of $t$ and the coefficients in the last equation to obtain

$$\beta = \beta_I = \frac{\alpha+2}{3}; \quad (4.15)$$

$$K = K_I = \left(\frac{Fr^2}{\beta_I^2}\right)^{1/3} [g'q]^{1/3}. \quad (4.16)$$

Here the subscript $I$ indicates that the flow is dominated by inertial (or inviscid) effects.

We summarize: the propagation of the 2D inviscid box-model gravity current is (in dimensional form)

$$x_N(t) = \left(\frac{Fr^2}{\beta_I^2}\right)^{1/3} [g'q]^{1/3} t^{(\alpha+2)/3}. \quad (4.17)$$

The ease and simplicity of this result is of course pleasing, but we must keep in mind not to confuse it with accuracy. In particular, we note that little is known about the value of $Fr$ for currents dominated by inflow conditions. We can argue that the accepted result for the $\alpha = 0$ case is still relevant in a deep ambient, and there are indications that $Fr \approx 0.8$ yields fair agreements with available experiments when $\alpha$ is about 1 (see below), but this issue is also a part of the uncertainties of the box-model approximation.

Substitution of (4.15) into (4.12) and use of (4.8) yields

$$\frac{F_I}{F_V} = \frac{1}{\nu}\beta_I \left(\frac{q}{K_I}\right)^2 t^{(4\alpha-7)/3} = \frac{1}{\nu}\beta_I \left(\frac{q}{K_I}\right)^2 \left(\frac{x_N}{K_I}\right)^{(4\alpha-7)/(\alpha+2)}. \quad (4.18)$$

These results reduce satisfactorily to the previously derived self-similar solution when $\alpha = 0$. In particular, we see that $F_I/F_V$ decays like $x_N^{-7/2}$, in accord to the calculation in § 2.7. It is left for an exercise to show that, for $\alpha = 0$, (4.18) yields the same decay of $\theta$ as (2.86), but with $C_V = Fr$.

For $\alpha > 0$, the rate of propagation is in agreement with the available similarity results discussed by Grundy and Rottman (1985b) and Gratton and Vigo (1994), as mentioned in § 2.6.1. However, the similarity results display discontinuities for values of $Fr$ larger than some critical value (which is of the order of 1, but decreases with $\alpha$). The inflow conditions at $x = 0$ are inconsistent with the box-model simplification $u = u_N x/x_N$; see (4.7b) . We assume that this discrepancy has a negligible effect on the global balances (4.9)-(4.11). A more realistic postulate for the velocity in the box-model volume is

$$u(x,t) = \beta_I K t^{\beta_I - 1} \left[ m + (1 - m) \frac{x}{x_N} \right], \tag{4.19}$$

where $m = \alpha/\beta_I$. The boundary condition at the source, $u(0,t)h_N(t) = \alpha q t^{\alpha-1}$, is now satisfied. The approximation (4.19) can be used *a posteriori*, together with the previous results for $\beta_I$ and $K$. The trend provided by this result is in good agreement with the continuous solutions of Grundy and Rottman (1985b). In particular, note that for $\alpha = 1$ we obtain $\beta_I = 1$, $m = 1$, and hence $u(x,t) = K$. The careful reader will notice that in this case $u_x = 0$, and wonder if this really means zero inertial terms and force. No, because an attempt to force a significant change of the motion of this current (by a viscous effect, say) will produce $u_t$ of the order of magnitude $u_N^2/x_N$. The estimate (4.18) remains valid.

Substituting the result (4.15) for $\beta_I$ into the thickness ratio requirement (4.13), we conclude that the present 2D inviscid box model is restricted to $\alpha < 4$. Otherwise, the dense fluid is expected to accumulate as a thick bulk.

Due to the complexity of the similarity solutions for $\alpha > 0$ flows, the present box-model approximation is a practical alternative analytical tool at the present state of knowledge. The price is a loss of accuracy which is difficult to estimate *a priori*. Numerical solutions of the SW equations are expected to provide more reliable results.

## 4.2.2   Viscous-buoyancy balance

When the inertial terms are negligibly small, the buoyancy is counteracted by viscous drag. For $F_V = F_B$ equations (4.9) and (4.11) yield

$$u_N = \frac{g' h_N^3}{\nu x_N}. \tag{4.20}$$

In the context of the present text, this is an interesting and novel result for the speed of propagation of a gravity current. (We shall see later a more rigorous derivation.) The difference with the speed of propagation of the inertial current, $Fr(g'h_N)^{1/2}$, is very significant.

Now we substitute the assumed behavior (4.7) into (4.20) to obtain

$$\beta K t^{\beta-1} = \frac{g'}{\nu} \frac{q^3}{K^4} t^{3\alpha - 4\beta}. \tag{4.21}$$

The solution is

$$\beta = \beta_V = \frac{3\alpha + 1}{5}; \tag{4.22}$$

$$K = K_V = \left( \frac{1}{\beta_V} \right)^{1/5} \left[ \frac{g'}{\nu} q^3 \right]^{1/5}. \tag{4.23}$$

Here the subscript $V$ indicates that the flow is dominated by viscous effects.

We summarize: in dimensional form, the box-model propagation of a viscous current is

$$x_N(t) = \left(\frac{5}{3\alpha+1}\right)^{1/5} \left[\frac{g'}{\nu}q^3\right]^{1/5} t^{(3\alpha+1)/5}. \tag{4.24}$$

Like in the inertial-buoyancy case, the easy derivation and simplicity of the results lends appeal to the box-model approximation. This, however, is not a proof of accuracy or applicability to a particular problem.

Substitution of (4.22) into (4.12) and use of (4.8) yield

$$\frac{F_I}{F_V} = \frac{1}{\nu}\beta_V \left(\frac{q}{K_V}\right)^2 t^{(4\alpha-7)/5} = \frac{1}{\nu}\beta_V \left(\frac{q}{K_V}\right)^2 \left(\frac{x_N}{K_V}\right)^{(4\alpha-7)/(3\alpha+1)}. \tag{4.25}$$

The previous result for $\beta_V$ satisfies the thickness ratio requirement (4.13) for any positive $\alpha$. In this respect the viscous current is also different from the inviscid current solution (which displays the $\alpha < 4$ restriction).

The viscous box-model propagation results are in good agreement with the similarity solution which will be derived later in Chapter 11. However, we shall see that the more rigorous similarity solutions are quite simple and can be easily implemented for any practical value of $\alpha$. This reduces the motivation for using the box-model results.

## 4.2.3 Critical $\alpha$

Intuitively, we expect that the viscous current spreads at a lower rate than the inertial one. This implies $\beta_V < \beta_I$. However, our "intuition" is based mostly on currents of fixed volume, $\alpha = 0$, and constant filling rate, $\alpha = 1$. Huppert (1982) made the interesting observation that there is a "critical" value of $\alpha$, $\alpha_c = 7/4$, for the behavior of the solutions. This is depicted in Fig. 4.1. We see that at this point $\beta_I = \beta_V = 5/4$, and for larger values of $\alpha$ the rate of spread of the viscous current is larger than that of the inertial current. An interesting branching concerning the possible change of flow regime during propagation appears.

Consider the dependency of the ratio $F_I/F_V$ on $\alpha_c = 7/4$, for both inertial and viscous dominated flows. Indeed, both (4.18) and (4.25) show that: (a) For $\alpha < \alpha_c$, and in particular for the fixed-volume current ($\alpha = 0$), the ratio of inertial to viscous forces decays continuously during the propagation. This means that in this domain of $\alpha$ an inertial gravity current is bound to become viscous. The change of regime of propagation means a *decrease* of the power of $t$ from $\beta_I$ to $\beta_V$. (b) For $\alpha > \alpha_c$ the ratio of inertial to viscous forces increases continuously during the propagation. This means that in this domain of $\alpha$ a viscous gravity current is bound to become inertial. (This also implies the increase of the Reynolds number and the possibility of associated instability effects.) The change of regime of propagation means a *decrease* of the power

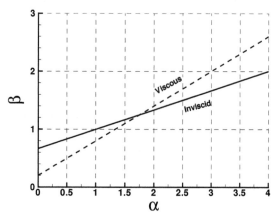

**FIGURE 4.1:**   The power $\beta$ as a function of $\alpha$ for the inertial and viscous 2D current.

of $t$ from $\beta_V$ to $\beta_I$. (c) For $\alpha = \alpha_c$ the ratio of inertial to viscous forces does not change with time or distance of propagation. For this $\alpha$ there is no transition between viscous and inertial regimes.

Consider the 2D current with $\alpha = \alpha_c$. It turns out that the type of this current is conveniently determined by the magnitude of the dimensionless parameter

$$J = \nu^3 \frac{g'^2}{q^4}, \qquad (4.26)$$

which has been defined as the "Julian number" by Huppert (1982). Inertia is always dominant if $J \ll 1$ and viscous drag is always dominant if $J \gg 1$. The justification is as follows. We go back to (4.12). For $\alpha = \alpha_c$ the ratio $F_I/F_V$ is time independent and hence, for the entire flow process, we can write

$$\frac{F_V}{F_I} = \frac{4}{5} \nu \left( \frac{K}{q} \right)^2. \qquad (4.27)$$

First, for the inviscid current we substitute (4.16), and use $Fr \approx 1$; we find that $F_V/F_I \approx J^{1/3}$. Second, for the viscous current we substitute (4.23); we find that $F_V/F_I \approx J^{1/5}$. We conclude that a large (small) value of $J$ corresponds to a large (small) ratio $F_V/F_I$. (Recall that the dimension of $q$ is $cm^2 \, s^{-4/7}$ for the present $\alpha = 7/4$ case. )

These results are sufficiently wide and intriguing to prompt experimental investigations into their validity.

## 4.2.4   Comparisons and experimental verifications

From the theoretical aspect, we can compare these box-model results with more reliable solutions of the equations of motion. Evidently, the present behavior corresponds to self-similar solutions. Thus, in general, it suffers from the same deficiencies: the problem of compatibility with realistic initial conditions, and the possibility of a "virtual origin." In particular, the initial "current" is expected to be a small volume under the influence of the inlet conditions rather than compatible with the standard hydrostatic and thin-layer assumptions.

There is, however, a general encouraging outcome: the powers $\beta_I$ and $\beta_V$ derived from box-model approximations of the type used here (§ 4.2) are in full agreement with self-similar results of more rigorous thin-layer approximate equations for the corresponding $\alpha$ (when available). There is also good qualitative agreement for the coefficients $K_I$ and $K_V$. The quantitative agreement of the coefficients is not so good. We must keep in mind that we assumed the shape of the interface, the linear dependency of $u$ on $x$, and the dependency of the viscous shear on $1/h_N$. These are very strong simplifications, and hence deviations of 10–50% from more accurate solutions should not surprise us. This has been illustrated for the $\alpha = 0$ inviscid case in § 4.1, and will be considered in Chapter 11 for the viscous case.

Interestingly, the viscous current with inflow is significantly more amenable to similarity solutions than the inviscid counterpart. The reason is that the thin-layer viscous approximation (essentially, the lubrication theory) is governed by parabolic PDEs (as shown later in Chapter 11). On the other hand, the hyperbolic nature of the SW inviscid equations allows for various discontinuities and boundary-condition control which complicate the solutions of currents with inflow. This is outside the scope of the present book, and the interested reader is referred to the papers of Grundy and Rottman (1985b) and Gratton and Vigo (1994). (For easy comparison, we mention that the notations $\delta$ and $\beta$ in these papers correspond to our $\beta$ and $Fr$ (assumed constant).)

As previously mentioned, the most intriguing results of the box-model solution are concerned with the critical behavior. A pertinent experimental investigation for 2D currents was presented by Maxworthy (1983b). The currents were created in a tank of dimensions $300 \times 50 \times 20\,\text{cm}^3$ (length×depth×width) filled with freshwater to a depth of 40 cm. The density of the currents was set by addition of salt in the Boussinesq range $\epsilon \approx 0.001 - 0.06$, and the filling rate $\alpha$ was set by a pump controlled by a flowmeter and a valve. The viscosity was $\nu = 0.01\,\text{cm}^2\,\text{s}^{-1}$ in all cases. The dyed current was photographed against a bright background with a length scale and watch. The position of the nose of the current, $x_N(t)$, was obtained.

Most experiments of Maxworthy (1983b) were for the critical $\alpha_c = 7/4$. For this $\alpha$, six values of $q$ were used in the experiments, in the range $0.018 - 0.088\,\text{cm}^3\,\text{s}^{-7/4}$. The various $\epsilon$ and $q$ produced both inviscid and viscous currents, with $J$ typically in the range of 0.5-5. A certain combination of $\epsilon$ (i.e., $g'$) and $q$ produced a current with a fixed $J$. For each current, the plot of $x_N$ vs. $t$ on log-log axes yields the slope of the corresponding $\beta$. In all the tested cases, the propagation of the current was with a constant slope (except for some short initial times which were discarded from the data processing). The measured values of $\beta$ were typically $1.16-1.19$, slightly below the theoretical value $1.25$. The experiments confirmed the major prediction that when $\alpha = \alpha_c$ the current in both the inviscid and viscous regimes (i.e., with small and large values of $J$) spreads with $t$ at the same power. (The discrepancies of the spreading data were within approximately $10\%$.)

But how do we know that these currents really covered both the viscous and inviscid cases? To answer this question, the experimental values of $x_N$ at a fixed $t$, for currents of fixed $g'$, were plotted against the various values of $q$ used in the experiments, on log-log scales; recall that $\nu$ was also fixed in these experiments. This is expected to reproduce the dependency of $K$ on $q$; see (4.7). But since $K_I \propto q^{1/3}$ while $K_V \propto q^{5/3}$, see (4.16) and (4.23), a significant difference can be expected in the graph of $\log x_N$ vs. $\log q$ between the inviscid and viscous regimes. Indeed, Maxworthy (1983b) reported (see inset in Fig. 1 there) a change of slope from $0.33$ (for smaller values of $J$) to $0.6$ (for larger values of $J$). The transition occurred at the average value of about $J_{TR} = 3.2$ (the subscript stands for transition).

The experimental value of $K_I$ was by about $25\%$ smaller than the value calculated from (4.16) with $Fr = 1.19$. We have no clear-cut explanation to this behavior, but we recall that the box model is a crude approximation and it would be exaggerated to expect that both $\beta$ and $K$ are in good agreement with measurements. We note in passing that the empirical $Fr = 0.80$ will remove the discrepancy. The experimental value of $K_V$ was by about $20\%$ smaller than the value calculated from (4.23). This is consistent with the prediction of the similarity solution discussed in Chapter 11.

Maxworthy (1983b) also presented experimental results for $\alpha = 1.5, 2$, and $3$. The experimental values of $\beta$ were, again, in good agreement with the box-model results. Some discrepancies were detected concerning the propagation during initial time, and the dependency of $K_I$ on $q$. It is possible that these reflect limitations of the experimental apparatus. The reader is referred to the original paper for details.

# Chapter 5

## Two-layer SW model

### 5.1  Introduction

For definiteness, we consider a dense (bottom) current released from a standard, rectangular lock of height $h_0$; in the Boussinesq case the light counterpart is the mirror image with respect to the boundary on which propagation occurs. Let $H$ be the dimensionless height of the ambient, scaled with $h_0$ of the lock.

The one-layer SW model discards the effect of the return flow in the ambient above the current. This approximation is expected to lose accuracy in flow domains where the height ratio of ambient to current is not large. In particular, when this ratio is smaller than 2 the return flow is expected to be (locally) more significant than that of the current. This suggests that the dam-break and slumping phases in a configuration with $H < 2$ are different from the predictions of the one-layer model. Indeed, laboratory observations show two clear different features of the $H \approx 1$ case (as compared to the $H > 2$ case): (a) the slumping distance of motion with constant $u_N$ is significantly larger; and (b) the smooth backward moving rarefaction wave from the lock into the reservoir develops a steep jump (shock), which is subsequently reflected from the backwall toward the nose. The contrast between the deep and shallow ambient currents is illustrated by Figs. 2.8 and 5.1. The displayed laboratory systems differ in the depth of the ambient: $H = 7$ and $H = 1$, respectively. The frames display the situations: (a) the fluid near the backwall is still stationary, while the wave from the lock moves backward; (b) the wave is reflected; and (c) the reflected wave moves away. The deep current has a smoother interface in all three frames. It is evident that in the shallow ambient ($H = 1$) it takes a longer time for the motion (wave) to reach the backwall. The one-layer model does not capture this interesting feature.

The understanding and accurate prediction of the slumping stage behavior is important from both academic and practical aspects. This is a fundamental ingredient in the development of the motion in most experimental devices and a possible stringent test for theoretical prediction tools. These interesting flow field features and modeling assessment considerations motivate the introduction of the two-layer SW model. As we shall see later in this chapter, this model predicts well the above-mentioned behavior. Moreover, this model also

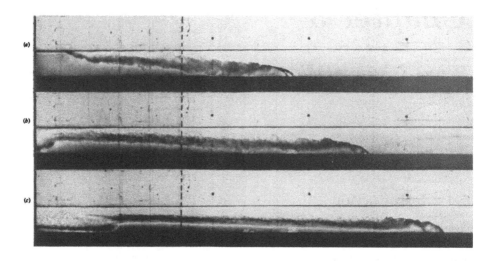

**FIGURE 5.1:** Shadowgraph of a gravity current with $h_0 = 7\,\text{cm}$, $x_0 = 50\,\text{cm}$, $g' = 47\,\text{cm}\,\text{s}^{-2}$ at $t = 4.7, 7.7$, and $10.7\,\text{s}$ (in dimensionless units, $t = 1.7, 2.8$, and $3.8$) after release. The current is shallow: $H = 1$ (dimensionless). The dotted vertical line indicates the position of the lock gate, and the thin lines are at $10\,\text{cm}$ intervals. (Rottman and Simpson 1983, with permission.) Here the reference speed and time are $U = 18\,\text{cm}\,\text{s}^{-1}$ and $T = 2.8\,\text{s}$.

provides the theoretical link between the realistic lock-release problem and the idealized steady-state current considered by Benjamin (1968). For large $H$ (and practically for $H > 3$) the differences between the two- and one-layer models are insignificant. Here we follow the papers of Rottman and Simpson (1983), Klemp, Rotunno, and Skamarock (1994), D'Alessio, Moodie, Pascal, and Swaters (1996), and Ungarish and Zemach (2005).

Even before we master the details, we can postulate the message: We must keep in mind that there are qualitative and quantitative differences between the shallow and deep ambient cases. This alert is relevant because the available data are strongly biased toward the former configuration. Laboratory experiments for $H = 1$ are preferred because it is easier to set up the lock and to follow the current than in a large $H$ configuration. Similarly, numerical simulations require simpler grids for $H = 1$ than for large $H$. The temptation is to extrapolate the more abundant results to the domain of scarce data.

## 5.2 The governing equations

Here we use dimensional variables. The flow is 2D, in a channel of depth $H$. The current is initially in a lock of length $x_0$ and height $h_0$. We use a two-dimensional Cartesian coordinate system $\{x, z\}$ with the gravity acceleration $g$ acting in the $-z$ direction.

We denote by the subscript $a$ the ambient fluid. In some cases we shall also use the subscripts 1 and 2 to denote the dense and the ambient fluid variables, respectively. For example, $h_a$ and $u_a$ (and/or $h_2$ and $u_2$) are the thickness and the $x$-velocity in the ambient. The variables for the current are typically without subscript, but the subscripts $c$ or 1 are used in some places for clarity. We consider the domain $0 \leq x \leq x_N(t)$. The rest of the fluid in the channel is stationary ambient fluid.

The inclined interface $z = h(x, t)$ (for $0 \leq x \leq x_N(t)$) is a key factor in the derivation of the two-layer model. The matching conditions between the lower (current) and the upper (ambient) layers are: continuity of the normal velocity, and continuity of the pressure. We can also write the integrated continuity relationship:

$$h_a = H - h; \quad \bar{u}_a = -\frac{h\bar{u}}{h_a}, \tag{5.1}$$

where $\bar{u}_a$ is the averaged $x$-velocity in the upper layer,

$$\bar{u}_a(x, t) = \frac{1}{h_a} \int_h^H u_a dz. \tag{5.2}$$

We recall that the cases of interest are when the upper and lower layers of moving fluid are of comparable thickness. In these circumstances the shallow-water approximations, introduced for the flow of the lower layer of dense fluid in Chapter 2, are relevant to the upper layer of ambient fluid too.

The motion of the interface is given by (2.3b), which is repeated below

$$\frac{\partial h}{\partial t} + u_h \frac{\partial h}{\partial x} - w_h = 0.$$

In this equation $u_h$ and $w_h$ are the velocity components of the dense fluid at the interface. Now we can also use the velocity on the upper side, and rewrite

$$\frac{\partial h}{\partial t} + u_{ah} \frac{\partial h}{\partial x} - w_{ah} = 0. \tag{5.3}$$

The fact that the sum of the last two terms on the LHS is the same for both the lower and upper layers expresses the continuity of the normal velocity at the interface. The tangential velocity at the interface must be discontinuous;

the average motion must be in opposite directions to satisfy volume continuity, as indicated by the second equation in (5.1).

The previously derived continuity equations for the current, (2.5) and (2.24), remain valid. The flow in the layer of ambient fluid has a dynamic effect which appears in the momentum equations, as follows.

We use, again, the reduced pressure $p = P + \rho_a g z$. The $z$-momentum equations are approximated by the hydrostatic balances. This is because the vertical accelerations are small as compared with the horizontal ones, and our objective is to derive a pressure term of the order of magnitude of the $x$-accelerations. We write these balances as

$$\frac{\partial p_a}{\partial z} = 0; \quad \frac{\partial p_c}{\partial z} = -\Delta \rho g, \tag{5.4}$$

where $\Delta \rho = \rho_c - \rho_a$. The integration of these equations is subjected to the pressure continuity conditions $p_a = p_c$ at $z = h(x,t)$. We obtain

$$p_a(x,z,t) = \Phi(x,t); \quad p_c(x,z,t) = \Phi(x,t) - \Delta\rho g z + \Delta\rho g h(x,t), \tag{5.5}$$

where $\Phi(x,t)$ is an unspecified function. The meaning of this function is the reduced pressure at the inclined interface and in the ambient fluid above this interface. For definiteness, we set $\Phi(x_N,t) = 0$ and $p_a = 0$ for $x \geq x_N(t)$.

The consequence of (5.5) is

$$\frac{\partial p_a}{\partial x} = \frac{\partial \Phi(x,t)}{\partial x}; \quad \frac{\partial p_c}{\partial x} = \Delta\rho g \frac{\partial h(x,t)}{\partial x} + \frac{\partial \Phi(x,t)}{\partial x}. \tag{5.6}$$

The interesting result is that the pressure $x$-driving terms are $z$ independent.

We compare (5.6) to (2.8) and (2.15) of the one-layer model. We realize that the function $\Phi(x,t)$ reflects the motion in the upper layer. Equation (2.22), which expresses the integrated form of the inviscid $x$-momentum balance in the current, is still valid. However, now the pressure term contains also the unknown $\Phi$. To determine it, we must use the $x$-momentum equation in the upper layer, in addition to the $x$-momentum equation (2.22).

We start with the inviscid $x$-momentum equation of the upper layer in conservation form

$$\frac{\partial u}{\partial t} + \frac{\partial u^2}{\partial x} + \frac{\partial}{\partial z}(uw) = -\frac{1}{\rho_a}\frac{\partial p_a}{\partial x}, \tag{5.7}$$

where the subscript $a$ for $u$ and $w$ has been omitted for simplicity. We integrate across the ambient layer, $h(x,t) \leq z \leq H$, to obtain

$$\int_h^H \left[\frac{\partial u}{\partial t} + \frac{\partial u^2}{\partial x}\right] dz - u_{ah}w_{ah} = -\frac{1}{\rho_a}\int_h^H \frac{\partial p_a}{\partial x} dz, \tag{5.8}$$

where we used the boundary condition $u_a w_a = 0$ on $z = H$. After some algebra similar to that employed for (2.18)-(2.21) we obtain

$$\frac{\partial \bar{u}_a h_a}{\partial t} + \frac{\partial \, \varsigma_a \bar{u}_a^2 h_a}{\partial x} = -\frac{1}{\rho_a} \int_h^H \frac{\partial p_a}{\partial x} dz, \tag{5.9}$$

where

$$\varsigma_a = \frac{1}{h_a \bar{u}_a^2} \int_h^H u_a^2 dz = 1 + \frac{1}{h_a} \int_h^H \left(1 - \frac{u_a}{\bar{u}_a}\right)^2 dz. \tag{5.10}$$

We assume that the velocity shape factors in both layers satisfy $\varsigma = \varsigma_a = 1$. The justification for the lower layer is as discussed in the text following (2.22). The justification for the ambient fluid layer is more problematic. The ambient fluid into which the current propagates must be elevated during the propagation, and some $z$ dependency of $u_a$ is, in general, expected. We use $\varsigma_a = 1$ as a simplifying assumption. In addition, we substitute (5.6) for the pressure terms on the RHS of the integrated momentum equations (2.22) and (5.9). The resulting two-layer model momentum equations are

$$\rho_a \left[\frac{\partial \bar{u}_a h_a}{\partial t} + \frac{\partial \bar{u}_a^2 h_a}{\partial x}\right] = -h_a \frac{\partial \Phi}{\partial x}; \tag{5.11}$$

$$\rho_c \left[\frac{\partial \bar{u} h}{\partial t} + \frac{\partial \bar{u}^2 h}{\partial x}\right] = -h \frac{\partial \Phi}{\partial x} - \Delta \rho g \frac{1}{2} \frac{\partial h^2}{\partial x}. \tag{5.12}$$

The next obvious step is to eliminate the $\partial \Phi / \partial x$ term from these equations. Furthermore, we can eliminate $h_a$ and $u_a$ in terms of $h$ and $u$ (and the constant $H$); see (5.1). We drop the overbar; it is clear that the following analysis is concerned with the averaged velocities. The result is one combined $x$-momentum equations in terms of the variables $h(x,t)$ and $u(x,t)$ of the current

$$\left(1 + \frac{\rho_a}{\rho_c} \frac{h}{H-h}\right) \frac{\partial u}{\partial t} + \left[1 - \frac{\rho_a}{\rho_c} \frac{h(H+h)}{(H-h)^2}\right] u \frac{\partial u}{\partial x}$$
$$+ \left[g_c' - \frac{\rho_a}{\rho_c} \left(\frac{H}{H-h}\right)^3 \frac{u^2}{H}\right] \frac{\partial h}{\partial x} = 0, \tag{5.13}$$

where, again, $g_c' = (\Delta \rho / \rho_c) g$. The details of the derivation of (5.13) from (5.11), (5.12), and (2.24) are left for an exercise. Note that this equation reduces simply to the one-layer momentum equation when $\rho_a / \rho_c \approx 1$ and $H \gg h$, or when $\rho_a / \rho_c \ll 1$, as expected.

The general two-layer formulation requires the solution of the system of the continuity equation (2.24), momentum equation (5.13), and nose condition (2.25). This is a difficult task and the results are awkward for interpretation. For progress, we introduce again the Boussinesq simplification: we substitute $\rho_a / \rho_c = 1$ into (5.13), and also use $g_c' = g_a' = g'$.

## 5.3   Boussinesq system in dimensionless form

We use the same scaling as for the one-layer model. We switch all dimensional variables, denoted below by an asterisk, to dimensionless variables as follows

$$\{x^*, z^*, h^*, H^*, t^*, u^*, p^*\} = \{x_0 x, h_0 z, h_0 h, h_0 H, Tt, Uu, \rho_a U^2 p\}, \quad (5.14)$$

where

$$U = (g'h_0)^{1/2}; \quad T = \frac{x_0}{U}. \quad (5.15)$$

The relevant shallow-water equations of continuity (2.24) and momentum (5.13) can be expressed in dimensionless form for the lower layer variables only. The only free parameter is $H$. In terms of $h$ and $q = hu$ this yields

$$\begin{cases} \dfrac{\partial h}{\partial t} + \dfrac{\partial q}{\partial x} = 0; \\[2mm] \dfrac{\partial q}{\partial t} + \left(1 - \dfrac{h}{H}\right) \dfrac{\partial}{\partial x} \left(\dfrac{q^2}{h} + \tfrac{1}{2}h^2\right) - \dfrac{h}{H} \dfrac{\partial}{\partial x} \dfrac{q^2}{H-h} = 0. \end{cases} \quad (5.16)$$

In characteristic form, the system reads

$$\begin{cases} \dfrac{\partial h}{\partial t} + h\dfrac{\partial u}{\partial x} + u\dfrac{\partial h}{\partial x} = 0; \\[2mm] \dfrac{\partial u}{\partial t} + (1 - 2A)u\dfrac{\partial u}{\partial x} + (1 - B)\dfrac{\partial h}{\partial x} = 0, \end{cases} \quad (5.17)$$

where

$$A = \frac{h}{H - h}; \quad (5.18)$$

$$B = \frac{h}{H} + \frac{u^2}{H} \left(1 - \frac{h}{H}\right)^{-2}. \quad (5.19)$$

The equations represent conservation of volume and horizontal momentum in the dense fluid layer (the term $\partial h/\partial x$ reproduces the pressure gradient). The equations are valid in the domain $0 \le x \le x_N(t)$, where $x_N(t)$ is the dimensionless position of the nose.

The two-layer system (5.17) is hyperbolic. Following the method of Appendix A.1, we obtain the characteristic relationships and paths

$$\frac{dh}{du} = \frac{1}{1 - B} \left[Au \mp \sqrt{(Au)^2 + (1 - B)h}\right], \quad (5.20)$$

on

$$\frac{dx}{dt} = c_\pm = u(1 - A) \pm \sqrt{(Au)^2 + (1 - B)h}. \quad (5.21)$$

These relationships show that, in the absence of shocks, $u/\sqrt{h}$ is a function of $h/H$ only. The derivation is left for an exercise.

Both equations (5.20) and (5.21) contain a complex dependency on $H$. This creates qualitative differences in the shape of the interface between the $H > 2$ ("deep ambient") and $H < 2$ ("shallow ambient") cases. Moreover, for $H < 1.25$ (approximately) the matching conditions at the nose are also affected. These difficulties are the mathematical reason for the differences between the two- and one-layer model, as discussed below. This can be interpreted as an increased mechanical rigidity of the system: the propagation of information and feasibility of motion in the current are restricted by the reaction of the ambient fluid when $H < 2$. Fluid domains of constant speed and height ($du = dh = 0$) are allowed by (5.20) for any $H$.

The task is to solve the system of equations (5.17) (or (5.16)) and obtain the variables $h$ and $u$ of the current. The variables $h_a(x,t)$ and $u_a(x,t)$ of the ambient are related with these of the current by

$$h_a = H - h; \quad u_a = -\frac{uh}{h_a}. \tag{5.22}$$

The initial conditions at $t = 0$ are $u = 0$, $h = 1$ ($0 \le x < 1$), and $x_N(0) = 1$. The boundary condition at the backwall $x = 0$ is $u = 0$.

The nose velocity condition is given, again, by (2.25). In Boussinesq, scaled form, we impose

$$u_N = Fr(a)\, h_N^{1/2}, \quad \text{and} \quad a \le a_{\max} \approx 0.5, \tag{5.23}$$

where $a = h_N/H$ and $Fr$ is a known function (see (2.26) and (2.27)). The justification is the same as for the one-layer model. Actually, the validity of this front condition in the two-layer system is higher than in the one-layer model. Benjamin's derivation uses a two-layer system, as explained in Chapter 3. As we shall see later in this chapter, an interesting equivalence between the two-layer time-dependent problem and Benjamin's steady-state current appears.

As expected, in the limit $H \to \infty$ the two-layer equations reduce to the one-layer model governing equations. In this limit $A = B = 0$.

What do these equations predict for $H$ close to 1? Before we proceed to interpretations, consider the finite-difference results for the initial propagation for $H = 1.2$ displayed in Fig. 5.2. The solutions of the two-layer SW system were obtained by the same Lax-Wendroff finite-difference method as used for the one-layer system, on a grid with 1000 intervals and time step $2 \cdot 10^{-4}$ (some fine tuning of the time step and of the numerical artificial viscosity (dissipation) was necessary to avoid numerical instabilities). We see that the behavior is very different from the $H = 3$ counterpart problem shown in Figs. 2.3 and 2.7. In the $H = 1.2$ case the interface displays a backward-moving jump in the initial dam-break stage (seen at $t = 1$ in the figure), which is reflected as a forward-moving jump (seen at $t = 3$ and 10 in the figure). The propagation is with constant $u_N$ over a long distance (the figure shows this effect up to $t = 10$, $x_N \approx 5.6$; the computations indicate that the deceleration starts at

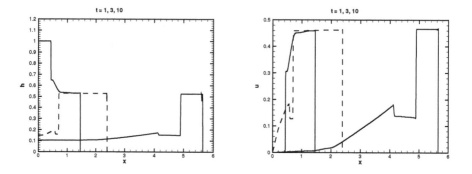

**FIGURE 5.2:**   SW two-layer results for $H = 1.2$: $h$ and $u$ as functions of $x$ for various times. The HS $Fr$ function was used.

$t \approx 15.5$, $x_N \approx 8$.)   The speed of propagation is significantly smaller than in the $H = 3$ configuration (0.46 as compared to 0.63). These features are, qualitatively at least, in good agreement with laboratory observation. This indicates that the two-layer model captures the correct underlying mechanism. We shall therefore proceed to a deeper interpretation and understanding of the predictions of this model.

### 5.3.1   "Critical" nose region for $H < 1.25$

Consider the characteristics of the two-layer system, (5.20)-(5.21). We notice that the value $B = 1$ indicates a "critical" change of behavior for the characteristics. For $B < 1$ we obtain $c_+ > u$, which is qualitatively like in the one-layer system. However, for $B > 1$ we obtain $c_+ < u$. This possibility is problematic for the solution close to the nose. It can be argued that the fluid there cannot travel faster than the perturbation that determines the motion. (This can be regarded as a cavitation restriction.) To overcome this non-physical situation, we subject the solutions to the condition $B \leq 1$ at the nose. In other words, we require that for $h = h_N$ and $u = u_N$ the result of (5.19) does not exceed 1. For a given $H$, this can be expressed as an upper limit constraint on the velocity of the nose

$$u_N \leq u_{N\,max} = \left[H(1 - h_{Nc}/H)^3\right]^{1/2};  \qquad (5.24)$$

here $h_{Nc}$ is the critical height of the nose associated to the $B = 1$ and $u_{N\,max}$ conditions. The restriction (5.24) can be rewritten as

$$u_N \leq u_{N\,max} = \left[\frac{(1-a)^3}{a}\right]^{1/2} h_{Nc}^{1/2},  \qquad (5.25)$$

where $a = h_{Nc}/H$.

On the other hand, we have the front-shock velocity equation $u_N = Fr(a)h_N^{1/2}$. Combining this equation with the limitation (5.25), we obtain

$$aFr^2(a) \le (1-a)^3. \tag{5.26}$$

This is clearly fulfilled for small values of $a$ (say 0.1 and below). Otherwise, the solution of (5.26) determines the value of the largest (critical) $a$, denoted by $a_c$.

It is left for an exercise to show that $a_c = 0.347$ for Benjamin's $Fr$ and $a_c = 0.427$ for the HS $Fr$. (It is interesting that for Benjamin's $Fr$ the function $a^{1/2}Fr(a)$ has a maximum at $a = a_c = 0.347$.) Subsequently, we obtain

$$h_{Nc} = a_c H; \quad u_{N\max} = (1-a_c)^{3/2}H^{1/2}. \tag{5.27}$$

In particular, for $H = 1$, the values of $h_{Nc}, u_{N\max}$ are $0.347, 0.527$ for Benjamin's $Fr$ and $0.427, 0.434$ for the HS $Fr$.

We recall that $a_c$ is the largest $h_N/H$ allowed by the two-layer SW model in a dam-break problem. It is interesting to note that for Benjamin's $Fr$, the attainable $a_c$ is already below the energy restriction $a_{\max} = 1/2$ which was derived in Chapter 3. Furthermore, we observe that even for $H = 1$, i.e., full depth lock geometry, the nose of the current with Benjamin's $Fr$ will not attain the dissipationless $1/2$ channel thickness. These conclusions, however, are restricted to Boussinesq systems, which are under consideration here.

In general, it is necessary to subject the nose velocity calculations, and the flow field behind the nose, to (5.24). The implication is as follows.

In § 2.5 we presented a straightforward method for determining the initial speed of the current. We assume that the $c_+$ characteristics from the reservoir reach the nose. Thus, we can integrate the characteristic balance equation on a $c_+$ characteristic that propagates from the reservoir into the current, to obtain $u$ as a function of $h$, until the value of the nose condition $Frh^{1/2}$ is reached. The current behind the nose is a simple rectangular slug with $h = h_N$ and $u = u_N$. Now we must check the result of $u_N$ against (5.24) (or that $h_N$ is smaller than the critical $a_c H$). If the condition is not satisfied, it means that the assumption that $c_+$ reaches the nose is invalid in this case. The fluid in the reservoir lacks the necessary pressure (or energy) for maintaining that motion. Instead, the nose will propagate with $u_{N\max}$, and the critical value of $h_N = a_c H$, quite uncontrolled by the conditions in the reservoir. The details will be elaborated later. The prominent effect is that the current behind the nose has a curved interface with a negative slope; see Fig. 5.3.

It turns out that the $u_{N\max}$ restriction is relevant only for non-large values of $H$, and actually only for $H < 1.25$ approximately (the exact value depends on the $Fr$ correlation). This can be inferred from the expectation that, typically, $h_N = 0.5$. Consequently, the critical nose height restriction $h_N < a_c H \approx 0.4H$ is fulfilled in configurations with $H > 1.25$. Also (5.27) indicates that $u_{N\max}$ is about 0.52 for $H = 1.25$ and increases faster than $H^{1/2}$. This is above the values of the expected $u_N$. Here we see, again, that gravity currents in configurations with $H \approx 1$ differ in significant aspects from the main domain of investigation. These currents are an exception even in the range of shallow-ambient $H < 2$ configurations. It is evident that results for $H \approx 1$ cases cannot be generalized.

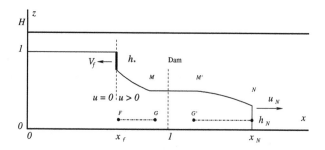

**FIGURE 5.3:**   Dam-break current for $H < 1.25$ case, with left-moving jump at interface (thick line) and $u_N = u_{N\,\text{max}}$.

## 5.4    Jumps of interface for $H < 2$

### 5.4.1    Backward-moving discontinuity

When $H < 2$ the interface displays a jump which moves from the gate to the backwall (to the left in our figures) during the dam-break stage. In Fig. 5.2 we see this jump of $h$ at $t = 1$. On the other hand, for large $H$ the interface displays a smooth left-moving rarefaction wave during the dam-break stage. To understand the formation of the left-moving discontinuity of the interface, let us go back to the dam-break problem solution predicted by the one-layer model. We learned in § 2.5 that the parabolic part of the interface behind the lock is an expansion fan, governed by the $c_-$ characteristics; see Fig. 2.5. This fan covers the transition from $h = 1$ to $h = h_N \approx 0.5$, the domain $LM$ in Fig. 2.5. It is clear that this smooth expansion behavior requires that the $c_-$ characteristics maintain a well-arranged velocity spectrum. A characteristic at a larger $h$ must propagate faster (in absolute value, into the reservoir) than a characteristic at a smaller $h$. If the opposite occurs, the characteristics will intersect. The one-layer model indeed satisfies the condition $d|c_-|/dh > 0$ in the expansion fan region.

Consider now the behavior expected for the two layer model. The leading left-moving characteristics (corresponding to point $L$ in Fig. 2.5) propagate into the unperturbed fluid where $u = 0$. The depression of the interface to the right of point $L$ is small, i.e., $h = 1 - \varepsilon$, where $0 < \varepsilon \ll 1$. According to (5.21) (with $u \approx 0$)

$$c_- \approx -\left[\left(1 - \frac{1}{H} + \frac{\varepsilon}{H}\right)(1 - \varepsilon)\right]^{1/2} =$$

$$-\left[1 - \frac{1}{H} + \left(\frac{2}{H} - 1\right)\varepsilon - O(\varepsilon^2)\right]^{1/2}. \quad (5.28)$$

We see that the dominant influence of the height decrement, $\varepsilon$, depends on the sign of $(2/H - 1)$. For $H < 2$ the perturbations from smaller $h$ (larger $\varepsilon$) tend to move faster to the left. The depressed interface continues to steepen until it forms a discontinuity (jump), as sketched in Fig. 5.4. This discontinuity propagates to the left with a larger speed than the $c_-$ result for $\varepsilon = 0$. The fluid in the reservoir is jumped into motion by this shock. On the other hand, when $H \geq 2$ a smooth expansion fan can develop, and the fluid in the reservoir is affected by a smooth rarefaction wave. This recovers the predictions of the one-layer model.

Indeed, Rottman and Simpson (1983) attempted to solve the two-layer model equations by the method of characteristics. They obtained satisfactory results for $H \geq 2$, but a multivalued unphysical behavior of the interface for $H < 2$ which indicates the formation of a shock.

The discontinuity is assumed to appear instantaneously in the SW framework. Actually, we shall see that it displays a constant height $h_*$ and speed $V_f$ (to the left), see Fig. 5.4, which prevail until the jump attains the backwall $x = 0$. The details are as follows.

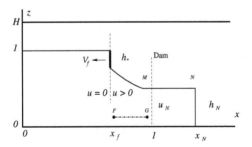

**FIGURE 5.4:** Dam-break current for $H < 2$ case, with left-moving jump at interface (thick line) and $u_N < u_{N\,\text{max}}$.

The analysis of this shock bears resemblance with that of the jump condition developed for the nose discontinuity (the derivation of Benjamin's condition

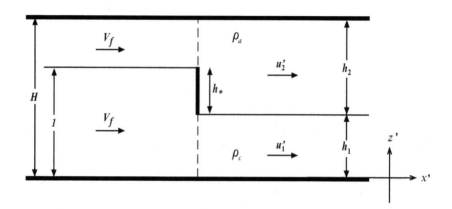

**FIGURE 5.5:** The flow about the left-moving jump for $H < 2$ (as seen by an observer moving with the shock).

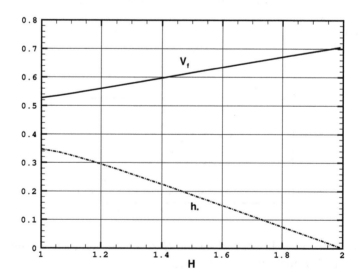

**FIGURE 5.6:** Speed $V_f$ and the height $h_*$ of the left-moving jump as functions of $H$. (Ungarish and Zemach 2005, with permission.)

in a control volume about the nose) in § 3.2. The key to the success of the analysis is to consider the balances in a frame of reference moving with the discontinuity. Let us call this frame $S'$; the axis $x'$ is on the bottom and the positive direction is to the right, like $x$; in the vertical direction, $z' = z$. In this frame the flow approaching and leaving the jump is in a quasi-steady state (over short $x$ distances, at least). We refer to Fig. 5.5. The frame $S'$ moves to the left in the laboratory frame with the absolute speed $V_f$. The transformation of the velocity is simply $u' = u + V_f$, where the prime denotes the $S'$ moving frame, and $u$ is the speed in the laboratory frame. We consider the balances for a thin control volume attached to the jump. The subscripts $l$ and $r$ denote the left and right boundaries of this volume in Fig. 5.4 and Fig. 5.5. The field with known (unperturbed) conditions into which the shock propagates is referred to as upstream (with respect to the jump), and the field with unknown $u$ and $h$, which develops after the arrival of the shock, is referred to as downstream.

The kinematic conditions relevant to the jump are as follows. On the left (upstream), there is unperturbed fluid with $u = 0$, and hence $u' = V_f$, in both the dense and ambient fluid. On the right, or downstream, with respect to the jump we obtain from geometry and volume continuity considerations

$$h_1 + h_2 = H; \quad V_f \cdot 1 = u_1' h_1; \quad V_f(H - 1) = u_2' h_2. \qquad (5.29)$$

We define the height of the jump as

$$h_* = 1 - h_1. \qquad (5.30)$$

The magnitude of $h_*$ is also an indication of the strength of the shock.

Consider the pressure field. The (reduced) pressure fields on the right and left of the jump are hydrostatic. Since only pressure differences are important, we can take the pressure in the ambient on the left as zero. Consequently, on the left the pressure field is given by

$$p_l = \begin{cases} 1 - z & (0 \le z \le 1) \\ 0 & (1 \le z \le H). \end{cases} \qquad (5.31)$$

To determine the pressure on the right of the shock, a jump condition for the pressure is needed. The boundary $z = H$ is a streamline. On this streamline Bernoulli's equation (again, in the $S'$ frame where the flow is in steady state) can be written as

$$\frac{1}{2} V_f^2 = p_R + \frac{1}{2} u_2'^2 + \Delta, \qquad (5.32)$$

where $p_R$ is the downstream pressure at the upper boundary, and $\Delta$ denotes a head loss due to irreversible dissipation. To simplify the analysis we shall assume $\Delta = 0$. (Calculations with more realistic values of $\Delta = 0$ show good agreement with the simplified results, as discussed in Klemp, Rotunno, and Skamarock 1994) However, this $\Delta = 0$ does not imply energy conservation in the two-layer system. The reader can check that Bernoulli's balance at the bottom of the shock, on the boundary $z = 0$, is, in general, not satisfied.)

Discarding the dissipation we obtain

$$p_R = \frac{1}{2}\left(V_f^2 - u_2'^2\right). \qquad (5.33)$$

The pressure distribution on the right of the jump can now be expressed as

$$p_r = \begin{cases} p_R + h_1 - z & (0 \le z \le h_1) \\ p_R & (h_1 \le z \le H). \end{cases} \qquad (5.34)$$

Finally, we employ the flow-force balance for the moving control volume

$$\int_0^H (u'^2 + p)_l\, dz = \int_0^H (u'^2 + p)_r\, dz, \qquad (5.35)$$

into which we substitute the expressions (5.29)-(5.31), and (5.33)-(5.34). After some algebra (which is left for an exercise) we obtain

$$V_f = \left[ \frac{\gamma^2 - \alpha^2}{\dfrac{2\gamma^2 - \alpha}{\alpha} + \left(\dfrac{1-\gamma}{1-\alpha}\right)^2 (1 - 2\alpha)} \cdot \frac{1}{\gamma} \right]^{1/2}, \qquad (5.36)$$

where $\gamma = 1/H$ and $\alpha = (1 - h_*)/H$.

Like Benjamin's formula for $u_N$, this is one equation for two unknowns: $V_f$ and $h_*$. This makes sense. The speed result (5.36) represents the kinematic and dynamic balances for a discontinuity which propagates in a general two-layer system, with known upstream conditions and unspecified downstream conditions. To obtain a particular solution for a given system, we must specify the boundary (or initial) condition which connects between this local discontinuity and the dam-break process which generated and supports this discontinuity. This will provide an additional equation for the unknown $V_f$ and $h_*$.

The second equation can be derived from the characteristics of the dam-break release. We argued that the jump develops because of mismatched dependencies of $c_-$ (from the right) on $h$. We therefore expect that the appropriate $V_f$ and $h_*$ replace the mismatch in such a manner that the resulting motion can be smoothly matched with the $c_-$ speed at the right of the jump. Otherwise, either a stronger shock is necessary, or a local expansion domain is possible. We therefore write the second equation for the jump as

$$|c_{-r}| = V_f. \qquad (5.37)$$

We emphasize that $c_{-r}$ is calculated using equation (5.21) with $u = u_1 = h_* V_f / (1 - h_*)$ and $h = h_1 = 1 - h_*$.

The numerical solution of the system (5.36)-(5.37) provides unique results for $h_*$ and $V_f$, as shown in Fig. 5.6. These results can be obtained for $1 \leq H < 2$ only. As expected, for $H > 2$ no left-moving discontinuity can appear and be sustained in the SW dam-break Boussinesq problem.

A fitted polynomial approximation to the shock results is

$$V_f \approx 0.1798H + 0.3475; \quad h_* \approx -0.093H^2 - 0.066H + 0.507. \qquad (5.38)$$

The velocities in the fluid just behind the jump (to the right, where $h_1 = 1 - h_*$), in the laboratory frame, are

$$u_1 = \frac{V_f h_*}{1 - h_*}; \quad u_2 = -\frac{V_f h_*}{H - 1 + h_*}. \qquad (5.39)$$

We see that for $H$ close to 1 the left-moving jump leaves behind (to the right) a very significant motion.

The left-moving shock is a time-independent effect, until it reached the backwall at the time $t = 1/V_f$. This is defined, again, as the end of the dam break phase. In the one-layer model this occurs at $t = 1$. Since $0.52 < V_f < 0.71$, we conclude that it takes a significantly longer time to set in motion a reservoir embedded in a shallow ambient than in a deep ambient. This can be attributed to the added inertia of the ambient. The inertia of the ambient behaves like $u_2^2 h_2 = (uh)^2/(H - h)$, and that of the current like $u^2 h$. The ratio is $(H/h - 1)^{-1}$, which indicates a very significant hindrance for $H$ close to 1 and a typical $h = 0.5$.

For $H = 2$ the height of the jump, $h_*$, vanishes, and the classical expansion fan (smooth rarefaction wave) is recovered. The speed of this rarefaction wave is $-1/\sqrt{2}$, while the one-layer model predicts $-1$, but otherwise there is good agreement between the two- and one-layer results.

We note in passing that Klemp, Rotunno, and Skamarock (1994) used a different argument from above to derive a "second equation" instead of (5.37). The results for $V_f$ and $h_*$ are the same.

## 5.4.2 The reflected bore

At $t = t_f = 1/V_f$ the left-moving discontinuity hits the wall at $x = 0$. The upper-layer ambient fluid in the region $x > 0$ has a considerable negative velocity $u_2$ at this time; see (5.39). The impingement on the wall forms immediately a shock-wave (bore, hydraulic jump), that propagates into the fluid with speed $V_b$, and brings the encountered fluid to an abrupt stop. The increased pressure in the stagnant ambient fluid pushes down the interface, and hence the dense fluid on the left of the shock is significantly thinner than in the moving current; see Fig. 5.7. This is consistent with the observations of Rottman and Simpson (1983); see Fig. 5.1.

The analysis of this phenomenon can be performed by the method used above for the left-moving discontinuity. We use again a frame of reference $S'$,

*An introduction to gravity currents and intrusions*

which is moving with the shock with speed $V_b$ in the positive $x$ direction (to the right), and an associated thin control volume. The configuration sketched in Fig. 5.7. The transformation of speed is simply $u' = u - V_b$.

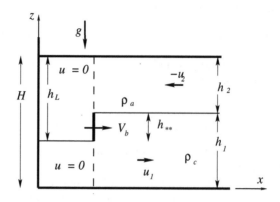

**FIGURE 5.7:** The reflected bore configuration. The flow field $u_1, h_1, u_2, h_2$ is known.

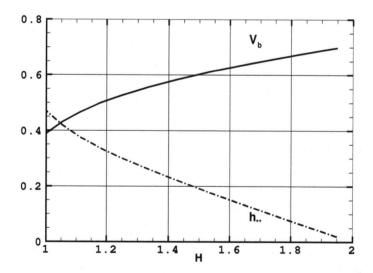

**FIGURE 5.8:** The reflected bore speed (solid line), and height (dash-dot line) as functions of $H$ at time of formation.

We denote by $h_L$ the thickness of the ambient fluid on the left (downstream) of the bore. The corresponding thickness to the right of the bore is $h_2$, which is known from the solution of the flow field in the previous phase. We introduce the presently unknown ratio

$$\chi = \frac{h_L}{h_2},$$ (5.40)

and the height of the jump

$$h_{**} = h_L - h_2 = (\chi - 1)h_2.$$ (5.41)

We expect $\chi > 1$. We shall see below that this is the only unknown that must be determined from the appropriate balances.

The kinematic conditions are as follows. On the left, or downstream, with respect to the bore, there is stagnant fluid with $u = 0$, and hence $u' = -V_b$, in both the dense and ambient fluid. On the right, or upstream, with respect to the bore, the thicknesses of the layers $h_1$ and $h_2$ and the velocities $u_1$ and $u_2$ are known. For further use we write, for the right side,

$$u_1 h_1 + u_2 h_2 = 0; \quad u_1' = u_1 - V_b; \quad u_2' = u_2 - V_b.$$ (5.42)

Continuity of volume fluxes $u'h$ that enter and leave our control volume, for both the dense and ambient fluid, yields

$$V_b = -\frac{u_2}{\chi - 1}.$$ (5.43)

Consider the pressure field. The hydrostatic pressure on the right of the bore is

$$p_r = \begin{cases} h_1 - z & (0 \le z \le h_1) \\ 0 & (h_1 \le z \le H). \end{cases}$$ (5.44)

For the pressure jump condition we use the streamline along the boundary $z = H$. Bernoulli's equation yields

$$\frac{1}{2}V_b^2 + p_L + \Delta = \frac{1}{2}u_2'^2,$$ (5.45)

where $p_L$ is the downstream pressure at the upper boundary, and $\Delta$ denotes an irreversible dissipation head loss. We again discard the head loss, and obtain

$$p_L = \frac{1}{2}\left(u_2'^2 - V_b^2\right) = \frac{1}{2}u_2^2 - u_2 V_b.$$ (5.46)

Using this result, the hydrostatic pressure on the left of the bore can be expressed as

$$p_l = \begin{cases} p_L + (H - h_L) - z & (0 \le z \le H - h_L) \\ p_L & (H - h_L \le z \le H). \end{cases}$$ (5.47)

Finally, we apply the flow-force balance (5.35). We substitute the foregoing results for $p_r$ and $p_l$, use equations (5.42), and eliminate $h_R$ and $V_b$ by (5.40) and (5.43). After some algebra we obtain

$$\frac{u_2^2}{\chi - 1}G = u_1^2(G - 1) + u_2^2(1 - \frac{1}{2}G) + \frac{1}{2}h_2\left[(G - 1)^2 - (G - \chi)^2\right], \quad (5.48)$$

where $G = H/h_2$. We recall that $u_1, u_2$, and $h_2$ are the known conditions of the flow into which the bore propagates. At the formation of the bore, these are exactly the conditions left behind by the left-moving jump which hits the wall; see (5.39).

We obtained one equation for the single unknown $\chi$. In this respect there is difference with the front and left-moving discontinuities where a similar analysis produces one equation for two unknowns. The reason is that here the velocities in the up- and down-stream sides of the discontinuity are specified, and the thickness of the layers upstream is also given. The only degree of freedom is the height partition between the stagnant fluids in the downstream domain.

The solution of (5.48) in the range $\chi > 1$, for given values of $u_1, u_2$, and $h_2$, is straightforward. In particular, we can use the results left behind by the left-moving jump. This yields the conditions at the formation of the bore. The velocity and height of the bore (at the time of formation) as a function of the total depth $H$ are given in Fig. 5.8. The height of the jump, $h_{**} = h_L - h_2$, which represents the strength of the bore, decays to zero as $H$ approaches 2, as expected.

The reflected bore will eventually reach the nose. Up to this time the propagation of the current is expected to be with constant speed $u_N$ (the slumping phase). We notice that when $H$ increases from 1 to 2, $V_b$ increases by a factor of 2, roughly. This explains the quite drastic decrease of the slumping distance in this range of $H$. Since the velocity $u_N$ of the nose increases only slightly with $H$, the relative velocity between the bore and the nose decreases strongly with $H$. In other words, the bore will reach the nose faster for larger $H$ than for $H = 1$.

However, the details of the journey of the bore toward the nose are not simple. The speed of the bore is not constant, because the upstream conditions are $x$-dependent. Indeed, when the left-moving jump hits the wall, the interface has an inclined portion near the backwall and a horizontal part following the nose. For $H > 1.25$ this horizontal part is of height $h = h_N$. For $H < 1.25$ the horizontal part is higher than $h_N$; then an additional decrease of height to $h_N$ occurs. The use of the local flow-field values in (5.48) is expected to provide the details of the time-dependent $V_b$. As a rough approximation for the averaged speed of the bore, we can take the mean between the values of $V_b$ at formation and at (almost) the position of the nose, where $h_1 = h_N$ and $u_1 = u_N$. For example, let us consider the $H = 1$ case. The speed of $V_b$ at formation is 0.39. The nose conditions $(h, u)$ are $(0.427, 0.434)$ and $(0.347, 0.527)$

for HS and Benjamin $Fr$ formulas. The values of $V_b$ for these conditions are 0.54 and 0.60, respectively. The mean value of $V_b$ is 0.47 and 0.54. The mean speed of the bore in the experiments of Rottman and Simpson (1983), for $H = 1$, was 0.55. These experiments suggest that the velocity of the bore is constant, but the theoretical variation of $V_b$ is sufficiently small for being in the range of the experimental uncertainties.

The transition from the jump-bore regime to the smooth interface regime is theoretically at $H = 2$. However, in practice this is not a sharp occurrence. When $H$ approaches 2 from below (say, for $H > 1.8$) the theoretical jumps $h_*$ and $h_{**}$ become sufficiently small to be blurred and smoothed out by the local eddies and shear instabilities at the interface. Moreover, in real fluids the discontinuities are weakened by head loss (which we assumed to be zero for both the backward and forward moving shocks). This explains why the experimental transition is at about $H \approx 1.7$, not 2.

### 5.4.3 Dam break and the speed of propagation during slump- ing

#### 5.4.3.1 Theory

In the one-layer SW model the value of $u_N$ during the initial slumping phase is given by (2.66). This simple solution resulted because the $c_+$ characteristic equation can be integrated analytically all the way from the reservoir to the nose. A similar result can also be obtained from the two-layer model equations, if we take $A = B = 0$ (justified as an approximation for a deep ambient by the fact that this is the limit of these variables for $H \to \infty$). However, we still use the finite value of $H$ in $Fr(a)$, and this leaves some inconsistency in the results.

The more rigorous approach, which is of course necessary for at least the domain $H < 2$, is to solve the dam-break problem using the two-layer model. The task is complicated by the possible presence of the left-moving jump and the $u_{N\,\text{max}}$ restriction. We attempt to follow the development of the variables $h$ and $u$ along a forward-propagating characteristic, $c_+$, starting from a point $F$ with known conditions in the reservoir domain, to a relevant point $G$ in the bulk of the fluid behind the nose. Using (5.20) we obtain, formally,

$$u_G(h_G; h_F, u_F) = \text{Solve} \left\{ \frac{du}{dh} = (1 - B)\left[Au - \sqrt{(Au)^2 + (1-B)h}\right]^{-1} \right\},$$
(5.49)

where $A$ and $B$ are functions of $h$, $u$, and of the parameter $H$, see (5.18)-(5.19), and $h_F, u_F$ are prescribed initial conditions. The integration must be performed numerically, e.g., by a Runge-Kutta method. No jumps are allowed on the trajectory $FG$. The function on the right-hand side is negative (for $B \neq 1$), consistent with the fact that for the initial slumping phase we are interested in a forwardly propagating characteristic along which $h$ decreases and $u$ increases (in the simplest case, from zero to $u_N$). We note that, typically, the value of $B$ increases on the $c_+$ characteristics from the reservoir forward.

Now we must distinguish between three cases:

1) When $H \geq 2$ the interface between the reservoir and nose is smooth. Points $F$ and $G$ can be conveniently taken like in the one-layer case. In other words, the integration starts in the unperturbed fluid in the lock, with $h = 1$ and $u = 0$, and ends in the rectangular region of constant height $h_N$ and velocity $u_N$ which trails the nose. In this case the height and velocity given by (5.49) at point $G$ are intersected with the nose jump condition (5.23) to obtain $h_N$ and $u_N$. The value of $B$ is smaller than 1 along the path of integration, and hence the $u_{N\,max}$ restriction is implicitly satisfied.

2) When $H < 2$ point $F$ must be taken at the position to the right of the left-moving jump of the interface; see Fig. 5.4. The values of $h$ and $u$ at this position are known from the solution described in § 5.4.1: $h = 1 - h_*$ and $u = u_1$ given by (5.39). Using these as initial values for the integration, the previous calculations for $h_N$ and $u_N$ can be repeated. Now we must also check the value of $B$ along the path of integration. For most of the $H < 2$ range, $B$ remains smaller than 1 until a point $G$ as above (i.e., where $u_G$ intersects the $u_N$ condition with the corresponding $h_G$) is reached. A domain of constant height and speed $(h_N, u_N)$ follows the nose. The position of point $M$ where the inclined and horizontal portions of the interface meet is given by

$$x_M = 1 + c_-(h_N, u_N) \cdot t. \qquad (5.50)$$

This is the locus of the fastest characteristic in the expansion fan, of the rarefaction wave; see (5.21).

3) When $H$ is close to 1 the $u_{N\,max}$ constraint, see (5.24), becomes relevant. The integration for the fluid behind the lock is as above. However, an additional matching domain is necessary between the nose and the horizontal interface domain; see Fig. 5.3. Indeed, for $H < 1.25$ (the exact value depends slightly on the specific $Fr$ function) the value of $B$ in (5.49) may exceed 1 on the path $FG$. A direct matching between the fluid in the reservoir and the nose along this path could produce a non-physical behavior (e.g., a nose velocity larger than $u_{N\,max}$). This is avoided by means of an expansion-fan zone (of the $c_+$ characteristics) which follows the nose.

In this case the nose propagates with $u_N = u_{N\,max}$ and the corresponding $h_N = h_{Nc}$, as specified in § 5.3.1. The domain left behind the nose is typically a region of non-constant $h$ and $u$. To be more specific, in this region we expect $\partial h / \partial x < 0$ and $\partial u / \partial x > 0$. We estimate the behavior in this region by integrating (5.20) along the $c_-$ characteristic from the nose backwards. The path of integration is displayed schematically from the nose to point $G'$ in Fig. 5.3. The matching assumption is that the conditions at points $G$ and $G'$ are the same. The corresponding diagram in the $h - u$ plane of the characteristics and other velocities, for the case $H = 1$, is illustrated in Fig. 5.9.

After obtaining the matched values of $h_G, u_G$, we can calculate the boundaries of the domain $MM'$, see Fig. 5.3, in which $h$ and $u$ are constants. The speeds of these points are constant, and given by the $c_-$ and $c_+$ values for $h_G, u_G$, and the initial position is that of the dam, $x = 1$.

It is interesting to put all this together for the $H = 1$ case with Benjamin's $Fr$ condition. We leave the details for an exercise. The results are a perfectly symmetric motion from and into the reservoir, or between the current and ambient. In particular, $h_N = h_* = 0.347$, $u_N = V_f = 0.527$, and $h_G = 0.500$. The points $M$ and $M'$ spread out symmetrically with speeds $\pm 0.157$. This is a good test for the theory. Physical considerations show that in a Boussinesq inviscid system the bottom and the top currents should be mirror images, and in the $H = 1$ geometry there is no clear distinction between the current and ambient. The solution reproduces these expectations. However, this symmetry is violated when a different $Fr$ function is used. The reason is that now some additional momentum loss (say by friction) is introduced at the nose, and this makes a distinction between the left-moving jump and the nose. This difference can be justified when the current propagates on a solid bottom while the return flow in the ambient is along a free surface.

### 5.4.3.2 Some comparisons

The $H = 1$ full-depth lock release, or lock-exchange, motion has received considerable attention in the analysis of gravity currents. We obtained a fairly complete SW description of the $H = 1$ dam-break motion. This stage ends at $t = 1/V_f = 1.90$ when the jump hits the backwall. There is in general a fair agreement between the two-layer theory and observations. However, the rich details of the SW theoretical prediction are difficult points for accurate experimental and numerical verifications. The experimental interface is usually so blurred and contaminated by mixing and local shear instabilities (see for example Fig. 5.1) that it seems quite impossible to obtain a direct confirmation of the predicted height decrease from the $h = 0.50$ at the gate to 0.35 at the nose. Pertinent high-resolution numerical simulations for a Boussinesq system were presented by Härtel, Meiburg, and Necker (2000). The numerical interface (see Figs. 3 and 11 in that paper) is sharper than the experimental one, but still contaminated by additional effects which prevent a conclusive comparison.

The speed of propagation, $u_N$, is a more reliable variable for comparisons. However, the evidence concerning $u_N$ for $H = 1$ is also not fully conclusive. The SW theory with Benjamin's $Fr$ predicts $u_N = 0.527$. In laboratory experiments, values of $0.40 - 0.46$ were obtained. The numerical simulations for Boussinesq systems presented by Härtel et al. (2000) reproduced well these experimental values, and prove that the viscous friction on the boundary reduces the speed of propagation, as follows. First, the speed increase from 0.40 to 0.46 is accompanied by an increase of $Re$ by a factor of about 50. Second, simulations with no-slip conditions produced $u_N = 0.49$. For details, see Fig. 4 in that paper (note that $Fr$ there is $u_N/\sqrt{2}$ in our notation, and $Gr = 8Re^2$). Apparently, the propagation corresponds to the non-dissipative $h = 1/2$, in which case Benjamin's $Fr$ yields exactly $u_N = 1/2$. However, this apparent interpretation of the measured speed for $H = 1$ is very problematic from the point of view of the SW two-layer model. For these values of $h_N$ and $u_N$ we obtain $B = 1.5$ and hence this flow displays the non-physical condition $u_N < c_+$; see (5.19)-(5.21). Such a serious internal inconsistency cannot be dismissed just in order to achieve a better agreement with data.

Consider the general dependency of the SW results on $H$, when we depart from the full-depth lock geometry. The two-layer SW model predicts that, during the dam-break phase, the interface has a backward-moving jump for $H < 2$ only. The SW height of this jump decreases like $0.35(2 - H)$, roughly. The experiments confirm the existence of a backward-moving jump for $H < 1.7$ (approximately; see Rottman and Simpson 1983). For $H > 2$ both the SW and experimental interface display smooth rarefaction profiles. There is also fair agreement between theory and experiment concerning the reflection (the bore) of the backward-moving jump from the $x = 0$ backwall. The one-layer SW model does not reproduce these jumps.

Sharper comparisons can be performed, again, for the speed of propagation. Here we actually can verify three features: (a) The prediction that the initial propagation is with constant speed over a significant time interval and length, which we call the slumping phase. This has been so widely confirmed, that it seems a self-understood property of the current. We must keep in mind that this is not a trivial issue. Both the one- and two-layer SW models predict initial motion with constant speed. (b) The value of $u_N$ during the slumping phase. (c) The extent of the slumping phase.

Following Ungarish and Zemach (2005), we consider results for four $Fr(a)$ functions. The first two are the usual Benjamin and HS formulas. Rottman and Simpson (1983) suggested a simple modification of Benjamin's formula: divide it by $\sqrt{2}$. This $Fr$, referred to as RS, is based on the empirical observation that it produced good agreement of $u_N$ between experiments and two-layer SW solutions for $H \geq 2$. The fourth function, referred to as UZ, has been suggested by Ungarish and Zemach (2005) as a simple compromise between the suggestions of HS and RS. It reads

$$Fr(a) = (1 + 3a)^{-1/2}. \tag{5.51}$$

These functions are displayed if Fig. 5.10. We recall that only Benjamin's formula has been derived theoretically with rigor; the three other functions contain empirical adjustments.

A summary of analytical results obtained by the above-mentioned two-layer model calculations is presented in Fig. 5.11. It displays the nose height $h_N$ and the nose velocity $u_N$ as *functions of* $1/H$ for the four suggested $Fr$ functions. In general, the predicted $u_N$ increases with $H$. The predicted $h_N$ first increases with $H$, in the domain of the critical $h_{Nc} = a_c H$, until $H \approx 1.25$.

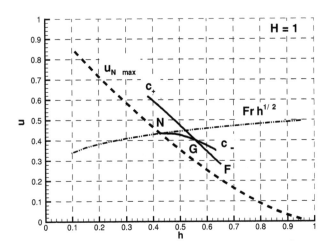

**FIGURE 5.9**: The characteristics and velocities in the $h - u$ plane for $H = 1$. Point $F$ is to the right of the backward-moving discontinuity. $h$ and $u$ in the current are given by the path $FGN$ along $c_+$ and $c_-$. In this diagram points $G$ and $G'$ coincide. (Ungarish and Zemach 2005, with permission.)

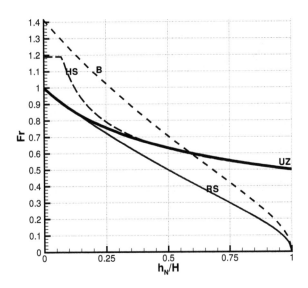

**FIGURE 5.10**: The $Fr$ functions considered by Ungarish and Zemach (2005).

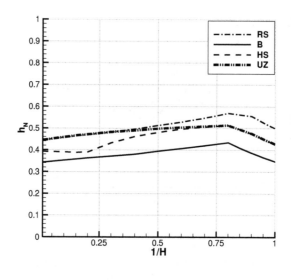

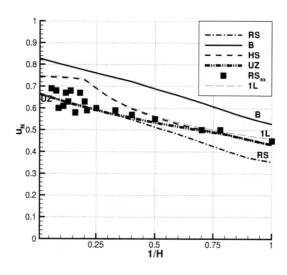

**FIGURE 5.11:** Two-layer slumping results for $h_N$ and $u_N$ as functions of $1/H$ for various $Fr$ functions. Also shown for $u_N$: one-layer (1L) results for HS $Fr$, and experimental results (symbols). (Ungarish and Zemach 2005, with permission.)

For larger $H$, $h_N$ decreases with $H$. (The value of $a_c$ is $0.347, 0.500, 0.427$, and $0.428$ for the $Fr$ function B, RS, HS, and UZ, respectively.) Experimental results of $u_N$, taken from the paper of Rottman and Simpson (1983), are also shown. There is fair agreement between theory and experiment. Benjamin's $Fr$ over-predicts the speed of propagation. The best agreement is obtained with the UZ $Fr$. The value of $h_N$ is not a clear-cut experimental variable, and hence comparisons are not useful.

The prediction of $u_N$ by the one-layer model with HS $Fr$ and no restriction on $h_N$ is also displayed in Fig. 5.11. The agreement with the two-layer model with the same $Fr$ function is excellent. The one-layer model slightly overpredicts the value of $u_N$ for $H < 2$. We therefore conclude that the major difference between the one- and two-layer models is the shape of the interface, not the speed of propagation during slumping. The shape of the interface, however, affects the length of the slumping stage.

### 5.4.4 The slumping distance: theory and experiments

At time $t_s$ after release, when the nose is at the position $x_s$, the slumping propagation with constant $u_N$ ends and the deceleration of $u_N$ begins. Laboratory measurements of $x_s$ were presented by Rottman and Simpson (1983). The only free parameter which affects the measured values of $x_s$ is $H$, in accord with the SW theory. The experiments showed that $x_s$ decreases when $H$ increases. This is displayed in Fig. 5.12 (note that the horizontal coordinate is $1/H$). First, there is a very significant decrease from about 10 to 5 when $H$ varies from 1 to 2. Further increase of $H$ caused milder reductions of $x_s$, to about $3 - 4$ for $H = 10$.

The corresponding SW results, calculated numerically by Ungarish and Zemach (2005), are also shown in Fig. 5.12. The two-layer SW solutions were performed for the four different $Fr$ functions mentioned above.

The available experimental points are quite scattered, and have been obtained indirectly from the intersection of the plotted bore and nose vs. time curves; Rottman and Simpson remarked that this procedure is expected to over-estimate $x_s$. In any case, the accuracy is low and it is difficult to draw strong quantitative conclusions from this comparison. The theoretical curve for the Froude function B seems to provide the best agreement with experiments, in particular for $H < 5$. However, the results of $Fr$ functions HS and UZ can be considered in similar agreement, on account of the expected over-prediction of the experimental evaluation. Additional and more accurate experiments and Navier-Stokes simulations are needed for a better resolution of this issue.

Here the performance of the one-layer model is not so good. In general, this model predicts that $x_s$ increases with $H$, in contradiction to the two-layer and experimental trends. For $H < 2$ the values of $x_s$ predicted by the one-layer model are significantly smaller than the real values. However, for $H > 3$ the values of $x_s$ predicted by the one-layer model are in acceptable agreement with the experiments and two-layer model.

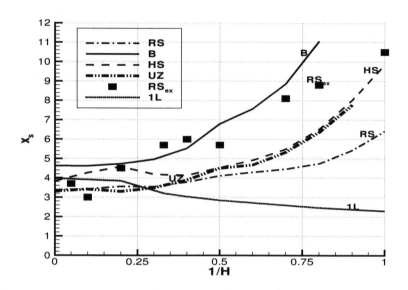

**FIGURE 5.12**:   Two- and one-layer slumping results for $x_s$ as function of $1/H$ for various $Fr$ correlations, and experimental points from Rottman and Simpson (1983). (Ungarish and Zemach 2005, with permission.)

To summarize: The two-layer model captures well the effect of the increasing slumping distance $x_s$ when $H$ decreases, in particular in the domain $H < 2$. This is a result of the increased rigidity of the system in this domain. The backward-forward propagation of information is performed by relatively slow jumps. The tail of the currents becomes thin and slow during the long slumping motion, but the head "does not know" about these changes.

The finite-difference solutions of the two-layer SW equations displayed in Fig. 5.12 were obtained with a special code. This code uses the analytical results of the left-moving jump as boundary conditions. The computation distinguishes between two phases of propagation. The first is the dam-break phase, from $t = 0$ to $t = 1/V_f$, when the left-moving jump reaches the left wall, $x = 0$; the domain of solution is $x_f(t) \leq x \leq x_N(t)$. For efficient solution this domain is mapped into $y \in [0, 1]$ by the transformation $y = [x - x_f(t)]/[x_N(t) - x_f(t)]$. The equations for the variables $h(y, t)$ and $u(y, t)$ are derived along the lines presented in § 2.3; the details are left for an exercise (see also Ungarish and Zemach 2005). In the second phase of propagation the domain of solution is the usual $0 \leq x \leq x_N(t)$, also mapped into $y \in [0, 1]$. The hyperbolic system for $y \in [0, 1]$ was solved with a finite-difference scheme of the type discussed in Appendix A.2.

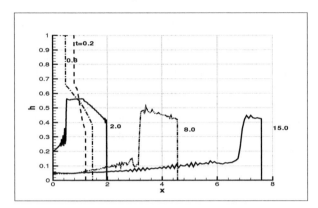

**FIGURE 5.13**: Finite-difference SW two-layer results of $h$ as a function of $x$ at various times, for $H = 1$. A special code was used, such that during the dam-break stage the position of the left-moving jump was calculated analytically. (Ungarish and Zemach 2005, with permission.)

An example of the SW results obtained with this code is shown in Fig. 5.13, for the $H = 1$ case with the HS $Fr$ function. (The numerical grid has 100 intervals and the time step is $2 \cdot 10^{-3}$.) In this solution the nose moves with the critical $u_{N\,max} = 0.43$ and leaves behind a downwardly-inclined interface. Initially, the interface in the region $x < 1$ has a jump of $h_* = 0.35$ which moves to the left, until it reaches the backwall $x = 0$ (at $t \approx 1.9$); these were imposed as boundary conditions. However, eventually the left-moving jump is reflected as a right-moving bore, that produces a deep penetration of the upper fluid into the lower region. Subsequently, the bore propagates from the backwall towards the nose which moves with its initial velocity and height. The bore arrives at the nose at time $t_s = 20.2$ and position $x_s = 9.8$. (The small oscillations in the profiles are a spurious numerical effect.)

## 5.5 Energy and work in a two-layer model

The exchanges of energy for an inviscid gravity current which is released from a lock and then propagates over a horizontal boundary are considered. We showed in the previous sections that the motion of the current can be described, with fair accuracy, by the averaged variables and the corresponding volume and momentum balances. The free parameter is $H$, and there is some flexibility in the choice of the front condition $Fr$ function. It makes sense to ask what do these results teach us about the energy of the flow field. Since we are concerned with an inviscid and incompressible idealization, the relevant energy is mechanical. Now a bell starts to ring: the analysis of the steady-state current and Benjamin's front condition in Chapter 3 indicated that pressure head loss and energy dissipation are a common feature of the gravity current. This needs clarification and quantification. Indeed, where does this energy come from, where does it go, how big is this effect (and compared to what)?

To answer these and related questions, we first focus attention on the energy budgets during the initial stage of propagation of a rectangular "dam-break" current configuration. We follow the presentation of Ungarish (2008). To avoid the complications of the left-moving jump we restrict the main discussion to the $H > 2$ cases. The configuration is sketched in Fig. 5.14.

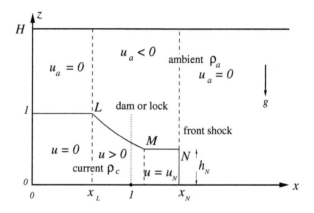

**FIGURE 5.14:** Schematic description of the two-layers system used for energy budgets.

We use dimensionless variables unless stated otherwise. We scale the usual variables according to (5.14)-(5.15). In addition, we scale the energy terms, and the rate of change of energy terms, as follows:

$$\{E^*, \dot{E}^*\} = \{(1/2)\rho_a U^2 h_0 x_0 E, (1/2)\rho_a U^3 h_0 \dot{E}\}, \qquad (5.52)$$

where, again, $U = (g'h_0)^{1/2}$, and the overdot denotes time derivative. The energy is scaled with the initial (reduced) potential energy of the body of dense fluid. In other words, the potential energy of a dense fluid particle is calculated with respect to the buoyancy force (per unit volume, dimensional) $(\rho_c - \rho_a)g$; this eliminates the passive effect of the embedding hydrostatic field, and renders the reduced potential energy of the ambient fluid equal to zero.

Using (5.4), (5.11), and (5.12), the momentum balances can be expressed as

$$\frac{Du_a}{Dt} = -\frac{\partial \Phi}{\partial x}; \qquad \frac{\partial p}{\partial z} = 0; \qquad (h < z \leq H) \qquad (5.53)$$

$$\frac{Du}{Dt} = -\frac{\partial \Phi}{\partial x} - \frac{\partial h}{\partial x}; \qquad \frac{\partial p}{\partial z} = -1; \qquad (0 \le z < h). \qquad (5.54)$$

We recall that $\Phi(x,t)$ is the reduced pressure in the ambient and on the interface $z = h(x,t)$, subject to the boundary condition 0 at $x = x_N(t)$. This set of SW, Boussinesq, inviscid equations is useful for the energy calculations.

The motion of the fluids is in the active domain $0 \le x \le x_N(t)$ only. There is no motion and no energy in the unperturbed ambient (passive domain) at $x > x_N(t)$.

The initial conditions are zero velocity and unit dimensionless height and length at $t = 0$. Also, the velocity at the wall $x = 0$ is zero.

We are interested in the energy during the propagation. Both the current and the ambient fluids contain kinetic energy, denoted $K_c$ and $K_a$, respectively. Since $u$ and $h$ are functions of $x, t$, and using (5.1), we obtain

$$K_c(t) = \int_0^{x_N(t)} u^2 h \, dx; \qquad K_a(t) = \int_0^{x_N(t)} \frac{(uh)^2}{H - h} \, dx. \qquad (5.55)$$

In the original definition of the kinetic energy a $1/2$ coefficient exists. This coefficient is not seen in the last equations because they are in scaled form. The scaling with $(1/2)\rho_a g' h_0^2 x_0$ eliminates the $1/2$ coefficient. The corresponding factor is also incorporated into the following definitions of potential energy and work.

In the reduced gravity field considered here the vertical displacement of ambient fluid particles of density $\rho_a$ is not influenced by the effective body force. Therefore, the (reduced) potential energy, denoted by $P$, is contained only in the lower layer (the current)

$$P(t) = \int_0^{x_N(t)} h^2 \, dx. \qquad (5.56)$$

Evidently, $P(0) = 1$.

The total energy of the two-layer system is

$$\text{total}(E) = P + K_c + K_a. \qquad (5.57)$$

The change of this total energy is defined as the dissipation. When the dissipation (or, to be more precise, the rate of dissipation) is zero, the system is energy conserving, or non-dissipative.

The pressure performs work on the moving parts of the interface between the fluids. Consider first the nose of the current ($x = x_N(t), 0 \le z \le h_N$). The SW model assumes that this boundary between the fluids is a sharp discontinuity between two domains of hydrostatic pressure. This is consistent with the $Fr$ jump conditions used in this model; see § 3.2. On the left of this front discontinuity we have the hydrostatic $p = -z + h_N$, and on the right the unperturbed $p = 0$, and the corresponding pressure forces are $(1/2)h_N^2$ and 0 (scaled with $\rho_a g' h_0^2$). Consequently, during the displacement of this boundary with speed $u_N$ only the dense fluid on the left performs work, and this is at the rate

$$\dot{W}_N(t) = u_N h_N^2. \tag{5.58}$$

The moving curved part of the interface ($LM$ in Fig. 5.14) displaces the pressure force associated with $\Phi(x, t)$. Here the pressure is continuous, and hence the same amount of work, but with opposing signs, is performed by the lower and upper fluids. The corresponding rate of performed work by the lower fluid (the current) at the interface can be expressed as

$$\dot{W}_i(t) = 2 \int_0^{x_N(t)} \Phi \frac{\partial h}{\partial t} dx, \tag{5.59}$$

where the subscript $i$ denotes the inclined portion of interface. The derivation of this result is left for an exercise.

### 5.5.1 Some SW energy calculations

The SW governing equations (5.16) with the appropriate boundary conditions were solved by a finite-difference scheme as explained before. The resulting discrete values of $h$ and $u$ were used to calculate the energy integrals (5.55) and (5.56), interfacial pressure $\Phi$, and work terms by standard finite-difference methods with second order truncation errors.

The runs were performed for various initial fractional depths $H$ in the range from 2 to 100. The $H = 100$ case reproduces well the approach to the one-layer model limit.

The second parameter of interest is connected with the boundary condition at $x_N(t)$. The usual approach is to specify that $u_N = Fr h_N^{1/2}$. We employ the $Fr$ functions of Benjamin and HS, (2.26) and (2.27). However, in the energy context, two additions have been suggested by Shin, Dalziel, and Linden (2004) with the purpose to reduce the dissipation: First, the function (labeled below S1)

$$Fr(a) = (1 - a)^{1/2}; \tag{5.60}$$

and second, instead of a condition for $u_N$, to use the explicit half-lock-height condition (labeled below S2)

$$h = \frac{1}{2} \quad (x = x_N(t)). \tag{5.61}$$

The four boundary conditions considered here are labeled as Be, HS, S1, and S2.

The three $Fr$ formulas are shown in Fig. 5.15 for $0 < a \leq 0.5$ which is the relevant range of gravity current flow. We emphasize that Benjamin's $Fr$ is the only result which has been derived with some rigor. The formula of HS is a result of the curve-fit of experimental data; the justification is that it represents a friction-adjusted Benjamin-type balance; see § 3.2. The S1 and

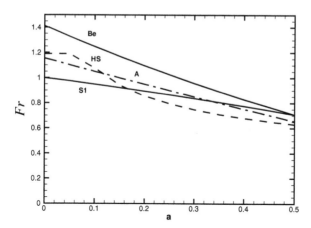

**FIGURE 5.15:**  Froude functions of Be, HS, and S1 for which energy behavior is calculated.  Curve A is a Benjamin-type condition with a friction
loss $\delta = 0.25$; see (3.16). (Ungarish 2008, with permission.)

S2 formulas were derived by box-model arguments in the attempt of reducing
the dissipation. We see that Benjamin's $Fr$ provides the largest value of $Fr$
for a given $a$. We shall see that Benjamin's Froude produces the smallest
dissipation among the four tested boundary conditions.

The solutions of the governing equations reveal the behavior of the two-layer
system as follows. The representative profiles for the interface and velocity
are shown in Fig. 5.16. The lower layer displays the typical dam-break pattern: the nose of the current moves with constant velocity $u_N$, followed by
a rectangular bulk of constant height $h_N$, which is then joined to the bulk
of stationary fluid by a domain of gradually-increasing height (expansion region). At a given time, the lengths of the rectangular and expansion regions
are roughly equal. The velocity of the dense fluid is $x$-independent in the
rectangular region, and increases with $x$ (almost linearly) from zero in the
expansion region. The velocity in the ambient is quite small, which indicates
that the (kinetic) energy of this fluid cannot play a significant role in the
energy balance of the whole system. The interfacial pressure $\Phi$, not shown,
is constant along the horizontal parts of the interface, but increases with $x$
on the curved (inclined) part. The one-layer model (for $H \to \infty$) produces
very similar profiles, but with $u_a = 0$, and the increase of $u$ below the inclined
interface is linear with $x$.

The SW results relevant to the energetic behavior of the system are displayed in Fig. 5.17. The curves are presented as functions of $1/H$ (=the depth
ratio of the lock to the ambient), for the various nose conditions. The various
rates of change and performance of energy and work, displayed in Fig. 5.17,
are constant with time during the considered initial slumping stage.

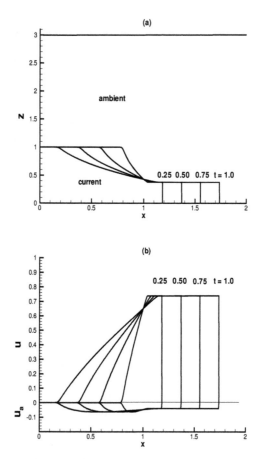

**FIGURE 5.16**: SW results of interface and velocity as functions of $x$ at times $t = 0.25, 0.50, 0.75, 1.00$, for $H = 3$. (Ungarish 2008, with permission.)

Fig. 5.17(a) shows that the rate of decrease (or release) of potential energy is about 0.3. This essential property of the system depends little (up to about 15%) on the nose boundary condition because it is dominated by the motion of the expansion wave into the ambient. The total energy of the system decays in all considered cases; see Fig. 5.17(b). This dissipation is at the typical rate of about 0.1. Here the nose condition may influence the results up to about 50%.

Fig. 5.17(c), which shows the ratio of rates of decrease of total energy (dissipation) to potential energy, is the most informative concerning the dissipative behavior of the system. We see that *the ratio total*$(\dot{E})/\dot{P}$ *is larger than* 0.2 *for all tested cases.* Here the influence of the nose condition is very significant, up to about 70%. The ratio of the rate of dissipation to $total(\dot{E})/\dot{P}$ is typically

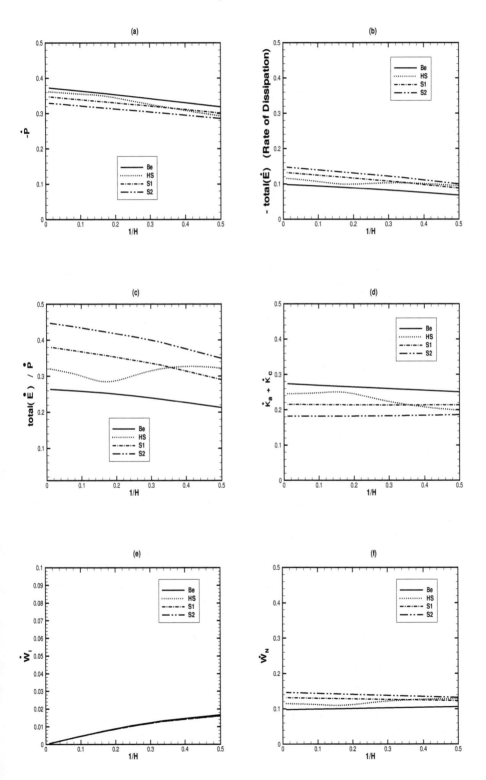

**FIGURE 5.17**: (continued on next page)

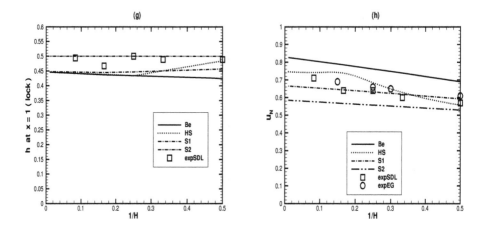

**FIGURE 5.17:** (continued). SW results as functions of the inverse fractional depth, $1/H$. The curves are for various boundary conditions at the nose. Frames (a)-(f) show rates of change of energy and work. Frames (g)-(h) show the height at the position of the lock, and velocity $u_N$. The square symbols in the last two frames are the experimental data of Shin et al. (2004); the circle symbols in the last frame are the experimental data of Ermanyuk and Gavrilov (2007) for the largest $Re$. (Ungarish 2008, with permission.)

the worst for the conditions S1 and S2 suggested by Shin et al. (2004), about 50% larger than for Benjamin's. The HS semi-empirical condition also yields a higher dissipation than Benjamin's.

The rate of increase of the kinetic energy of the system is typically 0.25, and also significantly influenced by the nose conditions; see Fig. 5.17(d). Benjamin's condition allows for the fastest development of kinetic energy. The contribution of the kinetic energy of the ambient is in any case quite small and practically negligible for $H > 5$. We recall that the SW formulation solves for the averaged $z$-independent velocity components. The kinetic energy results obtained here represent the contribution of these components only.

Fig. 5.17(e)-(f) show that the work rate of the inclined interface is positive, small, and negligible for $H < 5$, while the work rate of the nose is typically 0.1 and little affected by the value of $H$.

The last two frames of Fig. 5.17 show the height of the current at the lock ($x = 1$), and the velocity of the nose. The symbols are from some recent lock-exchange experiments.

We can summarize that, in general, for a fixed $H$ a smaller $Fr$ produces a smaller velocity $u_N$, a larger height $h_N$, and a larger rate of dissipation. This is compatible with our expectations and will be discussed further below. The condition $h_N = 1/2$ yields the smallest $u_N$, and the largest nose work and dissipation rates. This condition turns out to behave like the smallest $Fr$ number in our tests.

### 5.5.2 Analytical global energy balances

The volume and momentum balances in the SW results are clearly prescribed by the governing equations. The energy is a by-product which needs clarification. The dependency is on $H$ and $Fr$ nose condition.

General and versatile global energy balances can be derived and used in the interpretation and generalization of the foregoing results. Consider the current (dense fluid). In this closed-volume system the potential energy is transformed into kinetic energy and pressure work performed by pushing the upper inclined and side vertical interfaces. This is expressed as the energy conservation rule of the current

$$\dot{P} + \dot{K}_c = -\dot{W}_N - \dot{W}_i. \tag{5.62}$$

Obviously, $\dot{P}$ is negative while $\dot{K}_c$ and $\dot{W}_N$ are positive. The foregoing numerical solutions show that $\dot{W}_i$ is also positive.

Consider the ambient fluid. The (reduced) potential energy of this fluid was and remains zero. The kinetic energy develops (from zero) in the active layer above the current, $x_L(t) \le x \le x_N(t); h(x,t) \le z \le H$ (see Fig. 5.14). Outside this active layer the ambient is motionless, $u_a = 0$. Volume continuity shows that the volume of the layer of ambient to the left of $x_N(t)$ increases at the rate $\dot{V}_a = H u_N$. The "new" activated ambient fluid which is added above the moving nose has the velocity $u_{aN} = -u_N h_N/(H - h_N)$. Consequently, this new ambient fluid adds kinetic energy to the ambient at the rate $\dot{V}_a u_{aN}^2$. (On the other side, at position $x_L$ in Fig. 5.14, the expansion of the active ambient layer to the left is performed smoothly into a domain of $u_a = 0$, and hence there is no contribution to the kinetic energy balance.)

In addition, the upper layer receives work via the action of the inclined interface. Combining these contributions we obtain the energy balance rule of the ambient fluid

$$\dot{K}_a = \left( \frac{u_N h_N}{H - h_N} \right)^2 u_N H + \dot{W}_i. \tag{5.63}$$

Consider the sum of (5.62) and (5.63). It yields the total energy balance rule of the two-fluid system

$$total(\dot{E}) = \dot{P} + \dot{K}_c + \dot{K}_a = -\dot{W}_N + \left( \frac{u_N h_N}{H - h_N} \right)^2 u_N H. \tag{5.64}$$

Substituting $\dot{W}_N = u_N h_N^2$, this can be rewritten as

$$total(\dot{E}) = -u_N h_N^2 \left[ 1 - \frac{u_N^2 H}{(H - h_N)^2} \right], \tag{5.65}$$

and then using the definitions $u_N = Fr h_N^{1/2}$ and $a = h_N/H$, this can be expressed as

$$total(\dot{E}) = -Fr(a) h_N^{5/2} \left[ 1 - Fr^2(a) \frac{a}{(1-a)^2} \right]. \tag{5.66}$$

We emphasize that (5.62), (5.63), and (5.64) are fundamental mechanical energy balances in the framework of the shallow-water inviscid formulation in the absence of internal shocks. The energy results computed via the numerical solutions of the differential SW equations of motion confirm the global energy balances derived analytically. In the $a \to 0$ limit (the one-layer model) these results can be simplified; in particular, we know that $\dot{K}_a = \dot{W}_i = 0$, and $Fr$ is a constant.

### 5.5.3 Dissipation (formal and real)

Equation (5.64) or (5.66) expresses the (possible) imbalance between the decrease of potential energy of the current and the production of kinetic energy in both fluids, as viewed by the two-layer formulation. This imbalance represents the classical rate of dissipation of the system in the context of the inviscid modeling of the gravity current. Conversely, the system is energy-conserving when this rate of dissipation is zero. However, we wish to emphasize that the connection between this dissipation and the true irreversible loss of mechanic energy in a real flow is missing. As suggested by Benjamin, and as discussed below, there seems to be a qualitative similarity between these two effects, which makes the connection plausible, but the full correspondence is not clear or necessary. For sharpness, the energy deficiency effect which appears in our energy budgets could be termed "formal dissipation," or "averaged flow dissipation." Unless specified otherwise, this is the type of dissipation discussed here.

The compact expressions (5.65) and (5.66) provide some important results, as follows.

#### 5.5.3.1 General observations

It turns out that the dissipation is dominated by the values of $u_N$ and $h_N$ at the nose of the current. This is interesting, because the dissipation effect represents the energy changes of the entire two-fluid system.

We notice that only gravity currents which display non-positive $total(\dot{E})$ are physically acceptable. This is an interesting and non-trivial observation. We shall see below that the exact (and quantitative) interpretation of the dissipation effect is still missing. In any case, there is confidence that a positive $total(\dot{E})$ means an increase of the mechanical energy of the system, and this is impossible in normal conditions where no energy sources are given.

Equation (5.66) proves that, in general, a non-trivial SW current, of finite $h_N$ and $Fr$ while $a \to 0$, is significantly dissipative. This is exactly the one-layer model limit. We conclude that the one-layer SW system is significantly dissipative for all practical $Fr$.

### 5.5.3.2 Connection to Benjamin's result: the equivalence theorem

A useful clarification of a classical solution emerges. Let us substitute Benjamin's $Fr$ formula (2.26) into (5.66). We obtain

$$total(\dot{E})_B = -h_N^{5/2} \left[ \frac{(2-a)(1-a)}{1+a} \right]^{1/2} \frac{1-2a}{1-a^2} = -h_N^{5/2} Fr_B(a) \frac{1-2a}{1-a^2}.$$
(5.67)

Now compare this result with the rate of dissipation predicted by Benjamin's classical steady-state theory, equation (3.9). This is exactly the same result in dimensionless and dimensional form. We obtained here an interesting relationship between the time-dependent lock-release problem and the steady-state current. This can be expressed as an "equivalence theorem": If the two-layer SW dam-break time-dependent problem is solved using a nose boundary condition $u_N = Fr(a)u_N^{1/2}$ with Benjamin's $Fr(a)$, then the leading horizontal part of the current reproduces fully the flow-field properties of Benjamin's steady-state parallel current solution. (Formally, a semi-infinite reservoir $(-\infty < x \leq 1)$ and an ambient extending to infinity are assumed. Here we consider $H \geq 2$ cases only; the effect of the left-moving shock when $H < 2$, and the $u_{N\,max}$ restriction, are not incorporated in this analysis.)

This is not a trivial result. Benjamin's analysis assumes a steady-state flow in the frame moving with the current, without specifying how the horizontal interface was produced and whether the motion is compatible with realistic downstream conditions of a simple reservoir. In Benjamin's analysis the nose of the current is (roughly) a rounded wedge, while in the present SW formulation the transition region from the unperturbed ambient to the moving horizontal current is treated as a sharp vertical front. It is therefore interesting and important that the results coincide in a significant domain. Here we showed that the SW flow field to the right of point $M$ in Fig. 5.14 satisfies all the essential assumptions and results of Benjamin (shape of interface, velocity, rate of dissipation, pressure).

This equivalence strengthens considerably the relevance of Benjamin's local analysis to the solution of the lock-release problem. An observer moving with the nose at constant velocity $u_N$, and seeing a parallel flow ahead and behind, is likely to perform the steady-state, or local jump, analysis presented in Chapter 3. This observer may wonder about the source of the apparently disappearing energy as predicted by (3.9), perhaps define it as a paradox. This observer may also be puzzled about the condition which determines the thickness $h_N$ of the parallel stream behind (or rather the value of $a$). The present SW lock-release analysis indicates that the dissipated energy is simply supplied by the decrease of potential energy in the bulk of dense fluid far behind, and the value of $a$ is determined by the initial conditions in this reservoir (in particular, the value of $H$). The characteristics in the current, see (5.20)-(5.21), maintain a continuous connection between all parts of the dense fluid.

The only physically relevant zero of (5.67) is $a = 1/2$. For the non-trivial $a \to 0$ (i.e., finite $h_N$ and $H \to \infty$) equation (5.67) shows clearly that Benjamin's rate of dissipation is $-\sqrt{2}\, h_N^{5/2}$.

The interpretation of Benjamin's dissipation results for $a \to 0$ (i.e., finite $h_N$ and $H \to \infty$) requires some care. The difficulty may appear because of (a) improper scaling; and (b) confusion between the total energy imbalance, $total(\dot{E})$, and the local head loss, $\Delta$ (with respect to Bernoulli's balance on a streamline; see (3.10)). If the height of the ambient is used as a reference length (as done in some papers), one gets the wrong impression that the rate of dissipation vanishes for $a \to 0$. However, it is clear that the current is the driving phenomenon, and hence the proper length for scaling the rate of dissipation is $h_0$. With the proper scaling, as done here, the rate of dissipation remains $O(1)$ for $a \to 0$. This is clearly seen in (5.67) and in Fig. 5.17(b). The differences between the head loss and dissipation have been discussed in Chapter 3 following (3.12). The interesting result is that in a deep ambient the head loss $\Delta$ vanishes in spite of the fact that the dissipation of the system is large. Actually, we note that for Benjamin's $Fr$ the maximum ratio of dissipation rate to $\dot{P}$ is about 0.3 and occurs at $H \to \infty$; see Fig. 5.17(c).

### 5.5.3.3 The zero dissipation dilemma

Energy-conserving (non-dissipative) SW flows are theoretically possible. Using (5.66), we find that $total(\dot{E}) = 0$ is satisfied by the Froude number

$$Fr_{EC} = \frac{1 - a}{\sqrt{a}}, \qquad (5.68)$$

where the subscript means energy-conserving. On first impression, this is a welcome result. An inviscid flow is expected to be non-dissipative, and here is a simple way of imposing this property, so let us use it. This argument makes the plausible - but invalid - assumption that (5.66) reproduces the real dissipation of the system. A closer examination proves that the energy conserving results are of little practical importance (in the context of Boussinesq lock-release flows) and in general should not be used.

Consider the implications of $Fr_{EC}$. For $a = 1/2$ it reproduces Benjamin's value $Fr(1/2) = 1/\sqrt{2}$. This could be expected. We know already that a current propagating with Benjamin's $Fr$ is non-dissipative when $a = h_N/H = 1/2$. On the other hand, we showed that a lock-release current with Benjamin's $Fr$ is restricted to $a \le a_c = 0.347$. Moreover, we noticed that configurations with $H < 2$ develop internal shocks and bores that are unlikely to conserve energy. It therefore makes sense to consider the meaning of (5.68) for cases with $H > 2$, for which we expect $a < 0.25$ (because $h_N \approx 0.5$, typically).

Here the problematic behavior becomes evident: for $a = 0.25$ we obtain $Fr_{EC} = 1.5$, and as $a$ decreases further to 0 the value of $Fr_{EC}$ increases monotonically to $\infty$. The theoretical value 1.5 already exceeds the largest $Fr$

values provided by experimental and numerical evidence (for Boussinesq currents) for any $a$. The resulting non-dissipative-flow currents would display too large $u_N$ and too small $h_N$ as compared to experimental values (see Fig. 5.11). In other words, the use of the $Fr$ function that imposes zero dissipation produces SW currents with unrealistic speeds of propagation. On the other hand, we noticed that the $Fr$ functions that produce reasonable agreements of SW results with experimental observation also produce a significant dissipation. (The conditions S1 and S2 enter into this category, in spite of the fact that they were derived from a box model which is expected to be non-dissipative. The simplifications of that model are in general incompatible with the more realistic SW solutions. For further details see Ungarish 2008.)

An energy-conserving two-layer flow poses some additional inconsistencies with the observed behavior of gravity currents. Under this constraint the self-similar propagation in a deep ambient would be impossible, because in this stage both the potential and the kinetic energy of the current decrease, while the energy of the ambient is not affected ($u_a^2 H$ is practically zero). The details will be derived later in § 5.5.4. Moreover, it seems that a thinning long current will have to accelerate in the "energy conserving" regime. In the spirit of the box-model approximations, suppose that after the dam-break phase the current spreads out as a rectangle of height $h(t)$, with $h(t) < 0.5$. The representative velocity of the current in the box is $u(t)$, and the velocity of the ambient above this rectangle is $-uh/(H - h)$. The conservation of potential plus kinetic energy (in the current and ambient) yields

$$u^2(t) = [1 - h(t)]\left[1 - \frac{h(t)}{H}\right]. \tag{5.69}$$

The derivation of this result is left for an exercise. Equation (5.69) shows that when the thickness $h(t)$ decreases, the speed $u(t)$ increases. This is a very crude estimate, but the suggested trend clearly contradicts the well-known tendency of the current to develop a self-similar behavior of decreasing velocity with propagation.

The conclusion is that energy conserving (or non-dissipative) two-layer flows are incompatible with realistic gravity currents. The energy dissipation must be accepted as a by-product of realistic averaged-velocity models. Let us try to understand better the reasons and trends of this dissipation.

### 5.5.3.4  Physical interpretation

Equations (5.65) and (5.66) also provide insights into the reasons and trends of the dissipation. The fact that the kinetic energy of the SW model is determined by $z$-independent velocities plays a role. This is best observed for a typical current of finite $h_N$ and corresponding speed, both of the order of 1, in a very deep ambient, $H \to \infty$. What is the order of magnitude of the other variables with respect to the small $H^{-1}$? According to (5.65), in this case a finite dissipation rate of the order of 1 appears. The nose of the current performs the usual $O(1)$ work (for both small and large $H$); see (5.58). On the other hand, the $z$-averaged $u_a$ is $O(H^{-1})$, and hence the kinetic energy in the ambient is $u_a^2 H = O(H^{-1}) \to 0$. Thus, in the framework of averaged velocities, for large $H$ the upper layer of ambient fluid is unable to develop (and carry away) the kinetic $O(1)$ energy required to match the work of the nose. The SW result (5.66) informs us that there is an energy excess about the nose which must be re-deployed by a mechanism which is beyond the capability of the model. We can view this dissipation as a result of an expansion, in which the leading terms are $z$-independent (the averaged velocity field). These leading terms leave behind an imbalance which is satisfied by some higher order terms in the expansion (most likely associated with $z$ dependencies). But this imbalance does not mean that our solution for the $z$-averaged velocities is not accurate.

Indeed, consider the SW energy development in the expanding domain $0 \leq x \leq x_N(t)$. The model assumes that at the position $x = x_N(t)$, in some unspecified way, the work performed by the advancing nose is transfered into energy of the ambient. This energy should appear directly, in kinetic form, above the moving nose. Even in schematic terms, this requires a complex process: some ambient fluid must be displaced upward, integrated into the ambient which is already present at $z > h_N$, and accelerated from zero to $u_{aN}$. This must involve a strongly $z$-dependent velocity field in the ambient about the nose. (The case $h_N = (1/2)H$ is an exceptional configuration in which the volume of activated ambient fluid above the nose is equal to that displaced by the nose, which may explain the zero dissipation for $a = 1/2$.) The averaged SW approximation filters out some of these $z$-dependent effects. To illustrate this, let $u_a'(x, t, z)$ be the perturbation about the averaged $u_a(x, t)$, and denote by an upper bar $z$-averaged perturbations. Although $\overline{u_a'} = 0$, $\overline{(u_a')^2}$ contributes to the kinetic energy of the ambient. The superposition of the SW $u_a$ with a rather arbitrary wavy component $u_a'$ of amplitude $O(H^{-1/2})$ whose spectrum satisfies $\overline{u_a'} = 0$ could improve the energy conservation of the two-layer system (as compared with the SW results). This indicates that it is possible to correct the dissipation deficiency without changing the averaged flow field properties calculated by the SW approximation.

The dissipation result reflects a real need for distributing this energy imbalance inside the system by a means which is not included in the SW approximation. The missing agent is not necessarily the irreversible dissipation effect of a real system. Therefore, the energy which is dissipated according to the SW energy balances is not necessarily lost in an irreversible manner by the real system which is approximated by the SW formulation. However, the details concerning the propagation of this energy deficiency are beyond the resolution of a model which uses only $z$-independent horizontal velocities. Nevertheless, as noted above, the case of "positive" $total(\dot{E})$ is clear-cut; this can be safely dismissed because it requires the distribution of energy from an external source which does not exist in normal circumstances.

It is nevertheless interesting to note that the inviscid SW formal dissipation bears some qualitative similarities with a real irreversible dissipative process. This is inferred from a comparison between Figs. 5.17 (c) and (h). We see that, for a fixed $H$, a larger dissipation is associated with a smaller velocity of propagation. We can argue that the real dissipation produces a similar trend of speed reduction. To be more specific, we recall that empirical reductions of $Fr$ below Benjamin's value, as done by the HS formula, can be attributed to effects of friction. We therefore detect the following connection: the friction decreases the value of $Fr$, and the SW calculations show larger dissipation for smaller $Fr$. However, the quantitative connection of the SW dissipation with that in the real flow is an open topic for future investigation.

To summarize: we must distinguish between the irreversible loss of energy in a real system and the formal dissipation of the SW model. There is no clear-cut connection between these dissipations. While the former is expected to be small in a real large-Reynolds-number gravity current, the latter may be large (as compared to the decay of the potential energy during the considered time interval). In general, the dissipation increases when a value of $Fr$ smaller than Benjamin's is used in the SW model.

The presence of the formal dissipation is a deficiency of the SW model: the calculation of the energy of the ambient is unreliable. However, the SW model predicts fairly well the energy of the current. This will be shown later in Chapter 20.

## 5.5.4 Energy in the similarity stage

We showed in § 2.6 that in the self-similar stage of propagation in a deep ambient $(a \to 0)$ the SW flow is given by the exact solution

$$x_N(t) = At^{2/3}; \quad h = \dot{x}_N^2(c + \frac{1}{4}y^2); \quad u = \dot{x}_N y; \tag{5.70}$$

where $y = x/x_N$,

$$A = \left[ \frac{27}{2} \frac{Fr^2}{6 - Fr^2} \right]^{1/3}, \quad c = \frac{1}{Fr^2} - \frac{1}{4}. \tag{5.71}$$

(Here we use the notation $A$ for the constant defined as $K$ in § 2.6 to avoid confusion with the kinetic energy.)

We recall that for a deep ambient case $Fr$ is a constant, $K_a = 0$ and $W_i = 0$. Using (5.55) and (5.56) the energy of the system can be expressed as

$$total(E) = x_N \dot{x}_N^4 \int_0^1 \left[ y^2(c + \frac{1}{4}y^2) + (c + \frac{1}{4}y^2)^2 \right] dy = \frac{1}{Fr^4} \left( \frac{2}{3} \right)^4 A^5 t^{-2/3}. \tag{5.72}$$

The time derivative of the last result gives the rate of dissipation as

$$total(\dot{E}) = -\frac{1}{Fr^4}\left(\frac{2}{3}A\right)^5 t^{-5/3} =$$

$$-\left(\frac{2}{3}\right)^5\left(\frac{27}{2}\right)^{5/3}\left[\frac{1}{Fr^2(6-Fr^2)^5}\right]^{1/3}t^{-5/3}. \quad (5.73)$$

It is left for an exercise to verify that the rate of dissipation is equal to the rate of work of the nose, $-\dot{W}_N = -h_N^2 u_N$.

We conclude that the deep self-similar current is dissipative for any plausible $Fr$. This conclusion is valid for any initial shape of the current, because the similarity solution is independent of the initial conditions (except for a multiplication of $A$ by $\mathcal{V}^{1/3}$ when the initial volume $\mathcal{V}$ differs from 1; see (2.79)). This is an additional validation to the previous inferences that a deep current is dissipative, and that an energy-conserving flow contradicts the self-similar motion.

# Chapter 6

## Axisymmetric currents, SW formulation

Axisymmetric currents deserve a separate discussion. The axis of symmetry is the vertical $z$, and we assume that the propagation is above (or beneath) a horizontal plane. We use a cylindrical coordinate system $\{r, \theta, z\}$ in which the velocity components are $\{u, v, w\}$. The dependent variables are assumed to be independent of $\theta$, i.e., $\partial/\partial\theta = 0$. This type of flow appears when (a) The current is released from a cylinder in a laterally unbounded ambient (the fully axisymmetric case). In this case $v$ is not necessarily zero; and (b) The current is released in a wedge bounded by the vertical planes $\theta = 0, \Theta$, and $v = 0$.

Some of the properties derived for the 2D case can be easily extended to this configuration. However, there are important differences which are introduced by the diverging geometry and by the possible addition of rotation in the fully-axisymmetric case without violating the $\partial/\partial\theta = 0$ property. For example, currents propagating in the $r$ and $-r$ directions display contrasting behaviors, while in 2D propagation the sign of $x$ can be chosen arbitrary. Other notable differences with the 2D counterpart are the lack of constant-velocity propagation in the slumping phase, and a significantly shorter distance of inviscid motion for a given $Re$. The details will emerge below.

## 6.1  Governing equations

Similar to the 2D case, see § 2.1, the interface between the current and the ambient is considered to be a density kinematic shock, denoted by $\Sigma(\mathbf{r}, t) = 0$. The full differential and division by $dt$ yield

$$\frac{\partial\Sigma}{\partial t} + (\nabla\Sigma) \cdot \mathbf{V}_\Sigma = 0, \tag{6.1}$$

where $\mathbf{V}_\Sigma \equiv d\mathbf{r}/dt$ is the rate of displacement of the interface, which in the present problem is also the velocity of the fluid at the interface. Since $\nabla\Sigma$ is in the normal direction, it is clear that the displacement of the interface is governed only by the normal velocity component of the fluid at the interface.

The immediate implication is that at the transition between the current and
the ambient the normal velocity is continuous on the interface, but the tangential component may be discontinuous (this is an accepted simplification in
inviscid approximations).

To be specific, the interface under consideration in the present configuration
has two components. The first is a vertical front (also referred to as nose),
which is given by $\Sigma_N = r - r_N(t) = 0$ $(0 \leq z \leq h_N(t))$. The second, and major,
part of the interface is the inclined, sometimes piecewise inclined-horizontal,
surface which is given by $\Sigma_h = z - h(r,t) = 0$ $(0 \leq r \leq r_N(t))$. Here $r_N$ and
$h_N$ are the position and height of the nose. The velocity of the fluid is given
by $\mathbf{v} = u\hat{r} + w\hat{z}$. Using (6.1) ($\mathbf{V}_\Sigma$ is the value of $\mathbf{v}$ at the interface) we obtain
the motion of the interface in the axisymmetric case as

$$\frac{dr_N}{dt} = u_N; \tag{6.2a}$$

$$\frac{\partial h}{\partial t} + u_h \frac{\partial h}{\partial r} - w_h = 0, \tag{6.2b}$$

where the subscript $h$ denotes the inclined part of the interface. This equation
is usually supplemented by given initial conditions for $r_N$ and $h(r)$ at $t = 0$.

The continuity equation for the axisymmetric case in cylindrical coordinates
is

$$\frac{\partial w}{\partial z} = -\frac{1}{r}\frac{\partial ru}{\partial r}. \tag{6.3}$$

Using this equation, the no-penetration condition $w = 0$ on $z = 0$ and Leibniz's rule we can write

$$\frac{1}{r}\frac{\partial}{\partial r} r \int_0^h u\,dz = u_h \frac{\partial h}{\partial r} - w_h. \tag{6.4}$$

Combining (6.4) with (6.2b) we obtain

$$\frac{\partial h}{\partial t} + \frac{1}{r}\frac{\partial}{\partial r} r \int_0^h u\,dz = 0; \tag{6.5}$$

which we can also express as

$$\frac{\partial h}{\partial t} + \frac{1}{r}\frac{\partial r h \bar{u}}{\partial r} = 0. \tag{6.6}$$

Here $\bar{u}$ is, again, the $z$-averaged radial velocity of the current,

$$\bar{u}(r,t) = \frac{1}{h}\int_0^h u\,dz. \tag{6.7}$$

The result (6.5) (or (6.6)) is also referred to as the continuity equation of
the axisymmetric current. We keep in mind that in the derivation of (6.5)

we assumed that $z = 0$ is an impermeable boundary. Otherwise, a non-zero $w(r, z = 0)$ appears in the RHS of (6.5) and (6.6). This is consistent with the expectation that when the current gains (losses) fluid during propagation on a porous boundary its thickness increases (decreases) faster than usual.

Now we proceed to the momentum balances. We employ the one-layer model which has been considered in detail in § 2.2. We assume that the ratio of the vertical to horizontal length scales, $\delta$, is small, and we neglect viscous effects.

The horizontal balances are very similar to these of the 2D case. The main results, within errors of the order of magnitude of $\delta^2$ are as follows. First, the pressure obeys the hydrostatic balance in the $z$ direction. We employ the reduced pressure, $p = \mathcal{P} + \rho_a gz$, where $\mathcal{P}$ is the thermodynamic pressure. We recall

$$\Delta\rho = \rho_c - \rho_a. \tag{6.8}$$

We obtain for the ambient fluid

$$p_a = C; \tag{6.9}$$

and for the current

$$p_c(r, z, t) = -\Delta\rho gz + f(r, t) \quad (0 \leq r \leq r_N(t), 0 \leq z \leq h(r, t)). \tag{6.10}$$

Second, as a major component of the SW description, we use the pressure continuity condition

$$p_a = p_c \quad \text{at } z = h(r, t), \quad (0 \leq r \leq r_N(t). \tag{6.11}$$

The combination of (6.9)-(6.11) yields

$$p_c(r, z, t) = -\rho_c gz + \Delta\rho gh(r, t) + C \quad (0 \leq r \leq r_N(t), 0 \leq z \leq h(r, t)). \tag{6.12}$$

Finally, the longitudinal pressure gradient in the current can be written as

$$\frac{\partial p_c}{\partial r} = \Delta\rho g \frac{\partial h(r, t)}{\partial r}. \tag{6.13}$$

This is an important *fundamental* outcome which concerns the main internal driving force of the current. We confirmed that this effect is provided by the buoyancy (or reduced gravity) acceleration, expressed by the density difference times $g$, which justifies the term gravity (or density) current. This force decelerates the current where $(\partial h/\partial r)$ is positive, and accelerates it where $(\partial h/\partial r)$ is negative. We note that this effect is independent of $z$. This indicates that, in the present approximation, the speed $u$ of a current which is initially $z$-independent remains so during propagation. Moreover, this property also facilitates the development of the governing radial momentum equation to which we proceed now.

We consider here the case with no azimuthal motion, $v = 0$; this will be relaxed later (in Chapter 8). We start with the inviscid $r$-momentum equation of the current in conservation form

$$\frac{\partial u}{\partial t} + \frac{1}{r}\frac{\partial r u^2}{\partial r} + \frac{\partial}{\partial z}(uw) = -\frac{1}{\rho_c}\frac{\partial p_c}{\partial r}. \tag{6.14}$$

We integrate across the current to obtain

$$\int_0^h \left[\frac{\partial u}{\partial t} + \frac{1}{r}\frac{\partial r u^2}{\partial r}\right] dz + u_h w_h = -\frac{1}{\rho_c}\int_0^h \frac{\partial p_c}{\partial r} dz, \tag{6.15}$$

where we used the boundary condition $uw = 0$ on $z = 0$, and the subscript $h$ denotes values at the interface $z = h$. On the other hand, using Leibniz's rule, we obtain

$$\frac{\partial}{\partial t}\int_0^h u\,dz + \frac{1}{r}\frac{\partial}{\partial r}\int_0^h r u^2 dz - u_h\left(\frac{\partial h}{\partial t} + u_h\frac{\partial h}{\partial r} - w_h\right) = \text{LHS of (6.15)} . \tag{6.16}$$

The term in the brackets on the LHS of (6.16) vanishes on account of the kinematic conditions of the interface expressed by equation (6.2b).

We recall the definition of the shape factor

$$\varsigma = \frac{1}{h\bar{u}^2}\int_0^h u^2 dz = 1 + \frac{1}{h}\int_0^h \left(1 - \frac{u}{\bar{u}}\right)^2 dz. \tag{6.17}$$

The integrated form of the $r$-momentum equation, (6.16), can now be rewritten as

$$\frac{\partial \bar{u}h}{\partial t} + \frac{1}{r}\frac{\partial(\varsigma r \bar{u}^2 h)}{\partial r} = -\frac{1}{\rho_c}\int_0^h \frac{\partial p_c}{\partial r} dz. \tag{6.18}$$

We eliminate the pressure term by (6.13), and assume $\varsigma = 1$, using the same justification as in the 2D case. The result is the simplified SW inviscid $r$-momentum equation of the axisymmetric gravity current

$$\frac{\partial \bar{u}h}{\partial t} + \frac{1}{r}\frac{\partial r \bar{u}^2 h}{\partial r} = -\frac{\Delta\rho}{\rho_c}g\frac{1}{2}\frac{\partial h^2}{\partial r}. \tag{6.19}$$

The system (6.6) and (6.19) governs the behavior of $h(r,t)$ and $\bar{u}(r,t)$. We shall drop the overbar of $u$ in the following analysis. The initial conditions at $t = 0$ are known. The boundary conditions are (1) $u = 0$ at the axis $r = 0$ (or at some specified cylindrical boundary); and (b) at the propagating nose we apply again

$$u_N = \left(\frac{\Delta\rho}{\rho_a}g\right)^{1/2} h_N^{1/2}\, Fr(a), \quad \text{and} \quad a \leq a_{\max} \approx 0.5, \tag{6.20}$$

where $a = h_N/H$ and $Fr$ is the same Froude number as for the 2D case. The justification follows from the analysis of § 3.2. We showed that an observer attached to the moving front $x = x_N(t)$, see Fig. 3.1(b), could develop these results as "jump conditions" from global continuity and momentum balances, and energy constraints. The front is a thin vertical interface, and hence the curvature terms are unimportant (except for a small domain near the center). We argue that the jump conditions remain valid for an axisymmetric shock $r = r_N(t)$. We shall use the HS $Fr$, equation (2.27), unless stated otherwise.

To simplify the analysis, we introduce the Boussinesq assumption; see § 2.2.1. We define the reduced gravity as

$$g' = g'_a \approx g'_c. \qquad (6.21)$$

Next, we introduce the velocity and time scales

$$U = (g'h_0)^{1/2}; \quad T = \frac{r_0}{U}. \qquad (6.22)$$

We now switch all dimensional variables, denoted below by an asterisk, to dimensionless variables as follows

$$\{r^*, z^*, h^*, H^*, t^*, u^*, p^*\} = \{r_0 r, h_0 z, h_0 h, h_0 H, T t, U u, \rho_a U^2 p\}. \qquad (6.23)$$

An inspection of the continuity, $r$-momentum, and front condition equations confirms the consistency of this scaling for the variables of interest.

### 6.1.1 Summary 1

The dimensionless equations in conservation form can be written as

$$\frac{\partial h}{\partial t} + \frac{1}{r}\frac{\partial}{\partial r}(ruh) = 0, \qquad (6.24)$$

and

$$\frac{\partial}{\partial t}(uh) + \frac{1}{r}\frac{\partial}{\partial r}(ru^2 h) + \frac{\partial}{\partial r}\left(\frac{1}{2}h^2\right) = 0. \qquad (6.25)$$

In characteristic form, this becomes

$$\begin{bmatrix} h_t \\ u_t \end{bmatrix} + \begin{bmatrix} u & h \\ 1 & u \end{bmatrix} \begin{bmatrix} h_r \\ u_r \end{bmatrix} = \begin{bmatrix} -uh/r \\ 0 \end{bmatrix}. \qquad (6.26)$$

The characteristic curves and relationships provide useful information for the solution of the system, including a proper requirement of boundary conditions for the interface height $h$ at the ends of the current domain. The methodology for deriving the characteristics is summarized in Appendix A.1. Comparing (6.26) with (A.1) we readily identify that in the present case $\mathcal{H} = h$, $\mathcal{U} = u$, $A = 1$, $B = u$, $S_1 = -uh/r$, and $S_2 = 0$. Following the procedure outlined there, and with an obvious change of notation, we obtain the characteristic balances

$$dh \pm h^{1/2} du = -\frac{uh}{r}dt \quad \text{on} \quad \frac{dr}{dt} = c_\pm = u \pm h^{1/2}. \qquad (6.27)$$

The nose (front) condition needed at $r = r_N(t)$ is

$$u_N = h_N^{1/2}\, Fr(h_N/H), \quad \text{and} \quad \frac{h_N}{H} \le a_{\max} \approx 0.5, \tag{6.28}$$

$$Fr(h_N/H) = \begin{cases} 1.19 & (0 \le h_N/H < 0.075) \\ 0.5H^{1/3}h_N^{-1/3} & (0.075 \le h_N/H \le 1). \end{cases} \tag{6.29}$$

The initial and boundary conditions are provided by the physical situation under consideration. Here the fluid is released from rest from a vertical cylinder lock of unit length and height (in scaled form). At the axis $r = 0$ there is no source and hence the radial velocity there must vanish. All this is expressed as

$$u(r,t=0) = 0; \quad h(r,t=0) = 1; \quad (0 \le r \le 1) \tag{6.30}$$

$$u(r=0,t) = 0 \quad (t \ge 0). \tag{6.31}$$

This is supplemented by the kinematic condition

$$\frac{dr_N}{dt} = \dot{r}_N = u_N; \quad r_N(t=0) = 1. \tag{6.32}$$

The $r$ domain of solution is $[0, r_N(t)]$ which increases with time.

We compare the present one-layer SW formulation for the axisymmetric current with the 2D counterpart current governing equations, (2.37)-(2.43). The obvious difference with the 2D case is the presence of the curvature term proportional to $1/r$. This renders the first equation of (6.26) non-homogeneous, while the 2D counterpart is homogeneous. Consequently, the balance on the characteristics (6.27) is also non-homogeneous; moreover, the effect of this term is clearly time-dependent. While the (2.38) 2D balance admits a non-trivial solution with constant $h$ and $u$ (i.e., $dh = du = 0$), the axisymmetric counterpart (6.27) indicates that non-trivial solutions are time-dependent. The non-homogeneous term is proportional to $u$, which proves that the development of the current depends strongly on the direction of propagation. While a divergent outwardly-spreading axisymmetric current becomes thinner and slower, a converging current may become thicker and faster with time.

---

## 6.2   A useful transformation

The axisymmetric gravity current may spread out over a significant distance as compared to the initial $r_N(0) = 1$. It is convenient to keep track of it in a standard domain, $[0,1]$ say. This is achieved by introducing a stretched horizontal coordinate,

$$y = y(r,t) = \frac{r}{r_N(t)} \quad (0 \le y \le 1). \tag{6.33}$$

The reformulation of the governing equations for the original $h(r,t)$ and $u(r,t)$ variables as equations for the $h(y,t)$ and $u(y,t)$ variables is achieved by chain-rule derivatives. The necessary formulas are given by (2.45)-(2.48), in which $x$ is replaced by $r$.

### 6.2.1  Summary 2

We apply the transformation (2.44)-(2.47) (with $x$ replaced by $r$) to the system of SW governing equations (6.26). The result is

$$y = y(r,t) = \frac{r}{r_N(t)} \qquad (0 \le y \le 1). \tag{6.34}$$

$$\begin{bmatrix} h_t \\ u_t \end{bmatrix} + \frac{1}{r_N(t)} \begin{bmatrix} u - y\dot{r}_N & h \\ 1 & u - y\dot{r}_N \end{bmatrix} \begin{bmatrix} h_y \\ u_y \end{bmatrix} = \begin{bmatrix} -uh/r_N y \\ 0 \end{bmatrix}. \tag{6.35}$$

The characteristic balances for this system can be obtained directly, or by using (6.27) combined with (2.48). The derivation is left for an exercise. We obtain

$$dh \pm h^{1/2}du = -\frac{uh}{r_N y}dt \quad \text{on} \quad \frac{dy}{dt} = \frac{1}{r_N}\left(u - y\dot{r}_N \pm h^{1/2}\right). \tag{6.36}$$

The initial and boundary conditions (6.30)-(6.32) are not affected by the transformation, with the understanding that $y = 0$ is the left boundary, and that the nose condition (6.28) is applied at $y = 1$. Indeed, the $y$ domain of solution is now $[0, 1]$ for all times.

We compare with the 2D counterpart current governing equations, (2.49)-(2.51). As mentioned above, the obvious difference with the 2D case is the presence of the curvature term proportional to $1/y$. This renders the first equation of (6.35) non-homogeneous, while the 2D counterpart is homogeneous. Consequently, the balance on the characteristics (6.36) is also non-homogeneous; moreover, the effect of this term is clearly time-dependent. The obvious consequences are as mentioned in § 6.1.1.

---

## 6.3  The full behavior by numerical solution

We proceed to a global SW solution of the typical gravity current. This is achieved by a numerical finite-difference method. Here we use a Lax-Wendroff scheme and a grid with 1000 intervals in the domain $0 \le y \le 1$ (i.e., $0 \le r \le r_N(t)$); see Appendix A.2 for more details. The results are shown in Fig. 6.1.

As for the 2D counterpart, we distinguish between three stages: the main slumping and self-similar, with an intermediary transition between them. The first stage is a slumping motion in which the initial body of fluid changes its shape dramatically. The height and volume of fluid in the reservoir domain ($r < 1$) decreases rapidly. First, there is a fast backward rarefaction wave which reduces the height of the dense fluid in this domain, then further decrease occurs due to the reflected wave. The presence of the rarefaction wave and the formation of a forward moving current for $r > 1$ are similar to the 2D case. However, in the axisymmetric case both $h_N$ and $u_N$ decrease with time, while in the 2D case these variables are constant during the slumping interval. The difference is clearly a result of the diverging geometry. More details on the initial dam-break motion will be given below.

At $t \approx 2$ the interface of the fluid in the central domain ($r < 1$ say) becomes horizontal at $z \approx 0.2$, and subsequently the height of this horizontal tail decays like $t^{-2}$. The main volume of fluid becomes concentrated in an outer ring at $r > 1$ where the interface is curved like $(r - 1)^{1/2}$ (approximately). The radial velocity acquires a profile of double-linear increase with $r$, with a sharp transition (shock) at the position where the horizontal tail of fluid matches with the outer ring. (The profiles of $u(r)$ display some small oscillations near the shock. This is, again, a spurious numerical effect.)

Consider the behavior in the horizontal tail. It can be verified by substitution that the SW equations (6.26) subject to the boundary condition $u(0, t) = 0$ are satisfied by

$$h(r,t) = \frac{C_1}{(t + C_2)^2}; \quad u(r,t) = \frac{r}{(t + C_2)}, \tag{6.37}$$

where $C_1$ and $C_2$ are constants of the order of unity. This solution indeed appears for $0 < r < r_j$ for $t > 2$, where $r = r_j \approx 1.5$. A fit to the numerical results yields $C_1 = 0.50$ and $C_2 = -0.17$ for the results displayed in Fig. 6.1. At $r = r_j$ both the $h(r)$ and $u(r)$ profiles develop a discontinuity (shock); we see that $h$ jumps to a larger value, and $u$ to a smaller value (over a small $r$ increment). This jump is a result of incompatible conditions carried by the characteristics from the reservoir and from the nose. The information carried back from the nose is that of a self-similar behavior (to be discussed later), while the fluid in the tail is still under the influence of the reflected rarefaction wave. The inner solution will eventually disappear (i.e., $r_j$ will be pushed back to 0). This is a very slow process, because the speed of the characteristics in this region is small. However, the volume of fluid in the domain $r < r_j$ is very small as compared to the volume in the outer ring $r_j < r \leq r_N$.

The other major stage, seen for $t > 3$, corresponds to self-similar propagation. This will be discussed in detail below. The central nearly horizontal tail of fluid shrinks (eventually it is expected to disappear) and almost all of the current is beneath the upwardly inclined interface, and its velocity increases linearly with $r$. The development of the final self-similar shape is a long process, which has been studied in some detail by Grundy and Rottman (1985a) and Zemach and Ungarish (2007). What is important in the description of the flow is the fact that the leading part of the current, where most of the fluid is concentrated, starts to display the self-similar features after propagation to about $r_N = 3$. Indeed, the transition from the slumping stage to self-similar stage of the SW solution is quite sharply displayed by the change of the time-dependency of $h_N$ and $u_N$. On the log-log plots (the bottom frames in Fig. 6.1) we see that quite constant slopes appear for $t > 3$ (approximately), which indicate $h_N \propto t^{-1}$ and $u_N \propto t^{-1/2}$.

---

## 6.4 Dam-break stage

Consider an axisymmetric cylindrical configuration similar to that of Fig. 2.4. The axis of symmetry, $z$, replaces the backwall of the 2D reservoir. Also, $r$ replaces $x + 1$ of the 2D dam-break configuration (in dimensionless form).

The dense fluid in the reservoir is kept in equilibrium by the dam of radius $r_0$ (dimensional). The thickness of the dam is assumed negligibly small as compared to the radius. There is a significant pressure difference between the sides of the dam. The corresponding force is $(1/2)\Delta\rho g r_0 h_0^2$ (dimensional) per radian. We can say that this is the buoyancy force, $F_B$, of the fluid. This force is balanced by the dam in the static state. When the dam is suddenly removed, this force sets the fluid in motion in the radial direction, and a dynamic balance appears. Again, we assume that the viscous effects are unimportant, and the system is of Boussinesq type.

At time $t = 0$ the dam is instantaneously removed, and the current commences to spread out from $r = 1$. It is convenient to keep in mind that $\partial p_c / \partial r = 0$ in the reservoir when the dam is removed. This implies that the motion of the dense fluid develops quickly at $r \approx 1$, but the fluid inside the reservoir is not instantaneously affected.

A simple analytical solution of the type obtained for the 2D geometry is not available. The balances (6.27) on the characteristics are complicated by the time-dependent curvature term, and the boundary conditions at $r = r_N$ are also time-dependent. We therefore use the finite-difference solution also for the understanding of the dam-break stage. Results are presented in Fig. 6.2 for the same parameters as Fig. 2.7. The initial $u_N$ and $h_N$ (at $t = 0+$) are the same for both geometries and provided by (2.66), as displayed in Fig. 2.6. A comparison of Figs. 6.2 and 2.7 shows that at $t = 0.5$ the profiles of $h$ and $u$ in the 2D and axisymmetric problems are very close. Eventually the curvature terms introduce significant differences.

An interesting occurrence in the axisymmetric geometry is the fact that the velocity has a clear maximum inside the current during the slumping stage. The $h(r)$ profile has a converging-diverging throat at the same position. This has no counterpart in the 2D case. (At later times, $t > 3$ approximately, this maximum and throat evolve into jumps of $u$ and $h$.)

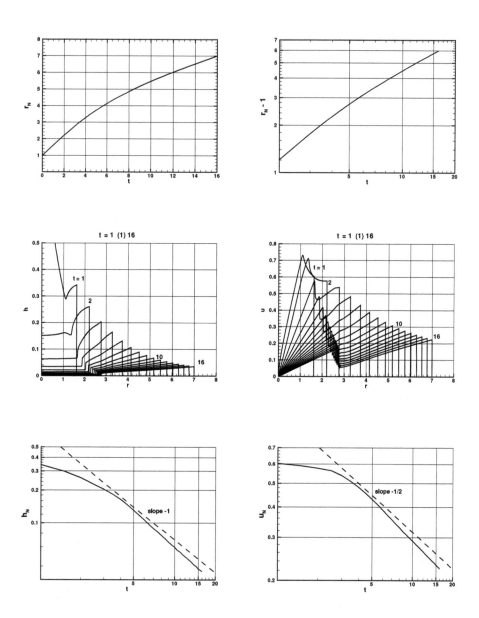

**FIGURE 6.1**: SW results for the axisymmetric current with $H = 3$ ($Fr$ given by HS formula). Top frames: the propagation $r_N$ vs. $t$ and $r_N - 1$ vs. $t$ on log-log axes. Middle frames: the $h$ and $u$ profiles as a function of $r$ for various $t$. Bottom frames: $h_N$ and $u_N$ vs. $t$, log-log axes. Lines (dashed) with slopes $-1$ and $-1/2$ are also shown to emphasize the self-similar behavior at large $t$.

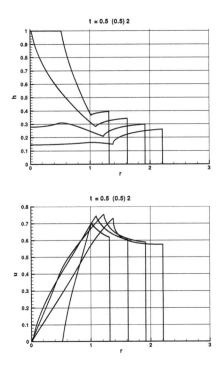

**FIGURE 6.2:** Axisymmetric case, SW numerical results of $h(r)$ and $u(r)$ at short times after release, for $H = 3$.

In general, in the diverging geometry $h_N$ and $u_N$ start to decrease after release, without the slumping delay typical to the 2D current. Ungarish and Zemach (2005) indicate that there is an exception, as follows. The two-layer model predicts that for $H < 1.25$ (approximately) $h_N$ and $u_N$ are controlled by the critical condition $u_N \leq c_+$, as discussed in § 5.3.1. As long as these limitations restrict the propagation of the nose, the speed $u_N$ is constant and smaller than the value predicted by straightforward calculations. This restriction, however, is relevant for a shorter time interval than in the 2D case. The restriction is relevant as long as the value of $h_N/H$ tends to be larger than some threshold $a_c$ value, typically 0.4. The nose of the divergent current attains after a short propagation a sufficiently small value of $h_N$, so that afterward $h_N/H$ is below the threshold, and the restrictions are no longer necessary.

We illustrate this as follows. The two-layer model shows that, for the HS $Fr$, $a_c = 0.427$ and hence the threshold values are $u_{N\max} = 0.434H^{1/2}$, $h_N \leq 0.427H$. Consider the most restricted case, $H = 1$. The one layer model yields, by (2.66), the corresponding initial $u_N$ and $h_N$ as follows: 0.458

and 0.594. If we calculate the SW one-layer propagation starting with these values without imposing any corrections, we find that after propagation to $t = 1.4$ and $r_N = 1.62$ the values of $h_N$ and $u_N$ will satisfy the above-mentioned restrictions (i.e., $u_N < 0.434$ and $h_N < 0.427$). The difference between $u_{N\,\text{max}}$ and the unrestricted speed is small. Propagation with that $u_{N\,\text{max}}$ during $t = 1.4$ would bring the current to $r_N = 1.61$. It is difficult to detect such differences in experiments and Navier-Stokes simulations.

---

## 6.5 Similarity solution

The axisymmetric current is also amenable to a self-similar solution for large times, when it "forgets" the initial conditions. The derivation follows the lines of the 2D case discussed in §2.6, but the diverging geometry introduces some important differences.

We use the transformation (6.34), $y = r/r_N(t)$, and the corresponding equations of motion (6.35), subject to the boundary conditions

$$u(y = 1, t) = \dot{r}_N = Fr\,[h(y = 1, t)]^{1/2}. \tag{6.38}$$

In addition, the similarity solution is subject to global volume conservation.

The current is thin and we therefore consider $Fr$ to be a constant.

We assume

$$r_N(t) = Kt^\beta; \quad h(y, t) = \varphi(t)\mathcal{H}(y); \quad u(y, t) = \dot{r}_N\mathcal{U}(y); \tag{6.39}$$

where $K$ and $\beta$ are some positive constants. The similarity coordinate is $y = r/Kt^\beta$ in the range $[0, 1]$. The obvious boundary condition is $\mathcal{U}(1) = 1$.

The form of $h$ is inspired by continuity consideration; we expect $\varphi = C\mathcal{V}/r_N^2(t)$, where $\mathcal{V}$ is the volume of the current, in general a time dependent variable, but a constant in the present case.

The objective is to determine $\varphi(t)$, $\beta$, $\mathcal{U}(y)$, $\mathcal{H}(y)$, and $K$.

Consider the boundary condition at the nose, (6.38). To satisfy it we use

$$\varphi(t) = \beta^2 K^2 t^{2\beta - 2} = \dot{r}_N^2; \tag{6.40}$$

and

$$\mathcal{H}(1) = Fr^{-2}. \tag{6.41}$$

Consider the integral volume continuity of the current, and use (6.39) and (6.40) to write

$$\mathcal{V} = \int_0^{r_N} h(r, t)r\,dr = \int_0^1 h(y, t)r_N^2 y\,dy = r_N^2 \varphi(t) \int_0^1 \mathcal{H}(y)y\,dy =$$

$$= \beta^2 K^4 t^{4\beta - 2} \int_0^1 \mathcal{H}(y)y\,dy. \tag{6.42}$$

This equation is satisfied by

$$\beta = \frac{1}{2} \qquad (6.43)$$

and

$$K = \left[ \frac{1}{4} \int_0^1 \mathcal{H}(y) y \, dy \right]^{-1/4} \mathcal{V}^{1/4} = K_1 \mathcal{V}^{1/4}. \qquad (6.44)$$

Next we use the equations of motion (6.35). Substitution of (6.39) and (6.40) yields, after some arrangement, the continuity equation as

$$(\mathcal{U} - y) \frac{\mathcal{H}'}{\mathcal{H}} + \mathcal{U}' + \frac{\mathcal{U}}{y} = 2 \left( \frac{1}{\beta} - 1 \right); \qquad (6.45)$$

and the momentum equation as

$$\mathcal{H}' + (\mathcal{U} - y)\mathcal{U}' + \left( 1 - \frac{1}{\beta} \right) \mathcal{U} = 0; \qquad (6.46)$$

where the prime denotes $y$ derivative. Recall that the boundary conditions at $y = 1$ are $\mathcal{U} = 1$ and $\mathcal{H} = Fr^{-2}$. In general, the system (6.45)-(6.46) must be solved numerically. However, for some values of $\beta$ the equations can be decoupled and simple solutions follow. This happens in the present case.

Indeed, for the present $\beta = 1/2$ the RHS of (6.45) is equal to 2. This equation admits the simple solution

$$\mathcal{U}(y) = y. \qquad (6.47)$$

As in the 2D case, this result satisfies the $u(y = 0, t) = 0$ boundary condition, and also simplifies the system of governing equations (6.35). (We note in passing that the $\mathcal{U}/y$ term in the LHS of (6.45) is incompatible with inflow at origin in self-similar flows. This is in contrast with the 2D counterpart.)

To calculate $\mathcal{H}(y)$, we substitute (6.47) and $\beta = 1/2$ into the momentum equation (6.46). We obtain

$$\mathcal{H}'(y) - y = 0. \qquad (6.48)$$

Integration, subject to the condition $\mathcal{H}(1) = Fr^{-2}$, gives

$$\mathcal{H}(y) = \frac{1}{Fr^2} - \frac{1}{2} + \frac{1}{2}y^2. \qquad (6.49)$$

Finally, we calculate $K$. With the result (6.49) we go back to the outcome of volume conservation (6.44), and obtain

$$K_1 = 2 \left( \frac{2Fr^2}{4 - Fr^2} \right)^{1/4}; \quad K = K_1 \mathcal{V}^{1/4}. \qquad (6.50)$$

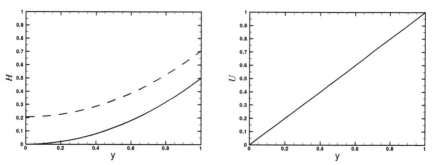

**FIGURE 6.3**: Similarity profiles for the axisymmetric SW current. $\mathcal{H}(y)$ is given for $Fr = 1.19$ (solid line) and $Fr = \sqrt{2}$ (dashed line).The similarity profile $\mathcal{U}$ does not depend on $Fr$.

To be specific $K_1 = 2.05$ for $Fr = 1.19$ and 2.378 for $Fr = \sqrt{2}$. Here $\mathcal{V}$ is the volume per radian (scaled with $r_0^2 h_0$). For a cylinder lock $\mathcal{V} = 1/2$, and for an ellipsoid lock ($h = (1 - r^2)^{1/2}$, $0 \leq r \leq 1$) we obtain $\mathcal{V} = 1/3$.

We note that for $Fr^2 \leq 2$ we obtain $\mathcal{H} \geq 0$ in $0 \leq y \leq 1$. Thus, for Boussinesq fluids the similarity result is acceptable over the full $y$ domain.

The similarity profiles are displayed in Fig. 6.3.

We summarize:

$$r_N = Kt^{1/2}; \quad h = \dot{r}_N^2 \left[ \frac{1}{Fr^2} + \frac{1}{2}(y^2 - 1) \right]; \quad u = \dot{r}_N y. \qquad (6.51)$$

Since $\dot{r}_N = (1/2)Kt^{-1/2}$, (6.51) can also be expressed as

$$r_N = Kt^{1/2}; \quad h = \frac{1}{4}K^4 r_N^{-2} \left[ \frac{1}{Fr^2} + \frac{1}{2}(y^2 - 1) \right]; \quad u = \frac{1}{2}K^2 r_N^{-1} y. \qquad (6.52)$$

Rescaling is necessary to finalize the similarity solution. We mentioned that in the self-similar propagation the current has forgotten the initial conditions. The results derived above are still dependent on $r_0$ and $h_0$ of the lock which were used to scale $r_N$ and $h$, and to determine the speed and time scales. To eliminate this difficulty, we introduce a new scaling length, $L^*$, for both vertical and horizontal lengths, and also new scales for time and speed

$$L^* = \mathcal{V}^{*1/3}; \quad T^* = (L^*/g')^{1/2}; \quad U^* = L^*/T^*; \qquad (6.53)$$

where $\mathcal{V}^*$ is the dimensional volume of the current (per radian).

Substituting this re-scaling into (6.51) we find that nothing changes, except for the fact that now $\mathcal{V} = 1$ in (6.50). The details are left for an exercise. This indeed proves that the similarity solution is independent of the shape of the lock.

The development of the self-similar solution in the entire $y$ domain from given initial conditions is a long process. Formally, we can take the SW equations with the given conditions, solve numerically, and compare the results at large $t$ with the analytical predictions. The difficulty is that the accuracy cannot be maintained when $h$ and $u$ become very small while the numerical errors accumulate. A transformation of variables resolves this deficiency; some details are presented later in § 16.3.3.1.

## 6.6 The validity of the inviscid approximation

The importance of the viscous friction on the motion of the current increases with time and distance of propagation. Even for quite large values of $Re$, the inviscid SW formulation may become invalid at moderate values of $r_N$. To monitor this effect, we use the previous results for $u$ and $h$ to estimate the time-dependent ratio of global inertial, $F_I$, to viscous, $F_V$, effects. The inertia per unit volume (dimensional) is well represented by $\rho_c u u_r$, and the viscous force per unit area (dimensional) is expected to be proportional to $\rho_c \nu u/h$. Switching to dimensionless variables with the aid of (6.23), we write the ratio of the inertial forces in the volume of the current to the viscous force on the contact area as

$$\frac{F_I}{F_V} = Re\frac{h_0}{r_0}\frac{\displaystyle\int_0^{r_N(t)} u u_r h r\, dr}{\displaystyle\int_0^{r_N(t)} (u/h) r\, dr} = Re\frac{h_0}{r_0}\theta(r_N). \tag{6.54}$$

The variable $\theta$ is expected to be of the order of unity at the beginning of propagation and decay to very small values as time (and $r_N$) increases.

For large times $\theta$ can also be estimated analytically using the self-similar profiles (6.52). To be specific, the integral at the numerator of (6.54) yields

$$I_I = \left(\frac{1}{2}K^2\right)^4 \frac{1}{r_N^3}\int_0^1 \left(c + \frac{1}{2}y^2\right)y^2\, dy, \tag{6.55}$$

and the integral at the denominator of (6.54) is

$$I_V = 2K^{-2}r_N^3 \int_0^1 \frac{y^2\, dy}{c + \frac{1}{2}y^2}, \tag{6.56}$$

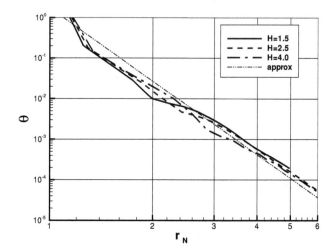

**FIGURE 6.4**:   Inertial to viscous effects ratio coefficient $\theta$, see (6.54), as a function of $r_N$, for various $H$ (log-log axes). Also shown the self-similar flow approximation $1.7r_N^{-6}$. (Ungarish 2007a, with permission.)

where $c = Fr^{-2} - 1/2$. Consequently,

$$
\theta = \left(\frac{1}{2}\right)^5 K_1^{10} \mathcal{V}^{5/2} \int_0^1 (c + \frac{1}{2}y^2)y^2 dy \left[\int_0^1 \frac{y^2 dy}{c + \frac{1}{2}y^2}\right]^{-1} r_N^{-6} =
$$
$$
C_V (2\mathcal{V})^{5/2} r_N^{-6}. \quad (6.57)
$$

The calculations give $C_V = 1.68$ for $Fr = 1.19$ and $C_V = 1.60$ for $Fr = \sqrt{2}$. (We note that the box-model approximation, to be discussed in Chapter 7, produces a very close estimate.)

The value of $\theta$, as a function of $t$ and $r_N$, can be easily calculated as a by-product of the finite-difference solution of the SW equations. Typical results of the calculated $\theta$ as a function of $r_N$ are shown in Fig. 6.4. The inviscid theory is expected to be relevant for, roughly, $F_I/F_V > 1$. The ratio $F_I/F_V$ is initially large, but decreases monotonically with the radius of propagation $r_N$. Let us denote by $r_V$ the radius of propagation at which $F_I/F_V = 1$ (i.e., where $\theta$ is already so small that it exactly counteracts the large $Re h_0/r_0$). When the current spreads to $r_N > r_V$ it enters into the domain of $F_V > F_I$ where viscous influence is important. This provides the estimate of $r_N < r_V$ as the domain of validity of the SW model. The present $\theta$ results, Fig. 6.4, provide a sharp practical means for evaluating $r_V$ for a realistic current of given $Re(h_0/r_0)$.

We can estimate $r_V$ using (6.57). The result is

$$r_V = \left[ \frac{1}{C_V(2\mathcal{V})^{5/2}} \right]^{1/6} \left( Re\frac{h_0}{r_0} \right)^{1/6} \approx \left( Re\frac{h_0}{r_0} \right)^{1/6}. \qquad (6.58)$$

## 6.7 Some comparisons

Here we discuss, following Ungarish (2007a), some comparisons of the one-layer SW axisymmetric model results with laboratory measurements (mostly) and Navier-Stokes (NS) simulation. We shall see that the model provides reliable insights and fairly accurate predictions (sometimes even better than obtained by full NS simulations) of distance of propagation and shape of interface, except for a very short initial phase. The experiments considered here are within the domain of validity of the inviscid approximation as defined in § 6.6.

The first set of data is from the experimental and numerical study of Patterson, Simpson, Dalziel, and van Heijst (2006). The observations concern the propagation in a wedge sector of angle 10° and length 2.35 m (see Table 6.1 for more details). The reported speed of propagation was divided into three phases: an "initial" very short one (to about $r_N \approx 1.1$), followed by a faster "secondary" phase up to $r_N \approx 2.7$; then a sharp reduction of velocity (the "tertiary" phase was limited to $r_N < 3.8$ by the outer boundary). The tests revealed a significant vortical motion above the head of the current, while its nose propagates to $r_N \approx 2.7$; afterward, the vortical motion disappears. Patterson et al. (2006) also presented NS simulations which predicted similar results, but with smaller speeds than recorded in the experiments.

The second set of relevant laboratory and numerical data is that of Hallworth, Huppert, and Ungarish (2001). They used the large tank at the Coriolis Laboratory in Grenoble, of about 13 m diameter, to create fully axisymmetric currents which propagate to $r_N \approx 6$ (see Table 6.1). (This experimental setup also allows the investigation of rotating currents, as discussed later in § 8.1.5.)

A comparison based on both observations of Hallworth et al. (2001) and Patterson et al. (2006), performed by different methods on very different geometries, provides a useful overview on the features of the axisymmetric gravity current and the means for predicting them.

The one-layer SW model is used in this comparison. Both Hallworth et al. (2001) and Patterson et al. (2006) performed some experiments with full-depth locks, $H \approx 1$. For these cases the initial speed is subjected to the cavitation restriction $u_N \leq u_{N\,max}$; see (5.25). As an alternative to this restriction one can impose the energy dissipation limitation $h_N/H \leq 0.5$. This produces a slightly larger bound for $u_N$. For full-depth lock release conditions, $H = 1$, the difference is about 3%. In any case these constraints

**TABLE 6.1:** The experiments of Hallworth et al. (2001) (labeled S1,S2,S3,S7) and of Patterson et al. (2006) (labeled P1-P5). The fluids were water with various amounts of dissolved salt (and some dye in the current). $r_V$ is the estimated radius of propagation beyond which viscous effects become more important than inertial effects. (Ungarish 2007a, with permission.)

| Expt | $r_0$ cm | $h_0$ cm | $H^*$ cm | $H$ | $g'$ $\mathrm{cm\,s^{-2}}$ | $Re$ $\times 10^4$ | $r_V$ | remark |
|------|-----|-----|------|-----|------|------|-----|---------|
| S1 | 100 | 41.1 | 50.1 | 1.2 | 4.91 | 6 | 5.8 | cylinder |
| S2 | 100 | 77.3 | 80.1 | 1.0 | 4.81 | 15 | 7.5 | cylinder |
| S3 | 100 | 45.8 | 79.8 | 1.7 | 19.2 | 14 | 6.8 | cylinder |
| S7 | 100 | 45.2 | 80.0 | 1.8 | 43.8 | 20 | 7.2 | cylinder |
| P1 | 60 | 30 | 30 | 1.0 | 13.2 | 6 | 6.0 | wedge |
| P2 | 60 | 22 | 30 | 1.4 | 13.2 | 4 | 5.4 | wedge |
| P3 | 60 | 17.5 | 30 | 1.7 | 13.2 | 3 | 4.9 | wedge |
| P4 | 60 | 9 | 30 | 3.3 | 13.2 | 1 | 3.7 | wedge |
| P5 | 60 | 7.5 | 30 | 4.0 | 13.2 | 0.7 | 3.2 | wedge |

are relevant only for shallow ambient cases $1 \le H < 1.25$. Since in the axisymmetric current the $c_+$ characteristics tend to decrease the value of $h_N$ all the time, see (6.27), the difficulties associated with $h_N/H \approx 0.5$ will in any case disappear a short time after release. In this aspect, when $H < 2$, the one-layer model is a more valid approximation for the diverging current than for the 2D current. The SW results used in the comparisons presented here were obtained by a finite-difference code, with grids of typically 500 intervals in the current domain; the HS $Fr$ function was used for the $u_N$ boundary condition.

The viscous friction is expected to be unimportant for $r < r_V$, as defined in § 6.6. The values of $Re$ and $h_0/r_0$ in the experimental systems used by Patterson et al. (2006) and Hallworth et al. (2001), and the corresponding $r_V$, are given in Table 6.1. We see that the inviscid theory is expected to be relevant for propagation to $r_N \approx 4$ and 6 in the respective tests; this is, approximately, also the limit of free propagation imposed by the outer wall of the experimental tanks (the domain close to the wall is anyhow excluded from the analysis due to possible boundary reflections and other interference effects). In other words, the experimental data considered here are expected to reproduce the inertia-dominated propagation and is consistent with the inviscid assumption of the SW model.

The only dimensionless parameter in the inviscid SW formulation is the initial height ratio $H$. The first interesting question is if the reduced experimental data obey this type of dependency. The experimental (and NS simulation) data considered here, after being reduced to dimensionless form with the scaling (6.22)-(6.23), confirm that the major features (speed and shape) of the propagating current are dominated by the dimensionless parameter $H$. This is an important support to the SW formulation.

Patterson et al. (2006) report measurements for the very early times of the process, which they call "initial phase." They report that, instantaneously after the lock is lifted, the current propagates with constant velocity over a distance $O(h_0/r_0)$. Then, a sharp increase of velocity occurs, and the "secondary phase" begins. During this "secondary phase" (which is significantly longer than the "initial phase") the propagation is to about $r_N = 3$. The initial formation of the head following the lift of the gate is a $z$-dependent viscous process, whose details are beyond the resolution of the inviscid SW theory. As explained in § 2.5, the SW formulation is not supposed to resolve behaviors on the radial scale $O(h_0/r_0)$. Thus, both physical and mathematical considerations indicate that this short "initial phase" after which a larger speed suddenly appears is consistent with the SW theory assumption that the motion starts instantaneously with the velocity $u_N$ predicted by the SW equations. Indeed, we find a good correspondence between the SW predicted $u_N(t = 0+)$ and the initial speed observed in the experiments (just after the short "initial phase"). To be specific: (a) for $H = 1$ SW predicts the initial $u_N = 0.43$, while experimental values are 0.43 (Hallworth et al. 2001) and 0.52 (Patterson et al. 2006); (b) for $H = 1.7$ the SW value of $u_N$ is 0.54 while the experiments give 0.50 (Hallworth et al. 2001) and 0.58 (Patterson et al. 2006); (c) for $H = 4$ the SW value of $u_N$ is 0.70, and the experiment gives 0.75 (Patterson et al. 2006). We see here good agreement between SW predictions and experiments concerning both the value and the dependency on $H$. The discrepancies are within the range of the scatter between the measurements of Patterson et al. (2006) and Hallworth et al. (2001).

The typical propagation predicted by the SW model is depicted in Fig. 6.1. We distinguished between two major stages (with a less clear-cut transition stage between them). The first major stage is a slumping motion in which the initial body of fluid changes its shape dramatically. The height and volume of fluid in the reservoir domain ($r < 1$) decreases rapidly. First, there is a fast backward rarefaction wave which reduces the height of the dense fluid in this domain, then further decrease occurs due to the reflected wave.

The SW prediction for the slumping shape of the current is consistent with the experiments and NS simulations. For example, in Fig. 4 of Patterson et al. (2006), shown here in Fig. 6.5, the frames (a) and (b) correspond, approximately, to the times $t = 0.5$ and 1.5. In both the experimental and SW figures we clearly observe: (1) the backward propagating rarefaction wave; and (2) the decay of the height of the dense fluid in the reservoir and the accumulation of the current in a propagating ring; subsequently, the ring spreads out and the curvature of the interface is changed.

The other major stage (observed for $t > 3$) corresponds to self-similar propagation. The central nearly horizontal portion of fluid shrinks (eventually it is expected to disappear) and almost all of the current is beneath the upwardly inclined interface, and its velocity increases linearly with $r$. We mentioned that the development of the final self-similar shape in the entire current is a long process. The intermediary stage between the slumping and self-similar

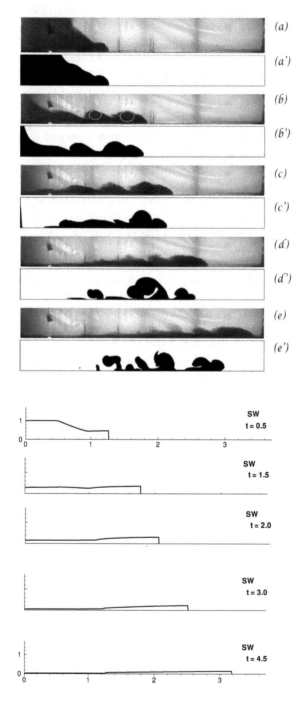

**FIGURE 6.5**:  Axisymmetric current released from a lock with $H = 1.7$. Side view from laboratory experiment in a wedge, photographs $a, b$, etc., at $t = 0.5, 1.5, 2, .3, 4.5$ (approximatively, dimensionless).  Below each photograph is the corresponding NS snapshot $a', b'$, etc. (Patterson et al. 2006, with permission.)  The last five frames are $h$ vs. $r$ SW solutions at these times.

stages is vague. What is important in the present context is the fact that the leading part of the current, where most of the fluid is concentrated, starts to display the self-similar features after propagation to about $r_N = 3$. Indeed, the change from the slumping behavior to the self-similar behavior of the SW solution is quite sharply displayed by the change of $h_N$ with $r_N$; see Fig. 6.6. The decay is first with the power of about $-1$, and then with the power $-2$ (as predicted by the similarity solution (6.52)). The transition of behavior in the leading part of the current occurs quite sharply, when propagation reaches about $r_N = 3$.

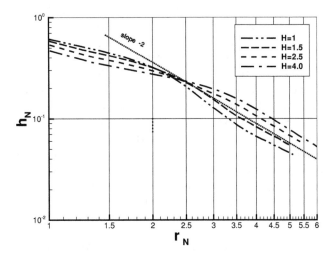

**FIGURE 6.6:** SW height of nose as function of $r_N$ for various $H$ (log-log axes, also shown the slope $-2$). (Ungarish 2007a, with permission.)

The foregoing results indicate that there is a good correspondence between the "phases" observed by Patterson et al. (2006) and the stages revealed by the SW solution, as follows: (1) The "secondary" phase ($1.1 < r_N < 2.7$, approximately) is modeled by the slumping stage (subject to an insignificant delay which covers the short "initial" phase); and (2) The "tertiary" phase ($r_N > 2.7$) is a manifestation of the self-similar stage. In the experiments of Patterson et al. (2006) the transition seems to occur instantaneously, while in both SW solutions and NS simulation the change occurs over a distance of about 0.5. The "tertiary" phase recorded by Patterson et al. (2006) was limited by the outer wall of the container to a quite short distance, from $r_N \approx$ 3 to 4. (Over short distances, or when the deceleration is weak, it is difficult to distinguish between straight and slightly curved $r_N$ vs. $t$. This explains why

Patterson et al. 2006 reported propagation with piecewise-constant speed, for the slumping and self-similar regime. More clarifications of this issue are given in Ungarish 2007a. The piecewise-constant speed simplification is not necessary, perhaps even misleading. The SW model gives accurate results within insignificant computer time, as shown below.)

Consider some comparisons of the distance of propagation as a function of time, in scaled form. Fig. 6.7 shows some experimental and SW results in the slumping stage. The agreement is good. We see that $H$, which is the only free dimensionless parameter in the SW model, differentiates correctly between the propagation results (in scaled form) obtained in various conditions.

The post-slumping propagation is displayed in Fig. 6.8 using log-log coordinates. We see that both experimental and SW results collapse to $r_N \propto t^{1/2}$ at larger times, in accord with the SW prediction of the self-similar propagation (6.51). It is clear that this "long time" behavior actually starts at the quite moderate values, $t = 2 - 3$, after a propagation to $r_N \approx 2.8$.

This is in good agreement with the transition radius from the "secondary" to "tertiary" phases detected by Patterson et al. (2006) in their tests. (In that study the variable $R_*$ denotes the distance $r_N - 1$ at which a velocity transition occurs, and the reported average value is 1.7.) The slope of the results in Fig. 6.8 is slightly larger than the theoretical $1/2$. This is because here $t$ is not truly large, and hence the motion is still under the influence of the initial conditions, and, moreover, $Fr$ is not a constant as assumed by the analytical self-similar solution. Again, we conclude that the agreement between theory and experiment is good; the scatter between the experiments of Hallworth et al. (2001) and Patterson et al. (2006) seems to be of the same order of magnitude as the disagreement between the experiments and SW theory.

In general, the experimental propagation reported by Patterson et al. (2006) is faster than that reported by Hallworth et al. (2001) for a similar $H$. The discrepancies are quite small, about 10%, but systematic. The SW results of $r_N$ vs. $t$ (for fixed $H$) are typically above these of Hallworth et al. (2001), which may be attributed to the unavoidable small viscous and mixing effects at the interface. On the other hand, the SW results (as well as the NS simulations presented by Patterson et al. 2006) are in many cases (in particular for $t > 1$) below the experimental data of Patterson et al. (2006). The discrepancy can be attributed to differences in the experimental settings and techniques. Hallworth et al. (2001) recorded the propagation of the dyed current from above, and averaged the radius of propagation over an angular sector. Patterson et al. (2006) recorded the propagation from a side projection. The first method is expected to filter out small azimuthal perturbations, while the second method is expected to represent the most forward point of the propagating volume of dyed fluid. The axisymmetric tank used by Hallworth et al. (2001) had a small inner rigid radius of $r_i = 0.21$, while the theory assumes a cylinder with no inner radius. The deviations from the theory with the idealized $r_i = 0$ geometry are estimated to be small (in the range of the experimental errors), as

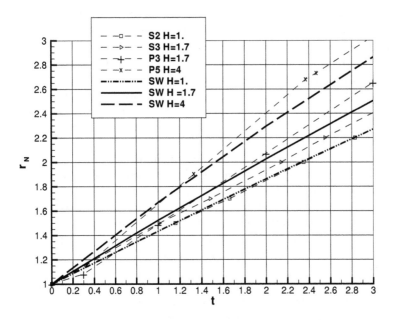

**FIGURE 6.7**: Radius of propagation as a function of time, for slumping stage (mostly) $t \leq 3$, experimental and SW results. (Ungarish 2007a, with permission.)

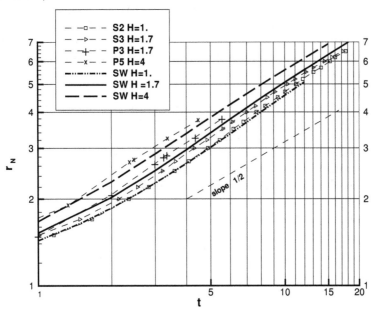

**FIGURE 6.8**: Radius of propagation as a function of time for transition and self-similar stage (mostly) $t \geq 1$, experimental and SW results (log-log axes, also shown the slope 1/2). (Ungarish 2007a, with permission.)

follows. After lifting the lock cylinder, the motion of the fluid starts at $r = 1$ (the dam-break problem) and there is no difference between an annular container and a full cylinder. At about $t = 0.8$ the backward-moving rarefaction wave first encounters the inner radius $r_i = 0.21$, and is then reflected to the front. We thus infer that the presence of $r_i$ shortens the time of propagation of the backward-forward rarefaction wave (by roughly $2r_i$ dimensionless time units), but does not influence the speed of the current during the slumping stage (until the reflected wave reaches the nose). In addition, this $r_i$ reduces by 4% the volume of fluid in the lock as compared with the ideal case, and this is expected to reduce the rate of the self-similar spread $Kt^{1/2}$ by about 1% (the coefficient $K$ is proportional to volume at the power 1/4; see (6.50)). The combined effect of this $r_i$ on the propagation is thus expected to be in the range of 1%. On the other hand, Patterson et al. (2006) used a $10°$ angle wedge container, while all theoretical predictions assume full axial symmetry. It is possible that the measured faster propagation is also a consequence of the geometry. A more rigorous understanding of this discrepancy requires more investigation.

We see that the SW theory provides a good prediction and interpretation of the axisymmetric gravity current phenomena. Since this is a $z$-averaged and asymptotic (small $h_0/r_0$, Boussinesq, large $Re$, laminar, one-layer) approximation, it has intrinsic resolution and accuracy limitations. Accounting also for experimental errors and gate-lift perturbations, agreements within 10% of speed and transition times, over a significant period of propagation, should be considered as satisfactory.

Some doubts concerning the suitability of the SW axisymmetric model are introduced by the fact that this model ignores the coherent vortical motion above the head of a real current which was reported by Patterson et al. (2006). Very clearly, this vortex is present during the "secondary" phase, but not later. Patterson et al. (2006) suggested that the presence-absence of this vortex is the dominant reason for the different velocity regimes in the "secondary" and "tertiary" phases. (Recall that we identified these phases with the slumping and self-similar SW regimes, respectively.) However, the comparisons show that the propagation in the initial slumping phase and the velocity transition features are well captured by the SW model, the presence of the vortex notwithstanding. It is obvious that the $z$-averaged velocity field predicted by the inviscid SW theory requires some shear-flow and Kelvin-Helmholtz corrections at the interface (where both a jump and change of sign of the horizontal velocity appear). Regarding the vortex as a component of this correction, it rather makes sense to claim that the influence of this (and similar two-dimensional velocity) mechanism is of the order of the disagreements between the experiments and SW averaged models, typically 10%. In other words, the vortex is a secondary by-product of the axisymmetric gravity current.

The SW model predicts accurately the position of the front for the full range covered by the considered experiments, which is about $r_N = 6$. Here the performance of the SW models is, strikingly, superior to that of reported

NS finite-difference solutions. Hallworth et al. (2001) (see Figs. 9-10 there) and more recently Patterson et al. (2006) (Figs. 3, 4, 6 there) obtained good agreement between NS and experiments for a limited radius of propagation, typically $r_N = 3$; afterward, the NS produce an incoherent interface and underpredict significantly the position of the nose.

Consider in particular the gravity current of experiment 3 of Patterson et al. (2006) with $H = 1.7$. In Fig. 6.5 we see snapshots of the experimental shape. For each time, the NS shape is displayed in the frame just below the photograph. In the lower part of Fig. 6.5 we see the SW results for the corresponding times. The experimental frames (d) and (e) for times 3 and 4.5 differ significantly with the NS results shown in frames (d') and (e'), respectively. On the other hand, these experimental results are in fair agreement with the SW profile at the appropriate times. This illustrates well the fact that the SW predictions are more reliable than the axisymmetric NS simulations for large time.

In any case, it is clear that the SW model is an advantageous prediction tool for the averaged properties of the current even when a NS solver is available. But, of course, the NS simulations can provide the smaller scale details of mixing and energy transfer which are beyond the resolution of SW models; moreover, the NS solutions are not restricted to Boussinesq systems like the present SW results.

The experiments confirm that viscous effects are unimportant during propagation to, at least, $r_V \approx 1.7Re(h_0/r_0)r_N^{-6}$. Significant viscous effects in the post-slumping stage would have reduced dramatically the slope of the experimental curves in Fig. 6.8 (from 1/2 to 1/8; see Chapter 11).

The typical axisymmetric current is divergent, i.e., the propagation is from the center outwardly, as considered here. However, we note that the SW model also provided correct insights and good agreements with experiments for axisymmetric convergent gravity currents (which propagate from the periphery to the center and display a decreasing-increasing velocity pattern). These problems are outside the scope of this book, and the interested reader is referred to Hallworth, Huppert, and Ungarish (2003) and Slim and Huppert (2004).

It is quite remarkable that in all these different circumstances the complicated flow of the axisymmetric gravity currents (with vortical regions, lobes and clefts, Kelvin-Helmholtz interfacial instabilities, etc.) is so well captured by a simple model of three equations (continuity, $r$-momentum, and the $Fr(a)$ function), and is governed by only one parameter, $H$, with no adjustable constants. This justifies the use of this model as the backbone in the interpretation and prediction of the phenomena. In this context we also recall that axisymmetric NS simulations did not provide, in general, better agreements with experimental data concerning propagation and shape as a function of time. Further study is necessary to find the reason and the remedy for these disappointing performances of the NS simulation. The SW models are not a substitute for the detailed NS solution, just an approximate tool for the prediction and interpretation of the averaged velocity and thickness of the current and related global variables (e.g., energy budgets; see Chapter 20). For an accurate and detailed prediction of the flow field of the two-fluid system, in particular concerning interfacial phenomena, the NS solutions must be employed.

# Chapter 7

## Box models for axisymmetric geometry

For some quick estimates of the expected behavior of the current we make a bold approximation. We assume that the current has a simple shape, a well defined "box." In the present axisymmetric case the box is a cylinder of radius $r_N(t)$ and height $h_N(t)$. The main objective is to determine the behavior of these two variables, which means that we need two equations. The first is provided by the obvious volume continuity requirement

$$\frac{1}{2}r_N^2(t)h_N(t) = \mathcal{V}. \tag{7.1}$$

The second equation must involve some dynamic considerations. In general, simplified forms for the velocity and reduced gravity fields are postulated, and integrals of the buoyancy, inertial, and viscous effects are obtained for the volume under consideration. Then some governing momentum integral balance is applied and an equation for $u_N$ is derived. The integration of this equation provides $r_N(t)$. The method is illustrated and discussed below for the axisymmetric current.

## 7.1 Fixed volume current with inertial-buoyancy balance

*In this section we use dimensionless variables.* In the fixed-volume particular case we can skip the integral balance and go directly to the boundary condition (2.39)

$$u_N = \frac{dr_N}{dt} = Fr(a)[h_N(t)]^{1/2}, \tag{7.2}$$

where $a = h_N(t)/H$. Combining (7.1) and (7.2) we obtain one equation for $r_N(t)$ (or for $h_N(t)$). Formally, for the case of a fixed volume $\mathcal{V}$, the solution is

$$t - t_I = \frac{1}{(2\mathcal{V})^{1/2}} \int_{r_I}^{r_N(t)} [Fr(a(r))]^{-1} r\,dr, \tag{7.3}$$

where $a = 2\mathcal{V}/(Hr^2)$. The initial condition is $r_N = r_I$ at $t = t_I$.

Consider the case of a deep current with a *constant Fr*. The result of (7.3) can be expressed as

$$r_N = \left[2Fr(2\mathcal{V})^{1/2}(t - t_I) + r_I^2\right]^{1/2} = (2Fr)^{1/2}(2\mathcal{V})^{1/4}(t + \gamma)^{1/2}. \quad (7.4)$$

Here $\gamma$ is a constant which combines the values of the initial $t_I$ and $r_I$.

This result (7.4) is remarkably close to the similarity solution (6.51). The propagation is with $t^{1/2}$, and the coefficient is proportional to $\mathcal{V}^{1/4}$ in both solutions; see (6.50). The ratio $2^{1/4}(2Fr)^{1/2}/K_1$ is 0.89 for $Fr = 1.19$ and 0.84 for $Fr = \sqrt{2}$. This means that the box model underpredicts the speed of propagation by about 15% (as compared to the similarity solution). As for the 2D case, this is explained by the fact that the self-similar current has a higher nose $h_N$ for a given $r_N$. Note that the box-model approximation $h(r, t) = h_N(t)$ implies a $r$-independent axial velocity and hence $u(r, t)$ is a linear function of $r$. This is, again, in agreement with the self-similar solution. Moreover, as in the self-similar result there is a time-shift constant $\gamma$ which must be determined by some initial conditions that may depend on the previous stage of propagation in which $Fr$ varies with $h_N$. This renders the box-model result also a quite weak prediction tool. We summarize that the box-model result (7.4) provides a plausible approximation for large times. However, if an analytical solution of the SW equations is available, there is actually no need for this model. It may still be convenient to use it for scaling arguments or for initial estimates of experimental data.

Equation (7.3) is amenable to analytical solutions also for non-constant $Fr$. In particular, consider the effect of the non-constant branch of HS $Fr$ formula (2.27) for a current with $\mathcal{V} = 1/2$ (standard cylinder lock of radius $r_0$ and height $h_0$). Substitution into (7.3) yields, after some algebra and use of the initial condition $r_I = 1$ at $t_I = 0$,

$$r_N = \left[\frac{2}{3}H^{1/3}t + 1\right]^{3/4} \quad (h_N = 1/r_N^2 \geq 0.075H). \quad (7.5)$$

Contrary to the long-time result (7.4), the present outcome lacks a clear-cut similarity with the initial behavior of a SW current. One may expect (see Huppert and Simpson 1980) that the use of (7.5) for the initial motion, matched with (7.4) for larger times, reproduces the behavior of a real current, i.e., a dam-break slumping which then evolves into the self-similar stage. This expectation is not fulfilled in general. A real axisymmetric current displays transition from slumping to self-similar propagation at about $r_N = 2.5$, while, according to the present box model, when $H^{-1} < 0.075$, the propagation is with $t^{1/2}$ from the start. The condition $a < 0.5$ cannot be imposed on (7.5). This makes the result especially unreliable for $H < 2$. The conclusion is that the box-model results should be used with great care. For $r_N > 2.5$ (approximately) the box model velocity regime is in fair agreement with the similarity results, but the motion to this position is not well predicted. Again, there is no good reason to prefer the box model when the analytical similarity results are available.

---

## 7.2 Inflow volume change $\mathcal{V} = qt^\alpha$

*In this section we use dimensional variables.* This provides a more direct insight into the balances and results which appear. For simplicity, we use the Boussinesq approximation, but this restriction can be relaxed with some care. Here we follow the discussions of Didden and Maxworthy (1982) and Huppert (1982). The 2D case is discussed in § 4.2.

We consider a gravity current of changing volume in time according to $qt^\alpha$ where $q$ ($> 0$) and $\alpha$ ($\geq 0$) are given constants. The $\alpha = 0$ case gives the simple fixed-volume current, and the $\alpha = 1$ case corresponds to a current supplied by a constant-rate influx (say, by an open tap). The $\alpha > 1$ cases may represent a leak or eruption which worsens with time. We assume that the source is located at the center $r = 0$.

The volume continuity requirement of the box model is

$$\frac{1}{2} r_N^2(t) h_N(t) = \mathcal{V} = qt^\alpha. \tag{7.6}$$

In the particular case $\alpha = 0$ we can replace $q = \mathcal{V}$, where $\mathcal{V}$ is the volume (per radian) and the dimensions are cm$^3$. In general, the dimensions of $q$ are cm$^3$ s$^{-\alpha}$.

The following simplified behavior is assumed:

$$r_N = Kt^\beta; \quad h_N = \frac{2\mathcal{V}}{r_N^2} = \frac{2q}{K^2} t^{\alpha - 2\beta}; \tag{7.7a}$$

$$u_N = \beta K t^{\beta - 1}; \quad u(r, t) = u_N \frac{r}{r_N} = \beta t^{-1} r, \tag{7.7b}$$

where $\beta$ and $K$ are constants to be determined by the solution. For later use we also rewrite the first dependency in (7.7a) as

$$t = \left(\frac{r_N}{K}\right)^{\frac{1}{\beta}}. \tag{7.8}$$

The major objective is to determine the constants $\beta$ and $K$.

The relevant forces can now be estimated as follows. The driving "buoyancy" force is the integral of $\partial p_c / \partial r$ over the volume of the current. This effect is represented by the pressure force on the periphery "dam" which bounds the current in the box (per radian)

$$F_B = \frac{1}{2} \rho_c g' r_N h_N^2. \tag{7.9}$$

We note that some care may be required when using this term in the box-model context. The pressure forces in a lock (box) tend to push the fluid at the boundary $r = r_N$ in the positive $r$ direction. However, the solution of the SW equations clearly shows that in some cases the pressure forces decelerate the fluid. The deceleration-acceleration is determined by the positive-negative inclination of the interface, an effect not properly incorporated in the box model. The inertial forces are

$$F_I = \rho_c h_N \int_0^{r_N} u u_r r dr = \frac{1}{3} \rho_c u_N^2 r_N h_N, \qquad (7.10)$$

and the viscous forces are

$$F_V = \rho_c \nu \int_0^{x_N} \frac{u}{h} r dr = \frac{1}{3} \rho_c \nu \frac{u_N}{h_N} r_N^2. \qquad (7.11)$$

The ratio of inertial to viscous forces can be expressed as

$$\frac{F_I}{F_V} = \frac{1}{\nu} u_N h_N^2 \frac{1}{r_N} = \frac{1}{\nu} \beta \left( \frac{2q}{K^2} \right)^2 t^{2\alpha - 4\beta - 1}. \qquad (7.12)$$

The thickness ratio $h_N/r_N$ behaves like $t^{\alpha - 3\beta}$; see (7.7a). The assumption of a thin current restricts the validity of the analysis to

$$\alpha - 3\beta < 0. \qquad (7.13)$$

### 7.2.1 Inertial-buoyancy balance

We start again with the result that the propagation of the inertial current is governed by the jump conditions at the nose. Using (7.7b) we write

$$u_N = Fr(g'h_N)^{1/2}, \quad \text{i.e.,} \quad \beta K t^{\beta-1} = Fr\, g'^{1/2} \left( \frac{2q}{K^2} \right)^{1/2} t^{(\alpha-2\beta)/2}. \qquad (7.14)$$

Assuming that $Fr$ *is a constant*, we equate the powers of $t$ and the coefficients in the last equation to obtain

$$\beta = \beta_I = \frac{\alpha + 2}{4}; \qquad (7.15)$$

$$K = K_I = \left( \frac{Fr}{\beta_I} \right)^{1/2} [2g'q]^{1/4}. \qquad (7.16)$$

Here the subscript $I$ indicates that the flow is dominated by inertial (or inviscid) effects. Comparing with the 2D case, we observe that the radial-divergent geometry entered into the behavior of $h_N$ in the equation for $u_N$. As a result, the propagation of the axisymmetric current is always with a smaller power $\beta$ than the 2D counterpart.

We summarize: the propagation of an inviscid axisymmetric box-model gravity current is (in dimensional form)

$$r_N(t) = \left( \frac{Fr}{\beta_I} \right)^{1/2} [2g'q]^{1/4} t^{(\alpha+2)/4}. \tag{7.17}$$

Substitution of (7.15) into (7.12) and use of (7.8) yields

$$\frac{F_I}{F_V} = \frac{1}{\nu}\beta_I \left( \frac{2q}{K_I^2} \right)^2 t^{\alpha-3} = \frac{1}{\nu}\beta_I \left( \frac{2q}{K_I^2} \right)^2 \left( \frac{r_N}{K_I} \right)^{4(\alpha-3)/(\alpha+2)}. \tag{7.18}$$

These results reduce well to the previously derived self-similar solution when $\alpha = 0$. In particular, we see that $F_I/F_V$ decays like $r_N^{-6}$, in accord to the calculation in §6.6. It is left for an exercise to show that, for $\alpha = 0$ and $q = \mathcal{V}$, the dimensionless form of (7.18) corresponds to the result (6.57) with $C_V = Fr$.

When $\alpha > 0$, the inflow conditions are inconsistent with the assumption that $u = u_N r/r_N$ at $r = 0$. We assume that this discrepancy is resolved in a small domain $r/r_N \ll 1$ which has a negligible effect on the global balances (7.9)-(7.11). Similarity solutions of the SW equations for axisymmetric configurations with inflow at the axis are not available (this restriction can be inferred from the presence of the $\mathcal{U}/y$ term in the RHS of (6.45); see Grundy and Rottman 1985b for a discussion). This leaves the box-model approximation as the only practical analytical tool at the present state of knowledge.

Substituting the previous result for $\beta_I$ into the thickness ratio requirement (7.13), we conclude that the present axisymmetric inviscid box model is restricted to $\alpha < 6$. Otherwise, the dense fluid is expected to accumulate as thick bulk.

## 7.2.2 Viscous-buoyancy balance

When the inertial terms are negligibly small, the buoyancy is counteracted by viscous drag. For $F_V = F_B$ equations (7.9) and (7.11) yield

$$u_N = \frac{3}{2} \frac{g'}{\nu} \frac{h_N^3}{r_N}. \tag{7.19}$$

In the context of the present text, this is an interesting and novel result. We shall see later a more rigorous derivation. The difference with the speed of propagation of the inertial current, $Fr(g'h_N)^{1/2}$, is very significant.

Now we substitute the assumed behavior (7.7) into (7.19) to obtain

$$\beta K t^{\beta-1} = 12 \frac{g'}{\nu} \frac{q^3}{K^7} t^{3\alpha-7\beta}. \tag{7.20}$$

The solution is

$$\beta = \beta_V = \frac{3\alpha + 1}{8}; \tag{7.21}$$

$$K = K_V = \left(\frac{12}{\beta_V}\right)^{1/8}\left[\frac{g'}{\nu}q^3\right]^{1/8}. \tag{7.22}$$

Here the subscript $V$ indicates that the flow is dominated by viscous effects.

We summarize: in dimensional form, the box-model propagation of the axisymmetric viscous current is

$$r_N(t) = \left(\frac{96}{3\alpha + 1}\right)^{1/8}\left[\frac{g'}{\nu}q^3\right]^{1/8}t^{(3\alpha+1)/8}. \tag{7.23}$$

Substitution of (7.21) into (7.12) and use of (7.8) yield

$$\frac{F_I}{F_V} = \frac{1}{\nu}\beta_V\left(\frac{2q}{K_V^2}\right)^2 t^{(\alpha-3)/2} = \frac{1}{\nu}\beta_V\left(\frac{2q}{K_V^2}\right)^2\left(\frac{r_N}{K_V}\right)^{4(\alpha-3)/(3\alpha+1)}. \tag{7.24}$$

The previous result for $\beta_V$ satisfies the thickness ratio requirement (7.13) for any positive $\alpha$. In this respect the viscous current is also different from the inviscid current solution (which displays the $\alpha < 6$ restriction).

The viscous box-model propagation results are in good agreement with the similarity solution which will be derived later in § 11.2. However, we shall see that the more rigorous similarity solutions are quite simple and can be easily implemented for any practical value of $\alpha$. This reduces the motivation for using the box-model results. When $\alpha > 0$, the inflow conditions are inconsistent with the assumption that $u = u_N r/r_N$ at $r = 0$. We assume that this discrepancy is resolved in a small domain $r/r_N \ll 1$ which has a negligible effect on the global balances. In the viscous case this difficulty is less severe than in the inertial case. The viscous diffusion smooths out large gradients which may develop about the source, while in the inviscid system the source conditions can be propagated along characteristics deep into the current.

### 7.2.3   Critical $\alpha$

As for the 2D current with inflow, a critical value of $\alpha$, $\alpha_c = 3$, for the behavior of the solutions exists (Huppert 1982). This is depicted in Fig. 7.1. We see that at this point $\beta_I = \beta_V = 5/4$, and for larger values of $\alpha$ the rate of spread of the viscous current is larger than that of the inertial current.

Consider the dependency of the ratio $F_I/F_V$ on $\alpha_c = 3$, for both inertial and viscous dominated flows. Indeed, both (7.18) and (7.24) show that: (a) For $\alpha < \alpha_c$, and in particular for the fixed-volume current ($\alpha = 0$), the ratio of inertial to viscous forces decays continuously during the propagation. This means that in this domain of $\alpha$ an inertial gravity current is bound to become viscous. The change of regime of propagation means a decrease of the power of

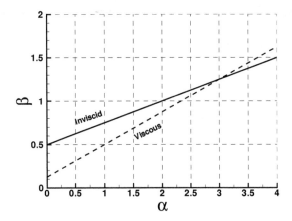

**FIGURE 7.1:** The power $\beta$ as a function of $\alpha$ for the inertial and viscous axisymmetric current.

$t$ from $\beta_I$ to $\beta_V$. (b) For $\alpha > \alpha_c$ the ratio of inertial to viscous forces increases continuously during the propagation. This means that in this domain of $\alpha$ a viscous gravity current is bound to become inertial. (This also implies the increase of the Reynolds number and the possibility of associated instability effects.) The change of regime of propagation means a decrease of the power of $t$ from $\beta_V$ to $\beta_I$. (c) For $\alpha = \alpha_c$ the ratio of inertial to viscous forces does not change with time or distance of propagation. For this $\alpha$ there is no transition between viscous and inertial regimes.

Consider the axisymmetric current with $\alpha = \alpha_c = 3$. It turns out that the type of this current is conveniently determined by the magnitude of the dimensionless number

$$J = \nu \frac{g'}{q}. \tag{7.25}$$

Inertia is always dominant if $J \ll 1$ and viscous drag is always dominant if $J \gg 1$. The justification is as follows. We go back to (7.12). For $\alpha = \alpha_c$ the ratio $F_I/F_V$ is time independent and, hence, for the entire flow process we can write

$$\frac{F_V}{F_I} = \frac{4}{5}\nu \left(\frac{K^2}{2q}\right)^2. \tag{7.26}$$

First, for the inviscid current we substitute (7.16), and use $Fr \approx 1$; we find that $F_V/F_I \approx J$. Second, for the viscous current we substitute (7.22); we find that $F_V/F_I \approx J^{1/2}$. The definition of the Julian number for the axisymmetric case is slightly different from the 2D counterpart (4.26) because the filling variables $q$ have different dimensions. Recall that the dimension of $q$ is $cm^3\,s^{-3}$ for the present $\alpha = 3$ case.

   To our knowledge, no experimental confirmation of the critical behavior has been reported for the axisymmetric current. This is a more challenging case than the 2D counterpart because (1) a significantly larger $\alpha$ is needed; (b) the difference between the slopes of $\beta_V$ and $\beta_I$ with respect to $\alpha$ is quite small and hence more difficult to detect experimentally.

# Chapter 8

## Effects of rotation

Here we consider a new effect, the rotation of the system. In addition to the gravity body force, the fluid is now also subjected to the action of the centrifugal body force. However, the more important effect is that of the Coriolis accelerations which appear in the direction perpendicular to that of the speed (in the rotating system). Rotating fluids are a complicated topic even for homogeneous fluid (see Greenspan 1968; Ungarish 1993) and hence we shall consider only some simple systems which, nevertheless, provide useful insights and illustrate the power of the available theory. These are: (1) the axisymmetric current which spreads out radially in a cylindrical geometry, and (2) the current in a rotating channel which propagates mostly in the longitudinal direction in a Cartesian geometry. In both cases the propagation is over (or under) a horizontal boundary, the rotation is about a vertical axis with constant angular velocity $\Omega$, viscous effects are negligible, and the Boussinesq simplification is valid.

A common practice and notation is to use

$$f = 2\Omega \tag{8.1}$$

which is sometimes called "Coriolis parameter." We shall not use this term, to avoid confusion with the dimensionless parameter $\mathcal{C}$ introduced later. In any case, the switch from $\Omega$ to $f$ is simple and will rarely be mentioned.

## 8.1 Axisymmetric case

The system under consideration is depicted schematically in Fig. 8.1. A layer of ambient fluid, of constant density $\rho_a$, above a solid horizontal surface at $z = 0$, is in solid-body rotation with angular velocity $\Omega$ about a vertical axis of symmetry $z$. At time $t = 0$, a fixed volume of co-rotating heavier fluid of density $\rho_c$, initially contained in a cylinder of height $h_0$ and radius $r_0$, is instantaneously released into the ambient fluid, generating a radially spreading rotating gravity current. We follow the analysis presented by Ungarish and Huppert (1998), Hallworth, Huppert, and Ungarish (2001), and Ungarish and Zemach (2003).

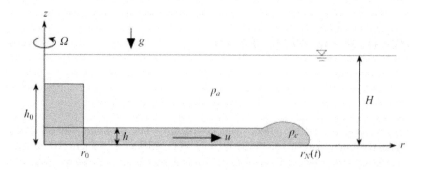

**FIGURE 8.1:**   A sketch of the axisymmetric current in the rotating system.
The azimuthal (angular) motion is not shown.

We use a cylindrical coordinate system $\{r, \theta, z\}$ rotating with angular velocity $\Omega$ about the vertical axis $z$ with a gravitational acceleration of $-g\hat{z}$.
The velocity vector (in the rotating system) is denoted by $\mathbf{v} = u\hat{r} + v\hat{\theta} + w\hat{z}$.

We are concerned with the prediction of the behavior of the current after
release, in particular, the radius of propagation, $r_N(t)$, the shape of the interface separating heavy and ambient fluid, $h(r,t)$, and the profiles of the radial
and azimuthal (or angular) velocities. To facilitate the theoretical description
we assume that the flow is axially symmetric, incompressible, inviscid, and
Boussinesq. We define

$$\epsilon = \frac{\rho_c - \rho_a}{\rho_a}; \quad g' = \epsilon\, g. \tag{8.2}$$

The importance of rotation in the gravity current flow is reflected by the
dimensionless parameter

$$\mathcal{C} = \frac{\Omega r_0}{(g'h_0)^{1/2}}. \tag{8.3}$$

This expresses the ratio of the typical Coriolis ($\Omega U$) to inertial ($U^2/r_0$) terms,
using the representative $U = (g'h_0)^{1/2}$. The inverse of this ratio is usually
defined as the Rossby number, $Ro$. A small value of $Ro$ roughly means a small
deviation from solid-body rotation, in particular concerning the azimuthal
velocity as compared to the local $\Omega r$. It is also evident that $\mathcal{C}$ is a ratio
of two lengths: $r_0$, which typifies the inertial terms, and a new one, which
characterizes the Coriolis terms.

The new length which appears in the problem, called the Rossby "radius" (of deformation, or adjustment), is given by

$$L_{Ro} = \frac{(g'h_0)^{1/2}}{2\Omega}. \tag{8.4}$$

An interpretation is provided by the following scaling argument. We estimate that the typical Coriolis term is of magnitude $2\rho\Omega U = 2\rho\Omega(g'h_0)^{1/2}$ and the pressure is $p \sim \rho g'h_0$. Suppose that the Coriolis term is the dominant dynamic effect. Then the pressure gradient over the Rossby distance, $p/L_{Ro}$, balances the Coriolis effect. Since in the gravity current there is close connection between $p$ and $h$, we can expect that major changes of thickness (including decrease to zero) occur over the Rossby interval $L_{Ro}$. However, it is important to keep in mind that this argument is valid only when the Coriolis terms dominate, i.e., when $Ro \ll 1$. The details will become more evident later.

Returning to (8.3) we can write

$$L_{Ro} = \frac{r_0}{2\mathcal{C}} = \frac{1}{2}Ro\, r_0. \tag{8.5}$$

The most interesting effects in our configuration appear for small values of $\mathcal{C}$, i.e., for large $Ro$. In this case there is a significant propagation after the release, and hence a significant deviation from the initial state of solid-body rotation. On the other hand, for large values of $\mathcal{C}$, i.e., $Ro \ll 1$, there is indeed little deviation from the initial solid-body rotation of the fluid in the lock. This small deviation implies a short propagation, which turns out to be, approximately, $L_{Ro}$. Such a restricted motion cannot produce a real gravity current, and hence we are only marginally interested in non-small values of $\mathcal{C}$.

We therefore note that in the problems of interest the typical ratio of centrifugal acceleration of the system to the gravitational acceleration is actually very small. This is because

$$\frac{\Omega^2 r_0}{g}\left(\frac{r_0}{h_0}\right) = \epsilon\, \mathcal{C}^2, \tag{8.6}$$

and we can safely restrict the analysis to negligibly small values of this parameter.

These considerations indicate the striking feature of the rotating axisymmetric inviscid current: it propagates to a finite radius only, owing to Coriolis effects. Our objective is to develop a reliable model for this flow. We shall see that the one-layer SW model provides, again, a good approximation.

To obtain a versatile incorporation of the new effects, we start with the momentum equation in a system rotating with constant $\Omega\hat{z}$, as introduced in § 1.2. This reads

$$\rho[\frac{D\mathbf{v}}{Dt} + 2\Omega\hat{z} \times \mathbf{v}] = -\nabla\mathcal{P} - \rho(g\hat{z} - \Omega^2 r\hat{r}) + \mu\nabla^2\mathbf{v}, \tag{8.7}$$

where $\mathcal{P}$ is the thermodynamic pressure. This equation is valid for both the current and ambient fluids, with the appropriate $\rho$. For more explicit expressions; see § C.2. We rearrange the body-force terms on the RHS as follows

$$\rho(g\hat{z} - \Omega^2 r\hat{r}) = (\rho - \rho_a)(g\hat{z} - \Omega^2 r\hat{r}) + \nabla\left[\rho_a(gz - \frac{1}{2}\Omega^2 r^2)\right], \qquad (8.8)$$

and use the term in the right brackets (the body-force potential $\mathcal{B}$ times $\rho_a$) to define the reduced pressure

$$p = \mathcal{P} + \rho_a(gz - \frac{1}{2}\Omega^2 r^2) + C. \qquad (8.9)$$

The momentum equation can be rewritten

$$\rho[\frac{D\mathbf{v}}{Dt} + 2\Omega\hat{z} \times \mathbf{v}] = -\nabla p - (\rho - \rho_a)(g\hat{z} - \Omega^2 r\hat{r}) + \mu\nabla^2\mathbf{v}. \qquad (8.10)$$

The one-layer SW model assumes that in the ambient $\mathbf{v} = \mathbf{0}$. Since $\rho = \rho_a$ in this domain, (8.10) is clearly satisfied by

$$p_a = C. \qquad (8.11)$$

This is the same result as in the non-rotating axisymmetric case. The volume continuity and $z$-component of the momentum equation are not influenced by the rotation. Hence we can apply exactly the same considerations as in the non-rotating inviscid case (§ 6.1): integrate the hydrostatic $\partial p_c/\partial z$ in the current, and match the pressures at the interface $z = h(r,t)$. We obtain, again,

$$p_c(r, z, t) = -\rho_c gz + \Delta\rho gh(r,t) + C \quad (0 \leq r \leq r_N(t), 0 \leq z \leq h(r,t)) \quad (8.12)$$

where $\Delta\rho = \rho_c - \rho_a$. Finally, the longitudinal pressure gradient in the current can be written as

$$\frac{\partial p_c}{\partial r} = \Delta\rho g\frac{\partial h(r,t)}{\partial r}. \qquad (8.13)$$

The interesting conclusion is that the driving pressure gradient in the rotating and non-rotating cases is the same. The difference between the currents emerges because of the reaction to this driving: the radial acceleration needed for propagation in a rotating system contains some terms which are not present in a non-rotating system. We keep in mind that a negative $\partial h/\partial r$ is necessary to provide a forward-driving pressure distribution.

Consider now the inviscid $r$-component of (8.10) for the current. We eliminate the $\partial p_c/\partial r$ by (8.13), use the continuity equation $\nabla \cdot \mathbf{v} = 0$, and obtain

$$\frac{\partial u}{\partial t} + \frac{1}{r}\frac{\partial ru^2}{\partial r} + \frac{\partial}{\partial z}(uw) - \frac{v^2}{r} - 2\Omega v = -\frac{\Delta\rho}{\rho_c}g\left[\frac{\partial h(r,t)}{\partial r} - \frac{\Omega^2 r}{g}\right]. \qquad (8.14)$$

The $\Omega^2 r/g$ term in the RHS is small as compared with the $\partial h/\partial r$ term, in accord with the observation (8.6), and will therefore be neglected. In other words, the dominant driving effect is the reduced gravity. The resulting radial momentum equation differs from the non-rotating counterpart (6.14) by the $-2\Omega v$ Coriolis and $-v^2/r$ centrifugal terms. We must distinguish between this local centrifugal term and the $\Omega^2 r$ centrifugal body force of the background rotation. The latter effect is negligible in our problem.

The azimuthal $\theta$-component of (8.10) for the current is relatively simple. In the axisymmetric and inviscid case considered here all the RHS terms are zero. We obtain

$$\frac{\partial v}{\partial t} + u\left(\frac{\partial v}{\partial r} + \frac{v}{r}\right) + w\frac{\partial v}{\partial z} + 2\Omega u = 0; \tag{8.15}$$

which can also be rewritten with the aid of the continuity equation $\nabla \cdot \mathbf{v} = 0$ as

$$\frac{\partial v}{\partial t} + \frac{1}{r}\frac{\partial}{\partial r}(ruv) + \frac{uv}{r} + \frac{\partial}{\partial z}(vw) + 2\Omega u = 0. \tag{8.16}$$

Next, we recast these equations for the $z$-averaged velocities. As suggested by the results in the previous cases, we can make a shortcut: just replace $u$ and $v$ in (8.14) and (8.15) by the $z$-averaged $u(r,t)$ and $v(r,t)$ (we drop the overbar). A more rigorous approach is to integrate (8.14) and (8.16) and employ Leibniz's rule, as done in § 6.1; the details are left for an exercise. We recall that the continuity equation is not influenced by the rotation. We obtain the SW system of three equations for the three variables: $h(r,t)$, averaged radial velocity $u(r,t)$, and averaged azimuthal velocity $v(r,t)$ in the rotating system. The corresponding angular velocity in the rotating system is defined by

$$\omega(r,t) = v(r,t)/r. \tag{8.17}$$

### 8.1.1 The scaled SW equations and boundary conditions

It is convenient to use different scalings for the radial and azimuthal velocities. Let us denote by asterisks dimensional variables. First, we apply the usual scaling

$$\{r^*, z^*, h^*, H^*, t^*, u^*\} = \{r_0 r, h_0 z, h_0 h, h_0 H, Tt, Uu\}, \tag{8.18}$$

where

$$U = (h_0 g')^{1/2} \quad \text{and} \quad T = r_0/U. \tag{8.19}$$

Next, we introduce

$$\{v^*, \omega^*\} = \{\Omega r_0 v, \Omega \omega\}. \tag{8.20}$$

We have two estimates for the importance of the viscous effects. The ratio of inertial to viscous effects is the usual $Re = Uh_0/\nu$, assumed large. The ratio of viscous, $\nu U/h_0^2$, to Coriolis, $\Omega U$, effects is defined as the Ekman number

$$\mathcal{E} = \frac{\nu}{\Omega h_0^2} = \left(CRe\frac{h_0}{x_0}\right)^{-1}, \tag{8.21}$$

assumed small.

In conservation form the dimensionless averaged balance equations of continuity, radial momentum, and azimuthal momentum can be written as

$$\frac{\partial h}{\partial t} + \frac{\partial}{\partial r}(uh) = -\frac{uh}{r}; \tag{8.22}$$

$$\frac{\partial}{\partial t}(uh) + \frac{\partial}{\partial r}\left[u^2 h + \frac{1}{2}h^2\right] = -\frac{u^2 h}{r} + C^2 vh\left(2 + \frac{v}{r}\right); \tag{8.23}$$

and

$$\frac{\partial}{\partial t}(vh) + \frac{\partial}{\partial r}(uvh) = -2uh\left(1 + \frac{v}{r}\right). \tag{8.24}$$

In characteristic form, this reads

$$\begin{bmatrix} h_t \\ u_t \\ v_t \end{bmatrix} + \begin{bmatrix} u & h & 0 \\ 1 & u & 0 \\ 0 & 0 & u \end{bmatrix} \begin{bmatrix} h_r \\ u_r \\ v_r \end{bmatrix} = \begin{bmatrix} -\frac{uh}{r} \\ C^2 v\left(2 + \frac{v}{r}\right) \\ -u\left(2 + \frac{v}{r}\right) \end{bmatrix}. \tag{8.25}$$

The initial conditions are zero velocity in both radial and azimuthal directions, and unit dimensionless height and length at $t = 0$. At $r = 0$ we impose $u = 0$ and a regularity of the angular velocity $v/r$. The boundary conditions at $r = r_N(t)$ will be discussed later.

The system is hyperbolic. It is useful to derive the characteristics (see § A.1). The corresponding eigenvalues and eigenvectors of the matrix in (8.25) are:

$$c_+ = u + h^{1/2}; \quad c_- = u - h^{1/2}; \quad \lambda_3 = u; \tag{8.26}$$

$$(1, h^{1/2}, 0); \quad (1, -h^{1/2}, 0); \quad (0, 0, 1). \tag{8.27}$$

The first two characteristic velocities of this system are like in the two-dimensional case, but the balance on these characteristics is modified by the curvature and Coriolis terms. We obtain

$$dh \pm h^{1/2}du = \left[-\frac{uh}{r} \pm h^{1/2}C^2 v(2 + \frac{v}{r})\right] dt, \tag{8.28}$$

on the characteristics

$$\frac{dr}{dt} = c_\pm = u \pm h^{1/2}. \tag{8.29}$$

This is the counterpart of (6.27) when rotation is present. The rotation contributes to the non-homogeneous term on the RHS of (8.28). Since $v$ is negative (see below), the effect of rotation enhances the effect of curvature on the

$c_+$ characteristic. The curvature causes the decrease of $h_N$ with propagation; the nose of the rotating current is expected to become even thinner.

The third characteristic is concerned with the azimuthal motion. It reads

$$dv = -\left(2 + \frac{v}{r}\right) u dt \quad \text{on} \quad \frac{dr}{dt} = u. \qquad (8.30)$$

This can be rewritten as

$$\frac{dv}{dr} + \frac{1}{r} v = -2 \quad \text{on} \quad \frac{dr}{dt} = u, \qquad (8.31)$$

subject to the initial condition $v = 0$ at the position $r = r_{init}$ where the characteristic starts the propagation. The solution is

$$\omega = \frac{v}{r} = -1 + \left(\frac{r_{init}}{r}\right)^2. \qquad (8.32)$$

The physical interpretation is that a particle of dense fluid that propagates to a larger radius will acquire more and more counter-rotation in the rotating frame of reference. This is the justification for the negative sign of $v$ used in the discussion above. The absolute angular velocity of this particle, in the laboratory frame, becomes smaller than $\Omega$ of the whole system. This could be anticipated due to conservation of angular momentum. But this also implies a positive radial acceleration, given by the $-2\Omega v$ (dimensional) Coriolis term.

*Boundary conditions* for the radial and azimuthal (angular) velocity components at the nose $r = r_N(t)$ are needed.

Concerning $u_N$. We shall use the same condition as for the non-rotating current, (6.28)-(6.29). The justification is as follows. We noticed that the pressure driving is like in the non-rotating case. When $\mathcal{C}$ is small, the dynamic reaction of the fluid in the jump about the nose is little affected by rotation. This is because a fluid particle which propagates a small radial distance $U\delta t$ is deflected an insignificant $\Omega r_0 \delta t$. The deviation from the non-rotating front jump balance is of the order of magnitude of $\mathcal{C}$. This conjecture is vindicated by the good agreement of the $r_N(t)$ predicted by the present SW formulation with full NS simulations and laboratory experiments, as discussed later in this chapter. On the other hand, when $\mathcal{C}$ is not small the distance of propagation is typically less than 1. The resulting flow is not truly a gravity current, but rather an adjusting bulk of rotating fluid. We are not really interested in the details of the time-dependent propagation in this case; we can still use (6.28) and (6.29) as an approximation for the qualitative behavior.

Concerning $v_N$ (or rather $\omega_N$). The result (8.32) provides $\omega$ for a fluid particle (actually, a ring) which was at $r = r_{init}$ at $t = 0$. The nose $r_N(t)$ is the ring which was at $r_{init} = 1$. For this initial condition (8.32) yields the necessary boundary condition

$$\omega_N(t) = -1 + \frac{1}{r_N^2(t)} \quad (r = r_N(t)). \qquad (8.33)$$

The SW formulation is now complete. The free parameters are $H$ and $\mathcal{C}$. For $\mathcal{C} = 0$ the non-rotating axisymmetric case is recovered.

### 8.1.1.1 Potential vorticity

The combination $(2\Omega + \hat{z} \cdot \nabla \times \mathbf{v})/h$ (in dimensional form) or $(2 + \hat{z} \cdot \nabla \times \mathbf{v})/h$ (in dimensionless form, scaled with $\Omega$) is defined as the potential vorticity (Greenspan 1968; Pedlosky 1997). This variable is governed by a compact equation which can be derived as follows.

The $z$-component of the local vorticity (i.e., in the rotating frame, in the direction parallel to the axis of rotation) is

$$\zeta = \hat{z} \cdot \nabla \times \mathbf{v} = \frac{1}{r}\frac{\partial}{\partial r}rv = \frac{1}{r}\frac{\partial}{\partial r}r^2\omega. \tag{8.34}$$

We apply the operator $\frac{1}{r}\frac{\partial}{\partial r}r$ to the last equation in (8.25), and, after some rearrangement and use of (8.34), obtain

$$\frac{\partial(2+\zeta)}{\partial t} + u\frac{\partial}{\partial r}(2+\zeta) + (2+\zeta)\frac{1}{r}\frac{\partial}{\partial r}ru = 0, \tag{8.35}$$

or

$$\frac{D}{Dt}(2+\zeta) = -(2+\zeta)\frac{1}{r}\frac{\partial}{\partial r}ru. \tag{8.36}$$

Note that the continuity equations (8.22) can be expressed as

$$h\frac{1}{r}\frac{\partial}{\partial r}ru = -\left(\frac{\partial h}{\partial t} + u\frac{\partial h}{\partial r}\right) = -\frac{Dh}{Dt}. \tag{8.37}$$

Dividing (8.36) by $h$ and combining with (8.37), we obtain the potential vorticity conservation result,

$$\frac{D}{Dt}\left(\frac{2+\zeta}{h}\right) = 0. \tag{8.38}$$

This is an interesting theoretical property of the rotating inviscid current. It shows that vorticity and height of a fluid particle (ring) are connected. A more explicit result can be derived by implementing the *initial condition* $h = 1$ and $\zeta = 0$ for all the particles of the current. In this case (8.38) can be integrated as $(2+\zeta)/h = 2$ for any particle. This can be rewritten

$$h(r,t) = 1 + \frac{1}{2}\zeta(r,t) = 1 + \frac{1}{2}\frac{1}{r}\frac{\partial}{\partial r}rv(r,t) = 1 + \omega + \frac{1}{2}r\frac{\partial\omega}{\partial r}. \tag{8.39}$$

Consider now the total volume conservation, combined with (8.39)

$$\frac{1}{2} = \int_0^{r_N} hr\,dr = \int_0^{r_N}[1 + \frac{1}{2}\frac{1}{r}\frac{\partial}{\partial r}r^2\omega]r\,dr = \frac{1}{2}\left[r_N^2 + r_N^2\omega(r_N,t)\right]. \tag{8.40}$$

This yields the condition

$$\omega_N(t) = -1 + \frac{1}{r_N^2(t)}. \tag{8.41}$$

This outcome is identical with (8.33), which was obtained from different considerations. The use of (8.39) for $r = r_N$, combined with (8.41), gives

$$h_N = \left(\frac{1}{r_N}\right)^2 + \frac{1}{2}r_N\left(\frac{\partial\omega}{\partial r}\right)_N \qquad (r = r_N(t)). \qquad (8.42)$$

We conclude that the potential vorticity of our current is conserved. This implies that at the nose there is counter-rotation (in the rotating frame); this counter-rotation approaches $-1$ when the current spreads significantly. Moreover, $h_N$ turns out to be significantly affected by the swirl at the nose. The first term on the RHS of (8.42) represents the decrease of thickness expected from continuity considerations; the second term, which turns out to be negative, imposes a significant additional reduction of $h_N$, which must be accompanied by a diminution of the speed of propagation. Indeed, even for small values of $\mathcal{C}$, the radial motion may stop when $r_N \sim O(\mathcal{C}^{-1/2})$ is reached. For large values of $\mathcal{C}$ the departure from the initial $r_N = 1$ is only one Rossby radius, $O(\mathcal{C}^{-1})$. How can the motion be stopped over such a small interval (compared to the radius of the lock)? The answer is indicated by (8.42): $\omega$ decreases (from zero) by a similarly small amount over this interval, such that $d\omega/dr = -2$. The details are elaborated below. The coupling between propagation, swirl, and thickness, which emerges from the conservation of potential vorticity, plays an essential role in the derivation of the detailed results.

## 8.1.2 Steady-state lens (SL)

The SW equations of motion (8.22)-(8.24) (or (8.25)), subject to the boundary conditions (6.28) and (8.33), admit a non-trivial steady-state solution with $u \equiv 0$ and $h_N = 0$. We call it steady lens (SL), as suggested by the shape of the interface. The relevance to the rotating gravity current is evident: we expect that in stable circumstances the time-dependent motion tends to a similar steady state (or quasi-steady state in more realistic circumstances, as explained below).

To obtain $h(r)$ and $\omega(r)$ at steady state we use the radial momentum balance (8.23),

$$\frac{dh}{dr} = \mathcal{C}^2 r\omega(2 + \omega), \qquad (8.43)$$

and the potential vorticity conservation in the form (8.39),

$$h = 1 + \frac{1}{2}\frac{1}{r}\frac{d}{dr}r^2\omega. \qquad (8.44)$$

Equation (8.43) means that the radial pressure gradient balances the Coriolis-centrifugal radial accelerations. Indeed, the term on the RHS of (8.43) expresses the combination of the radial Coriolis $-2v$ and centrifugal $-v^2/r$ acceleration; the minus sign has been canceled by the transfer from the LHS to the RHS of the simplified momentum equation.

The boundary conditions are

$$\omega_N = -1 + (1/r_N)^2 \quad \text{and} \quad h_N = 0 \quad (r = r_N). \tag{8.45}$$

The latter condition is necessary for consistency with $u_N = 0$. Note that (8.43) and (8.44) also imply $d\omega/dr = 0$ at $r = 0$. The determination of the constant $r_N$ is also a part of the problem; this enters implicitly via the boundary conditions. We recall that the boundary condition for $\omega_N$ was obtained from volume conservation considerations. Consequently, the SL that satisfies the equations (8.43)-(8.45) is bound to be of volume $1/2$. The total volume conservation can serve as a handy auxiliary equation in the verification and derivation of the solution.

We have a system for the unknown $\omega(r), h(r)$, and $r_N$. The free parameter is $C$. To facilitate the analysis and solution, we first substitute (8.44) into (8.43) to obtain a single equation for $\omega$

$$\frac{d}{dr}\frac{1}{r}\frac{d}{dr}r^2\omega = 2C^2 r\omega(2 + \omega). \tag{8.46}$$

Next, in the spirit of the usual transformation (§ 6.2) we introduce

$$y = r/r_N \quad (0 \leq y \leq 1) \tag{8.47}$$

and reformulate (8.46) and (8.44) as

$$\frac{d^2\omega}{dy^2} + \frac{3}{y}\frac{d\omega}{dy} - 2C^2 r_N^2 \omega(2 + \omega) = 0 \tag{8.48}$$

and

$$h = 1 + \omega + \frac{1}{2}y\frac{d\omega}{dy}. \tag{8.49}$$

The associated boundary conditions are

$$\frac{d\omega}{dy} = 0 \quad (y = 0); \quad \omega = -1 + 1/r_N^2 \quad (y = 1); \tag{8.50}$$

$$h = 1 + \omega + \frac{1}{2}\frac{d\omega}{dy} = 0 \quad (y = 1). \tag{8.51}$$

The (auxiliary) equation for the total volume of the SL is

$$r_N^2 \int_0^1 h(y)y\,dy = \mathcal{V} = \frac{1}{2}. \tag{8.52}$$

The numerical solution of the equations (8.48)-(8.51) which govern the SL is now straightforward. Some iterations are needed because of the non-linear Coriolis-centrifugal term in (8.48). We can use centered finite differences on a uniform grid to discretize $\omega(y)$ and $h(y)$ in the domain $0 \leq y \leq 1$. For a given $\mathcal{C}$ we estimate $r_N^2$ and solve (8.48) with conditions (8.50). Iterations are performed on the non-linear term $w(2 + w)$, starting with the value of $w(y = 1)$. For each iteration, we solve a simple tridiagonal linear system. We correct the estimate of $r_N$ until (8.51) is satisfied. Typical results are presented in Fig. 8.2.

Approximate analytical solutions for small and large values of $\mathcal{C}$ throw further light on the features of the SL.

### 8.1.2.1 Behavior for $\mathcal{C} \ll 1$

We expand $w(y)$, $h(y)$, and $r_N^2$ in powers of the small parameter $\mathcal{C}$. On account of the boundary conditions, we let

$$w = -1 + \mathcal{C}o_1(y) + O(\mathcal{C}^3); \quad r_N^2 = \frac{1}{\mathcal{C}}[R_1 + O(\mathcal{C}^2)]; \quad h = \mathcal{C}h_1(y) + O(\mathcal{C}^3).$$
$$(8.53)$$

The subscript 1 denotes the first unknown terms. Substituting into the equations (8.48)-(8.52) and solving for the matching powers of $\mathcal{C}$, we obtain the leading order behavior

$$r_N = \left(\frac{2}{\mathcal{C}}\right)^{1/2}; \tag{8.54}$$

$$w = -1 + \mathcal{C}(1 - \frac{1}{2}y^2); \tag{8.55}$$

$$h = \mathcal{C}(1 - y^2). \tag{8.56}$$

The jump in the powers of $\mathcal{C}$ in (8.53) is because the omitted terms turn out to be identically zero. Actually, more terms in the expansion can be derived, see Ungarish and Zemach (2003), but the leading terms given above provide the main insights.

Fig. 8.2 confirms the accuracy of the analytical approximation.

The striking outcome is the quite small radius of spreading of the "current," even for small $\mathcal{C}$ (recall that in a non-rotating frame a similar inviscid current will spread, theoretically, to infinity). We also observe that the current spreads significantly less than the Rossby radius of deformation, $1/(2\mathcal{C})$ (scaled with $r_0$). In other words, the Rossby radius is an irrelevant (actually, misleading) estimate for the distance of propagation of currents with small $\mathcal{C}$. As mentioned before, the Rossby radius arguments assume that the deviation from solid-body rotation is small; this assumption is invalidated by the large counter-rotation that develops for small $\mathcal{C}$ cases.

To estimate the time of propagation of the nose from release to the radius of the lens given by (8.54), we argue as follows. Since $r_N = (2/\mathcal{C})^{1/2} \gg 1$, the current is expected to be in the self-similar stage where $r_N \approx Kt^{1/2}$; see (6.51). Consequently, the relevant time is $1/\mathcal{C}$, approximately. In dimensional form, $T/\mathcal{C} = \Omega^{-1}$. Thus, such a lens-shape steady structure is expected to develop during the first revolution of the rotating frame.

## 8.1.2.2  Behavior for $\mathcal{C} \gg 1$

We mentioned that $\mathcal{C} \gg 1$ means a small Rossby number. The rotational effects are expected to dominate and hence the deviations from the initial solid-body rotation of the system remain small during the propagation. The radial displacement of the nose is also strongly restricted. Equations (8.48)-(8.49) are consistent with a small deviation from the initial conditions, i.e., $\omega \approx 0$ and $h \approx 1$, in the domain $0 \leq y < 1$. However, the boundary condition (8.51) requires a steep decrease of $h$ at the nose to zero, which stops the propagation. This indicates a boundary-layer structure about $y = 1$. The $\omega = 0$ and $h = 1$ approximation become the leading term of the "outer solution" of this singular perturbation problem.

In this case, an expansion in powers of $\varepsilon = \mathcal{C}^{-1} \ll 1$ is appropriate. Following the above-mentioned considerations, we let

$$\omega = \varepsilon o_1(y) + O(\varepsilon^2); \quad r_N^2 = 1 + \varepsilon R_1 + O(\varepsilon^2). \tag{8.57}$$

The need for $d\omega/dy = O(1)$, see (8.51), suggests the use of the stretched coordinate $\tilde{Y} = (y - 1)/\varepsilon$.

Letting $o_1 = o_1(\tilde{Y})$ we rewrite (8.48) as

$$\frac{d^2 o_1}{d\tilde{Y}^2} - 4 o_1 = 0 + O(\varepsilon) \quad (\tilde{Y} \leq 0) \tag{8.58}$$

subject to the corresponding leading order (8.50)-(8.51) conditions

$$o_1 = 0 \quad (\tilde{Y} \to -\infty); \tag{8.59}$$

$$o_1 = -R_1; \quad 1 + \frac{1}{2}\frac{do_1}{d\tilde{Y}} = 0 \quad (\tilde{Y} = 0). \tag{8.60}$$

After some algebra we obtain the leading terms

$$r_N = 1 + \frac{1}{2\mathcal{C}}; \tag{8.61}$$

$$\omega = -\frac{1}{\mathcal{C}}\exp[2\mathcal{C}(y - 1)]; \tag{8.62}$$

$$h = 1 - \exp[2\mathcal{C}(y - 1)]. \tag{8.63}$$

Evidently, when $\mathcal{C} \gg 1$ the SL is a bulk of fluid in solid-body rotation; the change from the initial state occurs only in a layer of thickness $1/(2\mathcal{C})$ about the periphery $r = 1$. Even in this thin layer the deviation from solid-body rotation is very small, $O(\mathcal{C}^{-1})$. In dimensional form the radius of the lens is $r_0 + L_{Ro}$; see (8.4). This provides the rigorous support to the definition of the Rossby radius of deformation (in the context of the axisymmetric current). The name "radius" is confusing here; $L_{Ro}$ is a radial interval.

The accuracy of the analytical approximation is confirmed by Fig. 8.2.

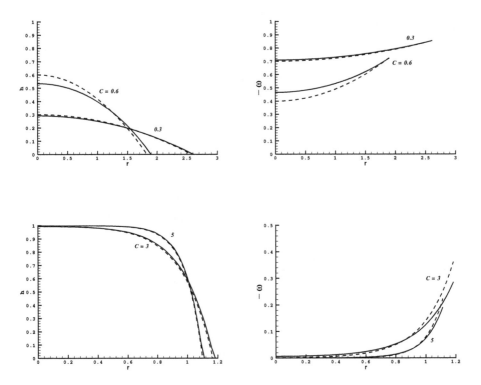

**FIGURE 8.2:** SL solutions for $h(r)$ and $-\omega(r)$ for small $\mathcal{C}$ (upper frames) and large $\mathcal{C}$ (lower frames). Numerical solution (solid line) and analytical approximations (dashed line).

We can now estimate the time of formation of the SL with $\mathcal{C} \gg 1$. Since a short propagation (compared to the initial $r_N = 1$) occurs, the current is in the dam-break phase and hence the speed of propagation can be approximated by 0.5. The propagation to $\Delta r = 1/(2\mathcal{C})$ is performed in a time interval of about $1/\mathcal{C}$. In dimensional form $T/\mathcal{C} = 1/\Omega$. Thus, the SL in the large $\mathcal{C}$ case develops in about the first revolution of the system. Interestingly, the same result for the time of formation was obtained for the SL with small $\mathcal{C}$, although the distances of propagation differ significantly between the large and small $\mathcal{C}$ cases.

### 8.1.3 The SW current and formation of the lens

The SW equations (8.22)-(8.24) and boundary conditions (6.28)-(6.29) and (8.33) are transformed to the domain $0 \le y \le 1$, and solved by the same method as the non-rotating current. Typical results (on a grid with 200 intervals) are shown in Fig. 8.3, for $H = 3$ and $\mathcal{C} = 0.4$. The difference with the non-rotating counterpart, see Fig. 6.1, is evident. A negative $\partial h/\partial r$ is seen already at $t = 1$; this means that the pressure gradient is used to supply the Coriolis-centrifugal acceleration in the radial direction. The propagation practically stops at about $t = 4$ ($\approx 1.7/\mathcal{C}$) where $h_N = 0$ and $r_N = r_{max}$.

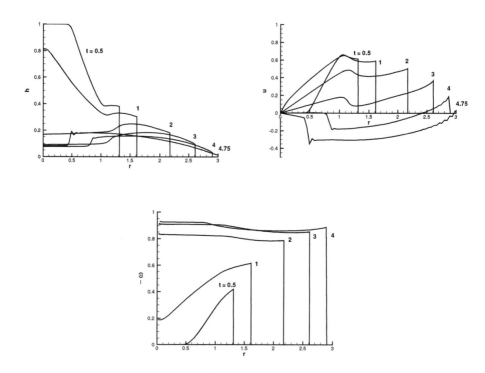

**FIGURE 8.3**:   The time development of a rotating SW gravity current with $H = 3$ and $\mathcal{C} = 0.4$.

We see that the current tends to develop into a SL as depicted in Fig. 8.2. The maximum radius of the current exceeds the radius of the SL with the same $\mathcal{C}$ by about 30%. However, at maximum propagation the radial speed $u(r)$ is negative. This surprising outcome indicates that the body of dense fluid will contract. The details are beyond the resolution of the present SW formulation, but we can speculate that oscillations about the SL solution will develop. Indeed, rotating fluids support inertial oscillations of period $\pi/\Omega$ (dimensional) in which the Coriolis effects are the restoring mechanism.

During the formation of the SL the current loses energy. In the final state, the dense fluid contains less mechanical energy than in the lock before release.

The discrepancy increases when $C$ decreases; some details will be presented later in § 13.1.2. Viscous effects, energy dissipation in various forms, and instabilities are expected to become relevant in the metamorphosis of the stopped current into the lens. These topics are a part of the complex "adjustment" process which has been widely treated in the literature. The interested reader will find more details in § 8.1.4 and references given therein.

Additional insights into the influence of $C$ on the propagation can be inferred from Fig. 8.4. In all the displayed cases the propagation practically stops at about $t_1 = 1.7/C$. In dimensional form, $t_1^* = 1.7/\Omega \approx \pi/(2\Omega)$. This corresponds to one-fourth revolution of the system. We see again that the maximum radius of the current exceeds the radius of the SL with the same $C$ by about 30%.

A laboratory visualization of this behavior is shown in Fig. 8.5. Propagation to a maximum radius (photo 3) followed by a contraction (photo 2) are seen.

The assumption is that the motion remains in the inviscid regime until the propagation is stopped by the Coriolis-centrifugal effects at $r \approx (2/C)^{1/2}$. This means that $C$ must be sufficiently large, as follows. In § 6.6 we defined $r_V$ as the radius where viscous effects begin to dominate in a non-rotating axisymmetric current. We can argue that when $r_V \gg (2/C)^{1/2}$ the viscous deceleration is unimportant during the spread to the maximum radius. Using (6.58) and after some algebra we obtain the condition

$$C \gg \left( Re \frac{h_0}{r_0} \right)^{-1/3}. \tag{8.64}$$

## 8.1.4 Two-layer models and more about the lens

A two-layer SW model for the rotating axisymmetric current is available; see Ungarish and Zemach (2003). As shown in Chapter 5, the non-rotating two layer model can be reduced to the solution of a set of equations for the variables of the current only, $h(r,t)$ and $u(r,t)$. The speed in the second layer is obtained by $u_2 = -u_1/(H-h)$. Here the subscripts 1 and 2 denote the current and the ambient, respectively. A key result in the rotating two-layer formulation is the fact that the angular velocity of the ambient, $\omega_2(r,t)$, can be expressed in a quite explicit manner in terms of the angular velocity of the current, $\omega_1(r,t)$. This allows the reduction of the SW problem to a set of equations for the variables $h, u, \omega$ in the current only. The system is hyperbolic and subject to the same initial and boundary conditions as the one-layer model. As expected, for large $H$ (and practically for $H > 2$) the two-layer results reduce to the one-layer results.

In general, in the upper layer above the current a positive $\omega_2$ appears, as a consequence of rings of rotating fluid which propagate from a larger to a smaller radius (just the opposite behavior of the spreading current); see Fig. 8.6. An important result is that the potential vorticity is conserved

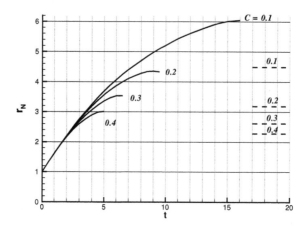

**FIGURE 8.4:** $r_N$ as a function of $t$ for various $C$ and $H = 3$. Also shown the radius of the SL (dashed lines).

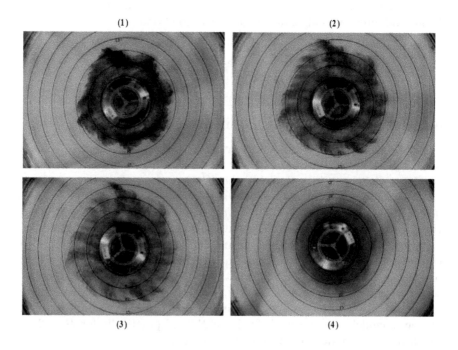

**FIGURE 8.5:** Photographs from above of the propagation of a saline current (made observable by dye) in a rotating frame, showing the expansion and contraction of the dense fluid. The marked rings are at 5 cm intervals. Time increases from photo 1 to photo 4. $C = 0.15$, $H = 1$, $h_0 = 10$ cm, $r_0 = 4.7$ cm, $g' = 20 \, \mathrm{cm \, s^{-2}}$. (Ungarish and Huppert 1998, with permission.)

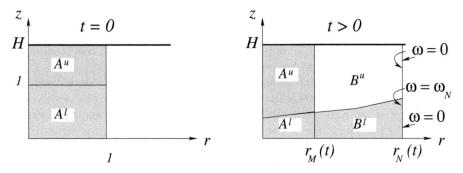

**FIGURE 8.6:** Sketch of upper regions $A^u$ (ambient fluid initially above the dense fluid in the lock) and $B^u$ (added ambient fluid to layer 2 during propagation of the current) and the lower regions $A^l$ and $B^l$ of dense fluid. At the initial time the region $B$ does not exist.

also in the upper layer of ambient fluid. However, the implementation of the potential vorticity balances is complicated by the fact that the fluid in the domain of interest ($0 \leq r \leq r_N$, $0 \leq z \leq H$) is composed of bulks which carry different values of potential vorticity, because of their different initial positions in the process. To sort this out, we split the domain into four regions, $A$ and $B$ each with an upper and lower part; see Fig. 8.6.

- Region $A^u$ : $h_1 \leq z \leq H$; $0 \leq r \leq r_M(t)$. Contains the volume of ambient fluid which at $t = 0$ was located above the "locked" dense fluid.

- Region $B^u$ : $h_1 \leq z \leq H$; $r_M(t) \leq r \leq r_N(t)$. Contains the volume of ambient fluid which entered the upper layer after the release of the current.

- region $A^l$ : $0 \leq z \leq h_1$; $0 \leq r \leq r_M(t)$. Contains the dense fluid below the ambient fluid of region $A^u$.

- region $B^l$ : $0 \leq z \leq h_1$; $0 \leq r_M(t) \leq r_N(t)$. Contains the dense fluid below the ambient fluid of region $B^u$.

Here $r_M(t)$ denotes the radius of the region $A$, which is provided implicitly by the volume conservation in the region $A^u$. Roughly, for the SL with small $\mathcal{C}$, $r_M \approx [(1 - 1/H)(1 + \mathcal{C}/H)]^{1/2}$.

Owing to the different initial position of the involved activated fluids, the initial potential vorticity has three different values: 2 for the dense fluid (originally in domain $A^l$); $2/(H - 1)$ for the ambient in domain $A^u$; and $2/H$ for the ambient in the domain $B^u$. These initial conditions are preserved in the values of $\omega$ in the subsequent motion and lens formation. Indeed, the calculations show that $\omega_2 = |\omega_1|/(H-1)$ for $r \leq r_M < 1$; then, for larger $r$, the ratio $|\omega_2/\omega_1|$ becomes smaller and it decays to zero like $1 - (r/r_N)^2$. The subscripts

1 and 2 denote the dense and ambient fluids. The boundary condition of $\omega_1$ at $r_N$ is still given by (8.33). Both $\omega_1$ and $\omega_2$ are continuous at $r = r_M$. The conclusion is that a large $\omega_2$ appears when $H$ is close to 1, but this is confined to a small volume of fluid bounded by $r_M$, approximatively.

The two-layer formulation also admits steady-state solutions with $u_1 = u_2 = 0$ and $\omega_{1,2} \neq 0$. This is the two-layer version for the SL. The positive $\omega_2$ in the upper layer enhances the Coriolis-centrifugal acceleration of the dense current in the lower layer. The local rotation of the ambient is larger than in the initial state. Consequently, the dense fluid "feels" a larger $C$ than the one calculated with the nominal $\Omega$. The result is a larger pressure gradient, i.e., $\partial h/\partial r$ is (locally) larger than predicted by the one-layer balance (8.43) in regions where $\omega_2/|\omega_1|$ is significant. Accordingly, the shape of the two-layer SL differs from the one-layer counterpart only when $H$ is close to 1, and mostly near the axis $r = 0$. Typical results are shown in Fig. 8.7. The rapidly-rotating spike of the lens near $r = 0$ when $H = 1$ and $C \ll 1$ is a striking feature which has been reproduced in laboratory. The released fluid maintains a "point of surface contact" (Dewar and Killworth 1990) at the original height 1. But we keep in mind that only a small portion of the volume is contained in this region, and the typical thickness of the lens is $C$. The radius of the SL is little affected by the height of the ambient. However, $H$ turns out to be an important parameter in stability considerations (see below). For $H > 2$ the differences between the one- and two-layer SL solutions are small, perhaps insignificant.

The "steady-state" status of the lens is an idealization. We showed that the lock-released current produces an approximation to this situation, which needs a further time-dependent adjustment. In a real fluid even the ideal SL structure is only a quasi-steady situation. The presence of a body of fluid with different density and angular velocity in a rotating ambient fluid triggers mechanisms which attempt to smooth out and annihilate the jumps at the interface.

First, it is clear that the $\omega < 0$ counter-rotation of the lens is incompatible with the $\omega = 0$ of the solid boundary and embedding fluid. This is mostly relevant to small $C$ cases in which the counter-rotation is very pronounced. Viscous Ekman layers develop at $z = 0$ and $z = h(r)$. The typical thickness of these layers is $(\nu/\Omega)^{1/2}$ (dimensional), and in dimensionless form, scaled with $h_0$, the thickness is $\mathcal{E}^{1/2}$. By an interesting coincidence, the Ekman layers also develop during about one revolution of the system, like the initial lens. Thus, after the current attains the maximum radius, it will be subjected to the torque of the Ekman layers from below and above. The thickness ratio of the lens to viscous layers is $A = C/\mathcal{E}^{1/2}$, which we assume to be large. The torque attempts to spin-up the lens, i.e., reduce the initial $\omega \approx -1$ to zero. This resembles a spin-up from rest process, and hence the spin-up time interval is, typically, $A/\Omega$ (see Greenspan 1968; Ungarish and Mang 2003). The circulation introduced by the Ekman layers, sketched in Fig. 8.8, and the tendency to diminish the counter-rotation, will support further spread out

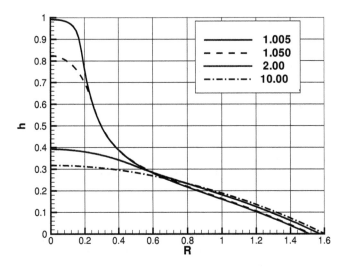

**FIGURE 8.7:** Two-layer SL results for $\mathcal{C}^2 = 0.1$ and various $H$: the thickness $h$ as a function $R$. Here $R$ is the radius scaled with $L_{Ro} = r_0/(2\mathcal{C}) = 1.58r_0$. (Ungarish and Zemach 2003 with permission.)

and thinning of the dense fluid. The relevant time scales are these of spin-up and viscous diffusion.

Second, there is experimental and theoretical evidence that instability waves may develop at the interface between the lens and the ambient fluid, in particular at the rim $r \approx r_N$. Indeed, the front of the real "axisymmetric" gravity current is rather ragged, as seen in Fig. 8.5. The instabilities depend on the azimuthal coordinate $\theta$. Eventually, due to the influence of these three-dimensional perturbations, the central core breaks into smaller, non-axisymmetric structures. The instabilities depend on the depth of the ambient, and are more pronounced for surface than for bottom currents. This topic has been widely investigated in the context of oceanographic lens-like structures (also called rings, eddies, or vortices). For more details see Saunders (1973), Flierl (1979), Csanady (1979), Griffiths and Linden (1981), Killworth (1992), Nof and Simon (1987), Hedstrom and Armi (1988), Verzicco, Lalli, and Campana (1997), Choboter and Swaters (2000), Rubino, Hessner, and Brandt (2002).

The mechanisms that dissipate the SL may require many revolutions of the system to accomplish their task, in particular when $\mathcal{C}$ is small. On the other hand, the propagation of the current and appearance of the SL are accom-

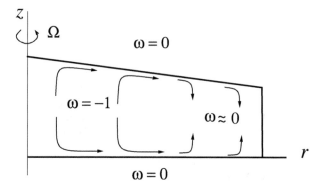

**FIGURE 8.8:** Sketch of the circulation introduced by the Ekman layers in spin-up from rest of a fluid embedded in boundaries co-rotating with the system.

plished in about one-fourth revolution of the system. This lends relevance to the present SW results and vindicates the (quasi) steady-state lens approximation. Strong support in this direction is provided by the work of Stegner, Bouruet-Aubertot, and Pichon (2004) discussed in § 8.1.5.2.

### 8.1.4.1 Lens with source at origin

Consider the time-dependent lens whose volume $2\pi\mathcal{V}^*(t^*)$ (dimensional) increases due to a confined source about $r = 0$. We assume that the situation is of the $\mathcal{C} \ll 1$ type. The added fluid enters the lens with zero azimuthal velocity, and hence the result $\omega \approx -1$ which typifies the SL remains valid in the main bulk of the dense fluid. It is justified to assume that, for a sufficiently slow filling rate, the shape of the dense fluid will be determined by the same mechanism as in the SL case: the Coriolis-centrifugal acceleration will be balanced by the pressure gradient represented by the $\partial h/\partial r$. The implication is that the geometry of the time-dependent lens will be similar to that of the SL.

As a first approximation, we can use the results (8.54) and (8.56) in which the volume $\frac{1}{2}r_0^2 h_0$ is replaced by $\mathcal{V}^*(t^*)$. The radius of the lens (dimensional) is predicted by (8.54)

$$r_N^*(t^*) = \sqrt{2}\frac{r_0}{\mathcal{C}^{1/2}} = \sqrt{2}\left(\frac{g'r_0^2 h_0}{\Omega^2}\right)^{1/4} = 2^{3/4}\left(\frac{g'}{\Omega^2}\right)^{1/4}[\mathcal{V}^*(t^*)]^{1/4}. \quad (8.65)$$

The corresponding height at center is obtained by (8.56)

$$h^*(r = 0, t^*) = \mathcal{C}h_0 = \left(\frac{\Omega^2 r_0^2 h_0}{g'}\right)^{1/2} = \sqrt{2}\left(\frac{\Omega^2}{g'}\right)^{1/2}[\mathcal{V}^*(t^*)]^{1/2}. \quad (8.66)$$

In particular, for lenses produced by constant flux $\mathcal{V}^*(t^*) = Q^*t^*/(2\pi)$, we obtain $r_N \sim t^{1/4}$ and $h(0) \sim t^{1/2}$. A more detailed analysis of this case is presented in Griffiths and Linden (1981). The associated experiments, performed with surface lenses of freshwater in saline, confirmed the above-mentioned trends with respect to time. However, the measured radius was larger and the height was smaller than the predictions of (8.65) and (8.66). The discrepancies, in the range of 30–100%, were attributed to the simplifications used in the theoretical model.

The behavior of the time-dependent volume lens can be derived by a different illuminating approach. Since $\omega \approx -1$, (8.43) can be rewritten as

$$\frac{dh}{dr} = -\mathcal{C}^2 r \quad \text{(dimensionless)}. \tag{8.67}$$

After the integration of this equation, subject to the obvious boundary condition $h(r_N) = 0$, we can calculate the volume of the resulting structure. It is left for an exercise to show that the previous results (8.65) and (8.66) are recovered. This derivation proves that: (a) The underlying mechanism of the lens is quite simple; and (b) Very little physics is involved in the results (8.65) and (8.66), and hence the discrepancy with experiments is not surprising.

Actually, in this section we used a variant of the box-model (or momentum integral) approach. For the constant-flux source the radius behaves like $Kt^\beta$. As in the other box-model solutions, the predicted power $\beta$ is in good agreement with experiments, but not the coefficient $K$.

## 8.1.5   Some experimental and Navier-Stokes results

### 8.1.5.1   The current

Here we consider in some detail the experimental observations and comparisons with Navier-Stokes simulations and SW results that have been reported in Hallworth, Huppert, and Ungarish (2001).

The experiments were conducted in the Coriolis laboratory at the Laboratoire des Ecoulements Géophysiques et Industriales, Grenoble. The turntable of this laboratory is a 14 m diameter circular platform capable of rotating about a vertical axis with a very accurate $\Omega$. The table was equipped with a 13.0 m diameter, 1.2 m deep circular tank, which could be filled with freshwater or brine supplied from large mixing tanks. (When the experiments were performed a central column of diameter 42 cm was present; this has been removed in later upgrades of this really formidable apparatus.)

The experiments involved lock-release gravity currents in which fixed volumes of dense saltwater, initially held behind a central cylindrical lock of radius 1 m and height 1 m, were allowed to intrude into freshwater surroundings following the rapid vertical removal of the lock. A schematic sketch of the experimental configuration is presented in Fig. 8.9. The lock could be raised and lowered by means of a pulley system connecting the uppermost hoop to a

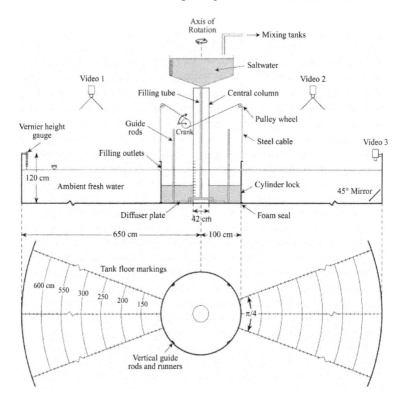

**FIGURE 8.9**: Elevation and plan schematic of the experimental tank. (Hallworth, Huppert, and Ungarish 2001, with permission.)

manually operated winding spool by four steel cables. The basal edge of the lock was seated on a 1 cm thick foam sealing ring on the tank floor.

The experimental procedure began by filling the entire tank with freshwater to the desired depth. Once the lock cylinder had been lowered and sealed on the tank floor, a known volume of red-dyed saltwater was then slowly fed into the base of the lock beneath a circular plate fixed 1 cm above the floor, designed to minimize mixing with ambient fluid upon entry. A sharp, stable interface separating the incoming saltwater from the overlying freshwater ambient inside the lock was thus formed, which slowly rose as the filling proceeded until the desired height was reached. The ambient fluid displaced during the filling of the lock escaped through a number of 1 cm diameter holes arranged around the perimeter of the lock wall at the height of the free surface. On completion of the filling, the final height, $h_0$, of the saltwater interface was read off a graduated scale on the central column, and the densities of both the saltwater and ambient freshwater were measured by hydrometers accurate to within $\pm 0.0001$ g cm$^{-3}$. To achieve efficient spin-up of the fluids, in the rotating experiments the filling process was performed while the whole table was in rotation with the chosen value of $\Omega$.

Each experiment was started by rapid vertical lifting of the lock cylinder. The vertical travel required was achieved in about 6 seconds. Upon release, the ensuing gravity current was observed to spread radially across the floor of the tank. The flow was recorded by two overhead video cameras covering diametrically opposite sectors of angular width $\pi/4$ marked on the tank floor. Each sector contained concentric arcs with radial spacings of 50 cm, in addition to 10 cm graduations along 3 radial lines (see Fig. 8.9). The rates of radial propagation of the currents were measured by subsequent analysis of the video recordings to determine the average time of arrival of the current front at fixed distances along all 6 radial lines. A third video camera was mounted on the perimeter wall of the tank above a submerged 45° mirror to record a vertical profile of the current along a radius. This view was illuminated by a vertical laser sheet in the plane of three vertical scale bars at fixed radii of 200, 250, 300 cm, and relied on either red dye or fluorescein to delimit the current profile.

A total of 17 experiments were conducted, exploring various combinations of initial conditions of the relevant parameters, namely: the rate of rotation $\Omega$; the ambient depth $H$; the initial height of the current $h_0$ (and hence the volume $\pi h_0(r_0^2 - r_i^2)$, where $r_0$, the radius of the lock, and $r_i$, the radius of the central column, were always fixed at 100 cm and 21 cm respectively); and the reduced gravity $g' = g(\rho_c - \rho_a)/\rho_a$. Four experiments (whose label is prefixed by the letter S) were performed with no rotation. The remaining thirteen experiments (label prefixed by the letter R) were conducted with rotation rates varying between 0.05 and 0.15 s$^{-1}$. The viscous effects were expected to be small. The initial Reynolds number was typically $10^5$, and the Ekman number in the rotating experiments was typically $10^{-4}$. (The Ekman boundary-layer length-scale $(\nu/\Omega)^{1/2}$ was 0.45 cm in the cases with the slowest rate of rotation.)

Table 8.1 summarizes five typical configurations which are discussed in detail in this section.

Each release in the experiment generated an almost perfectly axisymmetric

**TABLE 8.1:** Data of configurations in the experiments of Hallworth et al. (2001). In all cases $r_0 = 100$ cm. In the corresponding numerical simulations S3 and R5 the value of $Re$ was reduced by a factor of 10.

| | $h_0$ cm | $g'$ cm s$^{-2}$ | $\Omega$ s$^{-1}$ | $H$ | $\mathcal{C}^2$ | $Re$ | $U$ cm s$^{-1}$ | $T$ s |
|---|---|---|---|---|---|---|---|---|
| S3 | 45.8 | 19.2 | 0 | 1.74 | 0 | $4.3 \cdot 10^5$ | 29.7 | 3.37 |
| R5 | 46.3 | 19.0 | 0.100 | 1.72 | 0.114 | $4.3 \cdot 10^5$ | 29.7 | 3.37 |
| R11 | 77.0 | 9.61 | 0.064 | 1.06 | 0.057 | $2.1 \cdot 10^5$ | 27.2 | 3.68 |
| R14 | 44.0 | 9.72 | 0.050 | 1.84 | 0.059 | $3.1 \cdot 10^5$ | 20.7 | 4.84 |
| R16 | 16.7 | 23.9 | 0.078 | 4.99 | 0.150 | $5.0 \cdot 10^5$ | 20.0 | 5.01 |

gravity current which spreads radially across the floor of the tank. All the tested non-rotating currents reached the perimeter wall, whereupon they were reflected and observed to propagate back towards the center of the tank. A detailed discussion of the non-rotating currents is presented in § 6.7. With the compensation for the lock removal effect in mind, the conclusion is that the SW theory provides a very good prediction of $r_N(t)$ for an instantaneous release of an axisymmetric current in the parameter range tested by the experiments.

The behavior of the rotating currents, upon release from the lock, began in a very similar manner to those in a non-rotating system, with a decelerating front spreading radially across the floor of the tank. Beyond a certain radius, however, the deceleration became noticeably more pronounced than for the equivalent non-rotating currents. Overall, the propagation appeared nicely axisymmetric. Locally, however, the leading edge displayed some instabilities. Measured values of the propagation as a function of time are illustrated in Figs 8.10, 8.12, and 8.14. A noticeable effect is that, as opposed to the non-rotating cases, none of the rotating currents reached the outer wall of the tank. Evidently, at a certain radius defined here as $r_{max}$, the forward motion at the front ceased, at which point the height of the current was extremely thin. This is in agreement with the SW predictions. Moreover, for all tested cases, the SW model predictions of $r_N(t)$ are in fair agreement with the experimental points until the maximum radius is attained. The SW theory predicts that afterward a contraction occurs.

Indeed, in the rotating experiments, for a period several seconds prior to the arrest of the initial leading edge, a reverse flow in the ensuing tail was evident, and the bulk volume of the current contracted and increased in thickness back towards the center of the tank. Remarkably, thereafter this newly accumulated central body of fluid relaxed and generated a second outwardly propagating pulse of fluid with a clearly defined leading edge that reached and sightly exceeded the previous arrest radius at $r_{max}$. This behavior was repeated several times, and at least five discrete contractions followed by outward pulses were observed in each experiment. The radius of the initial front, and the leading edge of subsequent pulses, is plotted as a function of time in Fig. 8.10 for a typical experiment. The analysis of the pulses leads to the conclusion that the mean frequency is $\omega_p = 2.10\Omega$, independent of the initial conditions (the subscript $p$ denotes the pulse). This seems to be a manifestation of the inertial oscillations in rotating fluids. The more precise nature of these pulses could not be well understood from the experimental observation. The NS simulations (see below) suggest that after the contraction of the current, a ring of fluid separates from the bulk, forms a new head, and propagates again. This may be a manifestation of the wave breaking, suggested by Killworth (1992) as an essential mechanism for energy reduction between the initial state and the final SL - but this topic requires more investigation. (We note in passing that Holford 1994 performed experiments of the process of lock-release formation of a SL for large values of $\mathcal{C}$ and recorded oscillations

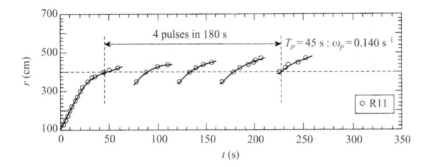

**FIGURE 8.10:** The radius of the leading edge of successive outward propagating fronts plotted as a function of time for experiment R11. The determination of the pulse frequency $\omega_p$ is shown. (Hallworth, Huppert, and Ungarish 2001, with permission.)

of the interface of frequency $1.9\Omega$ and of relatively small amplitude. In contrast, the amplitude in the experiments of Hallworth et al. 2001 was large, about one half the maximal radius.)

Let us consider in some detail the configurations labeled R5 and S3 in Table 8.1. The difference between the cases is the presence of rotation in the case R5. The standard inertial-buoyancy reference time is $T = 3.37\,\mathrm{s}$ and the period of revolution in R5 is $62.8\,\mathrm{s}$. The corresponding NS simulations were obtained with a numerical code of the type described in Appendix B. The numerical grid had typically $175 \times 240$ intervals. In dimensional form, the radial grid intervals were of uniform size $\delta r = 3.7\,\mathrm{cm}$, judged as acceptable because the typical length of the observed "head" was about 20 cm, and the estimated experimental error of its position is estimated as about 3–5 cm (the tank floor marks were at 10 cm intervals; see Fig. 8.9). The dimensional axial grid intervals changed from $\delta z = 0.26\,\mathrm{cm}$ near the bottom to $\delta z = 0.47\,\mathrm{cm}$ near the top free boundary; this is expected to provide a fair description of the thinning current during a considerable spread (the average thickness is about 4 cm when the container middle radius 3.25 m is reached at $t \sim 20\,\mathrm{s}$) and of the Ekman layer (whose thickness for water with $\Omega = 0.1\,\mathrm{s}^{-1}$ is approximately 1 cm). Fig. 8.11 displays the calculated shapes. Fig. 8.12 displays the behavior of the radius of propagation, $r_N(t)$ obtained by SW, NS, and experiments for these cases. The numerical code captures well the differences between the rotating and the non-rotating cases.

The NS simulations produce blurred and even patchy profiles of the interface (and other variables), which are quite different from the sharp and smooth $h(r,t)$ provided by the SW predictions. However, it seems that this type of

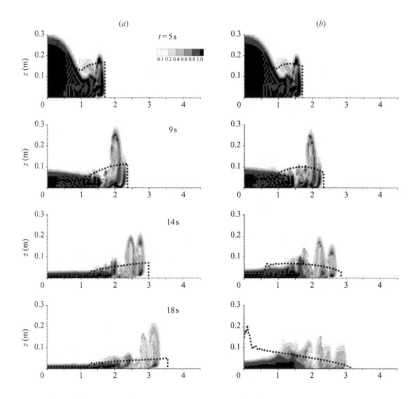

**FIGURE 8.11:** The NS results for the density function $\phi$ (contour lines and shading) and the SW results for the interface $h(r)$ (thick lines), at various times, for cases S3 (a) and R5 (b). $r$ and $z$ coordinates in m. (Hallworth, Huppert, and Ungarish 2001, with permission.)

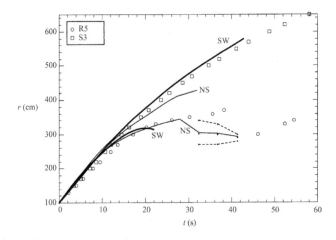

**FIGURE 8.12:** The radius of propagation as a function of time for experiments S3 and R5, compared with the corresponding NS and SW predictions. The dotted lines in the numerical R5 results indicate the thin spread of the head. (Hallworth, Huppert, and Ungarish 2001, with permission.)

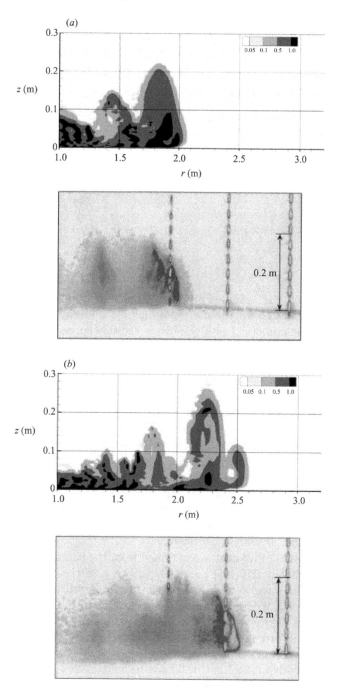

**FIGURE 8.13**: Experimental and numerical $rz$ profile for R14 at (a) $t \approx$ 10 s and (b) 16 s. (Hallworth, Huppert, and Ungarish 2001, with permission.)

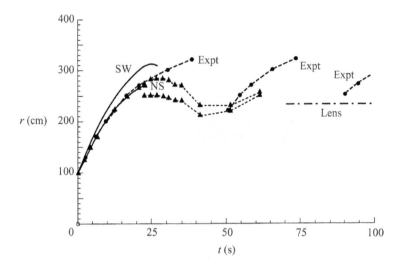

**FIGURE 8.14**:   The radius of propagation as a function of time for experiment R16, compared with the corresponding NS and SW predictions. The dotted lines in the numerical results indicate the thin spread of the head. The radius of the theoretical steady lens is also shown for $t \geq 75$. (Hallworth, Huppert, and Ungarish 2001, with permission.)

result indeed reflects the non-smooth behavior of a real gravity current. This is illustrated by the experiment R14, in which fluorescein and a laser sheet were used to visualize the $rz$ shape. Some video records are shown in Fig. 8.13 and compared with numerical results. The impression is that not only the position of the front, but also the shape of the head are in fair agreement. In particular, we notice that both the experiment and the computations show that the maximum height of the head increased by about 5 cm from $t = 10$ s to 16 s. (A more detailed comparison was not possible because the concentration of salt and fluorescein were not measured.)

Additional results of interest are obtained from the simulation corresponding to the experiment R16; see Fig. 8.14. In this case the current is initially deep, $H = 5$, in contrast with the previously discussed cases, S3 and R5, where $H = 1.7$ only. There is good agreement between the experimental and numerical results for $r_N(t)$ up to $t \approx 25$ s. (At this time the maximal expansion is achieved according to both the SW and NS results.) When the current is close to the maximum radius of propagation ($t = 20$ s) the nose region detaches from the body of the current. The latter remains roughly in the domain $0 \leq r \leq 250$ cm where, again, some contraction-expansion pulses appear. The second maximum of propagation is attained about 37 s after the first one, which suggests that the frequency of this occurrence is $0.17 \, \text{s}^{-1} \approx 2.2\Omega$, in agreement with the experimental observations.

The measurement of angular velocity inside the propagating current is a difficult task and reliable data are scarce. The NS simulations provide this information as a standard part of the solution. The results of Hallworth et al. (2001), see Fig. 12 in that paper, show that the dense fluid current has a very distinct "signature" as a region of strong counter-rotation (relative to the rotating system). This is in good qualitative agreement with the SW approximation. The strong counter-rotation in the core is also corroborated by the experiments of Stegner, Bouruet-Aubertot, and Pichon (2004) discussed below.

### 8.1.5.2 The lens

The study of Stegner, Bouruet-Aubertot, and Pichon (2004) presented convenient experimental data concerning the behavior of the lens. Boussinesq surface currents with $0.4 < \mathcal{C} < 1.8$ and $10 < H < 13$ were produced from locks of $3.45\,\mathrm{cm} \leq r_0 \leq 8.1\,\mathrm{cm}$ and visualized by fluorescein and laser sheets. The angular velocity was measured with particle image velocity (PIV) in both the lens and ambient fluid.

In all the tested cases $r_{\mathrm{max}}$ was achieved in about one-half revolution, then a contraction started. Overall, the lens structure was maintained for many revolutions of the system. The experiments were focused at the behavior after the first half-revolution. In general, the position of the interface, the azimuthal velocity, and the other properties were calculated from data averages over a period of time, usually one revolution.

Typical results are shown in Figs. 8.15 and 8.16. The mean motion (determined by particle trajectories averaged over one revolution) was axisymmetric. The measured thickness $h(0)$ was in very good agreement with the theoretical SL predictions, and the radius $r_N$ slightly larger. The angular velocity, measured with PIV technique, showed good agreement in the core. However, the maximum $|\omega|$ occurs in the interior, while at the edge of the lens $\omega$ decays to zero. (Note that $Bu$ and $\delta$ in that paper correspond to $(4\mathcal{C}^2)^{-1}$ and $H^{-1}$ in our notation.) Stegner, Bouruet-Aubertot, and Pichon (2004) also checked the potential vorticity of the SL. We recall that, theoretically, this variable is conserved at the value 2 in $[0, r_N]$ and decreases sharply to zero for $r > r_N$. In the experiments the decrease of the potential vorticity to zero is not sharp, and actually starts at a smaller radius than predicted by the theory. This "dissipation" of the potential vorticity becomes more pronounced and complex when $\mathcal{C}$ decreases. Also, the experimental dissipation of energy increased from about 12% to 50% when $\mathcal{C}$ decreased from 1.5 to 0.5. The mechanism of these dissipations is attributed to three-dimensional motions in the frontal region.

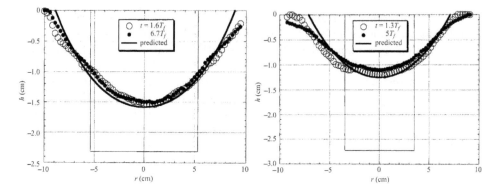

**FIGURE 8.15:** Profile of the density interface (averaged over one revolution) for: $\mathcal{C} = 0.8$, left frame; and $\mathcal{C} = 0.5$, right frame. The initial state confined within the cylinder is plotted with a thin solid line while the SL solution is plotted with a thick solid line. Here $T_f = \pi/\Omega$. (Stegner, Bouruet-Aubertot, and Pichon 2004 with permission.)

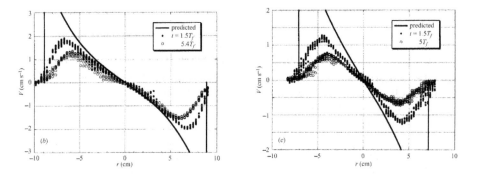

**FIGURE 8.16:** Angular velocity profiles within the lens (5mm below the free surface, averaged over one revolution) for: $\mathcal{C} = 0.8$, $\Omega = 0.88\,\mathrm{s}^{-1}$ (left frame); and (b) $\mathcal{C} = 0.5$, $\Omega = 0.92\,\mathrm{s}^{-1}$ (right frame). Other details as in the previous figure.

## 8.1.6 Summary

Overall, the analytical and numerical tools presented here provide a reliable description of the major behavior of the axisymmetric rotating current and associated (quasi) steady lens (SL). The understanding of the subsequent evolvement of this flow, and numerous fine details observed in practice, requires departure from the axisymmetric assumption. This is beyond the objectives of this text.

## 8.2   Rotating channel

In the previous section we considered the effects of rotation in the special case of axisymmetric currents. A significant feature of the axisymmetric flow is that the azimuthal motion is unrestricted by geometry; consequently, a significant angular velocity (in the rotating frame) accompanied by a strong Coriolis acceleration in the direction of propagation (radial) appear. The resulting radial spread is restricted.

Another important problem concerns the gravity current produced in a rectangular horizontal channel along the $x$ axis. The axis of rotation is the vertical $z$, and horizontal walls are placed at $y = 0$ and $y = D$. Suppose that the channel, of height $H$, is filled with co-rotating ambient fluid, into which a fixed volume of fluid of a slightly different density is released. The major propagation is in the $x$ direction while the lateral motion is restricted by the walls. In other words, there is a significant $u$ but a negligible $v$ speed component. Consequently, the major component of the Coriolis acceleration $2\Omega\hat{z} \times \mathbf{v}$ is now in the cross-channel direction $y$. To supply this acceleration, the current needs a $\partial p/\partial y$ pressure forcing, which is built up as follows: the current piles up, or leans, against the $y = 0$ wall and develops a significant inclination of the interface in the $y$ direction, $\partial h/\partial y$ . The result is a "boundary current" which has a restricted spread in the $y$ direction.

The typical flow arrangements that may appear in the leading domain of the current are sketched in Fig. 8.17. In this figure a surface current, $\rho_c < \rho_a$, is considered. We can distinguish between three cases: A, when the propagation is more or less like in the non-rotating case; B, when a significant inclination of the interface in the $y$ direction is present, and consequently the interface crops out at the top and the current is separated from the left channel wall $y = D$; and C, when the inclination of the interface is so steep that the current wets the bottom boundary. We argue that the particular case is determined according to the importance of the rotation. The dimensionless parameter which represents the importance of the rotation in the configuration of Fig. 8.17 can be estimated as follows.

Assume that the thickness of the current is $\sim H$ and the speed $\sim (g'H)^{1/2}$. The Rossby radius $L_{Ro}$ is defined by (8.4) with $h_0$ replaced by $H$. The dynamic argument associated with this variable remains relevant. In other words, the balance between Coriolis and pressure requires significant changes of the thickness of the current over a distance $\sim L_{Ro}$. However, in the present case the relevant geometrical length is the lateral width of the channel, $D$. Consequently, in analogy with (8.3), we expect that the importance of rotation is represented by

$$\mathcal{W} = \frac{2\Omega D}{(g'H)^{1/2}} = \frac{D}{L_{Ro}}. \tag{8.68}$$

The letter $\mathcal{W}$ can be attributed to the fact that this is a dimensionless width,

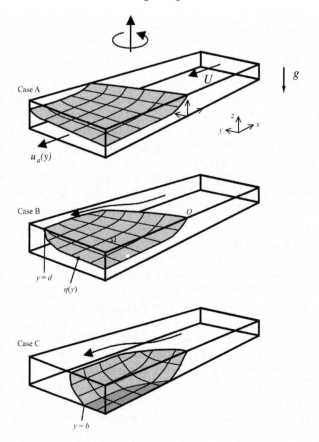

**FIGURE 8.17**: Typical flow geometries for a surface gravity current (shaded) in a rotating channel (of height $H$ and width $D$). The $\{x, y, z\}$ system is attached to the current, and hence the flow is steady. In the downstream domain the interface is parallel to the $x$ axis at $z = H - \eta(y)$. We denote $h_b = \eta(0)$. (Martin and Lane-Serff 2005 with permission.)

but this parameter has important dynamic significance which will be revealed soon. Again by analogy with the axisymmetric case, we expect that the Rossby number $Ro = 1/\mathcal{W}$. We keep in mind that the use of the channel depth $H$ in this definition may become problematic for a deep ambient; the use of the thickness of the current (in the lock, $h_0$, or at the right-hand-side wall, $h_b$) is more appropriate.

These flows received considerable experimental and theoretical attention, e.g., in Stern, Whitehead, and Hua (1982), Griffiths and Hopfinger (1983), Nof (1987), Hacker and Linden (2002), Martin and Lane-Serff (2005), and Helfrich and Mullarney (2005) where additional pertinent references are given. However, the details are complicated and compact analytical results are scarce. The discussion here attempts to highlight some basic effects and results.

## 8.2.1   Steady-state results

The problem of a *steady-state* current with rotation is an extension of Benjamin's analysis presented in Chapter 3. The flow field is sketched in Fig. 8.17, for a surface current ($\rho_c < \rho_a$). The current is assumed to move like a slug with constant speed $U$ with respect to the channel, and the interface of the tail is parallel to the $x$ axis. For definiteness, we denote the directions with respect to a point in the current: the upstream is the direction of unperturbed ambient (positive $x$ direction), and the right- and left-hand-side walls are with respect to an observer facing upstream.

The analysis is performed in a system of coordinates $\{x, y, z\}$ attached to the current. In this system the flow is in steady-state. This is like in Benjamin's analysis; see Chapter 3. However, in the present case the frame is not only translating, but also rotating and accelerating. The acceleration of the (origin) of this frame is $2\Omega U \hat{z} \times \hat{x} = 2\Omega U \hat{y}$. This is, simply, the Coriolis acceleration of a point which moves with speed $U$ in the $x$ direction in the frame rotating with $\Omega$ about axis $z$. With this addition to (8.10), we obtain the momentum equation for the motion in the channel as

$$\rho[\frac{D\mathbf{v}}{Dt} + 2\Omega\hat{z} \times \mathbf{v} + 2\Omega U \hat{y}] = -\nabla p - (\rho - \rho_a)[g\hat{z} - \Omega^2(x\hat{x} + y\hat{y})] + \mu\nabla^2\mathbf{v}, \quad (8.69)$$

where $p$ is the pressure reduced with $+\rho_a[gz - (1/2)\Omega^2(x^2 + y^2)]$. The volume and density conservation equations are as in the non-rotating case. In the steady-state the $\partial/\partial t$ terms are zero.

The simplifications of (8.69) invoke, again, the inviscid $Re \gg 1$ and Boussinesq $\epsilon \ll 1$ assumptions. In addition, we use the observation (or assumption) that in many cases of interest $\Omega^2 D/g \ll 1$, and hence the centrifugal force of the rotating frame can be neglected. (Roughly, this can be expressed as $\epsilon \mathcal{W}^2 \ll 1$. The validity of this simplification for a certain application with given parameters can be verified *a posteriori* using the results.) In other words, in the systems considered here the dominant driving effect in the momentum balances is the reduced gravity term, $-(\rho - \rho_a)g\hat{z}$. Next, the shallow-water approximations can be introduced, and the $z$ momentum equations are approximated by the hydrostatic balances. The averaged velocity components $u$ and $v$ are functions of $x$ and $y$ only.

We are now at the position to understand the formation of the three different types of flow seen in Fig. 8.17. We postulated that $\mathbf{v}_c = \mathbf{0}$ (the slug-like current is motionless in our frame of reference) and that, in the downstream (tail) domain, the flow is parallel, i.e., independent of $x$. We use dimensional variables. Let $\Delta\rho = \rho_c - \rho_a = -|\Delta\rho|$. The thickness of current in the downstream domain is $\eta(y)$, and $y = 0$ at the right-hand-side wall. We assume that the mean thickness is of the order of magnitude of $H$.

The simplified form of the momentum balances (8.69) for the motionless current of density $\rho_c$ yield

$$p_c = -\Delta\rho g z - \rho_c 2\Omega U y + C. \tag{8.70}$$

We focus attention on the downstream domain where the interface between the fluids is given by $z = H - \eta(y)$. Using (8.70) we find that the pressure at this interface is

$$p_{ci}(y) = -|\Delta\rho|g\eta(y) - \rho_c 2\Omega U y + C_1. \tag{8.71}$$

For the ambient fluid the $z$-momentum equation is simply $\partial p_a/\partial z = 0$, and the pressure continuity at the interface requires $p_{ai} = p_{ci}(y)$. The result, in view of (8.71), is

$$p_a(y) = -|\Delta\rho|g\eta(y) - \rho_c 2\Omega U y + C_1 \tag{8.72}$$

in the entire downstream domain.

The difference between the cases A, B, and C in Fig. 8.17 concerns mainly the lateral spread of the current, and hence the explanation is sought in the $y$-momentum equation; see (8.69). For the ambient fluid the $y$-balance reads

$$2\rho_a\Omega u_a + \rho_a 2\Omega U = -\frac{\partial p_a}{\partial y}. \tag{8.73}$$

Substitution of (8.72) and arrangement give

$$2\Omega u_a = g'\frac{d\eta}{dy} - \epsilon\,2\Omega U, \tag{8.74}$$

where $\epsilon = |\Delta\rho|/\rho_a$ and $g' = \epsilon g$. Since $u_a$ is of the order of magnitude of $U$, the last term in (8.74) can be discarded in the present $\epsilon \ll 1$ analysis.

We obtained a quite explicit expression for the slope of the interface. First, the slope $d\eta/dy$ is negative because the stream $u_a$ of the ambient is in the negative $x$ direction. This proves that the current tends to pile up adjacent to the right-hand-side wall, as depicted in Fig. 8.17. Finally, we consider the order of magnitude of the variables. $U$ and $u_a$ are $\sim (g'H)^{1/2}$, $\eta \sim H$, and $y \sim D$. The meaningful slope of the interface should be scaled by $H/D$. With these scalings we can write

$$\text{slope of interface} \quad = \frac{D}{H}\frac{d\eta}{dy} \sim \frac{2\Omega D}{(g'H)^{1/2}} = \mathcal{W}. \tag{8.75}$$

This proves that the parameter $\mathcal{W}$ reflects the importance of rotation in this problem and dominates the geometry of the current with respect to the channel. The parameter $\mathcal{W}$ is thus similar to $\mathcal{C}$ defined by (8.3), an inverse Rossby number, but concerning the lateral Coriolis acceleration. We showed that the interface profile $\eta(y)$ acquires the shape needed to supply the Coriolis acceleration by the pressure gradient $\partial p/\partial y$. If $\mathcal{W}$ is small, a small lateral inclination is needed and the current occupies the full width of the channel. This is case A of "weak rotation." In the "significant rotation" case B, the value of $\mathcal{W}$ is so large that the current occupies only part of the width of the channel. The interface outcrops at the top at $y = d$, but the current still occupies only part of the depth of the channel. In case C, for "strong rotation," large $\mathcal{W}$, the slope of the interface is so large that it outcrops on both the top (at $y = d$) and bottom (at $y = b$).

Another compact result can be derived from the conservation of the potential vorticity. The $z$-component of the vorticity is given by

$$\zeta = \hat{z} \cdot \nabla \times \mathbf{v} = \frac{\partial v}{\partial x} - \frac{\partial u}{\partial y}. \tag{8.76}$$

We follow a "particle" (column) of ambient fluid of height $h_a$ between far upstream and downstream position. Since there is no $x$ dependence at these positions, only the last term of (8.76) is relevant. We write the conservation of $(2\Omega + \zeta_a)/h_a$ as

$$\frac{2\Omega}{H} = \left(2\Omega - \frac{\partial u_a}{\partial y}\right) \frac{1}{H - \eta(y)}, \tag{8.77}$$

which yields

$$\frac{\partial u_a}{\partial y} = 2\Omega \frac{\eta(y)}{H}. \tag{8.78}$$

For the interpretation of this result we must recall that $u_a < 0$ and that the thickness of the current, $\eta$, decreases with $y$ (from right to left). The conclusion is that the speed of the ambient relative to the current, $|u_a|$, also decreases from right to left. This is consistent with volume continuity expectation: a column of fluid moves slower when the height is larger. Moreover, if we scale $u_a$ and $y$ as before, we obtain that the dimensionless slope of this speed with respect to the lateral coordinate is also proportional to $\mathcal{W}$. The details are left for an exercise.

Equations (8.74) and (8.78) can be combined into a single equation for the thickness

$$D^2 \frac{d^2\eta}{dy^2} - \mathcal{W}^2 \eta = 0 \tag{8.79}$$

(here the definition (8.68) was also used). The formal solution for $\eta(y)$ is straightforward, and then by use of (8.74) the solution $u_a(y)$ follows. The result is

$$\eta(y) = \eta(0)\cosh Y + u_a(0)\sinh Y; \tag{8.80}$$
$$u_a(y) = \eta_a(0)\sinh Y + u(0)\cosh Y; \tag{8.81}$$

where

$$Y = \mathcal{W}\frac{y-b}{D} \quad (0 \le y \le d). \tag{8.82}$$

In the $y$ domain not covered by (8.80), the thickness $\eta$ is known (either 0 or $H$) and $u_a$ is constant. Several other unknown constants are involved: $\eta(0); u_a(0);$

the width of the current, $d$ which is in the range $(0, D]$, and the width of the contact region with the bottom, $b$, which is in the range $[0, d)$. Additional conditions must be imposed to determine a more meaningful solution.

In the spirit of Benjamin's analysis, the main task is to determine the speed $U$ at which a steady-state can be maintained for given $\mathcal{W}$ and thickness of the current (downstream, in the tail). In the present configuration the thickness of the current varies with $y$. It is therefore convenient to use the thickness $h_b = \eta(0)$ as the representative thickness of the steady-state current. Here the subscript $b$ means boundary, namely the right-hand-side wall that supports (or pushes) the current. According to the previous results, $h_b$ is the maximum thickness of the current; in case C, $h_b = H$. The relative depth ratio of current to ambient we now express by $a_b = h_b/H = \eta(0)/H$.

The backbone of the subsequent analysis is the use of volume and $x$-momentum (flow-force) balances in a box control volume whose vertical boundaries ($yz$ planes) are at a far-upstream (say $x_2 > D$) and a far-downstream (say $x_1 < -D$) position, in the spirit of Benjamin's solution. The flow-force equation is an integral form of (8.69). Additional assumptions are that the potential vorticity in the ambient is materially conserved (as discussed above), and that the energy loss is uniform across the stream. (The relaxation of these assumptions does not appear to have a significant effect on the main flow results such as the speed and geometry of the interface; see Martin, Smeed, and Lane-Serff 2005.)

The derivation of the quantitative results involves manipulations and numerical processing which are beyond our scope; see Martin and Lane-Serff (2005). An important outcome is that, for given $\mathcal{W}$ and $a_b$ (denoted $\eta_0$ in that paper), there is a unique physical value of the Froude number

$$Fr = \frac{U}{(g'h_b)^{1/2}}. \tag{8.83}$$

This is displayed in Fig. 8.18. As expected, Benjamin's non-rotating two-dimensional results are recovered for $\mathcal{W} \ll 1$. Surprisingly, perhaps, the rotation increases the value of $Fr$. This can be attributed to the effect of the $y$-inclination; $h_b$ is defined at the right-hand-side wall, and $\eta(y)$ decreases with $y$, thus facilitating the flow of the ambient fluid. The influence of the parameter $\mathcal{W}$ decreases as $a_b$ becomes small. For $a_b \to 0$ the classical $Fr = \sqrt{2}$ is recovered for any final $\mathcal{W}$.

Broadly speaking, this steady-state theory indicates that the speed of propagation of a typical current in a rotating channel is given by $\beta(g'h_b)^{1/2}$ where $\beta$ is a numerical coefficient close to 1. Indeed, experimental studies which related the speed of propagation to the thickness on the sidewall reported values of $\beta$ in the range $1.0 - 1.3$ (Stern, Whitehead, and Hua 1982, Griffiths and Hopfinger 1983, Helfrich and Mullarney 2005). The experiments are obviously more complicated than in the non-rotating counterpart. In the real flows the interface is blurred by instability and even turbulence, and the

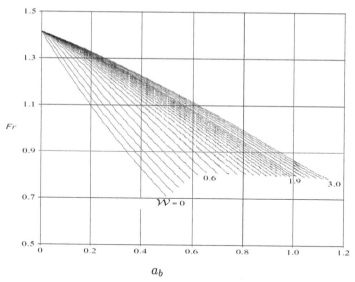

**FIGURE 8.18:** $Fr$ as a function of $a_b = h_b/H$ in a rotating channel for various $\mathcal{W}$. (Martin and Lane-Serff 2005 with permission.)

determination of the position (and value) of $h_b$ involves uncertainties. Moreover, we must keep in mind that the idealized steady-state result is only a non-rigorous approximation to the real flows produced in the experiments, typically by lock-release of a fixed volume of fluid. Indeed, the steady-state current is an abstraction, and we would like to know if a flow of this type will appear in a realistic current produced by dam-break, and if yes what is the connection between the local conditions and these in the reservoir which has generated the flow. This topic is discussed below.

## 8.2.2 Dam-break considerations

In the non-rotating case, the slumping stage of a rectangular dam-break two-layer problem reproduces the steady-state assumptions of Benjamin. This equivalence is discussed in Chapter 5 (§ 5.5.3 in particular). Moreover, Benjamin's results can also be derived directly as a jump condition. Similar two-layer results for the rotating current are yet unavailable. However, there are indications that a dam-break released current in a rotating channel contains a flow domain, behind the nose, where the foregoing steady-state analysis is relevant. This also implies that the speed of propagation can be correlated to the thickness of the current by a nose jump condition of the type $Frh_N^{1/2}$ (subject to proper scalings and incorporation of the parameter $\mathcal{W}$).

The one-layer Boussinesq SW model is, again, a useful theoretical tool for the investigation of the main effects. We shall consider some details of the formulation and results of the dam-break problem in this framework. We follow the study of Helfrich and Mullarney (2005). As in the previous section, a surface current ($\rho_c < \rho_a$) is considered. The coordinate system $\{x, y, z\}$ is attached to the channel and rotating with $\Omega\hat{z}$. The velocity in this system is $\mathbf{v} = u\hat{x} + v\hat{y} + w\hat{z}$. Let the origin of the coordinates be at the lower right point of the dam, and let $h(x, y, t)$ be the thickness of the current, i.e., $z = H - h(x, y, t)$ is the position of the interface (except for the vertical jump which is assumed to occur at $x_N(t)$). The fluid is released from a lock of length $x_0$, height $h_0$, and same width as the channel, $D$. See Fig. 8.19.

The formulation is in terms of dimensionless averaged variables defined as follows. We use $g' = |\rho_c/\rho_a - 1|g$. First, since the longitudinal and vertical motions are expected to be driven by the same mechanism as in the non-rotating case, we take

$$\{x^*, z^*, h^*, H^*, u^*, t^*, p^*\} = \{x_0 x, h_0 z, h_0 h, h_0 H, Uu, Tt, \rho_a U^2\} \quad (8.84)$$

where

$$U = (g' h_0)^{1/2}; \quad T = x_0/U. \quad (8.85)$$

Second, the lateral motion is expected to be dominated by Coriolis effects. The relevant length scale is the Rossby radius, and the speed scale follows from the continuity equation. The respective scales are

$$L_{Ro} = \frac{(g' h_0)^{1/2}}{2\Omega}; \quad V = \frac{L_{Ro}}{x_0} U. \quad (8.86)$$

We therefore use for the lateral ($y$) direction

$$\{y^*, D^*, d^*, v^*\} = \{L_{Ro}\, y, L_{Ro}\, \mathcal{W}, L_{Ro}\, d, V v\}. \quad (8.87)$$

Here

$$\mathcal{W} = \frac{D}{L_{Ro}} = \frac{2\Omega D}{(g' h_0)^{1/2}} \quad (8.88)$$

is a straightforward redefinition of (8.75). This parameter enters in a natural way in the lock release (dam-break) problem as the dimensionless width of the channel.

The derivation of the one-layer SW momentum equations is left for an exercise. The starting point for the momentum equations can be (8.69) (but without the translation acceleration term), and we recall that now $\mathbf{v}_a = \mathbf{0}$. The continuity equation is an extension of (2.5) for the case in which $h$ depends on both $x$ and $y$. The dimensionless governing equations are

$$\frac{\partial h}{\partial t} + \frac{\partial}{\partial x}(uh) + \frac{\partial}{\partial y}(vh) = 0; \quad (8.89)$$

$$\frac{\partial u}{\partial t} + u\frac{\partial u}{\partial x} + v\frac{\partial u}{\partial y} - v = -\frac{\partial h}{\partial x}; \quad (8.90)$$

$$\delta^2 \left[ \frac{\partial v}{\partial t} + u\frac{\partial v}{\partial x} + v\frac{\partial v}{\partial y} \right] + u = -\frac{\partial h}{\partial y}; \quad (8.91)$$

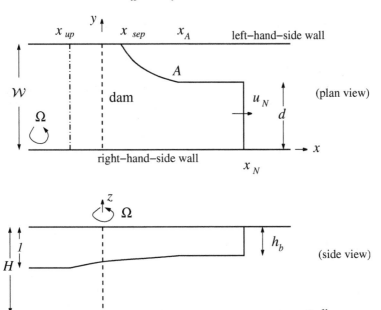

**FIGURE 8.19:** Sketch of the surface plan view and side view of a dam-break flow at some time after removal of the dam (the dashed line). The backward propagating wave is at $x_{up}$. The forward (from the reservoir) solution separates from the left-hand-side wall at $x_{sep}$ and meets the nose bulk at $x_A$. The quasi-steady nose is of width $d$ and terminates as a shock at $x_N$. The dimensional width of the channel is $D = \mathcal{W}L_{Ro}$. (Helfrich and Mullarney 2005 with permission.)

where

$$\delta = \frac{L_{Ro}}{x_0} = \frac{1}{\mathcal{W}} \frac{D}{x_0}. \tag{8.92}$$

The formulation is in terms of the variables of the current, $h$, $u$, and $v$ as functions of $x, y$, and $t$. The initial conditions are $u = v = 0$ and $h = 1$ in the lock at $t = 0$, and an obvious boundary condition is $v = 0$ at the right-hand-side wall $y = 0$.

The other boundary conditions are more complex. The current can separate from the left-hand-side wall at $x = x_{sep}(t)$ when $\mathcal{W}$ is sufficiently large. This creates a left-hand-side boundary where $h$ becomes 0, i.e., the interface crops out to the top. A more dramatic behavior appears at the nose which is created by the lift of the gate. The initial vertical discontinuity can (and will) propagate in the $x$ direction as a shock-front $x = x_N(t)$. The typical behavior is sketched in Fig. 8.19.

In general, the system (8.89)-(8.91) must be solved numerically. However, the limit $\delta \to 0$ (i.e., when the Coriolis term dominates the dynamic behavior in the $y$ direction) yields a simplified formulation, called the semigeostrophic (or long wave) theory. In this case analytical dam-break solutions can be derived by the method of characteristics. The details are omitted here; the reader is referred to Helfrich, Kuo, and Pratt (1999) and Helfrich and Mullarney (2005) (note that the present $u_N$ is denoted $c_b$ in these papers). These results justify and quantify the behavior sketched in Fig. 8.19. Moreover, the leading portion of this time-dependent current is in fair agreement with the results in the downstream domain of the steady-state current considered in § 8.2.1. The theory develops the necessary conditions of the jump at $x = x_N(t)$, and shows that the correlation

$$u_N = \beta h_b^{1/2}, \tag{8.93}$$

where $\beta$ is a constant of about 1.4, is a good approximation for both attached and separated flows. This is in fair agreement with the value $\beta \approx 1.2$ suggested by experimental data.

Like in the non-rotating two dimensional case, the SW current predicted by (8.89)-(8.91) displays a slumping stage in which the propagation is with constant speed, and the domain behind the front behaves like a steady-state slug parallel to $x$. The solution for the speed of propagation of the current, $u_N$, and the height of the domain trailing the nose can be obtained essentially like in the classical two-dimensional case: by intersecting the solution on the characteristics from the reservoir with the (quasi) steady-state conditions in the nose region. The steady-state results of § 8.2.1 become, again, a jump condition for the real current. In other words, the blunt nose of the current in Fig. 8.19 is a compression of the transfer from the upstream to downstream behavior in Fig. 8.17. For a given $\mathcal{W}$ the connection between the speed of propagation and $h_b$ (dimensional) is the same, but in the dam-break problem we are also able to determine $h_b$ as a function of $h_0$.

Some useful results are shown in Fig. 8.20 as functions of the dominant parameter $\mathcal{W}$ defined by (8.88). The front-shock speed $u_N$ and the corresponding height $h_b$ are smallest for $\mathcal{W} = 0$ and increase monotonically (but quite mildly) with $\mathcal{W}$. The interpretation of Fig. 8.20(b) requires some care. The current is attached to the left-hand-side wall (i.e., fills the channel) when $d = \mathcal{W}$. The separation occurs for $\mathcal{W} > 0.45$ (approximately). Once separated, $d$ grows only slightly with $\mathcal{W}$, but we must keep in mind that now $d$ is scaled with $L_{Ro}$. Thus, the width of the head of the separated current is approximately $0.7L_{Ro}$. The height of the ambient, $H$, does not enter into the results. In a one-layer model $H$ can influence via a nose condition with $Fr$ dependent on $h_N/H$, but here a constant $Fr$ was used.

The SW theory results are in fair agreement with experimental data. In general, the measured width of the current and the thickness $h_b$ are smaller than predicted. The experimental data show a slow decay of $u_N$ with time during the (theoretical) slumping interval, and this effect becomes more pronounced as $\mathcal{W}$ increases. We recall that the typical experiments in non-rotating chan-

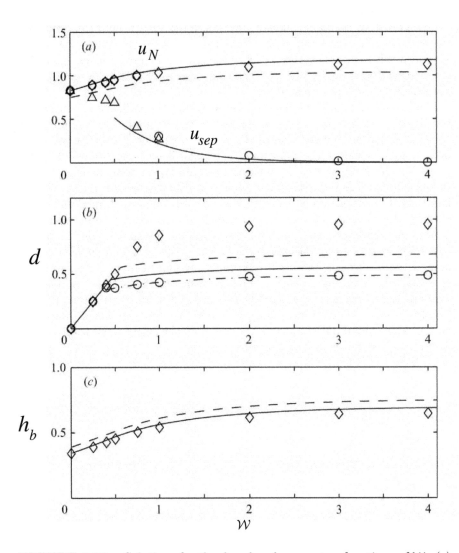

**FIGURE 8.20:** Solutions for the dam-break current as functions of $\mathcal{W}$. (a) speeds $u_N$ and $u_{sep}$ (scaled with $(g'h_0)^{1/2}$); (b) width of the head (scaled with $L_{Ro}$); and (c) thickness of head at the wall $h_b$ (scaled with $h_0$). The dashed and solid lines are solutions of the semigeostrophic model ($\delta \to 0$): the dashed line is for the closure $u_N = 1.2h_b^{1/2}$, and the solid line for a more rigorous formula suggested by Kubokawa and Hanawa (1984). The dash-dot line in (b) gives the width at which the gravity current depth is $h = 0.1$. The symbols are numerical results for the SW model with $\delta = 1$ which correspond to the semigeostrophic cases. (Helfrich and Mullarney 2005 with permission.)

nels reveal a constant $u_N$ for several lock-lengths. The reason for this discrepancy seems to be the lateral viscous friction on the right-hand-side wall, but the centrifugal acceleration of the system may also contribute to this effect. We finish this rather short discussion of the current in a rotating channel with the observation that the behavior after the dam-break and slumping stages are still largely un-investigated.

Finally, we conclude this chapter by stressing the difference concerning the Rossby number, $Ro$, between the rotating currents in the axisymmetric (free) propagation and the channel (boundary restriction) propagation. The interesting behavior of the axisymmetric current is for large $Ro$ (small $\mathcal{C}$); for small $Ro$, the stable motion is confined to a thin layer about the edge $r = r_0$. In the channel, the interesting behavior appears for small $Ro$ (large $\mathcal{W}$); for large $Ro$, the propagation is like in the non-rotating two-dimensional case. Clearly, the presence of the vertical wall is the reason for this remarkable dynamical difference: this wall enables a lateral pressure gradient, $\partial p/\partial y$, and thus supports the Coriolis acceleration needed for a significant radial flow, without affecting the basic solid-body angular velocity. (We can think of a blade in a turbine analogy.) On the other hand, in the axisymmetric case the lateral pressure gradient $\partial p/\partial \theta = 0$, and hence there is no agent to supply the lateral Coriolis acceleration needed to keep a significant radial flow on a straight radial path; a significant radial displacement must involve large deviations from the solid-body rotation.

# Chapter 9

## Buoyancy decays: particle-driven current, porous boundary, and entrainment

In ideal situations the gravity current fluid has a fixed density contrast to the ambient, $\rho_c - \rho_a$, and is enclosed in a well-defined, smooth, and impermeable envelope of volume $\mathcal{V}$. In practical flows the density contrast and volume of the current may be reduced due to various effects. Overall, the global buoyancy $|\rho_c/\rho_a - 1|g\mathcal{V}$ decays with time. Here we consider some typical circumstances in which this decay of buoyancy is so severe that the *inviscid* current "runs out" of (most of) the driving effect and actually stops (or vanishes) after a finite distance of propagation. The viscous friction plays a minor role in the pertinent momentum balances.

## 9.1  Particle-driven currents

If the current fluid is a mixture (suspension) of heavier particles in essentially the same interstitial fluid as the ambient, this is called a particle-driven current; while the current spreads, particles settle out and the effective driving strength of the current, as compared with a homogeneous current, decays. Particle-driven gravity currents are important in many geological and environmental settings; in these applications, they are usually called turbidity currents (when the interstitial fluid is a liquid, usually water) and pyroclastic currents (when the interstitial fluid is a gas); for details see Kneller and Buckee (2000), where additional references are given. The particle-driven current of fixed initial volume runs out of particles after a finite distance of propagation. Typically, the current stops and the interstitial fluid vanishes in the ambient fluid. However, the settled particles form a thin and persistent sediment (deposit) along the path of propagation which may prevail for a very long time after the current stopped. We follow the presentation of Bonnecaze, Huppert, and Lister (1993) and Ungarish and Huppert (1998).

### 9.1.1  Motion and concentration of particles

A small "heavy" particle of density $\rho_p$ and radius $a_p$, released from rest in a quiescent fluid of density $\rho_i$ and kinematic viscosity $\nu$, attains the Stokesian settling speed

$$W_s = \frac{2}{9}\varepsilon_p \frac{a_p^2}{\nu}g, \qquad (9.1)$$

in the $-\hat{z}$ direction, where

$$\varepsilon_p = \frac{\rho_p - \rho_i}{\rho_i}. \qquad (9.2)$$

The derivation of this simple drag-buoyancy result assumes a very small particle Reynolds number, $W_s a/\nu$, and a spherical particle. For details see Batchelor (1981).

For definiteness we assume $\varepsilon_p > 0$, i.e., "heavy" particles. A typical particle-driven current contains thousands of particles per $\text{cm}^3$. The presence of so many particles makes the current a two-phase fluid, or suspension. The concentration of the "dispersed" phase (the particles) is expressed by the volume fraction $\alpha(x, z, t)$ and the concentration of the "continuous phase" (the interstitial fluid) is $1 - \alpha$. For typical illustrative purpose, let us think of particles of radius $20\mu$ m, $\rho_p = 3.2\ \text{g cm}^{-3}$ (silicone carbide) suspended in water at $\alpha = 0.01$ volume fraction. It is left for an exercise to calculate $W_s$, the particle Reynolds number, and the number of particles in a reservoir of $20 \times 20 \times 20\ \text{cm}^3$.

In general, the phases move with different speeds and a separation process accompanies the main motion of the bulk. The relative speed of the dispersed phase with respect to the continuous phase is $\mathbf{v}_R$. This is similar to the diffusion in a binary gas, but driven by a macroscopic buoyancy-drag balance, not by the molecular Fick's law. Actually, in many cases of interest $\mathbf{v}_R \approx W_s(1-\alpha)^{-3.5}\hat{z}$, at least near the bottom boundary. In the presence of vigorous mixing this approximation must be adjusted. The term $(1-\alpha)^{-3.5}$ represents the mutual hindering effect between the particles, because (1) The effective viscosity increases like $(1 - \alpha)^{-2.5}$ (an extension of the famous formula of Albert Einstein); and (2) The effective buoyancy decreases like $(1 - \alpha)$ (the particles "feel" a fluid of effective density $\rho_c > \rho_i$; see (9.12)). A more detailed discussion can be found in Ungarish (1993).

The addition of the new variables $\alpha$ and $\mathbf{v}_R$ to the gravity current problem requires two more equations. Moreover, the original Navier-Stokes balances must be reconsidered, because differences between mass and volume velocities appear and the viscosity may also be affected by $\alpha$. These effects can be taken into account, see Ungarish (1993), but the complications are beyond the scope of most available studies on gravity currents. Fortunately, a simplified framework is available: the dilute ($\alpha \ll 1$) and Boussinesq ($\varepsilon_p\alpha \ll 1$) suspension. The necessary assumption is that $\alpha$ is sufficiently small. Note, however, that $\varepsilon_p$ may be large when the suspending fluid is a gas; in these cases the restriction of $\alpha$ may be so severe that the following approach, based on continuum equations, is physically irrelevant.

In the *simplified* framework the inviscid Navier-Stokes Boussinesq equations are valid, and in particular the continuity equation $\nabla \cdot \mathbf{v} = 0$. The speed of the dispersed (particle) phase is given by the rather intuitive

$$\mathbf{v}_p = \mathbf{v} + \mathbf{v}_R. \tag{9.3}$$

The conservation, or "diffusion," equation for the volume fraction in the suspension which is

$$\frac{\partial \alpha}{\partial t} + \nabla \cdot (\alpha \mathbf{v}_p) = 0 \tag{9.4}$$

can now be expressed as

$$\frac{\partial \alpha}{\partial t} + (\mathbf{v} + \mathbf{v}_R) \cdot \nabla \alpha = -\alpha \nabla \cdot \mathbf{v}_R. \tag{9.5}$$

The specification of $\mathbf{v}_R$ leads to two different SW models. The "laminar" (L) model argues that (9.1) is relevant everywhere in the propagating fluid (Ungarish and Huppert 1998). The "turbulent" (T) model assumes that in the propagating bulk of suspension there is vigorous mixing over the whole depth $0 < z < h(x,t)$. Consequently, particles at $z = h$ are actually moving with the fluid and settling with (9.1) occurs at the bottom only (in between $\mathbf{v}_R$ varies linearly with height). This model has been widely used (e.g., Einstein 1968, Bonnecaze, Huppert, and Lister 1993, Harris, Hogg, and Huppert 2002). We write

$$\mathbf{v}_R = \begin{cases} -W_s \hat{z} & (\text{L model}) \\ -W_s \left[1 - \dfrac{z}{h(x,t)}\right] \hat{z} & (\text{T model}). \end{cases} \tag{9.6}$$

We substitute (9.6) into (9.5). We obtain a relatively simple hyperbolic equation for $\alpha(x, z, t)$ which can be solved formally by the method of characteristics. The practical initial condition is $\alpha = \alpha_0$ for all the relevant characteristics, because the typical gravity current is released from a well-mixed reservoir. An inspection of the results reveals that: (1) In the L case $\alpha$ remains constant (this conclusion is valid for a general $\mathbf{v}(x, y, z, t)$ field); and (2) In the T case, for a SW $u(x, t)$ velocity of the fluid, the value of $\partial \alpha / \partial z$ remains zero. We summarize: the particle volume fraction in the 2D SW current is given by

$$\alpha = \alpha_0 \qquad (\text{L model}) \tag{9.7}$$

$$\frac{\partial \alpha(x,t)}{\partial t} + u(x,t)\frac{\partial \alpha(x,t)}{\partial x} = -\alpha(x,t)\frac{W_s}{h(x,t)} \qquad (\text{T model}). \tag{9.8}$$

Outside the current, $\alpha = 0$.

The verification and validation of there results is still an open question. Direct experimental measurements of particle concentration are difficult to perform. The available numerical simulations are also problematic. For example, Necker, Härtel, Kleiser, and Meiburg (2002, 2005) presented two- and three-dimensional high-resolution numerical simulations of a dilute Boussinesq suspension in a channel. The Navier-Stokes equations were supplemented with the particle conservation equation (9.5). However, the value of $\mathbf{v}_R$ itself was not calculated; the simulations used the closure $\mathbf{v}_R = -W_s \hat{z}$. In other words, the L model is implicitly incorporated in the results of these simulations. The simulations of deep currents reproduce the analytical $\alpha = \alpha_0$ result. However, for full-depth lock ($H = 1$ dimensionless) cases a significant departure from the sharp $\alpha_0$ structure develops in the numerical results after propagation to about $x_N = 7$ lock lengths. In the three-dimensional simulation this can be attributed to the fact that the interface displays a strong spanwise wavy structure and motion, and consequently the spanwise averaged $\alpha$ is not sharp. The influence of some small spurious numerical diffusion may also be present at large times. The numerical resolution between the L and T models requires additional investigation.

We keep in mind that the L and T models agree about the following property: in the current $\alpha$ does not depend on $z$ (under the assumption of an initially well-mixed suspension). Consequently, in the development of the SW momentum balance it is justified to use the simplification $\alpha = \alpha(x, t)$. Now the SW interpretation to (9.8) is evident: Take a control volume of length $dx$ in the current. The particle volume $\alpha h \, dx$ changes with time due to the horizontal fluxes $\alpha h u$ and the settling flux $\alpha W_s \, dx$ at the bottom. This balance, combined with the usual $\partial h/\partial t + u(\partial h/\partial x) = 0$, yields (9.8).

In both models the particle flux $\alpha W_s$ leaves the current at the bottom $z = 0$ and is assumed to settle as a motionless sediment. The volume fraction in the sediment is the maximum packing fraction, $\alpha_M$, close to 0.65 for spherical solid particles. The ratio of volumes of sediment to current is $\alpha_0/(1 - \alpha_M)$, and hence the thickness of the sediment, $h_s$, is negligibly small as compared with $h(x, t)$ in the dilute case. The consistent simplification is to apply the boundary conditions of the current at $z = 0$. In particular, the no-penetration condition for the fluid is expressed as $w = 0$ at $z = 0$.

## 9.1.2    The motion of the interface

We discussed the essential motion of the interface between the current and the ambient in § 2.1. The assumption was that there is a clear-cut density difference, but continuity of the normal velocity component, $\mathbf{v} \cdot \hat{n}$, at the interface. The intuitive understanding is that the particle-driven current is marked by the presence of the particles. In other words: the interface between the suspension and clear fluid, or the "last particle" seen from inside the suspension, provides the boundary of the current. This definition requires some care when $\mathbf{v}_R \cdot \hat{n} \neq 0$. This means that the particles and the interstitial fluid may encounter the ambient fluid with different speeds.

Since $\mathbf{v}_R$ has no component in the horizontal direction, this effect is irrelevant to the position of $x_N(t)$. Consequently the propagation of the interface at the nose is given like for the homogeneous fluid by (2.3a). In this respect there is no difference between the L and T models.

Consider now the inclined part of the interface, $z = h(x, t)$. When no particles are present, the motion is given by (2.3b), repeated here,

$$\frac{\partial h}{\partial t} + u_h \frac{\partial h}{\partial x} - w_h = 0, \tag{9.9}$$

where $u_h$ and $w_h$ are the velocity components of the fluid at the interface. Now the interface moves with the local components of $\mathbf{v}_p = \mathbf{v} + \mathbf{v}_R$. The necessary modification of (9.9) depends on the $\mathbf{v}_R$ model.

The settling models are sketched in Fig. 9.1. The T model, see (9.6), postulates that $\mathbf{v}_R$ vanishes at $z = h$. The encounter of the ambient fluid is like with a homogeneous fluid, represented by a single velocity for both interstitial fluid and particles. The motion of the interface is given again by the unchanged (9.9). Consequently, the SW continuity equations are like for a homogeneous current; in particular, (2.5) and (2.24) remain valid.

On the other hand, the L model postulates that the particles settle out from the interstitial fluid also at the interface; see (9.6). In (9.9) the horizontal velocity remains unchanged, but the component $-W_s$ must be added to the vertical $w_h$. This yields

$$\frac{\partial h}{\partial t} + u_h \frac{\partial h}{\partial x} - w_h = -W_s \qquad \text{(L model)}. \tag{9.10}$$

Subsequent use of $\nabla \cdot \mathbf{v} = 0$ and boundary condition $w(z = 0) = 0$ gives the continuity equation for the L model

$$\frac{\partial h}{\partial t} + \frac{\partial h\bar{u}}{\partial x} = -W_s. \tag{9.11}$$

The volume of the current in the L model is not conserved because there is detrainment of the interstitial fluid with respect to the upper envelope marked by the particles; see Fig. 9.1. The dynamic effect of the detrained fluid on the motion of the current will be incorporated later in the momentum equation. However, the dynamic effect of the detrained fluid on the ambient is neglected in the subsequent analysis because the detrained layer is initially very thin and can be quickly dispersed by the motion in the adjacent upper and lower layers. This justification is more acceptable when the interstitial and the ambient fluids have the same density, $\rho_i = \rho_a$. In this case the detrained fluid lacks buoyancy driving, and can be considered as a genuine part of the ambient in the hydrostatic pressure balance above the current. The difference between the L and T models is now more evident. In the L framework the particle volume fraction $\alpha$ in the suspension (current) is constant, but the volume of this suspension decreases with time because particles descend at $z = h$ and interstitial fluid is left behind. At the same time particles settle at the bottom.

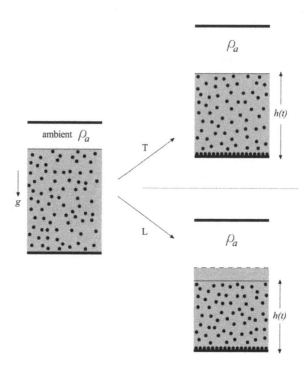

**FIGURE 9.1**: Schematic description of settling according to L and T models. The interstitial fluid of density $\rho_i$ is marked by the gray color.

In the T approach the volume of the suspension (current) is constant, but the volume fraction $\alpha$ dilutes with time. This is because particles settle at the bottom, and fewer particles are remixed in the same original volume. In both models the current will practically "run out" of particles during a finite time interval. If the settling speed $W_s$ is sufficiently large this will occur before the viscous effects become dominant. The details are derived below.

### 9.1.3 Effective reduced gravity and reversing buoyancy effect

The driving force is provided by the density difference between the ambient fluid, $\rho_a$, and the suspension, $\rho_c$. The latter is given by

$$\rho_c = \alpha\rho_p + (1-\alpha)\rho_i = \rho_i + \alpha(\rho_p - \rho_i) =$$
$$\rho_a(1 + \epsilon_i) + \alpha\varepsilon_p(1 + \epsilon_i)\rho_a, \quad (9.12)$$

where

$$\epsilon_i = \frac{\rho_i - \rho_a}{\rho_a} \tag{9.13}$$

and we used the definition (9.2) of $\varepsilon_p$.

It is convenient to use $\rho_a$ as a reference density. We also keep in mind that the Boussinesq approximation imposes $|\epsilon_i|, \alpha \varepsilon_p \ll 1$. The effective reduced gravity of the current is therefore

$$g'_e = \left(\frac{\rho_c}{\rho_a} - 1\right) g = (\varepsilon_p \alpha + \epsilon_i) g = \varepsilon_p \alpha_0 \left[\frac{\alpha(t)}{\alpha_0} + \tilde{\gamma}\right] g, \tag{9.14}$$

where $\tilde{\gamma} = \epsilon_i / (\varepsilon_p \alpha_0)$.

The speed boundary condition for the nose, in dimensional form, is given by a straightforward extension of (2.25)

$$u_N = (\varepsilon_p \alpha_0 g)^{1/2} \left[\frac{\alpha(t)}{\alpha_0} + \tilde{\gamma}\right]^{1/2} h_N^{1/2} Fr(a), \tag{9.15}$$

with the same $Fr$ as for the homogeneous current. The justification is that on the local time and thickness scales relevant to the nose jump, the suspension behaves like a homogeneous fluid of constant density $\rho_c$. The relevant condition is $\beta \ll 1$, to be elaborated below, where the dimensionless $\beta$ is defined by (9.24).

In general, two driving contributions are present. The density difference with $\rho_a$ appears because of $\varepsilon_p \alpha > 0$ and $\epsilon_i \neq 0$. When $\epsilon_i > 0$ (i.e., $\tilde{\gamma} > 0$) the heavy particles enhance the buoyancy of the interstitial fluid; this current will continue to propagate at the bottom after it runs out of particles.

When $\epsilon_i < 0$ (i.e., $\tilde{\gamma} < 0$) the two driving components of $g'_e$ act in opposite directions. The interstitial fluid tends to rise in the ambient. This effect can be strong enough for dragging up particles near the interface, where the light fluid is in direct contact with the heavier ambient. Some local instabilities may appear (see Ungarish 1996). The particles undergo a dilution which resembles the T model assumption, and hence the $\varepsilon_p \alpha$ contribution to the effective buoyancy decreases with time. Eventually, $g'_e = 0$ and the initial bottom current stops; then $g'_e$ becomes negative and the current lifts off the bottom boundary as a plume and forms a top or surface current which propagates in the same direction as the original current. A sequence of photographs of this fascinating behavior can be seen in Fig. 4 of Sparks, Bonnecaze, Huppert, Lister, Hallworth, Mader, and Phillips (1993); see also Hogg, Hallworth, and Huppert (1999).

This effect is termed "reversing buoyancy." The existence and importance of this effect in the particle-driven current are determined by the parameter

$$\tilde{\gamma} = \epsilon_i / (\varepsilon_p \alpha_0). \tag{9.16}$$

When $\tilde{\gamma} < 0$, $g'_e$ remains positive, even after the current runs out of particles. The case $\tilde{\gamma} = 0$ corresponds to the pure particle-driven current, where simple horizontal propagation takes place, and the current stops when it runs out of particles. When $-1 < \tilde{\gamma} < 0$ a reversing buoyancy behavior will appear at some lift-off point (while particles are still present in the current), and when $\tilde{\gamma} = -1$ the current is neutrally buoyant from the beginning, and will lift off shortly after release. (Some papers incorporate the contribution of the interstitial fluid with a negative sign from the beginning and use the parameter $\gamma = -\tilde{\gamma}$.)

Usually, the reduced gravity is defined as positive (absolute value). Here we keep the sign of $g'_e$ to emphasize the change of sign which may occur during propagation of this special current. (We note in passing that reverse buoyancy effects may also occur in compositional currents in some circumstances, typically due to entrainment of ambient fluid which is stratified (e.g. Fernandez and Imberger 2008) or near the density maximum (Leppinen and Kay 2006). These flows are beyond the scope of this book.)

In the following analysis, *we use $\tilde{\gamma} = 0$ unless stated otherwise.*

### 9.1.4 Momentum equation

We use the one-layer model. The SW momentum equations are developed along the lines presented in § 2.2. The pressure is reduced with $+\rho_a g$, and hence the hydrostatic approximation yields $p_a = C$. For the current we write

$$\frac{\partial p_c}{\partial z} = -\rho_a \varepsilon_p \alpha(x,t) g. \tag{9.17}$$

The integration, subject to $p_c = p_a$ at $z = h$, yields

$$p_c = \rho_a \varepsilon_p \alpha(x,t) g \left[ h(x,t) - z \right] + C. \tag{9.18}$$

Consequently, the longitudinal pressure gradient in the current can be written as

$$\frac{\partial p_c}{\partial x} = \rho_a \varepsilon_p g \left\{ \alpha(x,t) \frac{\partial h(x,t)}{\partial x} + [h(x,t) - z] \frac{\partial \alpha(x,t)}{\partial x} \right\}. \tag{9.19}$$

An interesting contrast with the classical compositional current appears. The fundamental result (2.15) for $\partial p_c / \partial x$ is $z$-independent. In the particle-driven current the pressure gradient depends on $z$ when $\alpha$ changes with $x$. According to (9.7)-(9.8), this $z$ variation is bound to occur in the T model; however, the L model will preserve the $z$-independence across the layer. For further use we calculate

$$\int_0^h \frac{\partial p_c}{\partial x} dz = \rho_a \varepsilon_p g \, h \left[ \alpha \frac{\partial h}{\partial x} + \frac{1}{2} h \frac{\partial \alpha}{\partial x} \right] = \rho_a \varepsilon_p g \, \frac{1}{2} \frac{\partial \alpha h^2}{\partial x}. \tag{9.20}$$

To derive the SW averaged momentum equations we start again with the inviscid $x$-momentum equation (2.16). The RHS pressure term is now given by (9.20) (but with $\rho_a$ instead of $\rho_c$, a valid substitution for the Boussinesq system). Concerning the subsequent manipulation of the LHS term of (2.16), we notice that (1) For the T model, the equation of motion of the interface is exactly like in the classical case. Consequently, the form (2.22) is recovered (again, with $\rho_c$ replaced by $\rho_a$). (2) For the L model, the equation of motion of the interface is given by (9.11). The third term on the LHS of (2.18) does not vanish: using (9.10) we find that an additional term, $W_s \bar{u}$, is contributed to the momentum fluxes. This is the momentum flux due to the fluid detrainment at the interface.

## 9.1.5 The governing SW equations

To summarize the results, we introduce dimensionless variables. The velocity and time scales are

$$U = (\varepsilon_p \alpha_0 g h_0)^{1/2}; \quad T = \frac{x_0}{U}; \tag{9.21}$$

note that this corresponds to the use of the reduced gravity $g' = \varepsilon_p \alpha_0 g$ in the previous scaling (2.33).

We now switch all dimensional variables, denoted below by an asterisk, to dimensionless variables as follows

$$\{x^*, z^*, h^*, H^*, t^*, u^*, p^*\} = \{x_0 x, h_0 z, h_0 h, h_0 H, Tt, Uu, \rho_a U^2 p\}. \tag{9.22}$$

In addition, we introduce the scaled volume fraction variable

$$\phi = \frac{\alpha}{\alpha_0}, \tag{9.23}$$

which is in the range $[0, 1]$.

The first dimensionless parameter of the SW formulation is the depth ratio $H$.

A new dimensionless parameter appears

$$\beta = \frac{W_s}{U} \frac{x_0}{h_0}. \tag{9.24}$$

Actually, this is a time-scale ratio which compares the propagation time of the current for a distance $x_0$ to the particle settling time for a height $h_0$. This parameter specifies the importance of the particle settling (phase separation) in the propagation process of the gravity current. We are interested in cases with small $\beta$. Otherwise, the particles settle out from the fluid during a relatively short propagation and no true gravity current develops. The use of (9.15) is thus vindicated. The limit $\beta = 0$ corresponds to a current of non-settling particles, i.e., particles whose radius $a_p$ is so small that $W_p$ is negligible. Regarding these particles as molecules, the $\beta = 0$ limit can be identified with a compositional current of fixed volume and constant $g' = \varepsilon_p \alpha_0 g$.

(Another possible parameter is $\tilde{\gamma}$, set here to zero.)

### 9.1.5.1   Model T

In conservation form the equations can be written as

$$\frac{\partial h}{\partial t} + \frac{\partial}{\partial x} uh = 0;$$
(9.25)

$$\frac{\partial}{\partial t} uh + \frac{\partial}{\partial x}\left(u^2 h + \frac{1}{2}\phi h^2\right) = 0;$$
(9.26)

$$\frac{\partial}{\partial t}\phi h + \frac{\partial}{\partial x} u\phi h = -\beta\phi.$$
(9.27)

In characteristic form, this becomes

$$\begin{bmatrix} h_t \\ u_t \\ \phi_t \end{bmatrix} + \begin{bmatrix} u & h & 0 \\ \phi & u & \frac{1}{2}h \\ 0 & 0 & u \end{bmatrix} \begin{bmatrix} h_x \\ u_x \\ \phi_x \end{bmatrix} = \begin{bmatrix} 0 \\ 0 \\ -\beta\phi/h \end{bmatrix}.$$
(9.28)

The characteristic paths and relationships are obtained as explained in Appendix A.1. The relevant eigenvalues and eigenvectors are

$$c_\pm = u \pm \sqrt{h\phi}; \quad \lambda_3 = u;$$
(9.29a)

$$\left(1, \pm\sqrt{\frac{h}{\phi}}, \frac{1}{2}\frac{h}{\phi}\right); \quad (0, 0, 1).$$
(9.29b)

The system is hyperbolic. The relationships between the variables on the characteristics are as follows

$$dh \pm \sqrt{\frac{h}{\phi}}du + \frac{1}{2}\frac{h}{\phi}d\phi = -\frac{1}{2}\beta dt \quad \text{on} \quad \frac{dx}{dt} = u \pm \sqrt{h\phi};$$
(9.30)

$$d\phi = -\frac{\beta\phi}{h}dt \quad \text{on} \quad \frac{dx}{dt} = u.$$
(9.31)

### 9.1.5.2   Model L

In conservation form the equations can be written as

$$\frac{\partial h}{\partial t} + \frac{\partial}{\partial x} uh = -\beta;$$
(9.32)

$$\frac{\partial}{\partial t} uh + \frac{\partial}{\partial x}\left(u^2 h + \frac{1}{2}\phi h^2\right) = -\beta u;$$
(9.33)

$$\frac{\partial}{\partial t}\phi h + \frac{\partial}{\partial x} u\phi h = -\beta\phi.$$
(9.34)

The conservation equation for $\phi$ is added as a generalization of the L model. For simple initial conditions $\phi = 1$ in a rectangular lock, the solution is $\phi = 1$

in the whole current. With $\phi = 1$ equation (9.34) becomes a duplicate of (9.32).

Note that the last term in (9.32) represents the detrainment of interstitial fluid with respect to the interface of particles which settle in the embedding fluid. The last term in (9.33) represents the loss of momentum due to this detrainment of interstitial fluid into the ambient.

In characteristic form, the system becomes

$$
\begin{bmatrix} h_t \\ u_t \\ \phi_t \end{bmatrix} + \begin{bmatrix} u & h & 0 \\ \phi & u & \frac{1}{2}h \\ 0 & 0 & u \end{bmatrix} \begin{bmatrix} h_x \\ u_x \\ \phi_x \end{bmatrix} = \begin{bmatrix} -\beta \\ 0 \\ 0 \end{bmatrix}.
\tag{9.35}
$$

A comparison between the governing systems for $u, h, \phi$, (9.28) and (9.35), reveals that the difference between the T and L model is only in the RHS source terms. The eigenvalues and eigenvectors of the coefficient matrix are given by (9.29). On characteristics the following relationships hold

$$
dh \pm \sqrt{\frac{h}{\phi}} \, du + \frac{1}{2} \frac{h}{\phi} d\phi = -\beta dt \quad \text{on} \quad \frac{dx}{dt} = u \pm \sqrt{h\phi};
\tag{9.36}
$$

$$
d\phi = 0 \quad \text{on} \quad \frac{dx}{dt} = u.
\tag{9.37}
$$

Again, the third equation of (9.35) can be replaced with $\phi = 1$ for standard initial conditions. The characteristics (9.36) are similarly simplified, and (9.37) is satisfied.

The initial and boundary conditions *for both models* are $u = 0, h = 1, \phi = 1$ at $t = 0$ in the lock $0 \le x \le 1$ and $u = 0$ at the backwall $x = 0$. At the front we use the condition (9.15); in dimensionless form, for $\tilde{\gamma} = 0$, this reads

$$
u_N = Fr(a)\phi_N^{1/2} h_N^{1/2},
\tag{9.38}
$$

where $a = h_N/H$ and the usual $Fr$ function (2.27) is used unless specified otherwise.

The formulation is complete. The particles enter into the formulation of the current due to two effects: their concentration, $\phi$, and their settling, $\beta > 0$. The particle concentration affects the internal pressure-driving (represented by the $\phi h_x$ term in the momentum equation) and the pressure jump at the nose via the $u_N$ condition.

It is interesting that the continuity and momentum equations of the T model do not contain explicit dependency on $\beta$. This allows a quite straightforward extension to the two-layer formulation. As shown in § 5.2, the ambient layer of fluid imposes on the lower layer an additional pressure gradient $-\partial \Phi/dx$ driven by (5.11). Combining this result with the steps used in this chapter to develop (9.25)-(9.26), we find that the two-layer form of these equations is given by (5.16) with one change: the pressure term is now $\frac{1}{2}\phi h^2$. The details are left for an exercise. The equation for $\phi$ is of course unchanged in the two-layer formulation.

We keep in mind that these equations were derived for a particles-driven current whose interstitial fluid is identical with the ambient, $\rho_i = \rho_a$ and $\tilde{\gamma} = 0$. The formulation can be extended for $\tilde{\gamma} \neq 0$ cases. Essentially, the influence of $\tilde{\gamma}$ enters into (1) The pressure term, which becomes $(\phi + \tilde{\gamma})h^2/2$ in LHS of the momentum equations in conservation form. (This will also affect the eigenvectors and the characteristics.) (2) The $u_N$ condition, which becomes proportional to $(\phi_N + \tilde{\gamma})^{1/2}$. This topic is outside the scope of this book. For more details see Bonnecaze, Huppert, and Lister (1993), Ungarish and Huppert (1998), and Hogg, Hallworth, and Huppert (1999).

### 9.1.6     Some SW results

The SW equations for the particle-driven current must be solved numerically. The type of equations and boundary conditions is like in the compositional case ($\beta = 0$ say) and hence the same finite-difference method can be applied. In the $\beta > 0$ case some numerical difficulties may appear in the tail after the slumping phase.

We recall that only $\beta \ll 1$ cases are of interest. For this parameter range the initial slumping is practically like in the classical case. Formally, a time decay of $u_N$ and $h_N$ appears from the beginning of propagation, but this is a small $O(\beta)$ effect. After the rarefaction wave (or left-moving shock when $H \approx 1$) hits the backwall $x = 0$, a quick decay of $h$ occurs in this region and a thin tail appears. Consider the T model: the characteristic (9.31) indicates that the decay of $\phi$ is enhanced when $h$ is small. Consider the L model: the characteristics (9.36)-(9.37) indicate that near the backwall, where $u \approx 0$, $h$ tends to become negative. The conclusion is that in long runs these effects may produce spurious results in the tail (such as negative $\phi$ and $h$). Consequently, some care is required to accommodate these trends and to prevent (even artificially) the propagation of possible errors into the other domains. The SW equations in the thin tail are anyway expected to be invalidated by viscous effects, and hence a loss of numerical accuracy in this domain is unimportant. A negative $h$ or very small $\phi$ can be interpreted as a bare region where the ambient is in direct contact with the bottom.

Now another question comes to our mind: for what time period shall we attempt the computation? If the current remains in the lock with fixed $h = 1$ and $u = 0$ then the particle runout time scale $t_{ro}$ is $1/\beta$. However, the propagation is bound to increase the settling area and decrease the thickness of the suspension. This enhances the sedimentation and is expected to reduce $t_{ro}$. Indeed, calculations that will be presented later show that $t_{ro} \approx 2/\beta^{3/5}$ and the corresponding distance of propagation is $x_{\max} \approx 2/\beta^{2/5}$.

Consider a typical current with $\beta = 2.5 \cdot 10^{-3}$ in a deep ambient $H = 10$. (A similar case has been presented by Bonnecaze, Huppert, and Lister 1993, but due to different scalings $x, t$, and $\beta$ in that paper are larger by factor 2.)

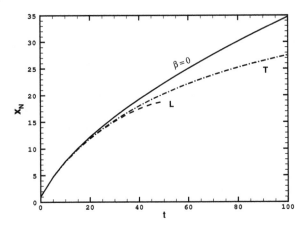

**FIGURE 9.2:** SW propagation for deep current. Particle-driven with $\beta = 2.5 \cdot 10^{-3}$, L and T models. Also shown the compositional-driven limit $\beta = 0$.

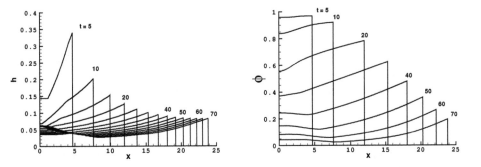

**FIGURE 9.3:** SW T model results for $h$ and $\phi$ as functions of $x$, at various $t = [5, 10 \ (10) \ 70]$. Here $\beta = 2.5 \cdot 10^{-3}$, $H = 10$.

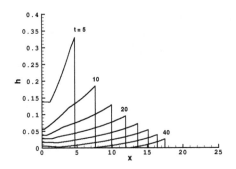

**FIGURE 9.4:** SW L model results for $h$ as a functions of $x$, at various $t = [5, 10 \ (10) \ 70]$. $\phi = 1$ in the current. Here $\beta = 2.5 \cdot 10^{-3}$, $H = 10$.

The SW propagation is shown in Fig. 9.2. The $\beta = 0$ limit that corresponds to a compositional gravity current with the same density excess, $\epsilon = \epsilon_p \alpha_0$, is also displayed.

The L and T models agree well in the $x_N(t)$ prediction, but the former model runs out of particles sooner than the latter one. More details on this effect are shown in Fig. 9.3 and 9.4. We see that, after a while, the main presence of the particle-driven current is concentrated near the front. The tail is thin and dilute in the T case, and practically absent in the L case.

A sharp experimental verification of the one-layer SW results is not available. The typical reported experiments were performed for a full-depth lock with the aspect ratio $h_0/x_0 = 2$ (Sparks, Bonnecaze, Huppert, Lister, Hallworth, Mader, and Phillips 1993, Bonnecaze, Huppert, and Lister 1993, Hogg, Hallworth, and Huppert 1999). Since the initial behavior of the current is little affected by the settling of the particles, we infer that the $H \approx 1$ case displays a special slumping behavior whose accurate description requires a two-layer model. Bonnecaze, Huppert, and Lister (1993) developed a two-layer formulation for the T model and confirmed this expectation. For these cases the predictions of the two-layer T model show good agreement of $x_N(t)$, sediment, and general shape of the current. The one-layer results also provide a fair approximation; the shorter slumping makes the overall propagation slightly slower, and also affects the thickness of the sediment (see Fig. 9.5).

In general, the particles are a source of errors and uncertainties in the formulation of the gravity current. The assumption of monodispersity (i.e., the same radius $a_p$ for all particles) is not easily satisfied in practice, and we see that the error in $a_p$ is amplified by the dependency of $W_s$ on $a_p^2$. This combines with the non-rigorous modification of the settling speed due to turbulence, effects of non-diluteness, and a quite unavoidable resuspension of the sediment. The fact that the emerging model for the particle-driven current can be solved with the same mathematical tools as the compositional-driven currents does not imply the same accuracy in the reproduction of the physical phenomena.

### 9.1.6.1   Sediment

The settled particles form a sediment. The mass per unit area is defined as the density of deposit. Since $\rho_p$ and the packing fraction, $\alpha_M$, are known, the thickness and the volume of the sediment layer can also be derived.

In theoretical calculations the thickness $h_s$ and the volume $\mathcal{V}_s$ (per unit width) of the sediment are conveniently scaled with $\alpha_0 h_0$ and $\alpha_0 x_0 h_0$, respectively. The behavior of these variables during the propagation of the current is given by

$$h_s(x,t) = \beta \int_0^t \phi(x,t')dt' \quad (0 \leq x \leq x_N(t)); \qquad (9.39)$$

$$\mathcal{V}_s(t) = \int_0^{x_N(t)} h_s(x,t)dx. \qquad (9.40)$$

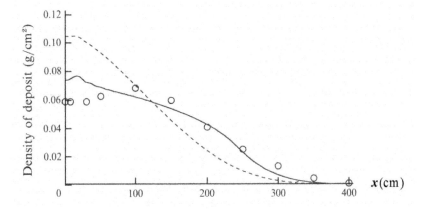

**FIGURE 9.5:** Density of deposit after the flow ends as a function of $x$, dimensional. Experiment (symbols) and SW T model predictions for one-layer (dashed line) and two-layer (solid line) formulations. Full-depth lock system with $x_0 = 15$ cm, $h_0 = 30$ cm, silicon carbide particles ($\rho_p = 3.22$g cm$^{-3}$, $a_p = 26.5\mu$m, $\alpha_0 = 0.0105$.). Here $\beta = 6.4 \cdot 10^{-3}$. (Bonnecaze, Huppert, and Lister 1993, with permission.)

In these definitions it is understood that $\phi = h_s = 0$ before the arrival of the nose at a given position $x$ at time $t_A(x)$. In other words, the integration in (9.39) actually starts at some $t_A(x)$. (The definition of $h_s$ assumes an idealized compact sediment with packing ratio $\alpha_M = 1$. The more realistic thickness is $h_s/\alpha_M$, with $\alpha_M \approx 0.65$.) The calculation of $h_s$ and $\mathcal{V}_s$ from SW results is straightforward.

The length of propagation is $\sim \beta^{-2/5}$, as shown later. Consequently, we infer that $h_s$ is of the order of $\beta^{2/5}$ (again, scaled with $\alpha_0 h_0$). The real-time observation and recording of this variable, when both $\alpha_0$ and $\beta$ are very small as assumed in the theory, is a difficult task. However, after the current stops it is possible to suck the sediment from selected subdomains of known area at the bottom, separate the particles from the fluid and weigh them, and calculate the density deposit (mass/area). A typical result of a laboratory experiment is shown in Fig. 9.5. Various verifications indicate that this technique gives good accuracy, of a few percent; however, the reconstruction of the time dependent process from the sediment left behind may contain significantly larger uncertainties. This is in particular relevant when applying the theoretical results to natural systems. This fascinating topic is outside the scope of this book, and the interested reader is referred to Dade and Huppert (1995a) and Kneller and Buckee (2000).

### 9.1.7 Similarity solution rudiments

Consider a deep current with $Fr = $ const. When $\beta$ is small the slumping phase occurs with insignificant particle settling. As a first approximation, the appearance of a self-similar behavior is thus expected, like in the classical case considered in § 2.6. (Note that the notation $\beta$ has different meanings in that and present section.) The classical self-similar solution is a function of $t$ and $y = x/x_N(t)$, with $x_N = Kt^{2/3}$. However, this self-similar solution can be a good approximation to the post-slumping stage of the particle-driven current for a restricted time only. The additional source terms in the equations of motion are small, $O(\beta)$, but their contribution is continuous and accumulates with time. Some corrections are necessary on a long-time scale to reconcile the self-similar tendency with the particle-settling buoyancy decay. This can be achieved by an expansion in powers of the small $\beta$, as demonstrated in Hogg, Ungarish, and Huppert (2000). The method is illustrated briefly for the T model.

To begin, the transformation

$$y = \frac{x}{Kt^{2/3}} \tag{9.41}$$

is introduced, and the governing equations are rewritten for the variables $h(y,t), u(y,t), \phi(y,t)$; see § 2.3. The system (9.28) in the new coordinates $t, y$ becomes

$$\frac{\partial h}{\partial t} + \left( K^{-1}t^{-2/3}u - \frac{2}{3}t^{-1}y \right)\frac{\partial h}{\partial y} + K^{-1}t^{-2/3}h\frac{\partial u}{\partial y} = 0; \tag{9.42}$$

$$\frac{\partial u}{\partial t} + \left( K^{-1}t^{-2/3}u - \frac{2}{3}t^{-1}y \right)\frac{\partial u}{\partial y} + K^{-1}t^{-2/3}\left( \phi\frac{\partial h}{\partial y} + \frac{1}{2}h\frac{\partial \phi}{\partial y} \right) = 0; \tag{9.43}$$

$$\frac{\partial \phi}{\partial t} + \left( K^{-1}t^{-2/3}u - \frac{2}{3}t^{-1}y \right)\frac{\partial \phi}{\partial y} = -\beta\frac{\phi}{h}. \tag{9.44}$$

Next, we seek an expansion in powers of the small $\beta$. The leading term is the classical self-similar solution; the next terms motivate the introduction of the new variable

$$\tau = \beta K^{-2}t^{5/3}. \tag{9.45}$$

The expansion, for $\tau \ll 1$, is conveniently written as

$$x_N = Kt^{2/3}[1 + \tau X_1 + \tau^2 X_2 + \cdots]; \tag{9.46}$$

$$u = \frac{2}{3}Kt^{-1/3}[U_0(y) + \tau U_1(y) + \tau^2 U_2(y) + \cdots]; \tag{9.47}$$

$$h = \frac{4}{9}K^2t^{-2/3}[H_0(y) + \tau H_1(y) + \tau^2 H_2(y) + \cdots]; \tag{9.48}$$

$$\phi = 1 + \tau\varphi_1(y) + \tau^2\varphi_2(y) + \cdots; \tag{9.49}$$

where $X_1, X_2$ are constants, $U_0(y) = y$, and $H_0(y) = \mathcal{H}(y)$; see (2.78). Evidently, for $\tau = 0$ ($\beta = 0$) the classical similarity solution (2.80) is recovered.

Finally, the next terms in the solution are obtained by substituting the expansion (9.47)-(9.49) into the governing equations (9.42)-(9.44) and balancing terms with the same power of $\tau$.

The leading $\tau^0$ terms are already in balance by virtue of the classical similarity solution. Consequently, the value of $K$ is given by (2.79) which implements the conservation of the initial volume.

The next terms provide facile equations for $U_1, H_1$, etc. as functions of $y$ only in the domain $[0, 1]$. The boundary conditions also require an expansion in powers of $\tau$. In particular, we note that due to (9.41) and (9.46), the nose of the current is at the $y$ position

$$y_N = 1 + \tau X_1 + \tau^2 X_2 + O(\tau^3), \tag{9.50}$$

which is expected to be smaller than unity. This is because $y$ is scaled with the propagation distance of the classical case, but the value of $x_N$ is reduced when $\beta > 0$. The details of the calculation of the terms in the expansion are given in Hogg, Ungarish, and Huppert (2000).

Some salient points of the results are as follows. First, the expansion method illustrated here for the T model works in the same way for the L model (with the simplification $\phi = 1$ instead of (9.44)). This strengthens the confidence that there is some essential compatibility between the models. Moreover, for $Fr = 1.19$ we find that the values of $X_1$ are $-0.181$ and $-0.197$ for the T and L models, respectively, and that both models give the same sediment volume to leading order in $\tau$.

Second, an important outcome is the rigorous derivation of the time and length scales of the combined propagation and settling process. As indicated by the expansion, and confirmed by the details of the solution, the relevant time scale for a dominant influence of the particle settling on the current, or runout of particles, is $\tau_{ro} \approx 1$, i.e., $t_{ro} \approx \beta^{-3/5} K^{6/5}$, where $K$ is given by (2.79). Let us express this runout time estimate in dimensional form

$$t_{ro}^* \approx 2 \frac{(\mathcal{V}_0^*)^{2/5}}{W_s^{3/5} (\varepsilon_p \alpha_0 g)^{1/5}}. \tag{9.51}$$

We note that the (estimated) time of propagation of a particle-driven current depends on the initial volume of the current, $\mathcal{V}_0^*$ (dimensional), irrespective of the lock aspect ratio and shape. This could be expected, because the current "forgets" the initial conditions while in the self-similar phase. The particles move with the fluid during the adjustment time, and thus also "forget" the initial conditions. This conclusion, however, must be treated with care when $\beta$ is not very small. We recall that the self-similar solution has a "virtual origin" which is affected by the shape of the lock, and this may introduce some associated settling before the beginning of the self-similar stage.

The relevant distance of propagation (particle runout, dimensionless) is $Kt_{ro}^{2/3}$. Use of (2.79) and some algebra give, to leading order,

$$x_{\max} \approx x_N(t_{ro}) \approx K_1^{9/5} \mathcal{V}_0^{3/5} \frac{1}{\beta^{2/5}}. \tag{9.52}$$

Here $\mathcal{V}_0$ is the initial dimensionless volume ($= 1$ in standard rectangular lock). It is left for an exercise to show that the dimensional form of this result, $x_{\max}x_0$, does not depend on the aspect ratio or shape of the lock. The correction factor $1 + X_1$, see (9.46), indicates that the propagation in the L model is shorter than in the T model.

The leading terms of this approximation provide good analytical support to the features observed in the numerical SW results. In particular, they confirm the tendency of the particle-driven currents to develop a thin tail, which is the first domain to run out of particles. Again, this happens sooner in the L model than in the T model.

### 9.1.8    Box models

We use dimensionless variables. The current is assumed to be a box of height $h_N$ and length $x_N$ whose volume is $\mathcal{V}(t) = h_N(t)\, x_N(t)$. The scaled particle volume fraction $\phi = \alpha/\alpha_0$ is assumed independent of $x$. The initial values at $t = 0$ are $x_N = 1$, $\mathcal{V} = \mathcal{V}_0$ ($= 1$ in the standard case), and $\phi = 1$. In general, $Fr$ is a function of $a = \mathcal{V}/(Hx_N)$.

For both T and L models we can write the total particle conservation

$$\frac{d(\mathcal{V}\phi)}{dt} = -\beta \phi x_N, \tag{9.53}$$

and the nose condition

$$\frac{dx_N}{dt} = Fr \, (\phi + \tilde{\gamma})^{1/2} h_N^{1/2} = Fr \, (\phi + \tilde{\gamma})^{1/2} \mathcal{V}^{1/2} x_N^{-1/2}. \tag{9.54}$$

The combination of (9.53) with (9.54) allows a convenient separation of the variables

$$x_N^{3/2} dx_N = -\frac{Fr}{\beta} \mathcal{V}^{1/2} \frac{(\phi + \tilde{\gamma})^{1/2}}{\phi} \, d(\mathcal{V}\phi). \tag{9.55}$$

Consider the T model. The volume of the current is constant, $\mathcal{V} = \mathcal{V}_0$, but $\phi$ decays. For simplicity, let $Fr = \text{const}$ and $\tilde{\gamma} = 0$, and integrate (9.55). While $x_N$ increases $\phi$ decreases, and we can calculate $x_{\max}$ when $\phi = 0$ (the current runs out of particles). We obtain

$$x_{\max} \approx (5Fr)^{2/5} \mathcal{V}_0^{3/5} \frac{1}{\beta^{2/5}}. \tag{9.56}$$

The details are left for an exercise. Further analytical elaborations (including the influence of $\tilde{\gamma}$ and non-constant $Fr$) and comparisons with experiments for this model were presented by Hallworth, Hogg, and Huppert (1998) and Hogg, Hallworth, and Huppert (1999). The qualitative behavior is well reproduced, but the runout and lift-off distances are under- and over-estimated, respectively.

Consider the L model. The volume fraction is constant, $\phi = 1$, but the volume of the suspension decays. For simplicity, let $Fr = \text{const}$, and integrate (9.55). While $x_N$ increases $\mathcal{V}$ decreases, and we can calculate $x_{max}$ when $\mathcal{V} = 0$ (the current runs out of particles). We obtain

$$x_{max} \approx \frac{1}{3^{2/5}}(1 + \tilde{\gamma})^{1/5}(5Fr)^{2/5}\mathcal{V}_0^{3/5}\frac{1}{\beta^{2/5}}. \tag{9.57}$$

For $\tilde{\gamma} = 0$ there is good agreement with the T model result (9.56). The reverse buoyancy for negative $\tilde{\gamma}$ is evident in the coefficient of $(1 + \tilde{\gamma})^{1/5}$. However, the experimental data of Hogg, Hallworth, and Huppert (1999) indicate that the reverse buoyancy effect increases faster with $-\tilde{\gamma}$ than suggested by this coefficient. Again, there is indication that the propagation in the L model is shorter than in the T model.

The results (9.56) and (9.57) suggest a rescaling of $x$ with $\beta^{-2/5}\mathcal{V}_0^{3/5}$. For consistency, $t$ should be rescaled with $\beta^{-3/5}\mathcal{V}_0^{2/5}$. The use of these rescaled variables will eliminate the dependency on $\beta$ and $\mathcal{V}_0$ from the governing balances (9.53)-(9.54). In other words, the box-model runout time is $\sim \mathcal{V}_0^{2/5}/\beta^{3/5}$ for both T and L cases. We conclude that the present $x_{max}$ and $t_{ro}$ runout length and time are consistent, and actually in fair agreement, with the more rigorous results derived from the self-similar solution in § 9.1.7.

## 9.2 Axisymmetric particle-driven current

Here we consider the counterpart problem of the 2D current discussed in the previous § 9.1 in a radially-diverging geometry. The container in which the current propagates is either a wedge or a full cylinder with a horizontal bottom at $z = 0$. The axis of symmetry is $z$, the gravity acts in the $-z$ direction, and the propagation is along the radial coordinate, $r$. The velocity components in this cylindrical coordinate system are $\{u, v, w\}$, and we assume that $v = 0$. The SW variables are the thickness $h(r, t)$, speed of current $u(r, t)$, and particle volume fraction $\alpha(r, t)$. Typically, the current is released from a cylinder of radius $r_0$ and height $h_0$, and contains particles of density $\rho_p$, radius $a_p$ at initial volume fraction $\alpha_0$. The simplification of dilute suspension and Boussinesq system are used again.

The settling speed $W_s$ and the relevant densities of the fluids and particles are subject to the same relationships as in the 2D case. The driving effective buoyancy is also not affected by the geometry, and the curvature terms have little influence on the nose jump condition. Consequently, equations (9.1) and (9.12)-(9.15) remain valid. We assume that the density of the interstitial fluid is equal to that of the ambient fluid, $\tilde{\gamma} = 0$, unless stated otherwise.

We use the one-layer SW approximation. A close inspection reveals that the derivation of the averaged horizontal pressure gradient in § 9.1.4 also remains valid, with $x$ replaced by $r$ in (9.17)-(9.20). Subsequently, the balances developed for the motion of the interface and the SW momentum equations can be carried over to the cylindrical axisymmetric flow taking into account the additional curvature terms.

## 9.2.1　The governing SW equations

We introduce dimensionless variables. The velocity and time scales are

$$U = (\varepsilon_p \alpha_0 g h_0)^{1/2}; \quad T = \frac{r_0}{U}; \tag{9.58}$$

note that this corresponds to the use of the reduced gravity $g' = \varepsilon_p \alpha_0 g$ in the previous scaling (6.22).

We now switch all dimensional variables, denoted below by an asterisk, to dimensionless variables as follows

$$\{r^*, z^*, h^*, H^*, t^*, u^*, p^*\} = \{r_0 r, h_0 z, h_0 h, h_0 H, Tt, Uu, \rho_a U^2 p\}. \tag{9.59}$$

In addition, we introduce the scaled volume fraction variable

$$\phi = \frac{\alpha}{\alpha_0}, \tag{9.60}$$

which is in the range $[0, 1]$.

The first dimensionless parameter of the SW formulation is the depth ratio $H$.

A new dimensionless parameter appears

$$\beta = \frac{W_s}{U} \frac{r_0}{h_0}. \tag{9.61}$$

The interpretation is as in the 2D case.

(Another possible parameter is $\tilde{\gamma}$, set here to zero).

### 9.2.1.1　Model T

In conservation form the equations can be written as

$$\frac{\partial h}{\partial t} + \frac{\partial}{\partial r} uh = -\frac{uh}{r}; \tag{9.62}$$

$$\frac{\partial}{\partial t} uh + \frac{\partial}{\partial r} \left( u^2 h + \frac{1}{2} \phi h^2 \right) = -\frac{u^2 h}{r}; \tag{9.63}$$

$$\frac{\partial}{\partial t}\phi h + \frac{\partial}{\partial r}u\phi h = -\beta\phi - \frac{u\phi h}{r}. \tag{9.64}$$

In characteristic form, this becomes

$$\begin{bmatrix} h_t \\ u_t \\ \phi_t \end{bmatrix} + \begin{bmatrix} u & h & 0 \\ \phi & u & \frac{1}{2}h \\ 0 & 0 & u \end{bmatrix} \begin{bmatrix} h_r \\ u_r \\ \phi_r \end{bmatrix} = \begin{bmatrix} -uh/r \\ 0 \\ -\beta\phi/h \end{bmatrix}. \tag{9.65}$$

The characteristic paths and relationships are obtained as explained in Appendix A.1. The relevant eigenvalues and eigenvectors are

$$c_\pm = u \pm \sqrt{h\phi}; \quad \lambda_3 = u; \tag{9.66a}$$

$$\left(1, \pm\sqrt{\frac{h}{\phi}}, \frac{1}{2}\frac{h}{\phi}\right); \quad (0,0,1). \tag{9.66b}$$

The system is hyperbolic. The relationships between the variables on the characteristics are as follows

$$dh \pm \sqrt{\frac{h}{\phi}}du + \frac{1}{2}\frac{h}{\phi}d\phi = -\left(\frac{uh}{r} + \frac{1}{2}\beta\right)dt \quad \text{on} \quad \frac{dr}{dt} = u \pm \sqrt{h\phi}; \tag{9.67}$$

$$d\phi = -\frac{\beta\phi}{h}dt \quad \text{on} \quad \frac{dr}{dt} = u. \tag{9.68}$$

### 9.2.1.2 Model L

In conservation form the equations can be written as

$$\frac{\partial h}{\partial t} + \frac{\partial}{\partial r}uh = -\frac{uh}{r} - \beta; \tag{9.69}$$

$$\frac{\partial}{\partial t}uh + \frac{\partial}{\partial r}\left[u^2 h + \frac{1}{2}\phi h^2\right] = -\frac{u^2 h}{r} - \beta u; \tag{9.70}$$

$$\frac{\partial}{\partial t}\phi h + \frac{\partial}{\partial r}u\phi h = -\beta\phi - \frac{u\phi h}{r}. \tag{9.71}$$

The conservation equation for $\phi$ is added as a generalization of the L model. For simple initial conditions $\phi = 1$ in a cylinder lock, the solution is $\phi = 1$ in the whole current. With $\phi = 1$ equation (9.71) becomes a duplicate of (9.69).

Note that the last term in (9.69) represents the additional settling of the interface due to the local sedimentation of the particles relative to the fluid. The last term in (9.70) represents the loss of radial momentum due to detrainment of interstitial fluid into the ambient.

In characteristic form, this becomes

$$\begin{bmatrix} h_t \\ u_t \\ \phi_t \end{bmatrix} + \begin{bmatrix} u & h & 0 \\ \phi & u & \frac{1}{2}h \\ 0 & 0 & u \end{bmatrix} \begin{bmatrix} h_r \\ u_r \\ \phi_r \end{bmatrix} = \begin{bmatrix} -uh/r - \beta \\ 0 \\ 0 \end{bmatrix}. \tag{9.72}$$

A comparisons between the governing systems for $u, h, \phi$, (9.65) and (9.72), reveals that the difference between the T and L model is only in the RHS source terms. The eigenvalues and eigenvectors of the coefficient matrix are given by (9.66). On characteristics the following relationships hold

$$dh \pm \sqrt{\frac{h}{\phi}}\, du + \frac{1}{2}\frac{h}{\phi}d\phi = -\left(\frac{uh}{r} + \beta\right) dt \quad \text{on} \quad \frac{dr}{dt} = u \pm \sqrt{h\phi}; \quad (9.73)$$

$$d\phi = 0 \quad \text{on} \quad \frac{dr}{dt} = u. \quad (9.74)$$

Again, the third equation of (9.72) can be replaced with $\phi = 1$ for standard initial conditions. The characteristics (9.73) are similarly simplified and (9.74) is satisfied.

The initial and boundary conditions *for both models* are $u = 0, h = 1, \phi = 1$ at $t = 0$ in the lock $0 \le r \le 1$ and $u = 0$ at the axis $r = 0$. At the front we use the condition (9.15); in dimensionless form, for $\tilde{\gamma} = 0$, this reads

$$u_N = Fr(a)\phi_N^{1/2} h_N^{1/2}, \quad (9.75)$$

where $a = h_N/H$ and the usual $Fr$ function (2.27) is used unless specified otherwise.

The formulation is complete. The particles enter into the formulation of the current due to two effects: their concentration, $\phi$, and their settling, $\beta > 0$. The particle concentration affects the internal pressure-driving (represented by the $\phi h_r$ term in the momentum equation) and the pressure jump at the nose via the $u_N$ condition.

## 9.2.2 Some SW results

The SW equations for the axisymmetric particle-driven current must be solved numerically. The type of equations and boundary conditions is like in the compositional case ($\beta = 0$ say) and hence the same finite-difference method can be applied. The numerical considerations are like in the 2D case discussed before.

Again, only $\beta \ll 1$ cases are of interest, and we ask for what time period shall we attempt the computation? If the current remains in the lock with fixed $h = 1$ and $u = 0$ then the particle runout time scale $t_{ro}$ is $1/\beta$. However, the propagation is bound to increase the settling area and decrease the thickness of the suspension. This enhances the sedimentation and is expected to reduce $t_{ro}$. Calculations that will be presented later show that $t_{ro} \approx 2/\beta^{1/2}$ and the corresponding distance of propagation is $r_{\max} \approx 1/\beta^{1/4}$. The runout time and distance are shorter than in the 2D counterpart with the same $\beta$. The settling in the axisymmetric case is more effective because the area of sedimentation increases like $r_N^2$.

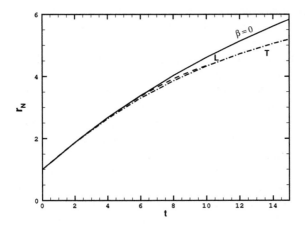

**FIGURE 9.6:** SW propagation for deep current. Particle-driven with $\beta = 10^{-2}$, L and T models. Also shown the compositional-driven limit $\beta = 0$.

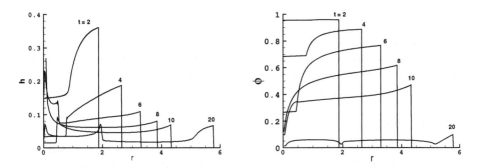

**FIGURE 9.7:** SW T model results for $h$ and $\phi$ as functions of $r$, at various $t$. Here $\beta = 10^{-2}$, $H = 1$.

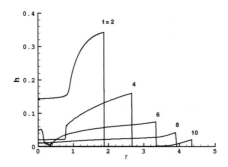

**FIGURE 9.8:** SW L model results for $h$ as a functions of $r$, at various $t$. $\phi = 1$ in the current. Here $\beta = 10^{-2}$, $H = 1$.

Consider a typical current with $\beta = 10^{-2}$ in a full-lock release ambient $H = 1$. (A similar case has been presented by Bonnecaze, Hallworth, Huppert, and Lister 1995, but due to different scalings $r, t$, and $\beta$ in that paper are smaller by factor 2.) The SW propagation is shown in Fig. 9.6. The $\beta = 0$ limit that corresponds to a compositional gravity current with the same density excess, $\epsilon = \varepsilon_p \alpha_0$, is also displayed.

The L and T models agree well in the $r_N(t)$ prediction, but the former model runs out of particles sooner than the latter one. More details on this effect are shown in Fig. 9.7 and 9.8. We see that, after a while, the main presence of the particle-driven current is concentrated near the front. The tail is thin and dilute in the T case, and practically absent in the L case. In the axisymmetric geometry, the difference between large $H$ and full-lock $H \approx 1$ is less pronounced than for the 2D case. The radial spread is bound to reduce $h_N$ and hence a significant slumping with constant $h_N$ is in any case not feasible. This carries over to the particle-driven current. Bonnecaze, Hallworth, Huppert, and Lister (1995) solved the one- and two-layer SW equations with the T settling model and found that the propagation and sedimentation results remained in good agreement for $H = 1$. That study also presented comparisons with experiments, performed mainly in a wedge container, with full-lock release ($H = 1$). The $r_N$ vs. $t$ results are in very good agreement with the SW predictions. The density deposit vs. $r$ was also measured, and there is only fair agreement with the SW predictions. The reasons for discrepancy are not clear. Further investigation is needed, including comparisons with the L model and effects of polydisperse particles. As mentioned before, the presence of particles introduces several uncertainties and complications that cannot be easily isolated from experimental data.

### 9.2.2.1 Sediment

The settled particles form a sediment. The mass per unit area is defined as the density of deposit. Since $\rho_p$ and the packing fraction, $\alpha_M$, are known, the thickness and the volume of the sediment layer can also be derived.

In theoretical calculations the thickness $h_s$ and the volume $\mathcal{V}_s$ (per radian) of the sediment are conveniently scaled with $\alpha_0 h_0$ and $\alpha_0 r_0^2 h_0$, respectively. The behavior of these variables during the propagation of the current is given by

$$h_s(r, t) = \beta \int_0^t \phi(r, t') dt' \quad (0 \le r \le r_N(t)); \tag{9.76}$$

$$\mathcal{V}_s(t) = \int_0^{r_N(t)} h_s(r, t) r dr. \tag{9.77}$$

In these definitions it is understood that $\phi = h_s = 0$ before the arrival of the nose at a given position $r$ at time $t_A(r)$. In other words, the integration in (9.76) actually starts at some $t_A(r)$. (The definition of $h_s$ assumes an idealized compact sediment with packing ratio $\alpha_M = 1$. The more realistic thickness is $h_s/\alpha_M$, with $\alpha_M \approx 0.65$.) The calculation of $h_s$ and $\mathcal{V}_s$ from SW results is straightforward.

Since the length of propagation is $\sim \beta^{-1/4}$ (as shown later), we infer that $h_s$ is of the order of $\beta^{1/2}$ (again, scaled with $\alpha_0 h_0$). The real-time observation and recording of this variable, when both $\alpha_0$ and $\beta$ are very small as assumed in the theory, is a difficult task. However, after the current stops it is possible to suck the sediment from selected subdomains of known area at the bottom, separate the particles from the fluid and weigh them, and calculate the density deposit (mass/area); see Bonnecaze, Hallworth, Huppert, and Lister (1995).

### 9.2.3 Similarity solution rudiments

Consider a deep axisymmetric current with $Fr = $ const. When $\beta$ is small the slumping phase occurs with insignificant particle settling. As a first approximation, the appearance of a self-similar behavior is thus expected, like in the classical case considered in § 6.5. (Note that the notation $\beta$ has different meanings in that and present section.) The classical self-similar solution is a function of $t$ and $y = r/r_N(t)$, with $r_N = Kt^{1/2}$. However, this self-similar solution can be a good approximation to the post-slumping stage of the particle-driven current for a restricted time only. The additional source terms in the equations of motion are small, $O(\beta)$, but their contribution is continuous and accumulates with time. Some corrections are necessary on a long-time scale to reconcile the self-similar tendency with the particle-settling buoyancy decay. This can be achieved by an expansion in powers of the small $\beta$, like in the 2D case discussed in § 9.1.7. The method is illustrated briefly for the T model.

To begin, the transformation

$$y = \frac{r}{Kt^{1/2}} \tag{9.78}$$

is introduced, and the governing equations are rewritten for the variables $h(y,t), u(y,t), \phi(y,t)$; see § 2.3. The system (9.65) in the new coordinates $t, y$ becomes

$$\frac{\partial h}{\partial t} + \left( K^{-1}t^{-1/2}u - \frac{1}{2}t^{-1}y \right) \frac{\partial h}{\partial y} + K^{-1}t^{-1/2}\left( h\frac{\partial u}{\partial y} + \frac{uh}{y} \right) = 0; \tag{9.79}$$

$$\frac{\partial u}{\partial t} + \left( K^{-1}t^{-1/2}u - \frac{1}{2}t^{-1}y \right) \frac{\partial u}{\partial y} + K^{-1}t^{-1/2}\left( \phi\frac{\partial h}{\partial y} + \frac{1}{2}h\frac{\partial \phi}{\partial y} \right) = 0; \tag{9.80}$$

$$\frac{\partial \phi}{\partial t} + \left( K^{-1}t^{-1/2}u - \frac{1}{2}t^{-1}y \right) \frac{\partial \phi}{\partial y} = -\beta\frac{\phi}{h}. \tag{9.81}$$

We define the variable

$$\tau = \beta K^{-2}t^2, \tag{9.82}$$

which will be utilized as a small parameter in the analysis which follows. We introduce the expansions

$$r_N = Kt^{1/2}[1 + \tau R_1 + \tau^2 R_2 + \cdots]; \tag{9.83}$$

$$u = \frac{1}{2}Kt^{-1/2}[U_0(y) + \tau U_1(y) + \tau^2 U_2(y) + \cdots]; \tag{9.84}$$

$$h = \frac{1}{4}K^2 t^{-1}[H_0(y) + \tau H_1(y) + \tau^2 H_2(y) + \cdots]; \tag{9.85}$$

$$\phi = 1 + \tau \varphi_1(y) + \tau^2 \varphi_2(y) + \cdots ; \tag{9.86}$$

where $R_1, R_2$ are constants, $U_0(y) = y$, and $H_0(y) = \mathcal{H}(y)$; see (6.49). Evidently, for $\tau = 0$ ($\beta = 0$) the classical similarity solution (6.51) is recovered.

As in the 2D case, we substitute these expansions into the governing equations (9.79)-(9.81) and balance terms of equal powers of $\tau$. The leading $\tau^0$ terms are already in balance by virtue of the classical similarity solution, and $K$ is given by (6.50) due to the conservation of initial volume requirement.

The next terms provide facile equations for $U_1, H_1$, etc. as functions of $y$ only in the domain $[0, 1]$. The boundary conditions also require an expansion in powers of $\tau$. In particular, we note that due to (9.78) and (9.83), the nose of the current is at the $y$ position

$$y_N = 1 + \tau R_1 + \tau^2 R_2 + O(\tau^3), \tag{9.87}$$

which is expected to be smaller than unity. The details of the calculation of the terms in the expansion are given in Hogg, Ungarish, and Huppert (2000).

Some salient points of the results are as follows. First, the expansion method illustrated here for the T model works in the same way for the L model (with the simplification $\phi = 1$ instead of (9.81)), but the details still require elaboration. For $Fr = 1.19$ we find $R_1 = -0.181$ (for the T model).

Second, an important outcome is the rigorous derivation of the time and length scales of the combined propagation and settling process. As indicated by the expansion, and confirmed by the details of the solution, the relevant time scale for a dominant influence of the particle settling on the current, or runout of particles, is $\tau_{ro} \approx 0.5$, i.e., $t_{ro} \approx \beta^{-1/2}K$, where $K$ is given by (6.50). Let us express this runout time estimate in dimensional form

$$t_{ro}^* \approx 2\frac{(\mathcal{V}_0^*)^{1/4}}{W_s^{1/2}(\varepsilon_p \alpha_0 g)^{1/4}}. \tag{9.88}$$

Here $\mathcal{V}_0^*$ is the dimensional volume of the current per radian. We note that the (estimated) time of propagation of a particle-driven current depends on the initial volume of the current irrespective of the lock aspect ratio and shape. This could be expected, because the current "forgets" the initial conditions while in the self-similar phase. The particles move with the fluid during the adjustment time, and thus also "forget" the initial conditions. This conclusion, however, must be treated with care when $\beta$ is not very small. We recall that the self-similar solution has a "virtual origin" which is affected by the shape of the lock, and this may introduce some associated settling before the beginning of the self-similar stage.

The relevant distance of propagation (particle runout, dimensionless) is $Kt_{ro}^{1/2}$. Use of (6.50) and some algebra give, to leading order, the maximum radius of propagation at particle runout time

$$r_{\max} \approx r_N(t_{ro}) \approx K_1^{3/2} \mathcal{V}_0^{3/8} \frac{1}{\beta^{1/4}}. \tag{9.89}$$

Here $\mathcal{V}_0$ is the initial dimensionless volume per radian, scaled with $r_0^2 h_0$ ($\mathcal{V}_0 = 1/2$ for the standard cylinder lock). It is left for an exercise to show that the dimensional form of this result, $r_{\max} r_0$, does not depend on the aspect ratio or shape of the lock.

The leading terms of this approximation provide good analytical support to the features observed in the numerical SW results. In particular, they confirm the tendency of the particle-driven currents to develop a thin tail, which is the first domain to run out of particles.

### 9.2.4   Box models

We use dimensionless variables. The current is assumed to be a cylinder of height $h_N$ and radius $r_N$ whose volume is $\mathcal{V}(t) = (1/2)h_N(t)r_N^2(t)$ per radian. The scaled particle volume fraction $\phi = \alpha/\alpha_0$ is assumed independent of $r$. The initial values at $t = 0$ are $r_N = 1, \mathcal{V} = \mathcal{V}_0$ ($= 1/2$ in the standard case), and $\phi = 1$. In general, $Fr$ is a function of $a = 2\mathcal{V}/(Hr_N^2)$.

For both T and L models we can write the total particle conservation

$$\frac{d(\mathcal{V}\phi)}{dt} = -\beta\phi\frac{1}{2}r_N^2, \tag{9.90}$$

and the nose condition

$$\frac{dr_N}{dt} = Fr\,(\phi + \tilde{\gamma})^{1/2}\,h_N^{1/2} = Fr\,(\phi + \tilde{\gamma})^{1/2}\,(2\mathcal{V})^{1/2}r_N^{-1}. \tag{9.91}$$

The combination of (9.90) with (9.91) allows a convenient separation of the variables

$$r_N^3\,dr_N = -\frac{Fr}{\beta}(2\mathcal{V})^{1/2}\frac{(\phi + \tilde{\gamma})^{1/2}}{\phi}\,d(2\mathcal{V}\phi). \tag{9.92}$$

Consider the T model. The volume of the current is constant, $\mathcal{V} = \mathcal{V}_0$, but $\phi$ decays. For simplicity, let $Fr = $ const and $\tilde{\gamma} = 0$, and integrate (9.92). While $r_N$ increases $\phi$ decreases, and we can calculate $r_{\max}$ when $\phi = 0$ (the current runs out of particles). We obtain

$$r_{\max} \approx (8Fr)^{1/4}(2\mathcal{V}_0)^{3/8}\frac{1}{\beta^{1/4}}. \tag{9.93}$$

The details are left for an exercise. Further analytical elaborations (including the influence of $\tilde{\gamma}$ and non-constant $Fr$) and comparisons with experiments for this model were presented by Dade and Huppert (1995b). This box model predicts well the radius and time of propagation. However, since the experiments were performed for a full-depth lock, the observed initial propagation is slower than predicted.

Consider the L model. The volume fraction is constant, $\phi = 1$, but the volume of the suspension decays. For simplicity, let $Fr = \text{const}$, and integrate (9.92). While $r_N$ increases $\mathcal{V}$ decreases, and we can calculate $r_{\max}$ when $\mathcal{V} = 0$ (the current runs out of particles). We obtain

$$r_{\max} \approx \left(\frac{8F}{3}\right)^{1/4} (1 + \tilde{\gamma})^{1/8} (2\mathcal{V}_0)^{3/8} \frac{1}{\beta^{1/4}}. \tag{9.94}$$

For $\tilde{\gamma} = 0$ there is good agreement with the T model result (9.93). Like for the 2D case, there is indication that the propagation in the framework of the L model is shorter than in the T model case.

The results (9.93) and (9.94) suggest to rescale $r$ with $\beta^{-1/4}\mathcal{V}_0^{3/8}$. For consistency, $t$ should be rescaled with $\beta^{-1/2}\mathcal{V}_0^{1/4}$. The use of these rescaled variables will eliminate the dependency on $\beta$ and $\mathcal{V}_0$ from the governing balances (9.90)-(9.91). In other words, the box model runout time is $\sim \mathcal{V}_0^{1/4}/\beta^{1/2}$ for both T and L cases. We conclude that the present $r_{\max}$ and $t_{ro}$ runout length and time are consistent, and actually in fair agreement, with the more rigorous results derived from the self-similar solution in § 9.2.3.

## 9.3    Extensions of particle-driven solutions

Theoretical and experimental extensions for gravity-driven currents with constant flux and in rotating systems have also been derived; see Bonnecaze, Hallworth, Huppert, and Lister (1995), Dade and Huppert (1995b), Ungarish and Huppert (1998, 1999).

Another important extension is concerned with polydisperse suspensions. The suspended particles belong to two or more species which are typified by different settling speeds, due to different sizes and/or densities. In dilute systems the settling of the different species of particles can be decoupled. The T approach is a versatile tool for this problem: we can assume that the various species dilute independently in the same body of interstitial fluid of thickness $h(x,t)$. For each species $i$ a conservation equation of the type (9.27) for $\phi_i$ driven by $\beta_i$ is added to the continuity and momentum equations of the current. In the more common systems the dispersed particles are made up of the same material, i.e., have the same $\varepsilon_p$, and the different $\beta_i$ are due to variations in size. The interesting outcome is that the runout length of the current phenomenon is actually dominated by the smallest particles, which remain suspended for the largest period of time. In other words, the runout length of a polydisperse current is larger than that of an equivalent current with an average settling speed. The flow sustained by the small (fine) particles carries along also larger (coarse) particles. The pertinent experimental observations (for bidispersions) and the theoretical models are in fair agreement, but the details are beyond the scope of this book. The interested reader is referred to Gladstone, Philips, and Sparks (1998) and Harris, Hogg, and Huppert (2002).

## 9.4   Current over a porous bottom

The lock-released gravity current which propagates over a horizontal porous boundary is another important example of continuous loss of buoyancy. In this case the system runs out of the denser component simply because this is absorbed into the "ground" below $z = 0$. Applications include currents impinging on coastal shelfs and accidental collapse of storage tanks surrounded by gravel beds. Here we briefly discuss and model the effects of a porous horizontal boundary on the propagation and shape of high-Reynolds-number, homogeneous currents resulting from the instantaneous release of a finite volume. We follow the presentations of Thomas, Marino, and Linden (1998) and Ungarish and Huppert (2000).

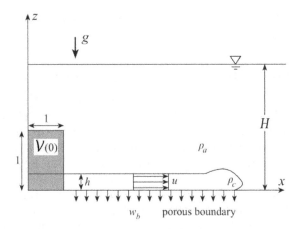

**FIGURE 9.9:** Schematic description of the system. The gray region represents the current at $t = 0$. (In the axisymmetric case $z$ is the axis of symmetry and the horizontal coordinate is the radius, $r$.)

It is emphasized the problems discussed in § 9.4 and and § 9.5 are concerned with propagation over a porous boundary. A gravity current which propagates *inside* a porous media over a solid boundary is an important, but different problem. The flow does not display the buoyancy-decay effects considered in this chapter; it is rather much closer to the viscous current, as discussed later in § 11.3.

The system under consideration is sketched in Fig. 9.9: a deep layer of ambient fluid, of density $\rho_a$, lies above a porous horizontal surface at $z = 0$. Gravity acts in the $-z$ direction. The porous substrate and the ambient fluid are in equilibrium. At time $t = 0$ a given volume of fluid, $\mathcal{V}(0)$, of density $\rho_c > \rho_a$ and kinematic viscosity $\nu$, initially at rest in a box of length $x_0$ and height $h_0$, is released into the ambient. For simplicity, we shall use the Boussinesq approximation. The reduced gravity of the current is

$$g' = \epsilon g; \quad \text{where} \quad \epsilon = \frac{\rho_c - \rho_a}{\rho_a}. \tag{9.95}$$

The initial total buoyancy, $g'\mathcal{V}(0)$, is like in the classical gravity current case. The special feature of the flow considered here is the lack of the no-penetration condition at $z = 0$. While the current propagates, fluid is also absorbed into the porous substrate with an expected double effect: the flow is slowed down due to loss of the effective buoyancy, and the practical distance of propagation is limited because the current runs out of fluid. The resulting flow bears some similarity with the particle-driven current discussed in the previous section. The resemblance with the L model is in particular close: in both cases the effective $g'$ of the current does not change, but the volume of fluid which displays this buoyancy excess over the ambient diminished with time.

The first task here is to incorporate the drainage effect of the porous substrate into the modeling of the flow.

### 9.4.1   The porous-boundary velocity condition

The porous boundary is conveniently incorporated into our analysis as a boundary condition for the vertical velocity component at the bottom, $w(x, z = 0, t)$, denoted $w_b$. This is of course equal to the rate of volume flux, or discharge, through the porous substrate. The underlying idea is as follows. The axial flow is slow, by which we mean that the Reynolds number of the flow through the porous media, $Re' = |w_b|d/\nu$, is smaller than about 10; here $d$ is the diameter of the pores. In this case the flow in the $z \leq 0$ domain is governed by Darcy's law (see Batchelor 1981). Since the relevant pressures are hydrostatic and the layer of dense fluid is relatively thin, the dominant pressure gradient is in the $-z$ direction. The resulting approximation, for a porous substrate of specific permeability $k$, is

$$\frac{\nu w}{k} = -\frac{1}{\rho_c}\frac{\partial p}{\partial z} - g' \quad (z \leq 0). \tag{9.96}$$

Suppose that the instantaneous thickness of the dense fluid layer in the porous domain is $l$. The reduced pressure in the substrate varies linearly from $\rho_c g' h$ at $z = 0$ (provided by the current above the floor) to zero at $z = -l$. For the corresponding $\partial p / \partial z$ the balance (9.96) yields

$$w = -g' \frac{k}{\nu} (\frac{h}{l} + 1) \quad (-l \le z \le 0).\qquad (9.97)$$

A detailed discussion of this derivation and experimental verification are presented in Acton, Huppert, and Worster (2002).

In general, $l$ is a function of $t$ (and $x$) whose initial condition is 0. This introduces a new dependent variable in the gravity current problem. The matching condition with the current is the volume conservation in the domain $0 < x < x_N(t)$ and $-l(x,t) \le z \le h(x,t)$. This provides the additional equation for the new variable. In this case the assumption is that the porous media is sufficiently thick ($> h_0$ say). The solution of a *viscous* current in such circumstances is presented in Acton, Huppert, and Worster (2002).

When the porous substrate is a thin layer of given thickness $l$ above a medium of higher permeability, the $h/l$ term in the RHS of (9.97) is dominant and $l$ becomes quickly a constant. This simplification was used by Thomas, Marino, and Linden (1998) to suggest (using dimensional variables)

$$w_b(x,t) = -\frac{1}{\tau} h(x,t),\qquad (9.98)$$

where $\tau = \nu l/(g'k)$. The time-constant $\tau$ can be measured in the laboratory with modest efforts (at least for representative systems as used by Thomas, Marino, and Linden 1998). Moreover, for a given porous substrate and with the effects of saturation of the pores and re-entrainment into the current from the substrate excluded, $\tau$ is a constant. This facilitates the analysis and interpretation of results, as shown below. The advantage of this simplification is that the flow of the current is decoupled from the details of the flow in the $z < 0$ domain.

We adopt the condition (9.98) as a prototype of the porous-boundary effect, and assume that the necessary value of $\tau$ is provided. This is a sufficient condition for the incorporation of the porous-boundary effect in the SW formulation. For simplicity (and also due to lack of more detailed information) we usually assume that $\tau$ is a constant on the porous portion of the bottom boundary. If an impermeable portion is present (usually in the lock domain) the boundary condition there is $w(x, z = 0, t) = 0$ which can be treated as the limit $\tau \to \infty$ of (9.98).

## 9.4.2 The shallow-water approximation

Consider a 2D configuration as sketched in Fig. 9.9. The velocity components are $\{u, w\}$. The motion of the interface is governed by the mechanism discussed in § 2.1. The nose is treated as a vertical shock with a prescribed $dx_N/dt$. The motion of the inclined (and vertical) portion of the interface, $z = h(x, t)$, is given again by (2.3b) repeated here,

$$\frac{\partial h}{\partial t} + u_h \frac{\partial h}{\partial x} - w_h = 0, \tag{9.99}$$

where $u_h$ and $w_h$ are the velocity components of the fluid at the interface. The vertical speed component is connected with $u$ via the continuity equation $\nabla \cdot \mathbf{v} = 0$. Subject to the boundary condition $w(z = 0) = w_b$, we thus write

$$\frac{\partial}{\partial x} \int_0^h u \, dz = u_h \frac{\partial h}{\partial x} + \int_0^h \frac{\partial u}{\partial x} dz = u_h \frac{\partial h}{\partial x} - w_h + w_b. \tag{9.100}$$

Consequently, for a 2D flow over a porous boundary (9.99) can be rewritten as

$$\frac{\partial h}{\partial t} + \frac{\partial}{\partial x} \int_0^h u \, dz = w_b = -\frac{h}{\tau}, \tag{9.101}$$

where (9.98) was also used. This is the "continuity equation" of the present SW model.

The momentum equation is derived for a one-layer model. We use again the thin-layer arguments of § 2.2 for $\delta = L_z/L_x \ll 1$. The additional assumption is that the porous flux $w_b$ is very small as compared with the typical horizontal $U$; to be precise, let $w_b = O(\delta U)$. The hydrostatic pressure approximations in the ambient and current remain valid, and the fundamental connection (2.15) between the buoyancy pressure driving and interface inclination can be applied. With $\Delta \rho = \rho_c - \rho_a$ this is expressed as

$$\frac{\partial p_c}{\partial x} = \Delta \rho g \frac{\partial h(x, t)}{\partial x} = \rho_a g' \frac{\partial h(x, t)}{\partial x}. \tag{9.102}$$

To determine the averaged momentum governing equation we start with the inviscid $x$-momentum balance of the current in conservation form

$$\frac{\partial u}{\partial t} + \frac{\partial u^2}{\partial x} + \frac{\partial}{\partial z}(uw) = -\frac{1}{\rho_c} \frac{\partial p_c}{\partial x}. \tag{9.103}$$

We integrate (9.103) to obtain

$$\int_0^h \left[ \frac{\partial u}{\partial t} + \frac{\partial u^2}{\partial x} \right] dz + u_h w_h - w_b u(z = 0) = -\frac{1}{\rho_c} \int_0^h \frac{\partial p_c}{\partial x} dz. \tag{9.104}$$

In the averaged formulation $u_h = u(z = 0)$. On the other hand, it can be verified that, on account of (9.99),

$$\frac{\partial}{\partial t} \int_0^h u \, dz + \frac{\partial}{\partial x} \int_0^h u^2 \, dz - u_h \left( \frac{\partial h}{\partial t} + u_h \frac{\partial h}{\partial x} - w_h \right) =$$

$$\text{LHS of (9.104)} + w_b u(z = 0). \tag{9.105}$$

Combining these equations with the definition of the average $u$, see (2.20)-(2.21), we conclude that the simplified SW momentum equation is like for the classical case, (2.23), with an additional $w_b \bar{u} = -\bar{u} h / \tau$ term on the RHS. This reads

$$\frac{\partial uh}{\partial t} + \frac{\partial u^2 h}{\partial x} = -\rho_a g' \frac{1}{2} \frac{\partial h^2}{\partial x} - \frac{uh}{\tau}. \qquad (9.106)$$

We used the Boussinesq assumption and dropped the overbar notation for the average $u(x,t)$.

We obtained a system of equations for the position of the interface, $h(x,t)$, and for the $z$-averaged longitudinal velocity, $u(x,t)$. The difference from the classical current over an impermeable boundary is the presence of source terms proportional to $1/\tau$ on the RHS of the continuity and momentum equations (9.101) and (9.106). These terms are negative and represent the loss of volume and momentum from the current into the porous boundary. The non-zero vertical flux on the boundary $z = 0$ is a simple function of $h(x,t)$, provided by the closure (9.98). For a given $\tau$ the dependent variables are $h$ and $u$, like in the classical case.

### 9.4.3 The SW equations

It is convenient to scale the dimensional (denoted by the asterisks) variables as follows

$$\{x^*, h^*, H^*, t^*, u^*, w^*\} = \{x_0 x, h_0 h, H h_0, T t, U u, U(h_0/x_0) w\}, \qquad (9.107)$$

where

$$U = (h_0 g')^{1/2} \quad \text{and} \quad T = \frac{x_0}{U}. \qquad (9.108)$$

This is the standard scaling of the 2D gravity current. The volume $\mathcal{V}$ (per unit width) is scaled with $x_0 h_0$.

This scaling of the problem produces the new non-dimensional parameter

$$\lambda = \frac{T}{\tau}, \qquad (9.109)$$

which simply reflects the ratio between the typical time of propagation of the nose to the typical time of descent of the interface due to the porosity of the boundary, the former over a length $x_0$ and the latter over a height $h_0$.

In the following discussion, we use *dimensionless variables* unless stated otherwise. The porous boundary condition (9.98) reads now

$$w_b(x,t) = -\lambda h(x,t). \qquad (9.110)$$

It can be anticipated that the interesting cases have small values of $\lambda$, because otherwise the fluid has drained into the porous boundary before any significant propagation occurs.

In conservation form the equations can be written as

$$\frac{\partial h}{\partial t} + \frac{\partial}{\partial x}(uh) = -\lambda h \tag{9.111}$$

and

$$\frac{\partial}{\partial t}(uh) + \frac{\partial}{\partial x}\left[u^2 h + \frac{1}{2}h^2\right] = -\lambda uh. \tag{9.112}$$

In characteristic form, this becomes

$$\begin{bmatrix} h_t \\ u_t \end{bmatrix} + \begin{bmatrix} u & h \\ 1 & u \end{bmatrix}\begin{bmatrix} h_x \\ u_x \end{bmatrix} = \begin{bmatrix} -\lambda h \\ 0 \end{bmatrix}. \tag{9.113}$$

We note that the porous-boundary condition appears in the equations as a source (rather sink) term. It is important to keep in mind that a non-porous section of the boundary, typically in the lock region $0 \leq x \leq 1$, may be present. This is readily accounted for by setting $\lambda = 0$ in the equations of motion for the impermeable region. Actually, we observe that $\lambda$ can be allowed to vary with $x$ and $t$ without affecting the foregoing formulation. This may model motion over a boundary of complex structure and prone to partial saturation. However, this flexibility still awaits implementation and confirmation.

The initial conditions are simply zero velocity and unit dimensionless height and length at $t = 0$.

The velocity variable $u$ is zero at the backwall $x = 0$. In addition, the usual boundary condition for the velocity at the nose can be used when $\lambda \ll 1$,

$$u_N = Fr h_N^{1/2}, \tag{9.114}$$

where $Fr$ is a function of $a = h_N/H$. The justification follows from the jump-condition analysis of § 3.2: when both $\lambda$ and the $x$-thickness of the jump control volume are small, the horizontal in- and out-flows dominate the continuity and force balances. We use again the HS $Fr$ formula (2.27).

The system is hyperbolic. It is left for an exercise to show that the relationships between the variables on the characteristics are

$$dh \pm \sqrt{h}\,du = -\lambda h dt \quad \text{on} \quad \frac{dx}{dt} = u \pm \sqrt{h}. \tag{9.115}$$

### 9.4.4 The global volume balance

Let the value of $\lambda$ of the porous domain be a constant. The negative flux is proportional to the local height $h$ of the dense fluid. As the current spreads out the height decreases but the area of absorption increases. We infer that the rate of volume loss is proportional to the volume. The details can be elaborated as follows.

Let $\mathcal{V}(t)$ denote the total volume of the dense fluid, and $\mathcal{V}_1(t)$ the volume above the impermeable boundary (typically, $0 \leq x \leq 1$). The respective expressions are

$$\mathcal{V}(t) = \int_0^{x_N(t)} h(x,t)dx; \quad \mathcal{V}_1(t) = \int_0^1 h(x,t)dx. \qquad (9.116)$$

On the other hand, for constant $\lambda$ and using (9.110), we can write

$$\lambda \mathcal{V}(t) = \lambda \int_0^1 h(x,t)dx + \int_1^{x_N(t)} \lambda h(x,t)dx = \lambda \mathcal{V}_1(t) - \int_1^{x_N(t)} w_b dx. \qquad (9.117)$$

The last term represents the rate of loss of volume into the porous boundary, $-d\mathcal{V}/dt$. Combining (9.116) and (9.117) we obtain the global volume decay behavior

$$\frac{d\mathcal{V}}{dt} = -\lambda \left[ 1 - \frac{\mathcal{V}_1(t)}{\mathcal{V}(t)} \right] \mathcal{V}. \qquad (9.118)$$

An upper limit for the rate of decay is obtained by the simplification $\mathcal{V}_1 = 0$, which implies that the boundary below the lock is also porous. In this case (9.118) yields

$$\mathcal{V}(t) = \mathcal{V}(0)e^{-\lambda t}. \qquad (9.119)$$

However, in the more realistic case the bottom below the lock reservoir is not porous. In the initial stages of propagation the ratio $\mathcal{V}_1/\mathcal{V}$ is close to 1 and hence only a slight decrease of volume occurs, while the flow field is expected to follow closely the classical dam-break pattern. Only for $t > 2$ (approximately), after the left-moving rarefaction perturbation reaches the backwall and is well reflected into the current, the volume $\mathcal{V}_1$ of fluid in the lock domain becomes small and an exponential-like decay of $\mathcal{V}$ appears. Overall, the rate of volume decrease is smaller than estimated by the simplified exponential decay equation (9.119). The discrepancy is more pronounced for larger values of $\lambda$.

### 9.4.5   Some results

The SW equations for the current on a porous boundary must be solved numerically. The finite-difference scheme used for the classical $\lambda = 0$ case can be straightforwardly extended to the $\lambda > 0$ case by the incorporation of the new terms in the discretized equations. These terms remain simple even when $\lambda$ is given as a function of $x$ and $t$. Only cases which correspond to small values of $\lambda$ are of interest; otherwise the current is drained before any significant propagation can occur. The computations can be stopped at $t \approx 3/\lambda$, when the current has practically run out of fluid (volume reduction to about $e^{-3} = 0.05$). The runout distance $x_{\max}$ is about $2/\lambda^{2/3}$ as shown later in § 9.4.6.

The SW numerical solutions presented here are for the full-depth lock release case $H = 1$. This was also the setup in the experiments of Thomas, Marino, and Linden (1998). The solutions were obtained on a numerical grid of 200 points and with a time step of $5 \cdot 10^{-3}$.

Fig. 9.10 gives a quick insight into the influence of the parameter $\lambda$ on the propagation distance and volume decay in configurations with an impermeable boundary in the lock region $0 \leq x \leq 1$. The porosity has little influence on the propagation during an initial period ($t < 12$). The explanation for this is given below. Afterward, the porosity slows the current down and shortens the effective distance of propagation. It is evident that these effects become more pronounced as the parameter $\lambda$ increases. Computations with non-zero $\lambda$ where stopped when either one of the following conditions was achieved: the volume decreased to 5% of the initial value, or $h_N$ became smaller than 0.01. The volume $\mathcal{V}(t)/\mathcal{V}(0)$ decreases slower than $e^{-\lambda t}$ because of the presence of the non-porous portion of the boundary; see (9.118).

More detailed results for the shallow-water equations in a rectangular configuration are presented in Figs. 9.11-9.12. Two cases are presented for comparison: in case A $\lambda = 0.1$ and the porous region extends for $x > 1$; and in case B $\lambda = 0.1$ and the porous region starts at the origin.

In the initial stages of the motion, for $t < 15$, and in particular for $t < 5$, the qualitative behavior of the interface is like in the classical case of propagation over a solid bottom. The fluid initially collapses from the front to $h_N \approx$ 0.5; in the classical case ($\lambda = 0$ say) the constant height of the head and speed persist during the slumping stage. The porosity causes a continuous decrease of $h_N$, as expected. The classical $\lambda = 0$ current tends to approach the similarity form of solution with a head-up, tail-down smooth profile and with $u$ a linear function of $x$. Here the porosity tends to strengthen the differences between the head and the tail regions, especially in case A where $u$ becomes a non-monotonic function of $x$. Quantitatively, up to $t \approx 10$ the distance of propagation is almost the same in the porous and non-porous cases, despite the fact that the height of the nose decreases considerably more in the porous cases. The reason is the behavior of $Fr$ with $h_N$, cf. (2.27), which makes the velocity of propagation a weak function of $h_N$ ($u_N \propto h_N^{1/6}$). In the deep stage of propagation, $h_N/H < 0.075$, the porous drainage has a more significant influence on the velocity of propagation.

Comparing cases A and B we notice that the non-porous portion $x < 1$ keeps the fluid at a higher level than in the tail which follows the head. In addition, the overall volume in case A is higher than the $e^{-\lambda t}$ behavior in case B, and therefore the propagation is slightly faster in the former case.

The comparisons of SW results with available experiments are not conclusive. A relevant set of experiments has been performed by Thomas, Marino, and Linden (1998). They considered the propagation of a saline current into freshwater in a rectangular container over a porous boundary (which they represented by a thin metallic grid below which a layer of freshwater was placed).

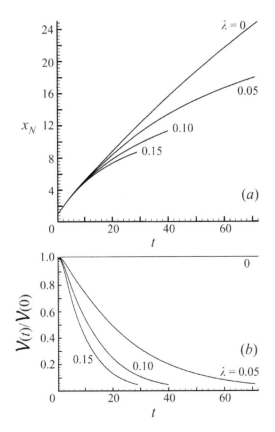

**FIGURE 9.10:**   SW results of $x_N$ and $\mathcal{V}(t)/\mathcal{V}(0)$ as functions of $t$ for various $\lambda$. The $\lambda > 0$ cases stopped when the volume reached about 0.05 of initial value. (Ungarish and Huppert 2000, with permission.)

The initial height of the current in the lock was $h_0 = 20$ cm, the same as that of the ambient fluid, i.e., $H = 1$. The length $x_0$ of the lock was between 15 and 37 cm, and the boundary in this region was non-porous. The width of the tank was 15.2 cm, and the length from the gate to the far-wall was 195 cm. Using $g' = 9.8$, 49, or 98 cm s$^{-2}$, the resulting values of $\lambda$ were in the range $0.02 - 0.24$. A small amount of dye was added to the current for visualization and video-tape image recording.

The small value of $\lambda$ in these experiments inhibits the rapid piling-up of a thick layer of saltwater in the region below the porous grid, and hence prevents the formation of a significant gravity current in this region. The fluid below the grid had a passive role during the propagation of the current above the grid.

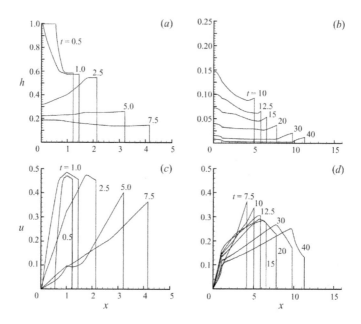

**FIGURE 9.11:** SW results for case A with $\lambda = 0.1$ for $x > 1$ and an impermeable wall for $x \leq 1$. (a),(b) $h$ as a function of $x$ at various $t$; (c),(d) $u$ as a function of $x$ at various $t$. (Ungarish and Huppert 2000, with permission.)

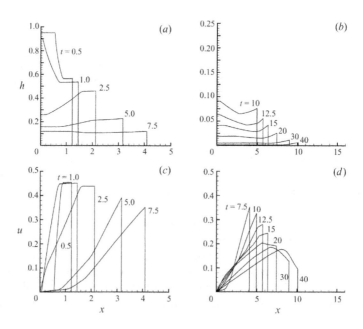

**FIGURE 9.12:** SW results for case B with $\lambda = 0.1$ for $x > 0$. (a),(b) $h$ as a function of $x$ at various $t$; (c),(d) $u$ as a function of $x$ at various $t$. (Ungarish and Huppert 2000, with permission.)

Data of propagation $x_N$ and volume $\mathcal{V}$ as functions of $t$ were obtained. However, the values of $\lambda$ in the experimental data are known only approximately. This parameter (actually, the dimensional $\tau$) was determined by an approximate post-processing of the $\mathcal{V}(t)$ under the assumption that the decay is exponential according to (9.119). As explained above, the real process is expected to be slower. This means that the fit to the exponential decay will produce an apparently larger $\tau$ than the real one. Attempts to improve these estimates by recovering the observed $\mathcal{V}(t)$ from SW calculations of $\mathcal{V}(t)$ according to (9.118) were made by Ungarish and Huppert (2000). The results indicate a typically 35% excess in the originally reported $\tau$. However, the accuracy of the $\lambda$ results obtained with this correction is still unknown.

In any case, the experimental propagation results are consistent with the SW predictions. The role of $\lambda$ as the governing dimensionless parameter of the porous-boundary effect has been confirmed. On the other hand, we must keep in mind that the guiding lines in the development of the theoretical model were provided by that experimental setup: the porous boundary behaves like a thin metallic grid that separates between the upper and the lower domains of the ambient fluid. The limits of applicability of this model are not clear.

### 9.4.6 Box models

The current is viewed as a control volume of rectangular cross-section, i.e., of height $h_N(t)$ and length $x_N(t)$. For the two unknowns $h_N$ and $x_N$ we need two equations; we use, again, the total volume balance and the nose-propagation jump condition. The initial conditions are $x_N = h_N = \mathcal{V}(0) = 1$ at $t = 0$.

For simplicity here we assume that *the porous floor starts at the origin*. The volume balance (9.119) yields

$$h_N(t) = e^{-\lambda t}/x_N(t), \qquad (9.120)$$

and the nose propagation condition can be written

$$\frac{dx_N}{dt} = Fr h_N^{1/2}, \qquad (9.121)$$

where $Fr$ is given by the HS function of $a = h_N/H$.

Consider the shallow stage of propagation with changing $Fr$; see (2.27). This appears only if the initial depth ratio $H < 1/0.075$. For this case, we substitute $Fr = (1/2)H^{1/3}h_N^{-1/3}$ into (9.121) and eliminate $h_N$ with the aid of (9.120). We obtain an equation for $x_N(t)$, which can be integrated to yield

$$x_N(t) = \left[\frac{7}{2\lambda}H^{1/3}(1 - e^{-\lambda t/6}) + 1\right]^{6/7} \quad (1 < x_N < e^{-\lambda t}/(0.075H)).$$
$$(9.122)$$

Suppose that the shallow propagation terminates at time $t_I$ when the nose is at $x_I$.

Consider the deep stage of propagation during which $Fr$ is constant (= 1.19). We use again (9.121), and eliminate $h_N$ with the aid of (9.120). The resulting equation, subject to the initial condition $x_N = x_I$ at $t_I$, yields

$$x_N(t) = \left[ 3Fr\frac{1}{\lambda}(e^{-\lambda t_I/2} - e^{-\lambda t/2}) + x_I^{3/2} \right]^{2/3} \quad (t \geq t_I). \qquad (9.123)$$

If the currents starts in the deep regime, we use in the last equation $x_I = 1$ and $t_I = 0$. The distance of propagation when the current runs out of fluid is

$$x_{\max} \approx \left( \frac{3Fr}{\lambda} \right)^{2/3}. \qquad (9.124)$$

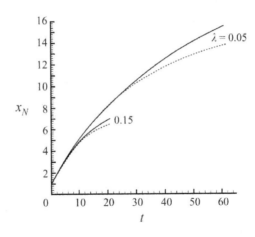

**FIGURE 9.13:** SW (solid line) and box-model (dotted line) 2D calculations for propagation as function of $t$ for two values of $\lambda$. The porous boundary starts at the origin and $H = 1$. (Ungarish and Huppert 2000, with permission.)

A comparison between the predicted propagation by the box-model results to those obtained by the SW equations for two values of $\lambda$ is shown in Fig. 9.13. The agreement is good. The box-model approximation predicts a slower propagation than the SW equations. This could be anticipated because in the SW (and also real) profiles the value of $h_N$ becomes quickly larger than the average $h$ and hence the velocity of the nose is larger than that obtained under the box-model assumption that $h = h_N$.

## 9.5 Axisymmetric current over a porous bottom

Consider now the axisymmetric counterpart of the 2D current over a porous boundary. The current is released from a cylinder lock of radius $r_0$ and height $h_0$. The volume is $\mathcal{V}$ (per radian). The propagation is in the radial direction in either a fully circular domain or a wedge.

The flux into the porous boundary is unaffected by the geometry of the horizontal flow, and is given (in dimensional form) by

$$w_b(r, t) = -\frac{1}{\tau} h(r, t). \tag{9.125}$$

The typical time constant $\tau$ is the same as in the 2D case, proportional to $\nu/g'$ and also depends on the porous properties of the boundary. We assume that $\tau$ is a known property of the bottom boundary.

We scale the variables like in (9.107)-(9.108), in which $x$ is replaced by $r$. The volume $\mathcal{V}$ is scaled with $r_0^2 h_0$. The porous effect is represented by the dimensionless parameter

$$\lambda = \frac{T}{\tau} = \frac{r_0}{(g' h_0)^{1/2} \tau}. \tag{9.126}$$

The dimensionless flux into the boundary porous boundary is given now by

$$w_b(r, t) = -\lambda h(r, t). \tag{9.127}$$

In the following discussion, we use *dimensionless variables* unless stated otherwise.

### 9.5.1 The SW equations

The derivation of the equations is similar to the 2D case. We use the one-layer model. The pressure is hydrostatic in both the current and ambient, and the driving pressure $\partial p_c / \partial r$ is connected to the inclination of the interface $\partial h / \partial r$ like in the classical case. The mathematical details are similar to these of § 9.4.2 and § 6.1 and are left for an exercise. The result is a system of volume and radial momentum balances, and the dependent variables are the thickness $h(r, t)$ and radial average speed $u(r, t)$ in the domain $0 \le r \le r_N(t)$.

In conservation form the SW equations can be written as

$$\frac{\partial h}{\partial t} + \frac{\partial}{\partial r}(uh) = -\frac{uh}{r} - \lambda h \tag{9.128}$$

and

$$\frac{\partial}{\partial t}(uh) + \frac{\partial}{\partial r}\left[u^2 h + \frac{1}{2}h^2\right] = -\frac{u^2 h}{r} - \lambda uh. \tag{9.129}$$

In characteristic form, the equations are

$$\begin{bmatrix} h_t \\ u_t \end{bmatrix} + \begin{bmatrix} u & h \\ 1 & u \end{bmatrix} \begin{bmatrix} h_r \\ u_r \end{bmatrix} = \begin{bmatrix} -uh/r - \lambda h \\ 0 \end{bmatrix}. \tag{9.130}$$

The presence of a non-porous section of the boundary, typically in the lock region $0 \leq r \leq 1$, can be accounted for by setting $\lambda = 0$ in the equations of motion for the impermeable region. We note that $\lambda$ can be allowed to vary with $r$ and $t$ without affecting the foregoing formulation.

The initial conditions are simply zero velocity and unit dimensionless height and radius at $t = 0$. The velocity variable $u$ is zero at the center $r = 0$. In addition, the usual boundary condition for the velocity at the nose, (9.114), can be used.

The system is hyperbolic. It is left for an exercise to show that the relationships between the variables on the characteristics are

$$dh \pm \sqrt{h}du = -\left(\frac{uh}{r} + \lambda h\right) dt \quad \text{on} \quad \frac{dr}{dt} = u \pm \sqrt{h}. \tag{9.131}$$

## 9.5.2   The global volume balance

Let $\mathcal{V}(t)$ denote the total volume of the dense fluid, and $\mathcal{V}_1(t)$ the volume above the impermeable boundary (typically, $0 \leq r \leq 1$). The respective expressions (per radian) are

$$\mathcal{V}(t) = \int_0^{x_N(t)} h(r,t)rdr; \quad \mathcal{V}_1(t) = \int_0^1 h(r,t)rdr. \tag{9.132}$$

We assume that $\lambda$ is constant for the porous boundary. It is left for an exercise to show that, like in the 2D counterpart, the global volume decay behavior is given by

$$\frac{d\mathcal{V}}{dt} = -\lambda\left[1 - \frac{\mathcal{V}_1(t)}{\mathcal{V}(t)}\right]\mathcal{V}. \tag{9.133}$$

An upper limit for the rate of decay is obtained by the simplification $\mathcal{V}_1 = 0$, which implies that the boundary below the lock is also porous. In this case (9.133) yields

$$\mathcal{V}(t) = \mathcal{V}(0)e^{-\lambda t}. \tag{9.134}$$

However, in the more realistic case the bottom below the lock reservoir is not porous. In the initial stages of propagation the ratio $\mathcal{V}_1/\mathcal{V}$ is close to 1 and hence only a slight decrease of volume occurs, while the adjustment is expected to follow the classical dam-break pattern. Only for $t > 2$ (approximately), after the left-moving rarefaction perturbation reaches the center and is well reflected into the current, the volume $\mathcal{V}_1$ of fluid in the lock domain becomes small and an exponential-like decay of $\mathcal{V}$ appears. Overall, the rate of volume decrease is smaller than estimated by the simplified exponential decay equation (9.134). The discrepancy is more pronounced for larger values of $\lambda$.

### 9.5.3 Some results

The SW equations must be solved numerically. Again, only cases which correspond to small values of $\lambda$ are of interest; otherwise the current is drained before any significant propagation can occur. The computations can be stopped at $t \approx 3/\lambda$, when the axisymmetric current has practically run out of fluid (volume reduction to about $e^{-3} = 0.05$). The runout radius $r_{\max}$ is about $2/\lambda^{1/2}$ as shown later in § 9.5.4. The SW numerical solutions presented here are for the full-depth lock release case $H = 1$, as in the 2D case discussed above.

Fig. 9.14 gives a quick insight into the influence of the parameter $\lambda$ on the propagation distance and volume decay in configurations with an impermeable boundary in the lock region $0 \le r \le 1$.

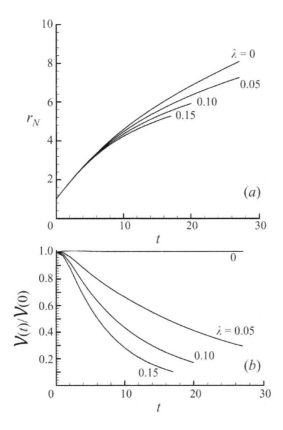

**FIGURE 9.14:** Axisymmetric SW results of $r_N$ and $\mathcal{V}(t)/\mathcal{V}(0)$ as functions of $t$ for various $\lambda$. The $\lambda > 0$ cases stopped when $h_N \approx 0.01$. (Ungarish and Huppert 2000, with permission.)

Further SW results for the axisymmetric configuration are presented in Figs. 9.15-9.16. Again, in case A $\lambda = 0.1$ and the porous region extends for $r > 1$; and in case B $\lambda = 0.1$ and the porous region starts at the origin. The qualitative behavior is as in the rectangular case, but the reduction of $h_N$ is proportional to $r_N^2$. Some oscillations appear near the center for $t > 5$ which seem to be introduced by the reflection (from the axis) of the backward-moving rarefaction wave. The porous drainage tends to amplify these oscillations, and it is difficult to decide if this is a numerical perturbation or an indication of a physical instability. However, since the amount of fluid in the relevant region is relatively small this is not expected to be of significance to the propagating current.

### 9.5.4 Box models

The current is assumed to be a cylinder of radius $r_N(t)$ and height $h_N(t)$. The initial conditions are $r_N = h_N = 1$ at $t = 0$.

For simplicity here we assume that *the porous floor starts at the origin*. The volume balance (9.134) yields

$$h_N(t) = e^{-\lambda t}/r_N^2(t), \tag{9.135}$$

and the nose propagation condition reads

$$\frac{dr_N}{dt} = Fr h_N^{1/2}, \tag{9.136}$$

where $Fr$ is given by the HS function of $a = h_N/H$; see (2.27).

We proceed as done above for the rectangular case. For the shallow stage with changing $Fr$ we obtain

$$r_N(t) = \left[4\frac{1}{\lambda}H^{1/3}(1 - e^{-\lambda t/6}) + 1\right]^{3/4} \quad (1 \le r_N \le \left(\frac{e^{-\lambda t}}{0.075H}\right)^{1/2}). \tag{9.137}$$

Suppose that the shallow propagation terminates at time $t_I$ when the front is at radius $r_I$.

For the subsequent deep stage of propagation we obtain

$$r_N(t) = \left[\frac{4Fr}{\lambda}(e^{-\lambda t_I/2} - e^{-\lambda t/2}) + r_I^2\right]^{1/2} \quad (t \ge t_I). \tag{9.138}$$

If the currents starts in the deep regime, we use in the last equation $r_I = 1$ and $t_I = 0$. The radius of propagation when the current runs out of fluid is

$$r_{\max} \approx \left(\frac{4Fr}{\lambda}\right)^{1/2}. \tag{9.139}$$

A comparison between the predicted propagation in axisymmetric circumstances of the box-model results to those obtained by the SW equations for

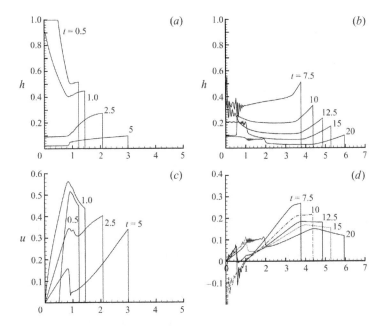

**FIGURE 9.15**: Axisymmetric SW results for case A with $\lambda = 0.1$ for $r > 1$ and impermeable wall for $r \leq 1$. (a), (b) $h$ as function of $r$ at various $t$; (c),(d) $u$ as a function of $r$ at various $t$. (Ungarish and Huppert 2000, with permission.)

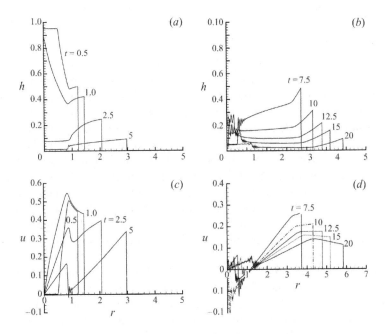

**FIGURE 9.16**: Axisymmetric SW results for case B with $\lambda = 0.1$ for $r > 0$. (a), (b) $h$ as function of $r$ at various $t$; (c),(d) $u$ as a function of $r$ at various $t$. (Ungarish and Huppert 2000, with permission.)

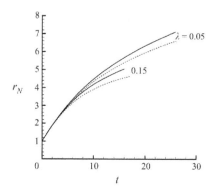

**FIGURE 9.17:** SW (solid line) and box-model (dotted line) axisymmetric calculations for propagation as function of $t$ for two values of $\lambda$. The porous boundary starts at the origin and $H = 1$. (Ungarish and Huppert 2000, with permission.)

two values of $\lambda$ is shown in Fig. 9.17. The agreement is good. As for the rectangular case, the box-model approximation predicts a slower propagation than the SW equations.

## 9.6   Entrainment

Entrainment is an intriguing effect. In the ideal situation we can distinguish sharply between a bulk of known volume of "current" of density $\rho_c$, and the "ambient" of fixed volume and density $\rho_a$. For definiteness let $\rho_c > \rho_a$. In practical situations there is some mixing between the fluids. Domains of intermediate density develop. This effect is more pronounced and important in the region of the propagating head which, according to our models, governs the speed of propagation. Small viscous effects and Kelvin-Helmholtz instabilities shear away dense fluid from the upper interface. Consequently, the volume of the current decreases with distance. On the other hand, the nose overrides less dense fluid. (Accurate numerical illustrations of the head structure can be seen in Härtel, Meiburg, and Necker 2000 and Härtel, Carlsson, and Thunblom 2000). The overridden thin layer of $\rho_a$ is buoyant with respect to the current, and thus the fluid captured in this layer is expected to rise into the body of dense fluid above. Evidently, the density contrast of current to ambient and hence the effective reduced gravity $g'_e$ are bound to decrease. The term "entrainment" covers, roughly, the changes of volume and density of the current due to the mixing effects. Here we follow the presentation of Hallworth, Huppert, Phillips, and Sparks (1996).

Rigorous theoretical derivations and predictions concerning the entrainment are lacking, and the relevant knowledge has been derived mostly from experimental measurements, scaling arguments, and *ad hoc* models. In the experiments it is necessary to measure the diluted volume and effective density as functions of time. This can be done by optical contrast visualization of the entire current (e.g., comparing photographs of the cross section of the dyed current) and analysis of selectively withdrawn samples. Hallworth, Huppert, Phillips, and Sparks (1996) used a chemical method as follows. The current contained a base of known concentration and some universal pH indicator, and the ambient was acidified. The acid entrains the moving current head and is instantaneously mixed. At a distance from the gate which depended on the initial excess concentration of the alkali, the amount of entrained acid has become sufficient to neutralize the fluid. This point was accurately detected by an abrupt change of color of the pH indicator. Experiments with different initial concentrations of alkali, but otherwise identical, mapped out accurately the proportion of the entrained volume as a function of the distance of propagation. The experiments were done for the full-depth lock release only.

The experiments show that the form of the entrainment in the horizontal flow of a gravity current is different from that used for a vertically rising plume. Only in the "head" of the current does a significant entrainment effect show up.

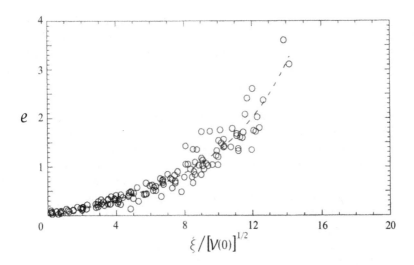

**FIGURE 9.18:** Measured entrainment ratio $e$ for 2D full-depth-lock currents over a smooth floor as a function of $\xi = (x_N - x_s)/[\mathcal{V}(0)]^{1/2}$. For $x_N < x_s$ no entrainment was observed. The dotted line is a curve-fit to the data. (Hallworth, Huppert, Phillips, and Sparks 1996, with permission.)

The dilution of the head can be expressed by

$$\tilde{\rho} - \rho_a = \frac{\rho_c - \rho_a}{e + 1}, \qquad (9.140)$$

where $\tilde{\rho}$ is the effective density of the head of the current and $e$ is the dimensionless "entrainment ratio." More explicitly, the entrainment ratio is given by

$$e = \frac{\rho_c - \rho_a}{\tilde{\rho} - \rho_a} - 1. \qquad (9.141)$$

A summary of experimental values of $e$ is shown in Fig. 9.18.

The first intriguing experimental result is that $e = 0$ during the slumping phase. In other words, right to the end of the slumping phase the head remains essentially undiluted. This is surprising, because the typical structure of the head (with lobe-and-cleft instabilities and overridden light fluid, as mentioned above) is present during the slumping stage; some mixing definitely occurs through shear instabilities at the interface of the current, and fluid is lost from the head. In the 2D geometry the position $x_N = x_s$ where the slumping ends is well defined as the end of the constant-speed propagation. In the axisymmetric case this is less rigorously defined as the radius of the current when the reflected rarefaction wave (or jump) reaches the nose. It is not clear why there is such a sharp transition of the entrainment mechanism at the end of the slumping. Experimental evidence and scaling analysis indicate that $e$ is independent of the initial $g'$ (i.e., $\Delta\rho$) of the current.

Hallworth, Huppert, Phillips, and Sparks (1996) developed a fairly comprehensive kinematic description of the entrainment. The resulting $e$ as a function of distance of propagation from the end of the slumping, $\xi = x_N - x_s$, displays the correct trends, in agreement with observations. However, these models contain three adjustable dimensional parameters which must be determined empirically: two entrainment constants $\alpha$ and $k$ (typical values 0.07 and 0.15, respectively) and a head shape factor $S_H$ (typical value 0.8). Roughly, the head entrains ambient fluid at the rate $\alpha h_H u_N$ and drains into the tail the instantaneously present fluid at the rate $k h_H u_N$. Here $h_H$ is the height of the head, which depends on the volume of the head and the shape factor. The kinematic model indicates that the importance of entrainment is proportional to the value of the lock aspect ratio parameter, $h_0/x_0$. (These results were also extended to the axisymmetric geometry.)

The obvious implication of (9.140) is that the effective reduced gravity which drives the current decays like $(1 + e)^{-1}$. According to Fig. 9.18 a very significant loss of buoyancy is expected after the slumping stage. Moreover, the current also losses volume at the interface, as a mixed wake. (According to the model of Hallworth, Huppert, Phillips, and Sparks 1996 the head of the current will practically run out of fluid at some finite $\xi$.) Overall, a large deviation from the ideal situation is indicated by the entrainment scenario presented above.

The second intriguing fact is that, in spite of these apparently severe buoyancy losses due to entrainment, there is in general good agreement between experimental propagation data and theoretical predictions based on models which completely discard the entrainment effects. Observed high-$Re$ two-dimensional gravity currents display constant slumping speed, enter into the self-similar phase where propagation is like $t^{2/3}$, and next enter into the viscous stage where propagation is like $t^{1/5}$. A similar robustness of the SW results is evident for the axisymmetric current. Experiments of Huq (1996) show typical dilution of about 2 for parameters that correspond to the cases discussed in § 6.7. However, in that discussion we noticed very good agreements between the SW theory (without entrainment effects) and experiments.

The reconciliation of these observations and the derivation of the entrainment behavior from first principles require further investigations. In this book we shall assume that the currents propagate in circumstances in which the entrainment effects are negligibly small; in particular, this can be justified for small values of $h_0/x_0$ (or $h_0/r_0$ in the axisymmetric geometry).

On the other hand, for currents on a slope the entrainment effect seems to be a rather essential component of the flow field. The topic of gravity currents along a sloping bottom (or incline) is beyond the scope of this book, except for this short digression. We assume a small angle of inclination, $\gamma$, a Boussinesq system, and continue to use the horizontal-vertical coordinate system $\{x, z\}$, with speed components $\{u, w\}$; see Fig. 9.19. For a thin layer current the $z$-accelerations in the fluid are small and hence the approximations of hydrostatic pressures and SW simplification are valid, along the arguments given in § 2.2.

The first important difference due to the inclination of the bottom is that an acceleration term $g' \tan \gamma$ appears in the $x$ momentum balance. This is a result of the pressure continuity at the interface. The reduced pressures in the ambient and current are

$$p_a = C; \quad p_c(x, z, t) = -(\rho_c - \rho_a)gz + f(x, t). \tag{9.142}$$

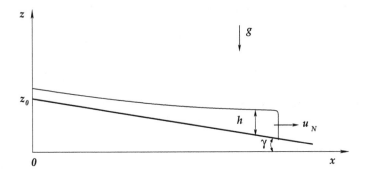

**FIGURE 9.19:**   Schematic description of a current on the inclined bottom.

We impose $p_c = p_a$ at the interface whose position is now $z = z_0 - x \tan \gamma + h(x,t)$. Here $h$ is the height of the current (not the "thickness" measured perpendicular to the boundary). After some algebra we obtain

$$-\frac{\partial p_c}{\partial x} = -\Delta \rho g \left[ \frac{\partial h(x,t)}{\partial x} - \tan \gamma \right], \qquad (9.143)$$

where $\Delta \rho$ is the density difference between the current and ambient.

The tendency of the $\Delta \rho g \tan \gamma$ component of $-\partial p_c / \partial x$ to accelerate the current is counteracted by the dilution of $g'$ by entrainment, and by a drag term $\propto u^2$ (which is usually unimportant on a horizontal bottom). The entrainment in this case is assumed to be distributed over the current, and causes an increase of the local $h$. This, in turn, causes a dilution of $\rho_c$ to a local $\tilde{\rho}(x,t)$, and a reduction of the effective $g'_e$. The calculation of this $g'_e$ is thus a part of the problem. The equations, in dimensional form, are

$$\frac{\partial h}{\partial t} + \frac{\partial}{\partial x}(uh) = \alpha |u|; \qquad (9.144)$$

$$\frac{\partial}{\partial t}(uh) + \frac{\partial}{\partial x}\left( u^2 h + g'_e \frac{1}{2} h^2 \right) - g'_e h \tan \gamma = -C_d u^2; \qquad (9.145)$$

and

$$\frac{\partial g'_e h}{\partial t} + \frac{\partial}{\partial x}(g'_e uh) = 0. \qquad (9.146)$$

The typical experimentally-determined value of the entrainment coefficient $\alpha$ is 0.1, and of the drag coefficient $C_d$ is 0.05. The last equation can be called "conservation of buoyancy", but it clearly is a consequence of the conservation of the initial $\rho_c \mathcal{V}$ of the dense fluid and (9.144). The initial value of $g'_e$ is $(\rho_c - \rho_a)g/\rho_a$. Evidently, when entrainment and continuous mixing in the current are present, $\rho_c$ must be replaced by the instantaneous value of $\tilde{\rho}$ in (9.142) and (9.143).

The front condition is $u_N = Fr(g'_e h_N)^{1/2}$. The usual $Fr$ can be used when $\gamma$ and $\alpha$ are small, but the dilution enters again via $g'_e$.

The extension to the axisymmetric geometry (a current flowing down a sloping cone) can also be obtained; see, for example, Ross, Dalziel, and Linden (2006). Experimental, numerical, and analytical investigations of this problem indicate that entrainment is an essential component in a realistic modeling of the flow on an inclined bottom (see Kunsch and Webber 2000 and Ross, Dalziel, and Linden 2006). In other words, the use of an appropriate $\alpha > 0$ in the foregoing equations seems to be even more important than the value of the drag coefficient $C_d$ for obtaining good agreements with experimental data. However, the derivation of the necessary coefficients from first principles requires additional study.

Finally, consider the combination of entrainment, sloping bottom, and continuously stratified ambient ($\rho_a$ increases with depth). The dense current propagates to lower levels of more and more dense ambient. The effective

$g'_e$ decreases due to both entrainment and stratification, and eventually the nose reaches a level of neutral buoyancy (local $g'_e = 0$) where the current begins to propagate horizontally as an intrusion. For more details on these complex flows see, for example, Baines (2001) and Fernandez and Imberger (2008) where other useful references are given. The flow of an intrusion in a stratified ambient will be considered later in Chapter 15.

# Chapter 10

## Non-Boussinesq systems

### 10.1 Introduction

In this chapter we attempt to elucidate the salient features of non-Boussinesq inviscid currents. This is a less investigated topic than the Boussinesq counterpart. The brief account presented here follows the paper of Ungarish (2007b), where additional relevant references are given.

We consider the propagation of a gravity current of density $\rho_c$ from a lock of length $x_0$ and height $h_0$ into an ambient fluid of density $\rho_a$ in a channel of height $H^*$ (dimensional). When the Reynolds number is large, the resulting flow is governed by two parameters: the depth ratio $H = H^*/h_0$, to which we became familiar in the foregoing discussions on the Boussinesq systems, and the density ratio $\rho_c/\rho_a$. The Boussinesq systems assume $\rho_c/\rho_a = 1$ plus or minus a small perturbation, say up to 10%. Now we ask what happens, qualitatively and quantitatively, when this perturbation is not very small.

It becomes important to distinguish between positive and negative deviations of $\rho_c/\rho_a$ from the value 1. When $\rho_c/\rho_a > 1$ we refer to a heavy (dense, bottom) current and when $\rho_c/\rho_a < 1$ we refer to a light (ceiling, top) current. We emphasize that the light or heavy term is with respect to the ambient. The current is released from a lock (of length $x_0$ and height $h_0$) adjacent to the horizontal boundary on which it will spread out. The configuration is illustrated in Fig. 10.1. We assume that the Reynolds number is large and hence viscous effects can be discarded.

Our previous results clearly show that in the Boussinesq case $\rho_c/\rho_a \approx 1$ the light and heavy currents display the same behavior. To be specific, in the configuration of Fig. 10.1 the bottom and top Boussinesq currents would appear as mirror images with respect to the horizontal boundary. But note that our figure emphasized the lack of symmetry. As shown below, this lack of symmetry is predicted by theory and also has strong experimental and numerical support (Fanneløp and Jacobsen 1984, Keller and Chyou 1991, Gröbelbauer et al. 1993, Lowe et al. 2005, Birman et al. 2005 and Etienne et al. 2005).

Theoretically, the density ratio $\rho_c/\rho_a$ is in the range $(0, \infty)$. However, it makes sense to leave out the extremes from our direct discussion, because such systems usually involve two-phase flow and some special effects. (The

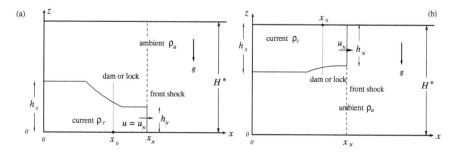

**FIGURE 10.1:** Schematic description of the (a) heavy and (b) light current released from a lock of length $x_0$ and height $h_0$ into an ambient of height $H^*$. (Ungarish 2007b with permission.)

$\rho_a/\rho_c = 0$ case, which reproduces a liquid propagating freely into a gas or vacuum, can be analyzed separately, and more accurately, than discussed here; see Stansby et al. 1998; Hogg and Pritchard 2004. On the other extreme, $\rho_c/\rho_a \to 0$, we find cases like the motion of a long bubble in a liquid, which may involve some surface tension effects which are not considered here.) Let us keep in mind that $\rho_c/\rho_a$ values of 0.5 and 2 can be considered small and large, respectively, because they represent very significant deviations from the 0.9-1.1 (say) limits of the Boussinesq domain. In this context we notice that the setup of experiments or simulations with fluids of significantly different densities requires more resources, and is prone to more undesirable side effects, than the relatively simple water-saline systems used for the Boussinesq cases. This yields practical restriction on the range of $\rho_c/\rho_a$ for which reliable data has been, or can be, obtained. For example: (a) for liquid-liquid systems Lowe et al. (2005) covered experimentally the $\rho_c/\rho_a$ range $0.6-1.7$ (approximately); (b) Gröbelbauer et al. (1993) used a combination of gases to cover the range $0.05-22$ (approximately); and (c) the numerical simulations of Birman et al. (2005) and Etienne et al. (2005) cover a fairly large range of density ratio, but for $H = 1$ only.

Here we attempt to apply the knowledge gained about the one-layer SW model in Chapters 2 and 3 to non-Boussinesq systems.

## 10.2 Formulation

We consider incompressible, immiscible fluids, and assume that the viscous effects are negligible (in both the interior and the boundaries). The density of the current and of the ambient are $\rho_c$ and $\rho_a$, respectively. The current propagates into the positive $x$ direction, and the gravitational acceleration $g$

acts into the negative $z$ direction. The bottom and the top of the half-infinite channel are at $z = 0$ and $z = H^*$. The propagating fluid is originally in a reservoir of length $x_0$ and height $h_0$ located adjacent to the boundary on which propagation occurs, i.e., the bottom for the heavy current and the top for the light current; see Fig 10.1. The $x = 0$ boundary is a rigid wall.

We use dimensional variables for the time being. The thickness of the current is $h(x,t)$ (non-negative) and the horizontal velocity ($z$-averaged) of the current is $u(x,t)$. Initially, at $t = 0$, $h = h_0$ and $u = 0$. We assume a shallow current. The derivation of the SW equations of motion is parallel to that of § 2.2, but we pay more attention to the possible differences between the light and heavy currents.

The continuity equation is not affected by the non-Boussinesq effects. This is a kinematic relationship concerning volume fluxes. The assumption here is that there is no volume entrainment from the horizontal boundary or through the interface. The variable $h(x,t)$ is the thickness of the current (non-negative). The continuity equation is

$$\frac{\partial h}{\partial t} + u\frac{\partial h}{\partial x} + h\frac{\partial u}{\partial x} = 0. \tag{10.1}$$

Let $p$ denote the pressure reduced with $+\rho_a g z$, and the subscripts $a, c$ denote the current and the ambient domains. The SW approximation implies that in the vertical direction the pressure obeys the hydrostatic balance. The one-layer model also assumes that the momentum flux in the ambient is negligible (compared to that of the current) and hence $p_a = C$. Pressure continuity $p_c = p_a$ at the interface ($z = h(x,t)$ for the heavy current and $z = H^* - h(x,t)$ for the light current) yields

$$p_c(x, z, t) = -\Delta\rho g z + |\Delta\rho| g h(x,t) + C_1, \tag{10.2}$$

where

$$\Delta\rho = \rho_c - \rho_a, \tag{10.3}$$

and $C_1 = C$ and $C + \Delta\rho H^*$ for the heavy and light currents, respectively.

Finally, the fundamental longitudinal pressure gradient in the current can be written, again, as

$$\frac{\partial p_c}{\partial x} = |\Delta\rho| g \frac{\partial h(x,t)}{\partial x}. \tag{10.4}$$

The $z$-averaged horizontal momentum equation is employed to express the balance between the inertial forces (proportional to $\rho_c$) and the $-\partial p_c/\partial x$ term, which is eliminated by (10.4). Here we assume again that the deviation of the real horizontal velocity component from the $z$-averaged $u$ is small, i.e., that the shape factor $\varsigma$ defined by (2.21) is close to 1. This is a reasonable approximation for low-viscosity fluids released from rest. We obtain

$$\frac{\partial u}{\partial t} + u\frac{\partial u}{\partial x} = -\frac{|\Delta\rho|}{\rho_c} g\frac{\partial h}{\partial x}. \tag{10.5}$$

The system (10.1) and (10.5) for $h(x,t)$ and $u(x,t)$ is hyperbolic. The characteristic equations, see Appendix A.1, can be expressed as

$$du \pm \left(\frac{|\Delta\rho|}{\rho_c}g\right)^{1/2}\frac{dh}{h^{1/2}} = 0 \quad \text{on} \quad \frac{dx}{dt} = u \pm \left(\frac{|\Delta\rho|}{\rho_c}g\right)^{1/2}h^{1/2}. \tag{10.6}$$

For obtaining realistic gravity current solutions the system of equations must be subjected to a boundary condition at the nose (or front) $x = x_N(t)$. The vertical plane which moves with the nose of the inviscid SW gravity current is treated as a discontinuity. We return to the jump conditions inspired by the analysis of Benjamin (1968). We argued in Chapter 3 that these results are not restricted to Boussinesq systems. For a quite general density difference, an observer attached to the discontinuity and imposing the balance of volume and momentum, and the condition of non-increase of energy, in a $x$-thin control volume finds a simple set of conditions. This is

$$u_N = \left(\frac{|\Delta\rho|}{\rho_a}g\right)^{1/2}h_N^{1/2}\,Fr(a), \quad \text{and} \quad a \leq \frac{1}{2}, \tag{10.7}$$

where $N$ denotes the nose (front), $a = h_N/H^*$ (here $h_N$ is dimensional), and $Fr(a)$ is Benjamin's Froude number function,

$$Fr(a) = \left[\frac{(2-a)(1-a)}{1+a}\right]^{1/2}. \tag{10.8}$$

We emphasize that the equations (10.1) and (10.5)-(10.8) are applicable to both heavy and light currents. This is the reason for using the absolute value of $\Delta\rho$. The non-Boussinesq effect is already evident: equation (10.7) indicates that the speed of the front is proportional to $|\Delta\rho|/\rho_a$, while according to (10.5) and (10.6) the intrinsic speed of the current is proportional to $|\Delta\rho|/\rho_c$. This apparent conflict of speeds is accommodated by the thickness (representing the pressure distribution) which thus also becomes a function of $\rho_c/\rho_a$. Since in the Boussinesq case that conflict is resolved, we can expect that the non-Boussinesq system will tend to attain solutions close to these displayed by the Boussinesq systems.

This interplay between speed and height is the backbone of the present model for the non-Boussinesq gravity current. In other words, the magnitude of $\rho_c/\rho_a$ affects both the speed and shape of the current; the tendency is to somehow "compensate" for the deviations from the Boussinesq $\rho_c/\rho_a \approx 1$ results. This is the reason for the non-symmetric behavior of the shape of light and heavy currents. A heavy current has a larger nose-driving buoyancy effect, $(|\Delta\rho|/\rho_a)g$, than a light current. To compensate for this bias, the height $h_N$ of the heavy current tends to be smaller than that of the light current. This behavior is sketched in Fig. 10.1. The details are complicated by the non-linear behavior of $Fr$, the energy restriction $a \leq 1/2$, tendency to self-similarity, and more. This will be elucidated below.

Estimates of the effect of the return flow suggest that the momentum balance (10.5) is restricted to, roughly, $H(\rho_c/\rho_a) > 2$, but there are indications, see below, that the model is useful beyond this bound. The other two equations of the model (continuity and front condition) have a wider range of relevance and apparently restrain the global error.

An advantage of the one-layer SW model is the possibility to obtain simple analytical solutions for the initial and fully-developed phases of propagation, as described below.

## 10.3   Dam-break and initial slumping motion

The analysis of the system (10.6)-(10.8) indicates that after release from the rectangular lock the current will enter into a slumping phase of propagation with constant velocity over a significant distance (several lock-lengths). This constant velocity feature is shared by both Boussinesq and non-Boussinesq currents, as confirmed by experiments and numerical Navier-Stokes simulations (Gröbelbauer et al. 1993; Lowe et al. 2005; Birman et al. 2005; Etienne et al. 2005; Bonometti and Balachandar 2009).

Following the procedure of § 2.5, we conclude that during this slumping stage, the domain of fluid trailing the nose is a rectangle of constant height, $h_N$. In the absence of shocks, the velocity of propagation and the thickness $h_N$ can be obtained analytically by matching the solution on a forward-propagating characteristic, see (10.6), with the front condition (10.7).

It is convenient to express the results in *dimensionless form*. We first introduce a sharper definition of the reduced gravity

$$g' = \frac{|\Delta\rho|}{\max(\rho_c, \rho_a)}g = \min(g_c', g_a'). \tag{10.9}$$

This choice is bound to keep $g' < g$ for both heavy and light currents. Indeed, the driving-force acceleration cannot exceed $g$.

The horizontal and vertical lengths are scaled as usual with $x_0$ and $h_0$, respectively. The velocity and time are scaled with

$$U = (g'h_0)^{1/2}; \quad T = \frac{x_0}{U}. \tag{10.10}$$

We non-dimensionalize equations (10.6)-(10.7). We then integrate the first equation of (10.6) (on the + branch subject to $u = 0$ for $h = 1$) and require matching with (10.7) when $h = h_N$. This yields

$$2\left[\frac{\max(\rho_a, \rho_c)}{\rho_c}\right]^{1/2}\left(1 - \sqrt{h_N}\right) = \left[\frac{\max(\rho_a, \rho_c)}{\rho_a}\right]^{1/2}\sqrt{h_N}\,Fr(a), \tag{10.11}$$

where, again, $a = h_N/H$ and $Fr(a)$ is given by (10.8). Equation (10.11) is formulated here in this apparently cumbersome way for the following reason: now the LHS is the value of $u_N$ imposed by the characteristic from the reservoir, and the RHS is the value of $u_N$ imposed by the front conditions. This equation provides the value of $h_N$; then, $u_N$ can be calculated using either side of the equation. However, this result has physical validity only if it satisfies the restriction $a \leq 1/2$; otherwise, the RHS with $a = 1/2$ dominates, as discussed later. We expect that the LHS of (10.11) overestimates the value of $u_N$ for a given $h_N$ because the hindering effects of the return flow were not incorporated in the momentum equation.

The values of $h_N$ and $u_N$ at $(\rho_c/\rho_a) = 1$ are defined as the corresponding limit $(\rho_c/\rho_a) \to 1$ of the results. This is actually the Boussinesq limit. Note that for $(\rho_c/\rho_a) = 1$ we obtain the trivial $g' = 0$ and $U = 0$; however, the right and left limits $(\rho_c/\rho_a) \to 1$ of the solution coincide and are meaningful. As expected, slightly heavier or lighter currents move with the same nose-height and speed. We obtain a continuous function of the density ratio for $h_N$ and $u_N$.

Our objective now is to derive the behavior of $u_N$ and $h_N$ as functions of $\rho_c/\rho_a$ and $H$. We recall that this analysis is for the slumping stage, in which these variables do not change with $t$.

### 10.3.1   Asymptotes

We first derive analytical expressions and insights concerning very heavy $(\rho_c/\rho_a \gg 1)$ and very light $(\rho_c/\rho_a \ll 1)$ currents, and also for shallow and deep ambients ($H \approx 1$ and $H \gg 1$).

*Consider the heavy current.* We rewrite (10.11) as

$$2 \left( \frac{\rho_a}{\rho_c} \right)^{1/2} \left[ 1 - \sqrt{h_N} \right] = \sqrt{h_N} \; Fr(a). \qquad (10.12)$$

For a very heavy current $\rho_a/\rho_c \to 0$ the LHS becomes small, and in order to balance it $h_N$ on the RHS must also be small, which also implies that $Fr(a)$ tends to $Fr(0) = \sqrt{2}$. A formal expansion of $h_N$ in terms of the small parameter $(\rho_a/\rho_c)^{1/2}$ yields the leading terms approximation

$$h_N = 2 \frac{\rho_a}{\rho_c} \left( 1 - 2\sqrt{2 \frac{\rho_a}{\rho_c}} \right); \qquad u_N = 2 \left( 1 - \sqrt{2 \frac{\rho_a}{\rho_c}} \right); \qquad (\frac{\rho_a}{\rho_c} \ll 1).$$
$$(10.13)$$

Simply, as the density ratio $\rho_c/\rho_a$ increases, the speed of propagation increases and the thickness of the nose decreases. Interestingly, this result does not depend on $H$; this feature is a consequence of the fact that the nose of the very heavy current is thin, and hence the encountered ambient is, relatively, very deep. The leading coefficient of the scaled speed $u_N$ is 2 which indicates that the heavy current tends to propagate much faster than what could be anticipated from the Boussinesq counterpart value which is typically below 1.

For $\rho_a/\rho_c = 0$ the result (10.13) recovers the behavior of the classical dam-break problem of a liquid (water) in a passive gas (air) (see Stansby et al. 1998 and Hogg and Pritchard 2004 where other references are also given). This is a non-trivial outcome. An inspection of (10.7) indicates a possible infinite value of Benjamin's front condition for $\rho_a = 0$. This is because Benjamin's result is one equation for two unknowns, $u_N$ and $h_N$, and it is impossible to know *a priori* the behavior of $h_N$ for $\rho_a/\rho_c \to 0$. The present SW model provides the second equation. The characteristic from the reservoir shows that $u_N$ is bounded by 2, and hence $h_N$ behaves like $2\rho_a/\rho_c$ in this limit.

Our results must be treated with care in the shallow ambient configuration $H < 2$. In this case there is a significant return flow in the ambient, which seems to be inconsistent with the one-layer dominance. However, we can argue that what is important is momentum (not volume) flux, and when $\rho_c/\rho_a$ is large the momentum flux of the return flow is relatively small. In other words, we expect that the validity of the one-layer model improves as the heavy current departs from the Boussinesq case.

Moreover, the restriction $a = h_N/H < 1/2$ is usually satisfied when $\rho_c/\rho_a$ is large, even for $H$ close to 1. Otherwise, it is necessary to impose the constraint $h_N = H/2$, which means that the flow is controlled (blocked, or choked) by the front condition, i.e., by the RHS of (10.11). We show below that for a heavy current no choking appears for $H > 1.1$ (approximately). Even in the worst case $H = 1$, no choking develops for $\rho_c/\rho_a > 1.373$. In other words, choking is expected to be a rare and quite mild effect for heavy currents. The details will be presented below.

*Consider the light current.* We rewrite (10.11) as

$$\left[ 1 - \sqrt{h_N} \right] = \frac{1}{2} \left( \frac{\rho_c}{\rho_a} \right)^{1/2} \sqrt{h_N} \; Fr(a). \tag{10.14}$$

For a very light current $\rho_c/\rho_a \to 0$, the RHS becomes small, and in order to balance it $h_N$ on the LHS must approach 1. A formal expansion of $h_N$ in terms of the small parameter $(\rho_c/\rho_a)^{1/2}$ yields the leading terms approximation

$$h_N = 1 - \sqrt{\frac{\rho_c}{\rho_a}} Fr(\frac{1}{H});$$

$$u_N = Fr(\frac{1}{H}) \left[ 1 - \sqrt{\frac{\rho_c}{\rho_a}} \left( \frac{1}{2} Fr(\frac{1}{H}) - \frac{|Fr'|}{H} \right) \right]; \quad (\frac{\rho_c}{\rho_a} \ll 1); \tag{10.15}$$

where $Fr' = dFr/da$ is calculated at $a = 1/H$. The difference with the heavy current is evident. Here the results depend on the initial depth ratio $H$. For small $\rho_c/\rho_a$ we obtain the interesting collapse of $u_N$ to $Fr$ corresponding to the initial $h_0/H^* = 1/H$. The speed $u_N$ of the light current increases with the depth of the ambient $H$ and its maximum is $\sqrt{2}$. The height of the front of the current tends to remain close to that of the lock, $h_N \approx 1$.

The results (10.15) are evidently non-applicable to shallow ambients $H < 2$ because the $h_N \approx 1$ outcome violates the energy constraint $h_N \leq H/2$. In

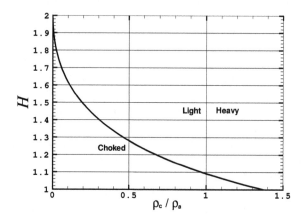

**FIGURE 10.2:**  Restricted flow curve (10.17). In the domain below this curve the motion is with $h_N/H = 1/2$, and $u_N$ given by (10.18).

this case the flow adjusts to $h_N = H/2$. In other words, the light current has a propensity for choked flow when $H < 2$. The details are presented below.

*The choked flow.* The restriction $a = h_N/H \leq 1/2$ can affect currents when the ambient depth is shallow, $1 \leq H < 2$. A current released in a deeper ambient will intrinsically satisfy this restriction. The restriction becomes relevant when the solution of (10.11) produces $h_N$ larger than $H/2$. This is energetically unacceptable, and we proceed as follows.

For both heavy and light currents, the nose condition takes control on the flow, and enforces propagation with the thickness of half-channel, $h_N = H/2$. Consequently, $Fr(a) = Fr(1/2) = 1/\sqrt{2}$. In this case the speed of the nose is smaller than that obtained when we follow the unperturbed characteristic from the reservoir to the height $H/2$. Using (10.14) (or (10.12)) this imbalance is expressed as

$$1 - \sqrt{\frac{H}{2}} > \frac{1}{2} \left( \frac{\rho_c}{\rho_a} \right)^{1/2} \sqrt{\frac{H}{2}} \sqrt{\frac{1}{2}}. \tag{10.16}$$

This provides the necessary conditions for a choked propagation,

$$\frac{\rho_c}{\rho_a} < 8 \left( \sqrt{\frac{2}{H}} - 1 \right)^2, \quad H < 2. \tag{10.17}$$

The condition (10.17) is displayed in Fig. 10.2. The choked flow domain is below the curve. We see that the heavy current, $\rho_c/\rho_a > 1$, occupies a very small portion of this domain. For $H > 1.1$ (approximately) the heavy currents are not choked. Even for $H = 1$ the heavy current is unchoked when $\rho_c/\rho_a > 1.373$. (This is referred to as the "critical density ratio" in the literature. Keller and Chyou 1991 suggested that the value is 3.56, but later experiments and simulation, Lowe et al. 2005, Birman et al. 2005, indicate that the real value of the critical density ratio is close to 1. Actually, given the limited size to the domain, the experimental verification of the existence of the choked heavy current is a difficult task. Small viscous and mixing effects, which are unavoidable in real systems, are expected to render the distinction between the choked and unchoked motions of the heavy current rather inconclusive. Another difficulty is contributed by the two-layer model restriction for a Boussinesq flow with $H \approx 1$, discussed in § 5.3.1. This topic requires further investigation.)

On the other hand, the light current is more prone to the choking effect. For $H < 2$ and a sufficiently small $\rho_c/\rho_a$ the front discontinuity blocks the internal velocity of the current, i.e., the flow is choked at this half-channel thickness by the attainable speed of the nose. This is in full agreement with the results of Lowe et al. (2005) and Birman et al. (2005) for the full-depth lock.

Both heavy and light currents display $h_N = H/2$ when choked. The speed is now provided by the RHS of (10.11). We obtain the compact result

$$u_N = \left[ \frac{\max(\rho_a, \rho_c)}{\rho_a} \right]^{1/2} \sqrt{\frac{H}{2}} \sqrt{\frac{1}{2}} = \begin{cases} \left( \frac{\rho_c}{\rho_a} \right)^{1/2} \frac{1}{2}\sqrt{H} & \left( \frac{\rho_c}{\rho_a} \geq 1 \right) \\ \frac{1}{2}\sqrt{H} & \left( \frac{\rho_c}{\rho_a} \leq 1 \right). \end{cases} \tag{10.18}$$

These considerations predict the height $h_N$ and speed $u_N$ (approximately) in choked circumstances, but do not resolve the mismatch between the sides of (10.11). This indicates that the interface of the choked slumping current cannot be obtained directly by integration along the characteristics from the unperturbed reservoir. An upstream adjustment may be necessary. A more accurate solution requires a two-layer analysis, along the lines of the expansion-wave solution discussed by Lowe et al. (2005). The details are beyond the scope of this book. We keep in mind that this reproduces a real physical effect concerning a gravity current in a shallow ambient, not an artifact of the non-Boussinesq model. We recall that the Boussinesq two-layer model also predicts discrepancies with the one-layer model concerning the shape of the interface and the slumping distance when $H < 2$; see Chapter 5.

*Consider the $H \to \infty$ case, for any $\rho_c/\rho_a$.* Here $a \to 0$ and hence $Fr(a) = \sqrt{2}$. Using this value in (10.11), and denoting $\sigma = [\rho_c/(2\rho_a)]^{1/2}$ we obtain after some algebra

$$h_N = \frac{1}{(1+\sigma)^2}; \qquad u_N = \begin{cases} \frac{2\sigma}{1+\sigma} & \left( \frac{\rho_c}{\rho_a} \geq 1 \right) \\ \frac{\sqrt{2}}{1+\sigma} & \left( \frac{\rho_c}{\rho_a} \leq 1 \right). \end{cases} \tag{10.19}$$

This remarkably simple result is an exact solution of (10.11) for the entire range of $\rho_c/\rho_a$. The prediction is that in a deep ambient $h_N$ decreases from 1 to 0 as $\rho_c/\rho_a$ increases from 0 to $\infty$, while $u_N$ first decreases from $\sqrt{2}$ to the minimum 0.83 at the Boussinesq limit $\rho_c/\rho_a = 1$, then increases to 2.

## 10.3.2   General results

In general, (10.11) must be solved numerically for the unknown $h_N$, subject to $h_N \leq H/2$ as discussed above. Then the result $u_N$ can be calculated. The results for $\rho_c/\rho_a$ in the range $[10^{-2}, 10^2]$ and for various $H$ are displayed in Fig. 10.3. The symbols are data from Navier-Stokes simulations and experiments, as specified in the caption.

This figure confirms and further elaborates the asymptotic trends discussed above. Indeed, except for a narrow range of density ratio about 1, there is a significant difference between the heavy and the light current. The heavy current displays a larger speed and a thinner front domain, and its features are less influenced by the depth ratio. The speed of the light current increases quite significantly with the thickness of the ambient. For $H = 1$ the light current is choked to occupy half the thickness of the channel and moves with restricted speed, while the heavy current propagates mostly with unrestricted height and speed. (This feature is relevant to $H < 2$.) For a given $H$, $u_N$ attains the minimum: (a) at the Boussinesq $\rho_c/\rho_a \approx 1$ for $H \geq 2$ ; and (b) on the light-current choked branch for $H < 2$.

The agreement with the numerical and experimental points is in general good. Unfortunately, only a very restricted comparison could be made in Ungarish (2007b), because the one-layer model is expected to be mostly relevant to values of $H$ larger than 1, for which practically no data was available. The relatively abundant and most accurate data are for the full depth case $H = 1$. Gröbelbauer et al. (1993) presented some experimental data for $H = 6$ depth ratio release geometry. However, the currents were actually released by a partly lifted gate from a deep reservoir, not from a lock of part depth. Considering the motion of the backward-moving characteristics and rarefaction wave, we could expect a significant difference between the theory and this type of experiment concerning the heavy current. Moreover, these experiments were performed with gases and may be affected by diffusion. We find, however, a quite good quantitative agreement for $0.046 \leq \rho_c/\rho_a < 1.1$, and qualitative consistently for the other points.

It is interesting to note that, according to Fig. 10.3, the best support to the predictions of the model comes from parameter regimes where the momentum equation of the model is expected to be least valid: (a) the full-depth $H = 1$ high resolution numerical data (the triangle symbols in Fig. 10.3); and (b) some of the deep light currents experimental results (the $\times$ symbols in Fig. 10.3 for $0.0463 \leq \rho_c/\rho_a \leq 0.334$, where $0.28 \leq H\rho_c/\rho_a \leq 2.0$). Actually, the agreement is remarkable. The plausible explanation is as follows. First, the discrepancy between the data and the model can be attributed to errors in the data. The data for $H = 1$ is the most accurate. Second, this figure is

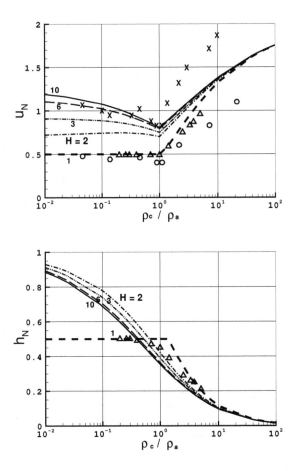

**FIGURE 10.3:** $u_N$ and $h_N$ as functions of $\rho_c/\rho_a$ for various depth ratio $H = 1, 2, 3, 6, 10$. Symbols show: experiments of Gröbelbauer et al. ($H = 1$ ○, $H = 6$ ×) and numerical simulations of Birman et al. ($H = 1$ △). Here $u_N$ is scaled according to (10.9)-(10.10). (Ungarish 2007b with permission.)

concerned only with speed and height of the nose. A comparison of additional features of the current is expected to reveal, in general, better agreements at larger values of $H$ (in particular for light currents which are prone to choking). Third, it is possible that the simple balances used in the one-layer model capture the governing physical mechanism better than suggested by the magnitude of the neglected terms (e.g., due to internal cancellation of effects the global deviation is smaller than the local error). This conjecture requires further verification; some support in this direction is provided by the recent numerical results of Bonometti and Balachandar (2009).

In any case, Fig. 10.3 indicates that the non-Boussinesq extension of the SW one-layer model is a useful tool, at least for predicting the initial behavior of the current. Note that the classical Boussinesq domain is just a very narrow strip about the line $\rho_c/\rho_a = 1$ in the parameter plane of Fig. 10.3.

We now proceed to the analysis of the non-Boussinesq effects on the developed motion.

---

## 10.4   The transition and self-similar stages

The non-Boussinesq SW model displays, after the slumping, a transition to self-similar motion, like the Boussinesq counterpart. Again, differences between the light and heavy current appear. To stress these differences it is convenient to use dimensionless variables as follows. We assume again a rectangular lock, and we scale $x, h$ with $x_0, h_0$. Now the speed and time are scaled with

$$U = \left( \frac{|\Delta\rho|}{\rho_c} gh_0 \right)^{1/2} = (g'_c h_0)^{1/2}; \quad T = \frac{x_0}{U}. \tag{10.20}$$

We emphasize that here the scaling of the speed and time is different from that used in the previous section, § 10.3 (practically, the difference appears only for the light current, $\rho_c < \rho_a$).

Using this scaling and the transformation of § 2.3, the governing equations (10.1), (10.5), and (10.6) become, again, the system (2.49)-(2.51) which was obtained for the Boussinesq current. Recall that $y = x/x_N(t)$. The initial conditions and the boundary condition $u = 0$ at $y = 0$ are also the same. This similarity between the systems of equations for the Boussinesq and non-Boussinesq current is not surprising. The density of the current is $\rho_c$, and the hydrostatic pressure buoyancy is $|\Delta\rho|g$ for both systems. The change, however, is in the nose boundary condition, (10.7), which predicts now

$$\dot{x}_N = u_N = \left( \frac{\rho_c}{\rho_a} \right)^{1/2} Fr(a)\sqrt{h_N}. \tag{10.21}$$

The details of the derivation are left for an exercise.

The numerical finite-difference solver used for the Boussinesq current works well in this case too (at least for moderate values of $\rho_c/\rho_a$). Typical results (using 200 grid points for $0 \le y \le 1$) for a light current of $\rho_c/\rho_a = 1/4$ and a heavy current of $\rho_c/\rho_a = 4$ are shown in Fig. 10.4. The differences are significant. The trends indicated by the slumping phase prevail: the heavy current is thinner and faster than the light current. In addition, a pronounced difference of the thickness profiles $h(x)$ appears. The difference is well detected analytically by the long-time similarity solution.

Indeed, after a significant propagation, the non-Boussinesq inviscid current becomes thin, "forgets" the initial conditions, and is amenable to a self-similar solution. Now the current is sufficiently thin (or deep) so that the $Fr$ at the front becomes constant. We define

$$\mathcal{F}^2 = \frac{\rho_c}{\rho_a} Fr^2(0) = 2\frac{\rho_c}{\rho_a}. \tag{10.22}$$

The boundary condition for the deep current is expressed as

$$\dot{x}_N = \mathcal{F}\sqrt{h_N}, \tag{10.23}$$

where $\mathcal{F}$ is a constant.

The similarity solution is derived by the method presented in § 2.6. The results are

$$x_N(t) = Kt^{2/3}; \quad h = \dot{x}_N^2(c + \frac{1}{4}y^2); \quad u = \dot{x}_N y; \quad (y_j \le y \le 1); \tag{10.24}$$

where

$$c = \frac{1}{\mathcal{F}^2} - \frac{1}{4}, \tag{10.25}$$

$y = x/x_N$, $y_j$, and $K$ are constants which will be calculated later. A novel feature is that the solution may be invalid in the inner part of the $y$ domain, therefore we introduced the $y_j$. The verification of the results is left for an exercise. Again, the similarity results are not sharp because initial conditions are not satisfied, and $t$ can be replaced by $t + \gamma$ without invalidating them. Moreover, in the similarity solution we can replace in the scaling both $x_0$ and $h_0$ by $L = \mathcal{V}^{1/2}$ where $\mathcal{V}$ is the volume (per unit width).

In the similarity stage, both light and heavy currents propagate with $t^{2/3}$. There is, however, a difference in the expected shape. This can be attributed to the fact that $\mathcal{F}$ can vary from small to large values. Consequently, the constant $c$ varies from large positive values to the negative $-0.25$, approximately.

For very light currents the small $\mathcal{F}$ produces $c \gg 1$, and hence the profile is a spreading rectangle, $h \approx (2K/3)^2 ct^{-2/3}$, with a relatively small contribution from the $y^2$ term. The opposite structure, of a sharp tail-to-head difference, is expected for the very heavy current, as follows.

First, we note that $c$ decreases when $\mathcal{F}$ increases. For $\rho_c/\rho_a > 2$ we obtain $\mathcal{F}^2 > 4$, and hence in this case $c < 0$ and $h$ predicted by (10.24) is negative for $y < y_1$ where

$$y_1 = \left(1 - \frac{4}{\mathcal{F}^2}\right)^{1/2} = \left(1 - 2\frac{\rho_a}{\rho_c}\right)^{1/2}. \tag{10.26}$$

The similarity solution (10.24) is invalid for negative $h$. We infer that $y_j = 0$ for $\rho_c/\rho_a \le 2$ and $y_j = y_1$ for heavier currents. However, what happens to the real heavy current in the domain $y < y_1$?

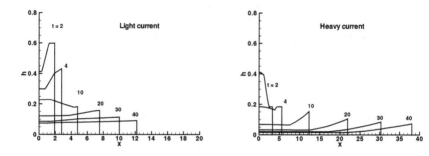

**FIGURE 10.4:** Propagation of light $\rho_c/\rho_a = 0.25$ and heavy $\rho_c/\rho_a = 4$ currents released from a rectangular lock with $H = 4$. Finite-differences solution of the SW equations. (Ungarish 2007b with permission.)

FannELøp and Jacobsen (1984) and Gratton and Vigo (1994) suggested that the negative $h$ of the similarity result means that the heavy current leaves behind a bare bottom, or an empty spot, in the region $x < y_1 K t^{2/3}$, approximately. However, the evolvement of this peculiar shape for a realistic current released from behind a lock needs clarification. We observe that the region where the classic similarity result (10.24) yields a negative $h$ is actually covered by a thin horizontal tail of the current. This corresponds to another branch of the long-time exact solutions of the SW equations (10.1) and (10.5), namely

$$u(x,t) = \frac{x}{t}; \quad h(x,t) = \frac{C}{t}; \quad (10.27)$$

where $C$ is a constant of the order unity. The validity of the result can be verified by direct substitution into (10.1) and (10.5), subject to the appropriate scaling. The analysis of the time-dependent behavior at small $x$ indicates that this solution develops about the backwall during the transition from slumping to similarity phases. But in many cases this solution is eventually suppressed by information which propagates from the nose backward. Only for a sufficiently heavy current this inner solution spreads out and prevails for large times, because the fast motion of the nose limits the backward zone of influence. Consequently, the self-similar profile for $\rho_c/\rho_a > 2$ combines the horizontal "tail" (10.27) for $0 \le y < y_m$, and the prominent head (10.24) for $y_m \le y < 1$, where $y_m \approx y_1$. This structure, and the evolvement of different shapes for the heavy and light currents, is confirmed by the numerical solutions of the SW equations displayed in Fig. 10.4.

The volume of the fluid in the tail is $(C/t)y_m x_N(t)$, which decays like $t^{-1/3}$. This justifies the simplification of a bare bottom left behind a very heavy current, but viscous effects will eventually slow down the decay of the tail.

Finally, the constant $K$ follows from the volume conservation in the $y$ domain $[y_j, 1]$ (recall that $y_j = 0$ for $\rho_c/\rho_a \leq 2$, and $y_1$ for a heavier current). This yields

$$K = \left[ \frac{4}{9} \left( c(1 - y_j) + \frac{1}{12}(1 - y_j^3) \right) \right]^{-1/3}. \qquad (10.28)$$

## 10.5 Summary

This chapter elucidates some salient features of the non-Boussinesq inviscid currents. The one-layer SW model captures well the major behavior for a wide range of $\rho_c/\rho_a$ and $H$. The advantage of this model is the mathematical power and simplicity. Essentially, this model provides a unified mathematical description for the Boussinesq and non-Boussinesq inviscid currents in the one-layer framework. Indeed, the underlying physical mechanism are the same for both systems, and the differences are mostly in the details of the resulting shape and motion. However, we must keep in mind that the accuracy and range of validity of this model are still under investigation. The theoretical predictions are in good agreement with available experimental and numerical simulations data. The comparisons cover a quite restricted parameter range due to lack of data, and in 2D configurations only. Extensions to axisymmetric and rotating configurations still await implementation. The plausible limitation of the model is $H(\rho_c/\rho_a) > 2$, but some valuable results were obtained for much smaller values of this parameter. Actually, we noted that the best agreements in Fig. 10.3 were obtained for parameter regimes outside that limitation, in which the model is expected to be least valid.

The interpretation and implications of this puzzling (but encouraging) better-than-expected performance requires further investigation. There is yet a great deal of missing knowledge concerning the inviscid non-Boussinesq currents. The influence of diffusion, entrainment, instabilities and surface tension may be more important in the non-Boussinesq systems than in the Boussinesq counterpart.

On the other hand, viscous currents are expected to be more robust concerning the effect of non-small density differences. We clarified that a major reason for the non-Boussinesq effects is the discrepancy between the reaction of the main current and of the nose to the driving force $|\Delta\rho|g$. The tendencies are to develop velocities proportional to $\rho_c^{-1}$ and $\rho_a^{-1}$, respectively, and hence both $g_c'$ and $g_a'$ play a role. The viscous current is not subject to a nose jump condition; the motion is rather controlled by a local balance between the driving force and the viscous shear. In these circumstances $g_c'$ is expected to dominate the motion. The details will be derived in the next chapter.

# Chapter 11

## Lubrication theory formulation for viscous currents

Here we consider the prototype thin viscous current which propagates over a horizontal solid boundary. Our main references here are the papers of Huppert (1982) (where other references are also given, in particular Fay 1969; Hoult 1986; Britter 1979) and Didden and Maxworthy (1982).

We assume that in the current the viscous forces, $F_V$, are dominant over the inertial forces, $F_I$. This means that the properly defined Reynolds number is small. This also implies that viscous diffusion smooths out velocity gradients in a very short time interval as compared to the typical time of propagation. For a current of fixed volume released from a rectangular lock we can use the results derived in § 2.7 and 6.6 to determine when the viscous dominance appears. That analysis makes it evident that the typical viscous current is a very thin layer of fluid. The deficiency of that estimate is the fact that it uses the inertial-buoyancy speed of propagation. A sharper estimate for the typical velocity in the viscous regime will be developed in this chapter. It turns out that there is fair agreement concerning the transition between the inviscid and viscous regimes. The box-model approximations considered in Chapters 4 and 7 also provide fairly good estimates about this transition.

## 11.1  2D geometry

### 11.1.1  The governing equations

We use a one-layer approximation. The objective is to derive and solve the governing equations for the thickness of the layer, $h(x,t)$, and the $z$-averaged speed of motion, $\bar{u}(x,t)$. This also provides the formula for the propagation, $x_N(t)$.

We assume a motionless hydrostatic ambient fluid of density $\rho_a$. The variables of the ambient are denoted by the subscript $a$. For definiteness, the current is more dense than the ambient, $\rho_c > \rho_a$, and propagates over a solid bottom. Usually the variables in this lower layer are without a subscript, and when emphasis is necessary we use the subscript $c$. For example, $\nu$ means the kinematic viscosity of the dense fluid current layer.

Consider the current. Suppose that for the horizontal motion the scales for length, velocity, and time are $L_x$, $U$, and $L_x/U$; while the vertical counterparts are $L_z$, $W$, and $L_z/W$. Here we introduce the thin-layer assumption that $(L_z/L_x) = \delta \ll 1$. The continuity equations shows that $W = (L_z/L_x)U = \delta U$. The thin layer configuration allows, again, the use of a powerful approximation. The lubrication theory, see for example Batchelor (1981), provides the viscous counterpart of the SW inviscid approach. In both cases the vertical velocity $W$ is negligibly small as compared to the horizontal velocity $U$. Using arguments similar to these applied in § 2.2, we conclude that, within a relative error $\delta^2$: (a) the pressure obeys the hydrostatic balance in the $z$ direction; and (b) the pressure is continuous on the interface $z = h(x,t)$. Consequently, the fundamental relationship (2.15) between the driving pressure gradient and the slope of the interface developed for the inviscid one-layer model is valid also for the viscous case. Again, the longitudinal pressure gradient in the current can be written as

$$\frac{\partial p_c}{\partial x} = \Delta \rho g \frac{\partial h(x,t)}{\partial x}, \tag{11.1}$$

where $\Delta \rho = \rho_c - \rho_a$.

In the $x$-momentum equation of the current the inertial terms of magnitude $\rho_c U^2/L_x$ are negligibly small as compared to the viscous $\rho_c \nu (\partial^2 u/\partial x^2 + \partial^2 u/\partial z^2)$ term of magnitude $\rho_c \nu U/L_z^2$. We of course assume that the effective Reynolds number, $(UL_z/\nu)\delta$, is small. Furthermore, since the $z$-shear is dominant when $\delta \ll 1$, we obtain the leading $x$-momentum balance as

$$0 = -\frac{1}{\rho_c}\frac{\partial p_c}{\partial x} + \nu \frac{\partial^2 u}{\partial z^2}, \tag{11.2}$$

which, using (11.1), is expressed as

$$0 = -\frac{\Delta \rho}{\rho_c}g\frac{\partial h}{\partial x} + \nu \frac{\partial^2 u}{\partial z^2}. \tag{11.3}$$

We see that the main dynamic balance in the viscous current is between the pressure-buoyancy driving provided by the inclined interface and the viscous $z$-shear hindrance. We note that the reduced gravity relevant to the viscous current is

$$g' = g_c' = \frac{\Delta \rho}{\rho_c}g. \tag{11.4}$$

The momentum equation (11.3) is readily integrated to yield

$$u(x,z,t) = \frac{g'}{\nu}\frac{\partial h}{\partial x}\left[\frac{1}{2}z^2 + Az + B\right] \quad (0 \le z \le h(x,t)), \tag{11.5}$$

where $A$ and $B$ are functions of $x,t$ to be determined by boundary conditions.

In the prototype problem, sketched in Fig. 11.1, the base of the current is a no-slip wall at $z = 0$ and hence $u = 0$ at this position. At the top $z = h$

the current is matched to a thick layer of ambient fluid. Here the boundary
condition is continuity of shear

$$\rho_c \nu \frac{\partial u}{\partial z} = \rho_a \nu_a \frac{\partial u_a}{\partial z} \quad (z = h). \tag{11.6}$$

We assume that the RHS is small as compared to $\rho_c \nu u(z = h)/h$. The justification is as follows. Consider the Boussinesq case with $\rho_a \nu_a \approx \rho_c \nu$. A
strong velocity gradient created at time $t_s$ spreads eventually over a thickness
$\delta = [\nu(t - t_s)]^{1/2}$. Take $t - t_s = x_0/U$, i.e., the typical time of propagation
of the current. We obtain $\delta/h_0 = (Re h_0/x_0)^{-1/2}$, where $Re = U h_0/\nu$. Thus,
for a typical Boussinesq viscous current, the shear of $u_a$ is spread over a much
thicker layer than that of $u$. Consider the non-Boussinesq case of practical
interest $\rho_a \ll \rho_c$ (for example, a liquid which spreads into the atmosphere).
Here the coefficient of the RHS term is very small. Consequently, in both
these typical cases, from the point of view of the flow in the current, the shear
stress at the top of the layer is very small as compared to the mean value
inside the layer, and can therefore be approximated by zero.

On account of the foregoing considerations, we apply to (11.5) the conditions

$$u(x, z = 0, t) = 0; \quad \frac{\partial u}{\partial z}(x, z = h, t) = 0; \tag{11.7}$$

to obtain

$$u(x, z, t) = \frac{g'}{\nu} \frac{\partial h}{\partial x} \left( \frac{1}{2} z^2 - hz \right). \tag{11.8}$$

Finally, the $z$-averaged velocity in the viscous current is

$$\bar{u}(x, t) = \frac{1}{h} \int_0^h u \, dz = -\frac{1}{3} \frac{g'}{\nu} h^2 \frac{\partial h}{\partial x}. \tag{11.9}$$

The last result reveals some fundamental features of the viscous current.
The speed of propagation decreases with $\nu^{-1}$, and forward propagation requires a negative $\partial h/\partial x$. The negative slope of the interface produces a higher
pressure in the back which pushes forward the dense fluid. This forcing is opposed by the viscous friction. The speed is fully determined by the local
behavior of the interface, so the next major objective is to calculate $h(x, t)$.

The behavior of $h(x, t)$ can be determined by the use of the continuity
equation (2.24), in which $\bar{u}$ is defined by (11.9). This combination of continuity
and momentum balances yields

$$\frac{\partial h}{\partial t} - \frac{1}{3} \frac{g'}{\nu} \frac{\partial}{\partial x} h^3 \frac{\partial h}{\partial x} = 0. \tag{11.10}$$

The governing equations of the viscous current are (11.9) and (11.10). The
boundary conditions will be discussed later. Contrary to the inviscid SW case,
(a) Here the equations for $h$ and $u$ can be easily decoupled. Actually, only a
single PDE must be solved; this is equation (11.10) for the unknown $h(x, t)$.

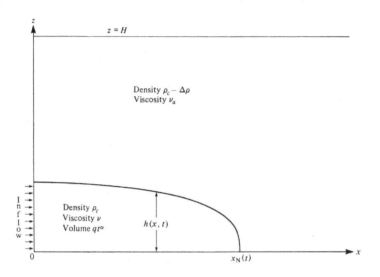

**FIGURE 11.1**: Schematic description of typical viscous gravity current with inflow. (Huppert 1982, with permission.)

(b) The dominant PDE (11.10) is parabolic (recall that the SW system is hyperbolic).

Before we proceed to the solution of (11.10), we reconsider the definition of $Re$. The appropriate typical velocity, $U$, is indicated by (11.9). Suppose that the typical length and thickness of the current are $x_0$ and $h_0$. We write

$$Re = \frac{U h_0}{\nu} = \frac{g'}{\nu^2} h_0^3 \left( \frac{h_0}{x_0} \right). \tag{11.11}$$

To illustrate the result, consider a layer of honey spreading on a plate or a pancake in the atmosphere. Using $x_0 = 10\,\text{cm}$, $h_0 = 0.1\,\text{cm}$, $\nu = 30\,\text{cm}^2\,\text{s}^{-1}$, and $g' \approx g = 980\,\text{cm}\,\text{s}^{-2}$ we obtain by (11.11) the value $Re \approx 1.1 \cdot 10^{-5}$. This, in accord with intuition, indicates a very viscous flow. Less intuitive is the fact that for $h_0 > 3\,\text{cm}$ that honey flows rather according to inviscid balances. Indeed, (11.11) gives $Re > 8$. Note that the use of the inviscid estimate $U = (g' h_0)^{1/2}$ in (11.11) produces $Re > 5$ for that layer of honey with $h_0 > 3\,\text{cm}$. This is a fair overlap. In other words, the inviscid and viscous estimates of the Reynolds number are consistent about the region of transition between the corresponding flow regimes. Moreover, (11.11) indicates that in a viscous current of fixed volume, $\mathcal{V}$, $Re$ decreases like $h^5$ (because $1/x_0 = h_0/\mathcal{V}$). Therefore, when such a current enters the viscous regime, the dominance of the viscous effects is quickly established.

To both generalize the results and simplify the manipulation, we perform two preliminary steps:

1. We assume that the volume of the current is given by

$$\mathcal{V} = qt^{\alpha}, \tag{11.12}$$

where $q$ ($> 0$) and $\alpha$ ($\geq 0$) are constants, with the particular case

$$q = \mathcal{V} \quad \text{for} \quad \alpha = 0. \tag{11.13}$$

The source is at $x = 0$. This covers a quite general variety of currents. In addition to the fixed volume current, we have these produced by constant discharge, $\alpha = 1$, by an open tap, say; and also currents produced by a worsening leakage, $\alpha > 1$.

2. We switch to dimensionless variables. The horizontal and vertical lengths are scaled with $x_0, h_0$, respectively, and volume (per unit width) is scaled with $x_0 h_0$. The velocity and time are scaled with

$$U = \frac{1}{3}g'\frac{1}{\nu}h_0^3\frac{1}{x_0}; \quad T = \frac{x_0}{U}. \tag{11.14}$$

Note that the coefficient $q$ is scaled with $x_0 h_0 T^{-\alpha}$, a quite unusual quantity. We also keep in mind that the proper choice of $x_0$ and $h_0$ may require some analysis of the particular problem. For example, for a fixed-volume current the intuitive choice for $x_0$ and $h_0$ is the dimensions of the lock. However, the initial motion may well be in the inviscid regime. Therefore, a more relevant choice will be the length and mean thickness of this current when transition to the viscous regime occurs.

The following analysis uses *dimensionless variables* unless stated otherwise. The system for solution is the dimensionless form of the equations (11.9) and (11.10), subject to the volume continuity (11.13). This is expressed as follows. The momentum equation is

$$\bar{u}(x,t) = -h^2\frac{\partial h}{\partial x}; \tag{11.15}$$

and the continuity equation (11.10) reads

$$\frac{\partial h}{\partial t} - \frac{\partial}{\partial x}h^3\frac{\partial h}{\partial x} = 0; \tag{11.16}$$

subject to

$$\int_0^{x_N(t)} h(x,t)dx = \mathcal{V} = qt^{\alpha}. \tag{11.17}$$

Clearly, we must concentrate on the solution of the PDE in this system for $h$, then the velocity follows. This is facilitated, again, by the transformation introduced in § 2.3. Letting $y = x/x_N(t)$ and using (2.46)-(2.47) we obtain the following equation for $h(y,t)$

$$\frac{\partial h}{\partial t} - y\frac{\dot{x}_N}{x_N}\frac{\partial h}{\partial y} - \frac{1}{x_N^2}\frac{\partial}{\partial y}h^3\frac{\partial h}{\partial y} = 0 \quad (0 \le y \le 1); \tag{11.18}$$

subject to

$$x_N(t)\int_0^1 h(y,t)dy = \mathcal{V} = qt^\alpha. \tag{11.19}$$

In general, the solution can be obtained by numerical methods. The behavior of $h$ poses a difficulty at the nose, when $y$ approaches 1. Since the equation is parabolic a nose jump condition of the type encountered for the SW inviscid hyperbolic formulation is not expected. The propagation of the viscous fluid is supported at any $y$ by a negative slope of $h$; see (11.15). We thus expect that $h$ decreases continuously with $y$, to $h = 0$ at $y = 1$. However, to keep this point moving we need to assume that $h^2(\partial h/\partial y)$ is finite when $h \to 0$ as $y \to 1$. This is consistent with the formulation. To work out the details, an analytical solution of (11.18) about $y = 1$ must be considered. The pertinent results are obtained below as a part of the similarity solution. A numerical finite-difference solution of (11.18) can be performed for the $y$ domain $[0, 1 - \Delta]$, with the boundary condition at $y = 1 - \Delta$ given by the analytical results, where $\Delta$ is reasonably small (say 0.005). The volume in this subdomain is negligibly small, as shown below. However, there is little incentive for calculating the numerical solution of the PDE for $h(y,t)$, because, as shown by Huppert (1982), a convenient analytical result is available. The initial conditions for this equation may also be difficult to specify, because in many cases of interest the current propagates first in an inertial-dominated regime where (11.18) is not valid.

## 11.1.2   Similarity solution

The analytical solution of (11.18)-(11.19) is obtained by self-similar assumptions. We seek a solution of the form

$$x_N = Kt^\beta; \quad h = \frac{\mathcal{V}}{x_N}\mathcal{H}(y) = \frac{q}{K}t^{\alpha-\beta}\mathcal{H}(y); \tag{11.20}$$

where $K$ and $\beta$ are some positive constants. The objective is to determine $\beta$, $\mathcal{H}(y)$, and $K$. Substitution into (11.18) and arrangement yields

$$(\alpha - \beta)\mathcal{H} - \beta y\mathcal{H}' - t^{3\alpha-5\beta+1}\frac{q^3}{K^5}(\mathcal{H}^3\mathcal{H}')' = 0. \tag{11.21}$$

Similarity requires that the power of $t$ is zero. This determines

$$\beta = \frac{3\alpha + 1}{5}. \tag{11.22}$$

Further simplification is achieved upon the substitution

$$\mathcal{H}(y) = \frac{K^{5/3}}{q}\lambda(y), \tag{11.23}$$

which reduces (11.21) to

$$(\lambda^3 \lambda')' + \beta y \lambda' - (\alpha - \beta)\lambda = 0. \tag{11.24}$$

The boundary condition for this equation is $\lambda(1) = 0$. The justification was mentioned above. The pressure gradient drives the current against the viscous forces, and this requires a decreasing $h$ with $x$. This implies that the self-similar profiles $\mathcal{H}$ (and $\lambda$) are positive monotonically decreasing functions of $y$ for $0 < y < 1$. This behavior must end with $h = 0$ at $x_N$, i.e., $\mathcal{H}(1) = \lambda(1) = 0$.

Equation (11.24) has, as expected, a singularity at $y = 1$. Using a Frobenius series expansion, $\lambda = \xi^\gamma(a_0 + a_1\xi + \cdots)$, where $\xi = (1 - y)$, we obtain

$$\lambda(y) = \left[\frac{3}{5}(3\alpha + 1)\right]^{1/3}(1 - y)^{1/3}\left[1 - \frac{3\alpha - 4}{24(3\alpha + 1)}(1 - y) + O[(1 - y)^2]\right]. \tag{11.25}$$

We can now calculate good approximations for the values of $\lambda$ and $\lambda'$ at $y = 1 - \Delta$ for some small $\Delta$ (0.005 say). These values can be used as boundary conditions for the numerical integration of (11.24) in the domain $[0, 1 - \Delta]$. (The solution of (11.24) can be treated now as an initial value problem. A Runge-Kutta, or similar method, can be used. Note that the integration is in negative $y$ direction.) This method determines uniquely the solution $\lambda(y)$ for relevant values of $\alpha$. The typical thickness profiles are shown in Fig. 11.2. This confirms the expectation that the viscous gravity current is a thin smooth layer, whose thickness decreases with the distance from the source or backwall.

Next, we calculate $K$ using the global volume balance (11.19), in which we substitute (11.20) and (11.23). After some algebra this yields

$$\int_0^1 \mathcal{H}(y)dy = \frac{K^{5/3}}{q}\int_0^1 \lambda(y)dy = 1, \tag{11.26}$$

and hence

$$K = q^{3/5}\left[\int_0^1 \lambda(y)dy\right]^{-3/5} = q^{3/5}\eta_N(\alpha). \tag{11.27}$$

Since in general $\lambda(y)$ is calculated numerically, so is also $\eta_N(\alpha)$; some results are displayed in Fig. 11.3. The values of $\eta_N$ are of the order of unity, which vindicates the scaling used in the derivation. The details of the derivation of (11.25) and (11.27) are left for an exercise.

Finally, we calculate the speed. Combining the previous results we can rewrite (11.15) as

$$\bar{u}(y, t) = -\frac{1}{x_N}\left(\frac{\nu}{x_N}\right)^3 \mathcal{H}^2\mathcal{H}'(y) = -Kt^{(3\alpha-4)/5}\lambda^2(y)\lambda'(y). \tag{11.28}$$

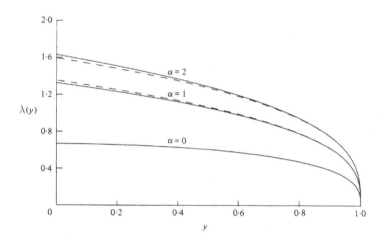

**FIGURE 11.2:** Similarity profiles of interface for 2D viscous current. $\lambda(y)$ is given for various values of $\alpha$. The dashed lines are the approximate solution (11.25), and the solid lines are the exact solution. (Huppert 1982, with permission.)

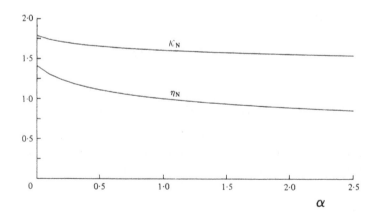

**FIGURE 11.3:** Coefficients for the self-similar viscous current as functions of $\alpha$. (a) $\eta_N$ for the 2D case; and (b) $\kappa_N$ for the axisymmetric case. (Huppert 1982, with permission.)

It is interesting that for the fixed volume $\alpha = 0$ case a simple analytical solution can be obtained. In this case $\beta = 1/5$. The thickness profile is given by

$$\lambda(y) = \left(\frac{3}{10}\right)^{1/3}(1 - y^2)^{1/3}, \tag{11.29}$$

which is easily verified by substitution into (11.24). With $q = \mathcal{V}$ (11.27) now yields

$$K = 1.411\,\mathcal{V}^{3/5}. \tag{11.30}$$

Then, using (11.23) and (11.28) we obtain

$$\bar{u} = 1.119\,\mathcal{V}^3\frac{1}{x_N^4}y = 0.282\,\mathcal{V}^{3/5}t^{-4/5}y. \tag{11.31}$$

We note that the boundary condition $\bar{u} = 0$ at $y = 0$ is satisfied. As in the inviscid case, the self-similar velocity in the fixed-volume current is linear with $y$. Otherwise, there are significant differences. In the inviscid case $\mathcal{H}$ is an increasing function of $y$ whose maximum is at the nose, $y = 1$, while in the viscous case $\mathcal{H}$ decreases with $y$ to $\mathcal{H}(1) = 0$ at the nose; see Figs. 2.11 and 11.2. This is, again, a manifest of the essential difference: the inviscid current is controlled by the pressure-jump conditions at the nose, while the viscous current is pushed from behind against the local shear. In the inviscid case the speed of propagation decreases like $x_N^{-1/2}$, while in the viscous case like $x_N^{-4}$.

Even for the simple $\alpha = 0$ case the transition process from the inviscid to the viscous solution is not clear. We can patch the two solutions as follows. In § 2.7 we derived an estimate for the position $x_N(t) = x_V$ where transition from inviscid to viscous regimes occurs. We assume that the inviscid SW results are valid for $x_N \leq x_V$, and that for $x_N > x_V$ the viscous formulation developed in this section becomes (instantaneously) relevant. The main matching conditions are, of course, continuity of $x_N(t)$ and of volume. This information is sufficient for switching from the inviscid SW results to the similarity solution derived in this section (some care is required with regard to the scaling of the variables). In many cases of interest the SW current is in the self-similar stage when transition occurs. In these cases we can switch from the similarity solution of § 2.6 to the similarity solution derived in this section (again, some care is required with regard to the scaling of the variables). We recall that the similarity solutions have a virtual origin which allow us to use $(t + \gamma_I)$ and $(t + \gamma_V)$ as the time variables in the inviscid and viscous domains, respectively. By the proper choice of $\gamma_I$ and $\gamma_V$ the initial conditions and the continuity at $x_N$ can be achieved. We also note that the inviscid solution of $h$ when $x_N = x_V$ may be a problematic "initial condition" for the viscous regime if a numerical solution of (11.16) is attempted. The inviscid profile has a positive slope which corresponds to negative $u$ in the viscous domain. Some non-trivial adjustment will be necessary. In addition, the boundary condition $h(y = 1) = 0$ is expected to introduce, again, a singularity of the type $(1 - y^2)^{1/3}$. The details still require investigation. These difficulties are circumvented by the above-mentioned patching, with the price of an unknown error.

For the more general $\alpha > 0$ case the situation is more difficult because we do not have simple similarity solutions for the inviscid regime. It is indeed emphasized that, in contrast to the inertial-buoyancy current, the viscous-buoyancy current admits smooth similarity solutions for a wide range of inflow conditions expressed by $\alpha$. As shown by Grundy and Rottman (1985b) and Gratton and Vigo (1994) the SW equations of the inviscid inertial current with inflow conditions admit a manifold of similarity solutions with different profiles of $u$ and $h$. Depending on the values of $\alpha$ and $Fr$, some of these profiles display discontinuities (shocks or hydraulic jumps). The details are beyond the scope of this book. In this book, the inviscid-inertial current with volume changing like $qt^\alpha$ is covered only by the imprecise box-model approximation. These differences can be related to the hyperbolic properties of the inviscid system, which support formation and propagation of steep changes of the dependent variables. On the other hand, the viscosity tends to smooth out gradients in a shorter time interval than the time of propagation. Therefore, the viscous current is intrinsically simpler than the inviscid-inertial current: it is governed by one parabolic PDE for $h$.

### 11.1.3   Summary

The behavior of the viscous 2D current of volume $\mathcal{V} = qt^\alpha$ is predicted by the similarity solution of the lubrication formulation as (in dimensionless form)

$$x_N(t) = Kt^{(3\alpha+1)/5}; \quad h(y,t) = K^{2/3}t^{(2\alpha-1)/5}\lambda(y); \tag{11.32a}$$

$$\bar{u}(y,t) = -Kt^{(3\alpha-4)/5}\lambda^2(y)\lambda'(y), \tag{11.32b}$$

where $K = q^{3/5}\eta_N(\alpha)$. $\lambda(y)$ and $\eta_N(\alpha)$ are displayed in Figs. 11.2 and 11.3. For $\alpha = 0$ the analytical results for $\lambda(y)$ and $\eta_N$ are given by (11.29) and (11.30). The scaling is given by (11.14).

The transformation of the foregoing dimensionless results to dimensional counterparts requires some work, but is straightforward. We must keep in mind the unusual scaling of $q$ with $x_0 h_0 T^{-\alpha}$. It is left for an exercise to show that

$$x_N^*(t^*) = \eta_N \left(\frac{1}{3}\right)^{1/5} \left[\frac{g^{*\prime}}{\nu^*}q^{*3}\right]^{1/5} (t^*)^{(3\alpha+1)/5}, \tag{11.33}$$

where dimensional variables are denoted by an asterisk. We also recall that for the fixed volume current, $\alpha = 0$, the value of $q^*$ is that of the volume.

### 11.1.4 Some comparisons

We compare the prediction (11.33) with the box-model approximation result (4.24) (both are in dimensional form). There is complete agreement concerning the power of $t$. The coefficient that multiplies $t$ has the same qualitative dependency on $g'$, $\nu$, and $q$ (this is the part in the square brackets). The only difference is in the value of the numerical coefficient. This numerical coefficient is larger in the box-model results (by about 50% for $\alpha = 0$ and 20% for $\alpha = 1$). This comparison gives some credence to the box-model approximation, but, on the other hand, the present similarity solution is more accurate and sufficiently simple for making it the preferred modeling tool.

Various comparisons with experiments were reported by Rottman and Simpson (1983), Didden and Maxworthy (1982), Huppert (1982), and others. Most experiments focused on the behavior of $x_N(t)$; see (11.33).

For $\alpha = 0$ this is a relatively simple task. The viscous-current behavior is expected to appear in the classical saltwater current released from a lock into freshwater in a 2D tank. The requirement is that the tank is sufficiently long as compared with $x_V$. The experiments of Rottman and Simpson (1983) were of this type. We note that this is a Boussinesq system. We emphasize again the contrast with the inviscid counterpart for the fixed volume $\alpha = 0$ current: the propagation is with $t^{1/5}$ as opposed to $t^{2/3}$, and the speed decay is with $t^{-4/5}$ as opposed to $t^{-1/3}$. Such differences are readily detected on a log-log plot of $x_N$ vs. $t$; see Fig. 2.13. The propagation with $t^{1/5}$ in the viscous regime has been confirmed.

The experiments for $\alpha > 0$ are in general more complicated. To create the desired inflow of saltwater into freshwater, an injector and pump are used. This allows for various $g'$, $q$, and $\alpha$.

Didden and Maxworthy (1982) reported results for $\alpha = 1$ for 2D viscous *surface* currents of this type. The dimensions length and width of the tank were 800 and 20.6 cm, respectively, the thickness of the ambient was 15 cm, and the length of the diffuser was only 10 cm. Food color was added to the injected fluid for flow visualization. The typical values of $g'$ and $q$ were 10 cm s$^{-2}$ and 0.6 cm$^2$ s$^{-1}$, and $\nu = 0.01$ cm$^2$ s$^{-1}$. The typical thickness of the current was 1 cm. A surprising observation was that the surface current is actually subjected to a no-slip condition on the surface (due to impurities on the interface) and hence comparison with the present theory is justified.

The ratio of inertial to viscous forces, estimated by (4.18) for $\alpha = 1$ and $Fr \approx 1$, is

$$\frac{F_I}{F_V} \approx \frac{1}{\nu} \frac{q^{4/3}}{g'^{2/3}} t^{-1}. \qquad (11.34)$$

We use dimensional variables. Didden and Maxworthy (1982) calculated the time $t_1$ at which this ratio is equal to 1, and concentrated on the behavior for longer times, for which the viscous forces are expected to be dominant. To be specific, in the experiments $t_1$ was typically 10 s, and the currents were followed for about 1000 s.

The theoretical prediction, see (11.33) for $\alpha = 1$, is

$$x_N(t) = 0.804 \left[ \frac{g'}{\nu} q^3 \right]^{1/5} t^{4/5}. \tag{11.35}$$

On a log-log plot, the recorded $x_N$ vs. $t$ collapsed well on straight lines with slope close to 0.8, in good agreement with the theoretical prediction 4/5. The other parameters also fitted well the theoretical dependency on $[g'q^3/\nu]^{1/5}$. The conclusion is that the experimental data fit well the formula (11.35), but with the numerical coefficient $0.73 \pm 0.03$. The deviation from the theoretical 0.80 was attributed to the friction on the side walls. Some data for the thickness of the current at the injector nozzle was also recorded, and showed consistency with the predicted dependency on $t$. Overall, the experiments provide good support to the theory for this configuration.

As mentioned above, in the experiments, the behavior of the viscous bottom current was mirrored by that of the "top" current under an open surface. This is surprising for a viscous current. There is a no-slip condition on the bottom, but, in ideal circumstances, a very different no-shear condition is relevant to the open surface. The visualization of the $z$-dependency of $u$ in the experiments indicated that at the open surface a very thin motionless layer was present during the propagation of the current. This was attributed to impurities on the interface of the ambient fluid. In practice, the top current was subject to a no-slip condition on the supporting boundary, and a no-shear condition at the interface with the deep ambient. This makes the surface current equivalent to a bottom current. (We recall that the experiments considered a Boussinesq system; in non-Boussinesq systems other effects, beside the boundary condition, may also be relevant to the difference between top and bottom currents.)

Similar experiments for $\alpha = 1.5, 1.75, 2$, and 3 but only with bottom currents, $\rho_c > \rho_a$, were performed by Maxworthy (1983a). Again, good agreement with the theory was reported. This was discussed in § 4.2.4.

## 11.1.5  Extensions to viscous surface currents and intrusions

The boundary conditions for the current in the foregoing analysis were (11.7), i.e., no-slip on the horizontal plane of propagation and no-shear at the contact interface with the ambient. These conditions were used to determine the functions $A$ and $B$ in (11.5). What happens when the viscous current spreads over an open surface where a no-shear condition prevails? What happens when the propagating fluid is an "intrusion" symmetric with respect to $z = 0$, instead of the current above $z = 0$ discussed above? A simple answer is obtained by interchanging the positions where the no-shear and no-slip conditions are applied. The arguments and results are briefly as follows.

Consider the upper half of the intrusion problem. A more detailed discussion is presented later in Chapter 15; here we just mention that $z = 0$ in an

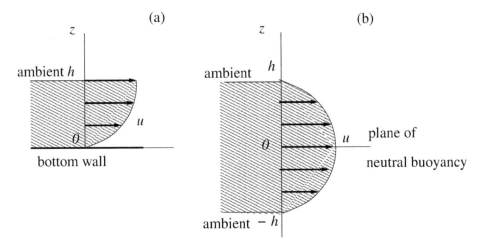

**FIGURE 11.4:** Schematic description of the change of the velocity profile from a gravity current configuration (a) to an intrusion configuration (b).

interface of symmetry between the upper and the lower halves of the intruding fluid. Essentially, $\rho_c$ is the mean density between the density of the ambient at $z > 0$ and that at $z < 0$, so that the intrusion floats at $z = 0$. The upper (and lower) half is a gravity current driven by the buoyancy force and hindered by the viscous friction. The momentum balance is the same as before and the general $u(x, z, t)$ result (11.5) is valid. However, now the particular $z$-shape of velocity profile must be different, because $z = 0$ is a plane of symmetry, not a solid boundary. The difference is sketched in Fig. 11.4. For the intrusion into a motionless ambient we thus impose the boundary conditions

$$\frac{\partial u}{\partial z}(x, z = 0, t) = 0 \quad \text{and} \quad u(x, z = h, t) = 0 \tag{11.36}$$

to obtain

$$u(x, z, t) = \frac{g'}{\nu} \frac{\partial h}{\partial x} \frac{1}{2}(z^2 - h^2). \tag{11.37}$$

The average speed, in dimensional form, is

$$\bar{u} = -\frac{1}{3} \frac{g'}{\nu} h^2 \frac{\partial h}{\partial x}. \tag{11.38}$$

This is the same result as for the flow above a plate (not really surprising given the symmetries shown in Fig. 11.4). The advantage of this *crude approximation* is that subsequent results for the thickness $h$ and position of nose $x_N(t)$ are identical with these developed for the flow above a plate. The lower part of the intrusion is just the mirror image. The drawback is that the zero velocity condition at $z = h$ is an oversimplification; in many practical cases the viscosity smooths out the speed differences with time, and hence the propagation is typically less restricted than estimated by this simplification. The details depend on the geometry and viscosity of the ambient fluid and are beyond the scope of this book; for a more comprehensive analysis and a compendium of spreading relations the reader is referred to the paper of Lister and Kerr (1989).

## 11.2   Axisymmetric current

### 11.2.1   The governing equations

Consider an axisymmetric configuration. The objective is to formulate and solve the governing equations for the thickness $h(r,t)$ and $z$-averaged speed, $\bar{u}(r,t)$, in the viscous current. This also provides the behavior of $r_N(t)$.

We again consider a one-layer system, and use the same type of simplifications as for the 2D geometry; see § 11.1. The horizontal coordinate $x$ of the 2D case is replaced with the cylindrical radius $r$. The momentum considerations used in the 2D case can be extended to the cylindrical propagation. To be specific, (11.1)-(11.9) and (11.11) are applicable when $x$ is replaced by $r$. Here the volumes and volume fluxes are per radian.

We start the analysis with the *dimensional form* of the averaged radial velocity

$$\bar{u}(r,t) = \frac{1}{h}\int_0^h u\,dz = -\frac{1}{3}\frac{g'}{\nu}h^2\frac{\partial h}{\partial r}, \tag{11.39}$$

and the corresponding continuity equation derived from (6.6), which is

$$\frac{\partial h}{\partial t} - \frac{1}{3}\frac{g'}{\nu}\frac{1}{r}\frac{\partial}{\partial r}rh^3\frac{\partial h}{\partial r} = 0. \tag{11.40}$$

Again, the equations for speed and thickness can be decoupled. The major task is to solve a single parabolic PDE for the unknown $h(r,t)$.

To both generalize the results and simplify the manipulation, we perform two preliminary steps:

1. We assume that the volume of the current (per radian) is given by

$$\mathcal{V} = qt^\alpha, \tag{11.41}$$

where $q\,(>0)$ and $\alpha\,(\geq 0)$ are constants, with the particular case

$$q = \mathcal{V} \quad \text{for} \quad \alpha = 0. \tag{11.42}$$

The source is, for simplicity, at $r = 0$. This introduces a non-physical behavior of $u$ near the source for the $\alpha > 0$ case, but we assume that this is a local effect which has minor influence on the main results.

2. We switch to dimensionless variables. The horizontal and vertical lengths are scaled with $r_0, h_0$, respectively, and volume (per radian) is scaled with $r_0^2 h_0$. The velocity and time are scaled with

$$U = \frac{1}{3} g' \frac{1}{\nu} h_0^3 \frac{1}{r_0}; \quad T = \frac{r_0}{U}. \qquad (11.43)$$

Note that the coefficient $q$ is scaled with $r_0^2 h_0 T^{-\alpha}$, a quite unusual quantity. We also keep in mind that the proper choice of $r_0$ and $h_0$ may require some analysis of the particular problem. For example, for a fixed-volume current the intuitive choice for $r_0$ and $h_0$ is the dimensions of the lock. However, the initial motion may well be in the inviscid regime. Therefore, a more relevant choice will be the radius and mean thickness of this current when transition to the viscous regime occurs.

The following analysis uses *dimensionless variables* unless stated otherwise. The momentum equation is

$$\bar{u}(r, t) = -h^2 \frac{\partial h}{\partial r}; \qquad (11.44)$$

and the continuity equation (11.40) reads

$$\frac{\partial h}{\partial t} - \frac{1}{r} \frac{\partial}{\partial r} r h^3 \frac{\partial h}{\partial r} = 0, \qquad (11.45)$$

subject to

$$\int_0^{r_N(t)} h(r, t) r \, dr = V = q t^\alpha. \qquad (11.46)$$

Clearly, we must concentrate on the solution of the PDE equation for $h$, then the velocity follows. This is facilitated, again, by the transformation introduced in § 2.3 and 6.2. Letting $y = r/r_N(t)$ and using (2.46)-(2.47) we obtain the following equation for $h(y, t)$

$$\frac{\partial h}{\partial t} - y \frac{\dot{r}_N}{r_N} \frac{\partial h}{\partial y} - \frac{1}{r_N^2} \frac{1}{y} \frac{\partial}{\partial y} y h^3 \frac{\partial h}{\partial y} = 0 \quad (0 \le y \le 1); \qquad (11.47)$$

subject to

$$r_N^2(t) \int_0^1 h(y, t) y \, dy = V = q t^\alpha. \qquad (11.48)$$

## 11.2.2 Similarity solution

We seek a similarity solution of the form

$$r_N = K t^\beta; \quad h = \frac{V}{r_N^2} \mathcal{H}(y) = \frac{q}{K^2} t^{\alpha - 2\beta} \mathcal{H}(y); \qquad (11.49)$$

where $K$ and $\beta$ are some positive constants. Substitution into (11.47) and arrangement yields

$$(\alpha - 2\beta)y\mathcal{H} - \beta y^2 \mathcal{H}' - t^{3\alpha - 8\beta + 1} \frac{q^3}{K^8} (y\mathcal{H}^3\mathcal{H}')' = 0. \tag{11.50}$$

Similarity requires that the power of $t$ is zero. This determines

$$\beta = \frac{3\alpha + 1}{8}. \tag{11.51}$$

Further simplification is achieved upon the substitution

$$\mathcal{H}(y) = \frac{K^{8/3}}{q}\lambda(y), \tag{11.52}$$

which reduces (11.50) to

$$(y\lambda^3\lambda')' + \beta y^2\lambda' - (\alpha - 2\beta)y\lambda = 0. \tag{11.53}$$

The boundary condition for this equation is $\lambda(1) = 0$. The justification is as in the 2D case. The pressure gradient drives the current against the viscous forces, and this requires a decreasing $h$ with $r$. This implies that the self-similar profiles $\mathcal{H}$ (and $\lambda$) are positive monotonically decreasing functions of $y$ for $0 < y < 1$. This behavior must end with $h = 0$ at $r_N$, i.e., $\mathcal{H}(1) = \lambda(1) = 0$.

Equation (11.53) has, as expected, a singularity at $y = 1$. Using a Frobenius series expansion $\lambda = \xi^\gamma(a_0 + a_1\xi + \cdots)$, where $\xi = (1 - y)$, we obtain

$$\lambda(y) = \left[\frac{3}{8}(3\alpha + 1)\right]^{1/3} (1 - y)^{1/3} \left[1 - \frac{1}{3(3\alpha + 1)}(1 - y) + O[(1 - y)^2]\right]. \tag{11.54}$$

In general, we use this approximation to obtain boundary conditions at $y = 1 - \Delta$ for numerical calculation of $\lambda(y)$, as explained for the 2D case. Results for typical values of $\alpha$ are displayed in Fig. 11.5. For $\alpha > 0$ a non-physical increase of thickness appears near the center $y = 0$. This is because the finite imposed influx $\alpha q t^{\alpha - 1}$ must be matched by $ruh$ with $r \to 0$ (theoretically). This introduces unrealistic large values of $u$ and $h$ near the center. The accepted remedy is to simply ignore the small region about the center ($y < 0.01$ say). The volume in this domain is small, and in any case a real source has a finite radius.

Next, we calculate $K$ using the global volume balance (11.48) in which we substitute (11.49) and (11.52). After some algebra this yields

$$\int_0^1 \mathcal{H}(y)y\,dy = \frac{K^{8/3}}{q}\int_0^1 \lambda(y)y\,dy = 1, \tag{11.55}$$

and hence

$$K = q^{3/8}\left[\int_0^1 \lambda(y)y\,dy\right]^{-3/8} = q^{3/8}\kappa_N(\alpha). \tag{11.56}$$

In general $\kappa_N(\alpha)$ is calculated numerically; some results are displayed in Fig. 11.3. (Note that the value of our $\kappa_N$ differs from $\xi_N$ of Huppert 1982. This is because our volumes are per radian, while the calculations in that paper are for a full $2\pi$ circumference.)

Finally, we calculate the speed. Combining the previous results we can write

$$\bar{u}(y,t) = -\frac{1}{r_N}\left(\frac{\mathcal{V}}{r_N^2}\right)^3 \mathcal{H}^2\mathcal{H}'(y) = -Kt^{(3\alpha-7)/8}\lambda^2(y)\lambda'(y). \qquad (11.57)$$

Consider the fixed volume case, $\alpha = 0$. An analytical solution of (11.53) can be obtained. It can be verified by substitution that

$$r_N(t) = Kt^{1/8}; \quad h = \left(\frac{16}{3}\right)^{1/3}\frac{8}{3}\frac{\mathcal{V}}{r_N^2}\lambda(y); \qquad (11.58)$$

$$\lambda(y) = \left(\frac{3}{16}\right)^{1/3}(1-y^2)^{1/3}; \qquad (11.59)$$

$$K = 1.781\mathcal{V}^{3/8}; \qquad (11.60)$$

$$\bar{u} = \left(\frac{2}{3}\right)\left(\frac{8}{3}\right)^3\frac{\mathcal{V}^3}{r_N^7}y = 0.223\mathcal{V}^{3/8}t^{-7/8}y. \qquad (11.61)$$

In the SW inviscid counterpart self-similar current the propagation is with $t^{1/2}$ and the speed decays with $r_N^{-1}$. Here the propagation is significantly slower, with $t^{1/8}$, and the decay of speed is remarkably faster, with $r_N^{-7}$. This is worth emphasizing: when the radius of a viscous gravity current of fixed volume doubles, its speed of propagation is reduced to less than 1% of the initial value. The similarity solutions for both the inviscid and viscous current predict a linear increase of $u$ with $y$. In the present case the boundary condition for $u$ at the center $y = 0$ is simply satisfied. For $\alpha > 0$ the behavior of $u$ for very small $y$ poses a difficulty, as discussed above.

### 11.2.3 Summary

The behavior of the viscous axisymmetric current of volume $\mathcal{V} = qt^\alpha$ (per radian) is predicted by the similarity solution of the lubrication formulation as (in dimensionless form)

$$r_N(t) = Kt^{(3\alpha+1)/8}; \quad h(y,t) = K^{2/3}t^{(\alpha-1)/4}\lambda(y); \qquad (11.62a)$$

$$\bar{u}(y,t) = -Kt^{(3\alpha-7)/8}\lambda^2(y)\lambda'(y), \qquad (11.62b)$$

where $K = q^{3/8}\kappa_N(\alpha)$. $\kappa_N(\alpha)$, and $\lambda(y)$ are displayed in Figs. 11.3 and 11.5, respectively. For $\alpha = 0$ the analytical results for the flow profiles are given by (11.59)-(11.61). The scaling is given by (11.43).

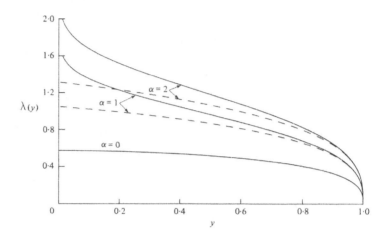

**FIGURE 11.5**:  Similarity profiles of interface for the axisymmetric viscous current. $\lambda(y)$ is given for various values of $\alpha$. The dashed lines are the approximate solution (11.54), and the solid lines are the exact solution. (Huppert 1982, with permission.)

The transformation of the foregoing dimensionless results to dimensional counterparts requires some work, but is straightforward. We must keep in mind the unusual scaling of $q$ with $r_0^2 h_0 T^{-\alpha}$. It is left for an exercise to show that

$$ r_N^*(t^*) = \kappa_N \left(\frac{1}{3}\right)^{1/8} \left[\frac{g^{*\prime}}{\nu^*} q^{*3}\right]^{1/8} (t^*)^{(3\alpha+1)/8}, \qquad (11.63) $$

where dimensional variables are denoted by an asterisk. We also recall that for the fixed volume current, $\alpha = 0$, the value of $q^*$ is that of the volume (per radian).

### 11.2.4   Some comparisons

We compare the prediction (11.63) with the box-model approximation results (7.23) (both are in dimensional form). There is complete agreement concerning the power of $t$. The coefficient has the same qualitative dependency on $g'$, $\nu$, and $q$. The only difference is in the value of the numerical coefficient. This numerical coefficient is larger in the box-model results (by about 13%). This gives some credence to the box-model approximation, but, on the other hand, the present similarity solution is more accurate and sufficiently simple for making it the preferred modeling tool.

Huppert (1982) reported experimental results for $\alpha = 0$ and $\alpha = 1$. Two silicone oils ($\nu = 13.2$ and $1110 \, \text{cm}^2 \, \text{s}^{-1}$) were used to create fully axisymmetric currents in the atmosphere over a flat Perspex surface. A ruled sheet of graph paper placed beneath the transparent surface allowed a direct measurement of the radius of propagation. The volumes involved were of the order of about one liter, and the currents propagated during several hours to a radius of about 50 cm. This is a non-Boussinesq system. In this case the zero-shear boundary condition at $z = h$ is well justified.

Britter (1979) and Didden and Maxworthy (1982) present data for $\alpha = 1$ inflow in saltwater Boussinesq systems, in a wedge. The experiments of Didden and Maxworthy (1982) for $\alpha = 1$ are the counterpart of the 2D configuration mentioned in § 11.1. Both surface and bottom currents were tested. The tank was a square with 245 cm sides. The typical depth of the ambient was 10 cm. The use of flow injectors of radius 10 and 15 cm placed in a corner of the tank created axisymmetric flows in a wedge of angle 90°. The typical flow rate was $q = 10 \, \text{cm}^2 \, \text{s}^{-1}$ (per radian of source angle).

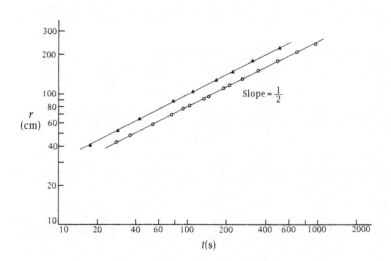

**FIGURE 11.6:** Experimental $r_N$ as a function of $t$ for bottom currents in a 90° wedge, and $\alpha = 1$. Log-log axes. $g' = 18.0 \, \text{cm} \, \text{s}^{-2}$, height of ambient about 11 cm, $q = 75/(2\pi) = 11.94 \, \text{cm}^3 \, \text{s}^{-1}$ ○, and $116/(2\pi) = 18.46 \, \text{cm}^3 \, \text{s}^{-1}$ ▲. (Didden and Maxworthy 1982, with permission.)

All these experiments give good support to the theoretical prediction (11.63). On log-log axes, the $r_N$ vs. $t$ data are straight lines, and the slopes are very close to the predicted 0.125 and 0.500 (for $\alpha = 0$ and 1, respectively). A typical result is shown in Fig. 11.6. The dependency on the other parameters collapses well in accord with the predicted $[g'q^3/\nu]^{1/8}$. The numerical coefficient is slightly bellow the theoretical value in the data of Didden and Maxworthy (1982) and Huppert (1982), but slightly above in the results of Britter (1979). The discrepancies are within a few percent and can be attributed to experimental errors and various deviations from the idealizations of the solution. After all, (11.63) contains many assumptions and simplification.

Measurements of the thickness at the center of an axisymmetric current of fixed volume are also presented in Huppert (1982). They show consistency with the theory. The scatter, however, is significantly larger than in the results for $r_N$. This is attributed to the practical difficulty of making the measurements.

An interesting observation of Didden and Maxworthy (1982) is that the surface and bottom currents displayed the same propagation. This quite surprising experimental outcome has also been obtained for the 2D case, and is discussed in § 11.1.4.

---

## 11.3    Current in a porous medium

An important related problem is that of a gravity current which propagates *inside* a porous media. The applications include injection of liquid or gas into the ground to enhance recovery of oil and heat in natural reservoirs. The typical gravity current in a porous medium bears some similarities with the viscous current. In both cases the inertial terms are unimportant and the governing momentum balance is between the buoyancy-pressure driving and a resistance to the motion proportional to $u$.

We can point out these similarities in a sketchy way as follows. Here we use dimensional variables. We assume a simple porous medium of constant permeability $k$ and porosity $\phi$ with a solid boundary at $z = 0$. The pressures in the ambient and current are hydrostatic. Consider the 2D case. The driving effect is given again by the fundamental

$$\frac{\partial p_c}{\partial x} = \rho_c g' \frac{\partial h(x,t)}{\partial x}. \tag{11.64}$$

Here we use $g' = g'_c$.

Darcy's law in the $x$ direction then gives simply

$$u(x,t) = -\frac{k}{\rho_c \nu} \frac{\partial p_c}{\partial x} = -\frac{kg'}{\nu} \frac{\partial h(x,t)}{\partial x}. \tag{11.65}$$

The lack of $z$ dependency facilitates the averaged-variables analysis.

The continuity equation is provided by (2.24) with a slight modification

$$\frac{\partial h}{\partial t} + \frac{1}{\phi}\frac{\partial h\bar{u}}{\partial x} = 0. \tag{11.66}$$

The correction is due to the fact that only the portion $\phi$ of the medium can contain fluid. Consequently, the axial displacement of the interface is faster for a given rate of volume accumulation.

The combination of (11.65) and (11.66) yields a single PDE for the variable $h(x,t)$

$$\frac{\partial h}{\partial t} - \frac{g'k}{\nu\phi}\frac{\partial}{\partial x}h\frac{\partial h}{\partial x} = 0. \tag{11.67}$$

The similarity with (11.10) is evident. The governing PDE is parabolic and admits the boundary condition $h = 0$ at $x = x_N(t)$. This resemblance becomes even more pronounced upon scaling height with $h_0$, length with $x_0$, speed with $U = g'kh_0/(\nu\phi x_0)$, and time with $x_0/U$. (Some care is necessary in the definition and interpretation of lengths and volume because of the apparent dilatation caused by $\phi$.) It is left for an exercise to show that (11.18) is recovered, but with $h$ instead of $h^3$ in the last term; and that this equation is again subject to the global volume continuity condition (11.19). Also consider the solution by the use of similarity variables, calculate the value of $\beta$, and formulate the boundary conditions.

This analogy can be carried over to the axisymmetric geometry.

The detailed derivation and assessment of the results involve effects and considerations that are beyond the scope of this book. The interested reader is referred to the papers of Huppert and Woods (1995), Woods and Mason (2000), and Lyle, Huppert, Hallworth, Bickle, and Chadwick (2005) where additional useful bibliography is given.

# Part II

# Stratified ambient currents and intrusions

# Chapter 12

## Continuous density transition

### 12.1  Introduction

Our primary aim is to extend the body of knowledge presented in the previous chapters to the case when the propagation of the gravity current is into a non-homogeneous ambient. The major interest is in a linear stratification, in particular over a layer whose thickness is larger than that of the current. This type of stratification is considered in this chapter. Density transition over a thin layer, including a sharp (jump) stratification of the ambient, will be considered later.

Before we consider the current, let us point out the prominent novel features of the ambient into which propagation will occur: a continuously stratified fluid supports internal oscillations and wave propagation. For simplicity we consider a Boussinesq inviscid system close to hydrostatic, stable, equilibrium.

To derive the frequency of the oscillations we consider a small fluid particle of density $\rho_A$, which is displaced slightly from its initial height $z_A$ to $z_A + \zeta$. The local density difference produces a buoyancy force that pushes back the particle. We can write the equation of motion

$$\rho_A \frac{d^2\zeta}{dt^2} = -[\rho(z_A) - \rho(z_A + \zeta)]g \approx \left(\frac{\partial \rho}{\partial z}\right)_A \zeta g. \tag{12.1}$$

The linear with $\zeta$ restoring force is justified when the displacement is small or the stratification is linear.

The solution $\zeta(t)$ of (12.1) behaves like $\cos(\mathcal{N}t)$, where

$$\mathcal{N} = \left(-\frac{1}{\rho}\frac{\partial \rho}{\partial z}\right)_A^{1/2} g^{1/2}. \tag{12.2}$$

This is defined as the buoyancy frequency. Clearly, a stratification with positive $\partial \rho / \partial z$ is unstable and produces an imaginary value of $\mathcal{N}$; in this case $\zeta(t)$ behaves like $\exp(|\mathcal{N}|t)$. For obvious reasons we shall consider propagation in stable setups only. In a linear Boussinesq stratification $\mathcal{N}$ is constant (within the usual $O(\epsilon)$ error).

The analysis of the internal waves is more complicated; see for example Baines (1995). A fundamental result is that in a channel of height $H$ filled with linearly-stably-stratified fluid, a small-perturbation long wave is associated with the stream-function $z$-dependency

$$\psi_j = \sin j\pi \frac{z}{H}, \quad j = 1, 2, 3 \ldots \tag{12.3}$$

and propagates with the speed

$$c_j = \frac{\mathcal{N}H}{j\pi}. \tag{12.4}$$

A gravity current which propagates at the bottom (or top) wall in the channel excites such perturbation. The most relevant effect is expected to be represented by waves with the leading mode $j = 1$, which propagate with the speed

$$u_W = \frac{\mathcal{N}H}{\pi}. \tag{12.5}$$

We can estimate the wavelength in the context of the gravity current as follows. The current propagating at speed $u_N$ displaces a particle (actually, an isopycnal) of the stratified ambient. The displaced particle is expected to perform an oscillation with time period $2\pi/\mathcal{N}$. An observer attached to the current will identify the wavelength with the relative $x$-propagation of the disturbed particle during this interval,

$$\lambda = 2\pi \frac{u_N}{\mathcal{N}}. \tag{12.6}$$

With these essentials in the background, let us consider the typical gravity current configuration. The prototype problem is, again, the propagation of high-Reynolds number currents resulting from the instantaneous release of a finite volume of fluid of constant density. The ambient fluid is linearly stratified over the full depth $H$.

The elementary system under consideration is sketched in Fig. 12.1. A deep layer of ambient fluid, of density $\rho_a(z)$, lies above a horizontal surface at $z = 0$. Gravity acts in the $-z$ direction. The density of the ambient increases from $\rho_o$ to $\rho_b$. In the rectangular case the system is bounded by parallel vertical smooth impermeable surfaces and the current propagates in the direction labeled $x$. At time $t = 0$ a given volume of homogeneous fluid of density $\rho_c$ ($\geq \rho_b$) and kinematic viscosity $\nu$, initially at rest in a rectangular box of height $h_0$ and length $x_0$, is instantaneously released into the ambient fluid. A two-dimensional current commences to spread. We assume that the appropriate Reynolds number of the horizontal flow is large and hence viscous effects can be neglected.

For analytical progress we shall use a one-layer SW model. The relevant theoretical papers are Ungarish and Huppert (2002, 2004, 2006) and Ungarish (2005a, 2005b, 2006). The major deficiency of the one-layer SW model is the fact that internal gravity waves are discarded. Our arguments are that

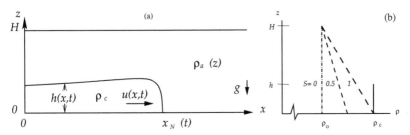

**FIGURE 12.1:** Schematic description of the system: (a) the geometry; and (b) the density profiles in the current (solid line) and unperturbed ambient, for various values of $S = (\rho_c - \rho_b)/(\rho_c - \rho_o)$ (dashed lines).

these waves have little influence on the motion of the current in certain practical circumstances, at least during a significant initial period of propagation, and that, for analytical progress, it is worthwhile, perhaps even necessary, to decouple the current and the waves. A closely related problem is the stratified flow over a fixed obstacle, a topic covered thoroughly by Baines (1995), from which useful insights can be gained. However, the lock-released gravity current is a time-dependent deformable "obstacle," whose shape interacts with the waves it produces in the ambient. The analytical study of this time-dependent flow field is evidently a formidable task, and our idea is to attempt the following decoupling: first solve for the propagation of the gravity current under the assumption of an unperturbed ambient; next consider the perturbations produced in an impulsively-started flow over an obstacle of prescribed height $h(x, t)$ over the bottom (the relative velocity far upstream is that of the front of the current, $u_N(t)$). In this book we consider mainly the solution of the first problem. This seems to be the more fundamental one, as reflected by the accurate predictions provided by the SW results for a considerable time interval. We note in passing that the waves are intrinsically incorporated in the Navier-Stokes simulations discussed in the text.

Another possibility is a steady-state motion, of the type considered in Chapter 3. In this case the changes produced in the ambient by the presence of the current can be taken into account analytically. This topic is postponed to Chapter 14.

## 12.2   The SW formulation

The configuration is sketched in Fig. 12.1. For the rectangular case we use a $\{x, y, z\}$ Cartesian coordinate system with corresponding $\{u, v, w\}$ velocity components, and assume that the flow does not depend on the coordinate $y$

and that $v \equiv 0$. The variables are in dimensional form.

Initially, the height of the fluid which will make up the propagating current is $h_0$, its length $x_0$, and the density $\rho_c$. The height of the ambient fluid is $H$ and the density in this domain decreases linearly with $z$ from $\rho_b$ to $\rho_o$, where the subscripts $b, o$ refer to bottom and top (or open) surface values, respectively. The density variation *in the ambient* is defined as

$$\widetilde{\Delta\rho} = \rho_b - \rho_o \tag{12.7}$$

(which should be distinguished from the $\Delta = \rho_c - \rho_a$ which we used for the homogeneous-ambient analysis). The linear variation of $\rho_a$ over the full depth of the ambient is taken here for simplicity of analysis. We shall see later that the SW analysis can be extended to more general forms of $\rho_a(z)$.

It is convenient to use $\rho_o$ as the reference density. In experimental sets, this is usually the density of freshwater; the stratification and the current are conveniently created by addition of salt.

We introduce the reduced density differences (a) with the current

$$\epsilon = \frac{\rho_c - \rho_o}{\rho_o}, \tag{12.8}$$

and (b) with the fluid at the bottom

$$\epsilon_b = \frac{\rho_b - \rho_o}{\rho_o} = \frac{\widetilde{\Delta\rho}}{\rho_o}. \tag{12.9}$$

The ratios between these density differences introduces the new important parameter

$$S = \frac{\rho_b - \rho_o}{\rho_c - \rho_o} = \frac{\epsilon_b}{\epsilon}. \tag{12.10}$$

The parameter $S$ represents the magnitude of the stratification in the ambient fluid. We consider only $0 \leq S \leq 1$. The homogeneous ambient is recovered by setting $S = 0$. The maximum stratification case $S = 1$ means that the density of the current is equal to that of the ambient at the bottom. The typical cases are illustrated in Fig. 12.1(b). In the Boussinesq system $\epsilon_b, \epsilon \ll 1$.

We also define the reference reduced gravity,

$$g' = \epsilon g, \tag{12.11}$$

where $g$ is the gravitational acceleration. We must keep in mind that now the real reduced gravity forcing changes with $z$ because the local buoyancy is proportional to $\rho_c - \rho_a(z)$.

Using the definitions above we can express the densities as

$$\rho_c = \rho_o(1 + \epsilon), \quad \rho_a = \rho_o \left[ 1 + \epsilon S \left( 1 - \frac{z}{H} \right) \right]. \tag{12.12}$$

The buoyancy frequency is obtained by substitution of (12.12) into (12.2)

$$\mathcal{N} \approx \left(-\frac{g}{\rho_o}\frac{d\rho_a}{dz}\right)^{1/2} = \left(\frac{Sg'}{H}\right)^{1/2}. \tag{12.13}$$

We recall that the leading, or mode one, linear internal waves in a closed 2D channel propagate with the speed

$$u_W = \frac{\mathcal{N}H}{\pi}. \tag{12.14}$$

In the ambient fluid domain we assume that $u = v = w = 0$ and hence the fluid is in purely hydrostatic balance and maintains the initial density $\rho_a(z)$. The motion is assumed to take place in the lower layer only, $0 \le x \le x_N(t)$ and $0 \le z \le h(x,t)$. As in the classical inviscid, shallow-water analysis of a gravity current in a homogeneous ambient, we argue that the predominant vertical momentum balance in the current is hydrostatic and that viscous effects in the horizontal momentum balance are negligibly small. Hence the motion is governed by the balance between pressure and inertial forces in this horizontal direction.

A relationship between the pressure fields and the height $h(x,t)$ is now obtained. In the motionless ambient fluid the pressure does not depend on $x$. The hydrostatic balances are $\partial p_i/\partial z = -\rho_i g$, where $i = a$ or $c$. Use of (12.12) then yields

$$p_a(z,t) = -\rho_o \left[1 + \epsilon S(1 - \frac{1}{2}\frac{z}{H})\right]gz + C \tag{12.15}$$

and

$$p_c(x,z,t) = -\rho_o(1 + \epsilon)gz + f(x,t). \tag{12.16}$$

As in the non-stratified case, pressure continuity between the ambient and the current on the interface $z = h(x,t)$ determines the function $f(x,t)$ of (12.16) and we obtain, after some algebra,

$$p_c(x,z,t) = -\rho_o(1 + \epsilon)gz + \rho_o g'\left[h - S(h - \frac{1}{2}\frac{h^2}{H})\right] + C. \tag{12.17}$$

Finally, the fundamental $x$-pressure driving gradient is given by

$$\frac{\partial p_c}{\partial x} = \rho_o g'\frac{\partial h}{\partial x}\left[1 - S(1 - \frac{h}{H})\right]. \tag{12.18}$$

(In the present derivation we did not use the reduced pressure. Since the hydrostatic pressure of the hydrostatic ambient is $x$-independent, the resulting (12.18) will remain unchanged if the pressures are reduced such that $p_a$ becomes a constant. For $S = 0$, (2.15) is recovered. The details are left for an exercise.)

We note several interesting points concerning the effect of stratification on the pressure terms. The linear with $z$ stratification of density introduces a non-linear, $z^2/H$, dependency. This has some non-trivial implications on the nose boundary condition. On the other hand, $\partial p_c/\partial x$ is not a function of $z$, and this renders the derivation of the averaged SW equations of motion as a straightforward extension of the homogeneous case. As expected, the effect of stratification increases as $S$ increases.

The stratification does not affect the volume continuity equation of the SW model. The next step is to consider the vertical average of the horizontal inviscid momentum equation in the dense fluid, and eliminate the pressure term by (12.18). The derivation is similar to that of § 2.2. We obtain a system of equations for $h(x,t)$ and for the averaged longitudinal velocity, $u(x,t)$.

It is convenient to scale the dimensional variables (denoted here by asterisks) by

$$\{x^*, z^*, h^*, H^*, t^*, u^*\} = \{x_0 x, h_0 z, h_0 h, h_0 H, Tt, Uu\}, \tag{12.19}$$

with

$$U = (h_0 g')^{1/2} = \mathcal{N}\sqrt{\frac{H}{S}} h_0 \quad \text{and} \quad T = \frac{x_0}{U} = \frac{1}{\mathcal{N}}\sqrt{\frac{S}{H}}\frac{x_0}{h_0}. \tag{12.20}$$

Here $U$ is a typical inertial velocity of propagation of the nose of the current and $T$ is a typical time period for longitudinal propagation for a typical distance $x_0$. We emphasize that $\mathcal{N}$ is dimensional; when we write $\mathcal{N} = (S/H)^{1/2}(g'/h_0)^{1/2}$ we use the dimensionless $H$, and this explains the difference with (12.13). Note that the horizontal and vertical lengths are scaled differently, which removes the initial aspect ratio $h_0/x_0$ from the SW analysis, like in the homogeneous circumstances. Indeed, the scaling here is the same as for the non-stratified case. However, there are some subtle differences that will become manifest in the results. In particular, we notice that in this problem the ambient introduces another intrinsic time scale, $\mathcal{N}^{-1}$, and speed, $u_W$; see (12.14). These scales may become important in the reaction of the ambient to the motion of the current, and eventually to the interaction between the flows. In addition, $g'$ is less representative than in the non-stratified case because the density difference (and the reduced gravity excess) varies with $z$.

Hereafter, *dimensionless variables* are used, unless stated otherwise. In any case, $x_0, h_0$, and $\mathcal{N}$ are dimensional.

In conservation form the SW equations can be written as

$$\frac{\partial h}{\partial t} + \frac{\partial}{\partial x}(uh) = 0, \tag{12.21}$$

and

$$\frac{\partial}{\partial t}(uh) + \frac{\partial}{\partial x}\left[u^2 h + \frac{1}{2}(1-S)h^2 + \frac{1}{3}S\frac{h^3}{H}\right] = 0. \tag{12.22}$$

In characteristic form, this becomes

$$\begin{bmatrix} h_t \\ u_t \end{bmatrix} + \begin{bmatrix} u & h \\ 1 - S + S\frac{h}{H} & u \end{bmatrix} \begin{bmatrix} h_x \\ u_x \end{bmatrix} = \begin{bmatrix} 0 \\ 0 \end{bmatrix}. \tag{12.23}$$

The system is hyperbolic in the range of parameters of interest. Following the standard procedure, see Appendix A.1, we obtain the characteristic balances

$$\mathcal{G}(h)dh \pm du = 0, \tag{12.24}$$

where

$$\mathcal{G}(h) = \left[ \frac{1 - S + (Sh)/H}{h} \right]^{1/2}, \tag{12.25}$$

on the characteristics

$$\frac{dx}{dt} = c_{\pm} = u \pm \left[ h(1 - S + S\frac{h}{H}) \right]^{1/2}. \tag{12.26}$$

The initial conditions are zero velocity and unit dimensionless height and length at $t = 0$. Also, the velocity at $x = 0$ is zero, and an additional condition is needed at the nose $x = x_N(t)$.

## 12.2.1 The nose condition

The nose is treated as a vertical discontinuity. The pertinent knowledge on the required condition is less developed than for the non-stratified counterpart. A major theoretical difficulty is that a two-layer SW model is unavailable. Both the density and the speed of the ambient fluid above the nose are expected to be $z$-dependent. An analysis of the type presented by Benjamin becomes a formidable task, and the results cannot be reduced to a simple analytical jump condition as done in Chapter 3. This will be elaborated in Chapter 14.

For progress, we adopt the heuristic approach of Ungarish and Huppert (2002). This is based on the arguments that, briefly, (1) the velocity of the nose is proportional to the square-root of the pressure head (per unit mass); (2) the factor of proportionality, defined as the Froude number $Fr$, varies in a quite narrow range with the ratio $h_N/H$; and (3) the behavior of $Fr$ in the stratified case is well approximated by the well known homogeneous situation. In other words, we go back to (3.15) which was inferred for the non-stratified case (in dimensional form)

$$u_N = Fr(a) \times \left[ \frac{p_c(z = 0) - p_a(z = 0)}{\rho_a} \right]^{1/2}, \tag{12.27}$$

where $a = h_N/H$. In the RHS the term in the brackets gives the pressure driving effect, and $Fr(a)$ represents the dynamic reaction effect. We argue that this separation of effects that dominate the speed of the discontinuity remains valid for the stratified ambient case.

The pertinent driving pressure difference at the nose is obtained from (12.15) and (12.17) as

$$p_c(z = 0) - p_a(z = 0) = \rho_0 g' h_N \left[1 - S(1 - \frac{1}{2}\frac{h_N}{H})\right] \quad \text{(dimensional)}.$$

(12.28)

We substitute (12.28) into (12.27), and use the Boussinesq approximation $\rho_a \approx \rho_0$. Then we scale the speed result with $(g' h_0)^{1/2}$.

The resulting dimensionless nose velocity condition is

$$u_N = Fr(a) \times \left[1 - S(1 - \frac{1}{2}\frac{h_N}{H})\right]^{1/2} \times h_N^{1/2}.$$

(12.29)

The term in the square brackets of (12.29) is equal to 1 in the non-stratified case ($S = 0$), and smaller than 1 for $S > 0$. This term expresses the explicit slow-down of the head due to the stratification effects.

The function $Fr(a)$ which is needed to close the present SW formulation is taken "off the shelf" from the non-stratified counterpart. As in these cases, unless stated otherwise, we use the simple HS formula

$$Fr(h_N/H) = \begin{cases} 1.19 & (0 \leq h_N/H < 0.075) \\ 0.5 H^{1/3} h_N^{-1/3} & (0.075 \leq h_N/H \leq 1). \end{cases}$$

(12.30)

A practical justification of the boundary condition (12.29)-(12.30) is that the SW results obtained with this closure are in good agreement with available experiments and simulations for a wide range of problems. The theoretical support will be discussed in Chapter 14.

The SW formulation is complete. There are two free parameters: the depth ratio, $H$, and the (relative) magnitude of the stratification, $S$. We see that for $S = 0$ the non-stratified case is identically recovered. In general, the stratification terms enter into the formulation as both $S$ and $Sh/H$. However, we notice that a significant simplification is obtained for the $S = 1$ case. We shall elaborate on this later.

For completeness, we use (12.14) to write the dimensionless speed of the fastest mode internal wave

$$u_W = \frac{(SH)^{1/2}}{\pi}.$$

(12.31)

A current is defined as super-critical when $u_N > u_W$ and sub-critical when $u_N < u_W$. An inspection of (12.29) and (12.31) indicates that the type of the current depends on the values of $H$ and $S$. Since $Fr$ is about 1 and $h_N$ smaller than 1 (typically 0.5 in the initial stages of motion), (12.29) indicates that $u_N$ is smaller than 1 and decreases with $S$, while, on the other hand, the forward wavespeed $u_W$ increases with $S$ and can attain values both smaller and larger than 1. Consequently, both super-critical ($u_N > u_W$) and sub-critical ($u_N < u_W$) currents are feasible; the former are expected for weak stratification and shallow ambients (small $S$ and non-large $H$) and the latter for strong stratification and/or deep ambients ($S$ close to 1 and/or large $H$). The quantitative critical separation curve $S_{cr}(H)$ will be developed in § 12.4.

Presently, let us consider some typical results and get an impression about the behavior of the current and the prediction power of the SW theory.

## 12.3  SW results and comparisons with experiments and simulations

Analytical solutions of the SW system can be obtained for some special cases, as shown in §§ 12.4, 15.6, and 16.3.2. In general the resulting SW system requires numerical solution. This is actually the situation also for the classical homogeneous case. We use the same finite-difference approach because the equations are of the same type. The current domain $[0, x_N(t)]$ is mapped into the fixed $y \in [0, 1]$ domain, see § 2.3, which is discretized in equal intervals. The typical grid for the SW results displayed here has 100 points, with time step $0.8 \cdot 10^{-2}$.

Data for comparisons are provided by Maxworthy et al. (2002). The reported 36 laboratory experiments were preformed in a tank 244 cm long, 20 cm wide, and 30 cm deep. The fluids were saline in the Boussinesq domain. Currents of $x_0 = 20$ or 40 cm and $1 \leq H \leq 3$ were released from a rectangular lock into a linearly stratified ambient of height 15 cm, and the motion visualized with dye. The typical values of the buoyancy frequency $\mathcal{N}$ were 1.4 and $2.0\,\mathrm{s}^{-1}$. The initial motion was in the buoyancy-inertial inviscid domain; see § 12.7 below. (The laboratory results were also supported by a two-dimensional high-resolution Boussinesq simulation.)

We now proceed to a more detailed solution of two typical cases.

Consider the case corresponding to Run 5 of Maxworthy et al. (2002): $H = 3$ and $S = 0.293$. To be specific, in this case $U = 23.82\,\mathrm{cm\,s}^{-1}$, $T = 0.840\,\mathrm{s}$, $h_0 = 5\,\mathrm{cm}$, $x_0 = 20\,\mathrm{cm}$, $\epsilon = 0.1156$, $\mathcal{N} = 1.48\,\mathrm{s}^{-1}$, $Re = Uh_0/\nu = 1.2 \cdot 10^4$.

The calculated SW profiles of $h$ and $u$ as functions of $x$ at various times are shown in Fig. 12.2.

We note that the initial SW propagation is with a constant speed $u_N = 0.54$, which is larger than $u_W = 0.30$. Indeed, Maxworthy et al. (2002) consider this configuration as typical to the super-critical domain, and emphasize that in this case no wave generation behind the head and no wave-head interaction is observed (at least during the initial propagation).

Figure 12.3 presents a comparison between the experimental results (taken from Fig. 5 of that paper) and SW predictions for $x_N$ as a function of $t$. The agreement is excellent for $t < 10$, and afterward the experimental results lag

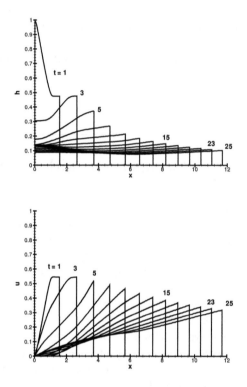

FIGURE 12.2: SW profiles of $h$ and $u$ as functions of $x$ for Run 5 case at $t = 1(2)25$. (Ungarish and Huppert 2004, with permission.)

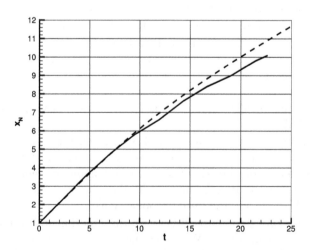

FIGURE 12.3: Comparisons of experiment (full line) and SW (dashed line) results for Run 5 of Maxworthy et al. (2002). (Ungarish and Huppert 2004, with permission.)

slightly behind the SW results. Overall, the agreement is quite satisfactory and the discrepancy at the later stage can be attributed to viscous and mixing effects which are not incorporated in the SW model. (The graph of $x_N$ vs. $t$ displays some weak oscillations after $t = 10$ which may be the result of internal waves in the ambient or of measurement errors, but their amplitude and mean contribution to the main propagation are insignificant.) Typical flow-field results of the NS simulations are shown in Fig. 12.4. The finite-difference numerical code is discussed in Appendix B. The numerical tank is of length $x_w = 8$, and a mesh of 320 horizontal and 200 vertical intervals was used.

Consider the case corresponding to Run 19 of Maxworthy et al. (2002): $H = 3$ and $S = 0.72$. To be specific, in this case $U = 19.86 \, \text{cm s}^{-1}$, $T = 1.01 \, \text{s}$, $h_0 = 5 \, \text{cm}$, $x_0 = 20 \, \text{cm}$, $\epsilon = 0.0804$, $\mathcal{N} = 1.94 \, \text{s}^{-1}$, $Re = Uh_0/\nu = 1.0 \cdot 10^4$.

The SW profiles of $h$ and $u$ as functions of $x$ at various times are shown in Fig. 12.5; the qualitative behavior is similar to that of Run 5, Fig.12.2, but the velocity is smaller in the present case for which $S$ is larger. The initial velocity of propagation predicted by SW, $u_N = 0.38$, is smaller than $u_W = 0.47$. This is in good agreement with the experimentally-detected sub-critical type of the current.

Comparisons for the propagation as a function of time are presented in Fig. 12.6. We observe that, initially, there is excellent agreement between SW predictions and experiment. However, at $x \approx 5$, $t \approx 12$, the propagation of the real current is strongly hindered by a new effect, which has no counterpart in the non-stratified case: the interaction between the nose and the waves. Let us denote by $x_2$ the position of the nose, $x_N$, when this effect is first pronounced.

We see that the speed decreases drastically at $x_2$, and the current almost stops. At the first glance, this resembles the transition position $x_V$ to the regime of viscous dominance. We can check that this is not the case. At $x_V$ the value of $F_I/F_V \approx Re(h_0/x_0)\theta$ becomes smaller than 1 (see § 12.7 and in particular Fig. 12.17). However, at $x_2$ the estimated $F_I/F_V$ is about 8. Moreover, we observe that the current in the experiment regains its speed after a while, in contrast with the behavior of a viscously-dominated motion which cannot accelerate. The speed reduction at $x_2$ is evidently an inviscid interaction with the internal waves. The interaction extends about 5 dimensionless time units, after which the previous speed is recovered. The SW solution evidently misses this interaction; the SW curve shows no special behavior in the pertinent time period, and, consequently, a significant discrepancy of $x_N$ is present for $t > 15$. An analytical prediction for the current during the phase of interaction is presently unavailable. The theoretical remedy is to use Navier-Stokes simulations.

Figs. 12.6 and 12.7 show results of such simulations with a finite-difference numerical code discussed in Appendix B using a mesh of 320 horizontal and 200 vertical intervals in a numerical tank of length $x_w = 8$. We see in Fig 12.6

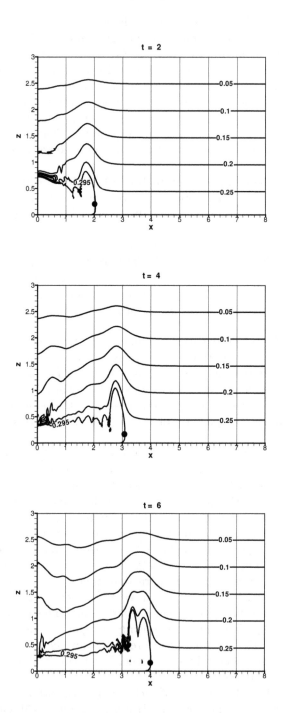

**FIGURE 12.4:** NS results. Density contours at various times for $S = 0.29$.
The dot marks the nose of the current, defined as the foremost forward point
of the dense fluid. (Recall that, initially, in the dense fluid $\phi = 1$, while in
the ambient $\phi = 0$ at the top and $\phi = 0.29$ at the bottom.) (Ungarish and
Huppert 2004, with permission.)

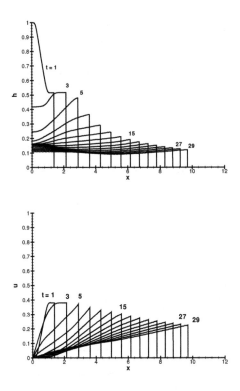

**FIGURE 12.5**: SW profiles of $h$ and $u$ as functions of $x$ for Run 19 case at $t = 1(2)29$. (Ungarish and Huppert 2004, with permission.)

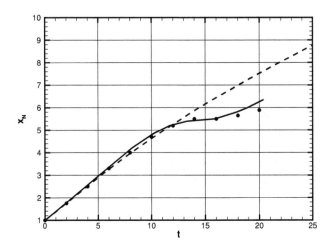

**FIGURE 12.6**: Comparisons of experiment (full line), SW (dashed line), and numerical NS (symbols) results for Run 19 of Maxworthy et al. (2002). (Ungarish and Huppert 2006, with permission.)

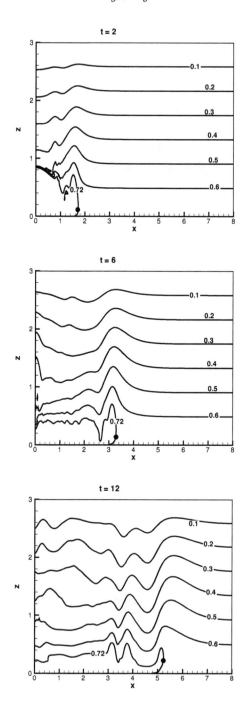

**FIGURE 12.7:** NS results: contours of the density function $\phi$ at various times for Run 19. The dot marks the nose of the current. (Recall that, initially, in the dense fluid $\phi = 1$, while in the ambient $\phi = 0$ at the top and $\phi = 0.72$ at the bottom.) (Ungarish and Huppert 2004, with permission.)

that the NS code reproduces well the $x_N(t)$ curve of the experiment, including the hindering of the interaction. This indicates that the interaction between the waves and the head is a quite robust property of the system, that occurs after a certain interval of propagation during which the SW approximation can be applied. The NS results shown in Fig. 12.7 display pronounced oscillations of the isopycnals in the region of the head. However, at $t = 12$ the head is very thin and almost detached from the main body of the current; this reduces the speed, and creates the pronounced wave-nose interaction hindrance.

We see that the SW theory provides reliable and useful information on the gravity current in a stratified ambient. We now seek to extend these insights by analytical solutions and interpretations.

## 12.4 Dam break

The numerical solutions of the SW equations show that the initial motion of the current is of the "slumping" type, with a constant speed $u_N$ and height $h_N$. The details of the initial motion, and in particular the values of $u_N$ and $h_N$, can be obtained analytically from the dam-break problem solution. The underlying ideas are like in the non-stratified problem solved in § 2.5, but the details are more complicated because now the solution depends on two parameters, $H$ and $S$.

We refer to Fig. 12.8. Here we shift the $x$ axis so that the gate is at $x = 0$. We use the equations of the characteristics (12.24)-(12.26) and the nose condition (12.29).

The balance on the characteristics (12.24) can be integrated to yield

$$u + \Upsilon(h) = \Gamma_+ \quad \text{on} \quad \frac{dx}{dt} = c_+; \qquad (12.32)$$

$$u - \Upsilon(h) = \Gamma_- \quad \text{on} \quad \frac{dx}{dt} = c_-; \qquad (12.33)$$

see (12.26). Here, see (12.25), $\Gamma_\pm$ are constants (to be determined by boundary conditions) and

$$\Upsilon(h) = \int_0^h \mathcal{G}(h')dh' = \int_0^h \left[ \frac{1 - S(1 - h'/H)}{h'} \right]^{1/2} dh'. \qquad (12.34)$$

The $WOL$ characteristic region is covered by characteristics that start at $t = 0$ in the domain $x < 0$ and satisfy the initial condition $u = 0$ and $h = 1$. Substitution of these conditions into (12.32)-(12.33) determines $\Gamma_+ = -\Gamma_- = \Upsilon(1)$. Further manipulation of (12.32)-(12.33) yields $u = 0$ and $h = 1$ in the entire $WOL$ domain. The characteristics that carry these values are straight

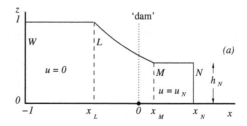

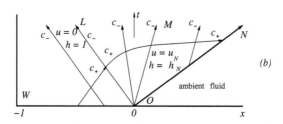

**FIGURE 12.8**: Schematic description of the current during the initial dam-break stage. (a) Geometry in $xz$ plane; note that the gate is at $x = 0$. (b) Characteristics in $xt$ plane (here point $O$ is the origin).

lines, and the largest $x$ of the pertinent domain is given by the line $OL$:

$$x_L(t) = t \cdot c_- = -[1 - S(1 - 1/H)]^{1/2}\, t. \qquad (12.35)$$

We conclude that the stratification decreases the speed of backward propagation of position $L$. The result (12.35) is valid for $t \leq [1 - S(1 - 1/H)]^{-1/2}$, until $L$ reaches the wall at $x = -1$. This occurrence can be considered the end of the dam-break stage.

Consider the solution of (12.32)-(12.33) in the $LOM$ domain of characteristics. This is an expansion fan with origin $x = 0, t = 0$. Assume that $c_+$ characteristics from the above-mentioned $WOL$ domain enter the present domain, i.e., on $c_+$

$$u + \Upsilon(h) = \Upsilon(1), \qquad (12.36)$$

while the $c_-$ characteristics are emanated from point $O$ as a fan with various initial $h$, i.e.,

$$u - \Upsilon(h) = \Gamma_- \qquad (h_N \leq h \leq 1) \qquad (12.37)$$

on

$$\frac{dx}{dt} = c_- = u - [h(1 - S(1 - h/H)]^{1/2}, \quad x(0) = 0. \qquad (12.38)$$

The intersection of (12.36) and (12.37) and use of (12.38) yield

$$u = \Upsilon(1) - \Upsilon(h) = C_1; \quad h = C_2; \qquad (12.39)$$

on the ray

$$\frac{x}{t} = u - \left(h[1 - S(1 - h/H)]\right)^{1/2}. \tag{12.40}$$

We expect that $(x/t)$ is a monotonic decreasing function of $h$. For $h = 1$ (12.39)-(12.40) recover the line $OL$, and the line $OM$ is given by (12.39)-(12.40) with $h = h_N$.

Finally, the above-mentioned $c_+$ characteristics are expected to enter the characteristic region $MON$ where $u = u_N$ and $h = h_N$. The intersection of (12.39) with (12.29),

$$u_N = \Upsilon(1) - \Upsilon(h_N) = Fr(h_N) h_N^{1/2} \times \left[1 - S(1 - \frac{1}{2}\frac{h_N}{H})\right]^{1/2}, \tag{12.41}$$

provides the values of $h_N$ and $u_N$. The $c_-$ characteristics in the $MON$ domain are lines parallel to $OM$ and carry the consistent information $u = u_N$, $h = h_N$.

To be more explicit, it is necessary to calculate $\Upsilon(h)$ and solve (12.41) for $h_N$ and $u_N$. It is convenient to distinguish between the following cases.

*Case $S = 0$.* The results reduce to the solution of the non-stratified case presented in § 2.5.

*Case $S = 1$.* The integration of the RHS of (12.34) is straightforward

$$\Upsilon(h) = \frac{1}{\sqrt{H}}h. \tag{12.42}$$

Consequently, (12.39) yields for the expansion fan domain

$$u = \frac{1}{\sqrt{H}}(1 - h) \qquad (h_N \le h \le 1). \tag{12.43}$$

Substitution into (12.40) and arrangement give

$$h = \frac{1}{2}\left(1 - \sqrt{H}\frac{x}{t}\right); \quad u = \frac{1}{2}\left(\frac{1}{\sqrt{H}} + \frac{x}{t}\right); \quad \left(-\frac{1}{\sqrt{H}} \le \frac{x}{t} \le \frac{1}{\sqrt{H}}(1 - 2h_N)\right) \tag{12.44}$$

in the region $LOM$.

The values of $h_N$ and $u_N$ are provided by (12.41)

$$u_N = \frac{1}{\sqrt{H}}(1 - h_N) = Fr(h_N)\frac{h_N}{\sqrt{2H}}. \tag{12.45}$$

The results are plotted in Fig. 12.9. For the HS $Fr$, we also have the particular $Fr = 1.19$ for $H \ge 7.4$. In this case (12.45) gives $h_N = 0.543$ and $u_N = 0.457/\sqrt{H}$.

*Case $0 < S < 1$.* The integral in (12.34) can be obtained analytically. Let us define the constant

$$K = (1 - S)\sqrt{\frac{H}{S}} \tag{12.46}$$

and the variable

$$\xi = \xi(h) = \left[ 1 + K \left( \frac{H}{S} \right)^{1/2} \frac{1}{h} \right]^{1/2}. \tag{12.47}$$

A manipulation of the integral, or a verification by substitution, proves that the result can be expressed as

$$\Upsilon(h) = K f(\xi) \tag{12.48}$$

where

$$f(\xi) = \frac{\xi}{\xi^2 - 1} + \frac{1}{2} \ln \frac{\xi + 1}{\xi - 1}. \tag{12.49}$$

Substitution of (12.48) into (12.39)-(12.40) gives, for the characteristic domain $LOM$,

$$u = u(\xi) = K \left[ f(\xi_L) - f(\xi) \right], \tag{12.50}$$

on the ray

$$\frac{x}{t} = K \left[ f(\xi_L) - f(\xi) - \frac{\xi}{\xi^2 - 1} \right] \quad (\xi_L \leq \xi \leq \xi_M), \tag{12.51}$$

where $\xi_L = \xi(1)$ and $\xi_M = \xi_N = \xi(h_N)$ (recall that $h_M = h_N$).

The values of $h_N$ and $u_N$ are provided by the combination of (12.50) and (12.41)

$$u_N = K \left[ f(\xi_L) - f(\xi_N) \right] = Fr(h_N) h_N^{1/2} \times \left[ 1 - S(1 - \frac{1}{2} \frac{h_N}{H}) \right]^{1/2}. \tag{12.52}$$

Here $\xi_N = \xi(h_N)$. The details of the manipulation concerning (12.48)-(12.52) are left for an exercise.

The outcome (12.52) is a non-linear equation for the unknown $h_N$. This is the stratified counterpart of the non-stratified (2.65). For a given pair $H, S$ we solve numerically for $h_N$, then calculate $u_N$. For plausible $Fr$ functions there are unique solutions in the physical $h_N$ domain $(0, 1)$.

The nose height and speed results for the HS $Fr$ function are displayed in Fig. 12.9. These values are valid for the slumping stage, for both sub- and super-critical currents. In general, we observe that: for a given $H$, $u_N$ decreases with the stratification $S$; for a given stratification $S$, $u_N$ increases with $H$. This qualitative behavior corresponds to our physical understanding, and we can proceed to a more detailed quantitative comparison.

Fig. 12.10 shows the comparison of the calculated slumping speed (with HS $Fr$ function) and the experimental data of Maxworthy et al. (2002). The display uses the same axes and parametric values as Fig. 7 of Maxworthy et al. (2002). The data cover $0.3 \leq S \leq 1$ and $H = 1, 1.5, 2$, and $3$, and hence a comparison is relevant only for this range. The qualitative agreement is excellent and the quantitative agreement can be considered very good. The

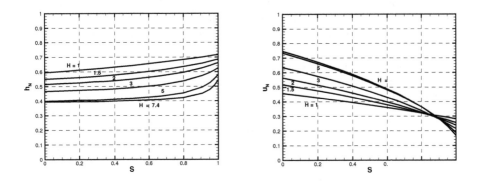

**FIGURE 12.9:** Nose height and speed in dam-break analytical solution as a function of $S$ for various $H$. Linear stratification, one-layer SW model, $Fr$ function of HS. (Ungarish and Huppert 2002, with permission.)

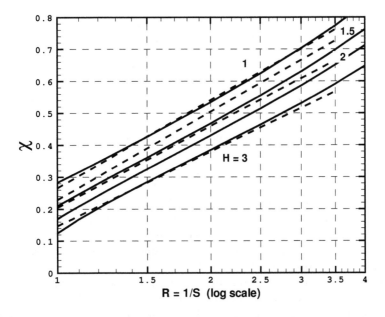

**FIGURE 12.10:** The behavior of $\chi = u_N^*/\mathcal{N}H^* = u_N/(SH)^{1/2}$ as a function of $1/S$ for various initial depth ratios. SW results (solid lines) and the experimental fit of Maxworthy et al. (2002) (dashed lines). (Ungarish and Huppert 2002, with permission.)

discrepancies of typically 5% can be attributed to the various simplifications introduced in the SW formulation, and in particular the use of the one-layer model for non-large values of $H$ and the neglect of internal waves in the ambient, and of mixing. Overall, this seems to be a good extension of the non-stratified one-layer SW model.

A versatile interpretation of the propagation speed results was suggested in Ungarish and Huppert (2002). In our scaling we used the formal reduced gravity $g'$ based on the $(\rho_c/\rho_o - 1)$ density difference; see (12.11). However, equation (12.41) indicates that the driving force in the presence of stratification is proportional to $[1 - S(1 - 0.5h_N/H)]^{1/2}$. This means that the effective reduced gravity that drives the propagation is actually proportional to the density difference between the current and the ambient at the position of half-height of the nose, $z = h_N/2$. In general $h_N$ is not known *a priori*, but a plausible estimate during the slumping stage is $h_N \approx 0.5$. We infer that the "effective" reduced gravity that drives the propagation of the current is given by

$$g'_e = \frac{\rho_c - \rho_a(z = 0.5h_N)}{\rho_o} g = g'A^2 \tag{12.53}$$

where

$$A = \left[1 - S\left(1 - \frac{h_N}{2H}\right)\right]^{1/2}. \tag{12.54}$$

Using the estimate $h_N \approx 0.5$ we obtain the approximation

$$A \approx B = \left[1 - S\left(1 - \frac{0.25}{H}\right)\right]^{1/2}. \tag{12.55}$$

Now we can define an "effective" Froude number for the current released from a lock

$$Fr_e = \frac{u_N^*}{(g'_e h_0)^{1/2}} = \frac{u_N}{A}. \tag{12.56}$$

In this formula, we use the slumping $u_N$ calculated from (12.52). It turns out, see Fig. 12.11, that this $Fr_e$ changes little with $S$; it is mostly a function of $H$. In other words, to get a first approximation for the speed of the current in a stratified ambient, we can use the non-stratified results for the same $H$, and multiply the corresponding $u_N$ by $A$.

The case $S = 1$ merits attention. In this case the effective gravity behaves like $g'(h_N/H)$. This indicates a significant reduction of speed in a deep ambient, and the need for a different scaling. Indeed, in the $S = 1$ case the density of the current is equal to that of the ambient at the bottom, $\rho_c = \rho_b$. The diving force is built only on the density differences from $z = 0$ to $z = h_N$, and hence is proportional to $h_N/H$. The speed of propagation is of the order of $\mathcal{N}h_0$. The case $S = 1$ is relevant to intrusions, and will be discussed in detail in Chapter 15.

The analytical predictions for the dam-break propagation for the typical geometry $H = 3$, cases $S = 1$ and $S = 0.5$, are compared with the numerical

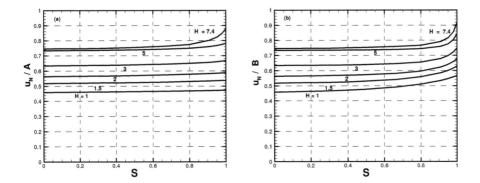

**FIGURE 12.11:** Rescaled $u_N$ as a function of $S$ for various $H$. (a) $u_N/A$ with $A = [1 - S(1 - 0.5h_N/H)]^{1/2}$; (b) $u_N/B$ with $B = [1 - S(1 - 0.25/H)]^{1/2}$. (Ungarish and Huppert 2002, with permission.)

solution of the SW equations in Fig. 12.12. The agreement between the analytical and numerical solutions is perfect before the backward-moving wave $L$ reaches the wall at $x = -1$. Afterward, the forward-moving reflected wave, not captured by the present analytical solution, influences the interface near the wall. The remarkable features for the linear stratification with $S = 1$ are (a) the profile of $h$ vs. $x$ in the $LM$ domain is linear, see (12.44), in contrast to the parabolic shape in the $S = 0$ case, see (2.64); and (b) position $M$ (i.e., the start of the rectangular region which trails the nose) propagates backward ($h_M = h_N > 0.5$ for all $H \geq 1$).

The SW results are an idealization. To get an impression of the more realistic behavior of the current in the circumstances depicted in Fig. 12.12 we consider results of Navier-Stokes simulations. Density contour plots at $t = 1$ and 1.5 are displayed in Fig. 12.13. The simulations are in fair global agreement with the SW predictions. The dense fluid displays, after a quick adjustment motion, flow-field domains that can be identified with the regions of Fig. 12.8: $MN$ following the head, $LM$ with inclined interface, and unperturbed $LW$. The nose is a prominent discontinuity and propagates with velocity close to the predicted $u_N$. Indeed, the SW predictions are that at $t = 1.5$ the nose is at $x_N = 0.33$ and 0.71 for $S = 1$ and 0.5, respectively (see Fig. 12.12), which is in very good agreement with the position of the nose obtained in the NS simulations, Fig. 12.13. The head is higher and slower for the larger $S$ case. The rarefaction wave of the interface reaches the wall $x = -1$ at (approximately) the time predicted by the SW theory. However, the shape of the interface in domain $LM$ differs from the SW predictions, perhaps as a result of the strong shear about the interface in the real flow. We emphasize that this discrepancy between the SW and NS results is also observed in the homogeneous $S = 0$ case.

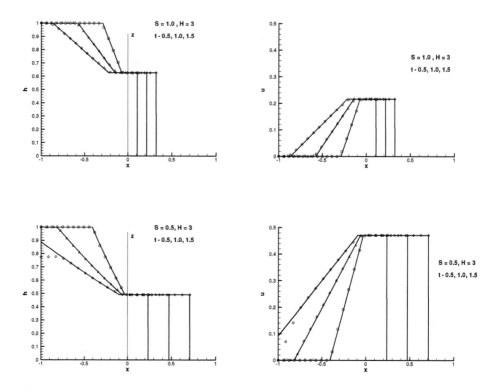

**FIGURE 12.12**:  Analytical (lines) and SW numerical (symbols) results, of $h(x)$ and $u(x)$ at $t = 0.5, 1.0, 1.5$, for linear stratification and $H = 3$. $S = 1$ (upper figures) and $S = 0.5$ (lower figures). (Ungarish 2005a, with permission.)

Contour lines of the simulated $u(x, z, t)$ fields at $t = 1$ and $2$ are shown in Fig. 12.14. The comparison with the SW results indicates, again, fair agreement in the $WM$ (practically, $x < 0$) domain; however, in the domain of fluid that trails the nose $u$ displays some fluctuations about the constant (average) value predicted by the SW theory. Indeed, the nose of a real gravity current is a complicated flow field, and the SW theory lumps this behavior in an idealized jump condition. This deviation from the SW theory is not a result of the stratification; in this respect, the non-stratified $S = 0$ simulations are not different from the $S = 1$ and $0.5$ cases.

Fig. 12.13 shows, again, that the release of the current in the stratified ambient produces significant perturbations in the ambient density field, in particular above the head. The $S = 1$ case is sub-critical and the $S = 0.5$ case is super-critical. In both cases, the major deviation of the density contours from the initial level is above the head - then the perturbed isopycnal returns

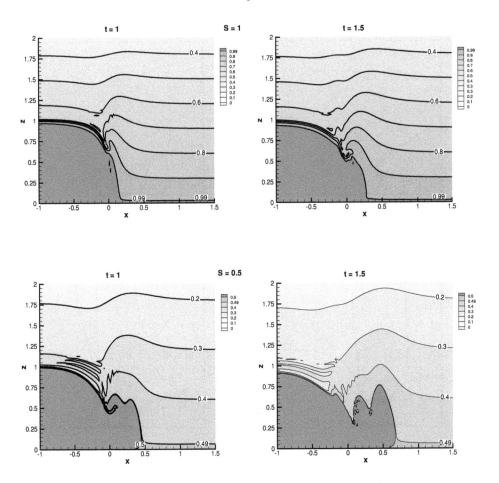

**FIGURE 12.13:** NS contours of the density function $\phi = [\rho(x,z,t) - \rho_o]/(\epsilon\rho_o)$ at two times, for $S = 1$ (upper figures) and 0.5 (lower figures). In both cases $H = 3$. (Recall that at $t = 0$: in the dense fluid $\phi = 1$, while in the ambient $\phi = 0$ at the top and $\phi = S$ at the bottom). (Ungarish 2005a, with permission.)

to its initial height, approximately, at $x = 0$. Also, we observe that the perturbation of the density field in front of the current head is small at these times. Consequently, the global pressure-force driving calculated from the unperturbed hydrostatic balance can be expected to remain valid in spite of the local density waves. This vindicates the use of (12.18) and (12.29) during the initial period of propagation for both sub- and super-critical $u_N$.

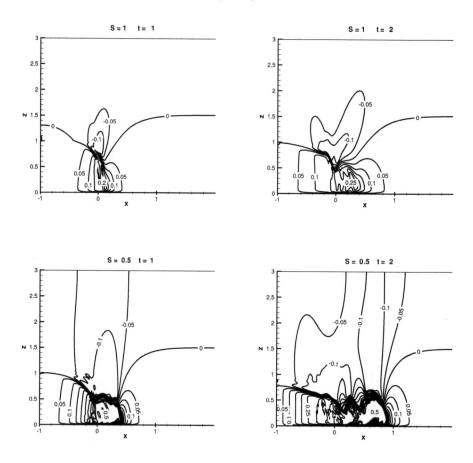

**FIGURE 12.14**:  NS contours of the $u$ at two times, for $S = 1$ (upper figures) and 0.5 (lower figures). In both cases $H = 3$. (Ungarish 2005a, with permission.)

## 12.5   Critical speed and nose-wave interaction

We are now in the position to provide a firm classification concerning the "criticality" of a current for given $H$ and $S$. We simply calculate $u_N$ from (12.52) and compare to $u_W$ given by (12.31). For a fixed $H$, the speed of the current decreases with $S$ while the speed of the wave increases with $S$. There is a critical $S_{cr}$ where these speeds intersect, and for larger $S$ the currents will be sub-critical. The critical curve, $S_{cr}(H)$, is displayed in Fig. 12.15 for the HS $Fr$.

t![h]

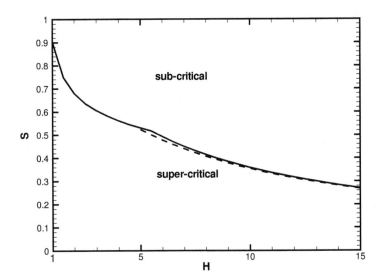

**FIGURE 12.15:** The sub- and super-critical sub-domains in the $S - H$ domain, as predicted by the SW theory with HS $Fr$. The dashed line is the approximation (12.58). (Ungarish and Huppert 2004, with permission.)

.

Using the previous results we can also derive an approximation to $S_{cr}$ for large $H$ (i.e., $a = h_N/H \to 0$) and $S$ not close to 1. We write

$$u_N(a, S) \approx u_N(a = 0, S = 0) \times (1 - S)^{1/2} = \frac{Fr(0)}{1 + Fr(0)/2}(1 - S)^{1/2}. \quad (12.57)$$

Here (2.66)(b) was used to express the speed in the non-stratified case, and (12.54) was used to incorporate the correction of the stratification. This approximate $u_N$ is equal to $u_W$ for values of $S$ given by

$$S_{cr} = \left[1 + \left(\frac{Fr(0) + 2}{2\pi Fr(0)}\right)^2 H\right]^{-1}. \quad (12.58)$$

In particular, the HS $Fr(0) = 1.19$; see (12.30).

Fig. 12.15 leads to the conclusion that a significant portion of the $S - H$ domain corresponds to super-critical currents. For these circumstances the wave-current interaction is expected to be unimportant for at least the SW slumping distance, $x_N \approx 3$. Eventually the velocity of the front decays and the sub-critical domain is attained, but viscous effects may also become important at this stage. We recall that the super-critical current Run 5 has been discussed in § 12.3.

On the other hand, the Run 19 current discussed in § 12.3 indicated that sub-critical currents are prone to a strong wave-nose interaction. Nevertheless, the initial motion is dominated by the inertia-pressure balance in the dense fluid, not by the waves in the ambient. We can therefore assume that the motion starts with the slumping values of $u_N$ and $h_N$. Then, using this information and the insights provided by the observations of Maxworthy et al. (2002), we can estimate the position $x_2$ where the first strong interaction between the waves and the nose occurs as follows. The nose propagates first two wavelengths with the wave locked to the head. Next, the wave is unlocked from the head and moves forward relative to the current until the crest reaches the nose (and thus slows it down). No mixing or instabilities were observed during this process.

We estimated the wavelength (dimensional) in (12.6). In dimensionless form ($\lambda$ scaled with $x_0$, speed scaled with $U$) this reads

$$\lambda = 2\pi \left(\frac{H}{S}\right)^{1/2} \frac{h_0}{x_0} u_N. \tag{12.59}$$

The propagation of the nose from $x = 1$ to the position of interaction is made up of $2\lambda$, plus the motion during the time in which the released wave catches up the distance $(1/4)\lambda$ to the nose. The relative velocity is $u_W - u_N$. We assume that $u_N$ is constant during this process. The resulting position is

$$x_2 = 1 + \lambda \left(2 + \frac{0.25}{u_W - u_N} u_N\right)$$

$$= 1 + 2\frac{h_0}{x_0} H \left(\frac{\sqrt{SH}}{\pi u_N}\right)^{-1} \left[2 + \frac{0.25}{\left(\dfrac{\sqrt{SH}}{\pi u_N}\right) - 1}\right]. \tag{12.60}$$

Using again our results for the initial $u_N$ as a function of $S$ and $H$ we obtain straightforward estimates of $x_2$. The denominator in the last term of (12.60) is positive for sub-critical currents, for which this result has been derived. We note in passing that the dimensionless ratio $u_N/\sqrt{SH}$ (expressed with dimensionless variables) can be rewritten as $u_N^*/(\mathcal{N}H^*)$ (expressed with dimensional variables). This ratio is sometimes referred to as the Froude number of the current; in this case, the value $1/\pi$ marks the transition from sub- to super-critical currents.

Maxworthy et al. 2002 provided the experimental distance (from the gate of the lock) $X_{tr}$ where significant deceleration of the nose occurs. A comparison with the present $x_2$ result, for sub-critical currents, is shown in Fig. 12.16.

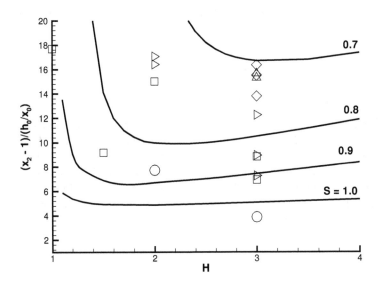

**FIGURE 12.16:** The distance of propagation where wave-nose interaction becomes important for sub-critical currents, as a function of $H$ for various $S$. The lines show the present SW estimate, the points are from Exp. 1, 4, 8-12, 15-16, 19, 21-25, 28, 29, 32 of Maxworthy et al. (2002). The symbols are for values of $S$: $\approx 1.0$ ○; $\approx 0.90$ □; $\approx 0.80$ ▷; $\approx 0.70$ ◇ ; and $\approx 0.60$ △. (Ungarish and Huppert 2004, with permission.)

There is some scatter of the experimental points, but overall the magnitude and the trend predicted by (12.60) are consistent with the data. In any case, this is a rough approximation, and further research is necessary for the elucidation of the interaction effect.

The aspect ratio $h_0/x_0$ has been scaled out from the SW formulation, but enters into the problem via (12.60) for the sub-critical current. This is because the wavelength of the perturbations in the ambient are proportional to $u_N^*/\mathcal{N}$. (We can compare the current with a moving solid obstacle which intrudes into a stationary stratified fluid. In the latter and better understood problem, see Baines 1995 Chapter 5, the instantaneously created perturbations display a local rather than global behavior. In this analogy, the dominant dimensionless parameter "$\mathcal{N} h/U$" in flows over an obstacle corresponds to the combination $(S/H)^{1/2}(h_N/u_N)$ of our dimensionless variables.)

Formally, the prediction power of the present SW approximation stops where the interaction begins, as far as sub-critical currents are concerned. However, this limitation may be less severe than it sounds here, because the interactions are not always so influential on the results as in the present $x_N(t)$

calculations. The geometries of the lock (i.e., not a rectangle) and of the ambient (e.g., axisymmetric) also play important roles in this phenomenon.

## 12.6 Similarity solution

Self-similar solutions for large $t$ of the type derived for the non-stratified ($S = 0$) current in § 2.6 are in general not available. The difficulty emerges because the combinations $1 - S + Sh(x, t)/H$ and $1 - S + Sh(x, t)/(2H)$ enter the formulation. These expressions are reduced to simple forms for $S = 0$ and $S = 1$, for which similarity solutions can indeed be derived. The $S = 1$ case corresponds to an intrusion, and the appropriate similarity solution will be considered in this context in § 15.6.

## 12.7 The validity of the inviscid approximation

As for the non-stratified current configuration, the importance of the viscous friction in the motion of the current increases with time and distance of propagation. Even for quite large values of $Re = Uh_0/\nu$, the inviscid SW formulation may become invalid at moderate values of $x_N$. This tendency is enhanced by stratification. As in the non-stratified case, the inertia per unit volume is well represented by $\rho_c uu_x$, and the viscous force per unit area is expected to be proportional to $\rho_0 \nu u/h$. The ratio of the total inertial to viscous forces, in dimensionless form, is therefore estimated by

$$\frac{F_I}{F_V} = Re\frac{h_0}{x_0}\frac{\int_0^{x_N(t)} uu_x h dx}{\int_0^{x_N(t)} (u/h)dx} = Re\frac{h_0}{x_0}\theta(t). \qquad (12.61)$$

Here $\theta$ is a function (of $t$ or $x_N$) whose value is expected to be of the order of unity at the beginning of the propagation and to become small with propagation. This function can be easily calculated from the SW results for $u(x, t)$ and $h(x, t)$. Typical results are shown in Fig. 12.17. In the stratified case $\theta$ depends on both $H$ and $S$.

We showed in § 2.7 that for the $S = 0$ case it is convenient to calculate $\theta$ analytically by using the self-similar results. However, no self-similar results are known for $0 < S < 1$. The $S = 1$ case will be considered in § 15.6. Here the analytical estimate can be obtained only using a box-model consideration as follows.

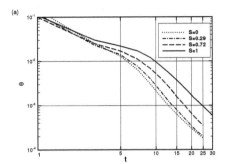
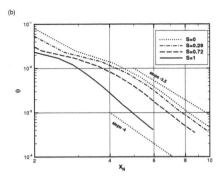

**FIGURE 12.17:** SW results. The coefficient $\theta$, see (12.61), for $H = 3$ and various values of $S$: (a) as a function of time; and (b) as a function of $x_N$, displaying also the slopes of approximation (12.62). (Ungarish and Huppert 2006, with permission.)

Assume that the current is a box of height $h_N(t)$ and length $x_N(t)$, and accordingly $u_x = u_N/x_N$. With these approximations in (12.61) we obtain $\theta \approx u_N h_N^2 x_N^{-1}$. The value of $u_N$ is estimated from (12.29)-(12.30) as: (a) $[(1 - S)h_N]^{1/2}$ for $S$ not close to 1; and (b) $h_N/H^{1/2}$ for $S \approx 1$. Finally, we use the volume conservation $h_N = 1/x_N(t)$. This yields

$$\theta \approx \begin{cases} (1 - S)^{1/2} x_N^{-7/2} & (S \text{ not close to 1}) \\ H^{-1/2} x_N^{-4} & (S \approx 1). \end{cases} \tag{12.62}$$

Using these results, as illustrated in Fig. 12.17, we can estimate the importance of the viscous terms, and also the limit of validity of the inviscid assumption, for a real gravity current with given $Re$ and $h_0/x_0$. We denote by $x_V$ the position where $F_I/F_V = 1$. The inviscid theory is expected to be relevant for, roughly, $F_I/F_V > 1$, i.e., for $x_N < x_V$. We see that the stratification reduces the value of $\theta$, which means that the transition to the viscously-dominated regime is achieved at a shorter $x_V$. This is because the stratification reduces the effective driving force, and hence the inertial effects, during the initial propagation. The enhancement of the viscous effects is in particular severe for the $S \approx 1$ case, which will be discussed in more detail later (in particular; see § 15.6.1).

We can summarize that the SW model in the stratified ambient is in general more prone to invalidation than in the non-stratified case. First, $x_V$ is shorter. In addition, the sub-critical currents are vulnerable to nose-wave interactions which restricts the SW propagation to $x_2$. In some ranges of parameters the second restriction is more severe than the first one. The details are left for an exercise, and we notice that the Run 19 case discussed above is an example of $x_2 < x_V$.

We conclude that the one-layer SW formulation for the stratified ambient $0 < S \leq 1$ is a versatile useful extension of the classical non-stratified model. The major limitation is due to the possible interaction of the neglected internal gravity waves with the motion of the head. When the nose velocity is sub-critical, this interaction will eventually hinder the propagation significantly below the SW predictions. However, this interaction occurs only after the nose has propagated about two wavelengths from the lock. We noticed that the SW results describe accurately the propagation in this initial period of motion. We have presently no analytical model for the behavior of the current during and after the interaction. However, the NS 2D simulations describe well the slow-down and the speed recovery observed in experiments. Whether and how these internal gravity waves can be incorporated in the SW formulation is a topic for future research.

# Chapter 13

## Axisymmetric and rotating cases

In this chapter we consider the current to be released from a cylindrical lock of height $h_0$ and radius $r_0$. The entire cylindrical system is in general rotating with a constant angular velocity $\Omega$ about the vertical axis $z$ (with $\Omega = 0$ as a particular case).

The stratification of the ambient is linear over the full depth $0 \leq z \leq H$, with $\rho_a(0) = \rho_b$ and $\rho_a(H) = \rho_o$. The density of the current is $\rho_c \geq \rho_b$. The values of $\epsilon, \epsilon_b, S, g'$, and $\mathcal{N}$ are defined as in the 2D configuration; see (12.8)-(12.13).

---

## 13.1  SW formulation

We use a cylindrical co-ordinate system, $\{r, \theta, z\}$, co-rotating with the ambient fluid; see Fig. 13.1. The velocity components in the rotating system are $\{u, v, w\}$ and we assume that the flow does not depend on the angular coordinate $\theta$. In the meridional plane $rz$ the current is similar to the 2D counterpart sketched in Fig. 12.1, but, in addition (a) the geometry diverges with $r$; (b) we allow rotation of the system about $z$, in which case Coriolis-centrifugal accelerations appear and an azimuthal velocity component develops.

It is convenient in this section to scale the dimensional variables (denoted

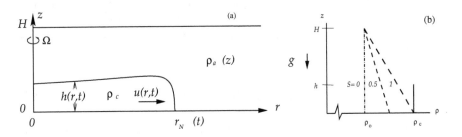

**FIGURE 13.1:** Schematic description of the system: (a) the geometry; and (b) the density profiles in the current (solid line) and unperturbed ambient, for various values of $S$ (dashed lines).

here by asterisks) by

$$\{r^*, z^*, h^*, H^*, t^*, u^*\} = \{r_0 r, h_0 z, h_0 h, h_0 H, T t, U u\}, \qquad (13.1)$$

where

$$U = (h_0 g')^{1/2} = \mathcal{N}\sqrt{\frac{H}{S}} h_0 \quad \text{and} \quad T = \frac{r_0}{U} = \frac{1}{\mathcal{N}}\sqrt{\frac{S}{H}} \frac{r_0}{h_0}. \qquad (13.2)$$

We also introduce the azimuthal and angular velocity (in the rotating system)

$$\{v^*, \omega^*\} = \{\Omega r_0 v, \Omega \omega\} \qquad (13.3)$$

where

$$\omega = v/r. \qquad (13.4)$$

The Reynolds number is defined as $Re = U h_0 / \nu$.

As compared to the previously considered 2D Cartesian case, two extensions of the SW equations of motion are necessary. First, the geometrical curvature terms must be incorporated, which is a quite straightforward task. Second, for $\Omega > 0$, the azimuthal momentum equation and Coriolis-centrifugal terms must be added to the formulation. The relevant dimensionless parameter, see § 8.1, is the typical Coriolis to inertial accelerations ratio

$$\mathcal{C} = \frac{\Omega r_0}{(g' h_0)^{1/2}} = \Omega T. \qquad (13.5)$$

This parameter does not take into account the stratification of the ambient. We shall see later that the effect of Coriolis forces is more pronounced when $S$ increases (and, actually, the inertia of propagation in the radial direction decreases.) We are interested in small values of $\mathcal{C}$; otherwise, the Coriolis effects restrict the propagation to a small distance and no real gravity current develops. On the other hand, when $\mathcal{C}$ is small the deviation of the current from the initial solid-body rotation is significant, i.e., the Rossby number of the flow is not small. We note in passing that an important parameter in stratified rotating fluids is the ratio of the inertial wave frequency, $f = 2\Omega$, to the usual buoyancy frequency of the ambient, $\mathcal{N}$. This ratio can also be expressed in terms of the previously defined parameters as follows:

$$\frac{f}{\mathcal{N}} = 2\Omega\sqrt{\frac{H}{S}} \frac{h_0}{(g' h_0)^{1/2}} = \frac{2}{\sqrt{S}} \frac{h_0}{r_0} \mathcal{C}\sqrt{H}. \qquad (13.6)$$

The one-layer model is used, again, for simplicity. As argued in § 8.1 the centrifugal term $\Omega^2 r^*$ is negligible as compared to $g$. The motionless ambient is governed by a hydrostatic pressure balance. The vertical momentum balance in the dense fluid is also (approximately) hydrostatic. After some algebra, it turns out that in the range of parameters considered here (in particular $\epsilon \ll 1$

and $C < 1$) the relationship between the lateral pressure gradient and inclination of the interface is similar to the two-dimensional case. The stratification of the ambient does not affect the $\theta$-component of the momentum equation of the current and the accelerations. Consequently, in the $z$-averaged equations, the azimuthal momentum balance and the contribution of the Coriolis terms and of the $(v^2/r)\hat{r}$ term are exactly like in the non-stratified case. The result is a set of equations which differs from the non-stratified counterpart only in the term which represents $\partial p/\partial r$.

In conservation form the averaged balance equations of continuity, radial momentum, and azimuthal momentum can be written as

$$\frac{\partial h}{\partial t} + \frac{\partial}{\partial r}(uh) = -\frac{uh}{r}, \tag{13.7}$$

$$\frac{\partial}{\partial t}(uh) + \frac{\partial}{\partial r}\left[u^2h + \frac{1}{2}(1-S)h^2 + \frac{1}{3}S\frac{h^3}{H}\right] = -\frac{u^2h}{r} + C^2vh\left(2+\frac{v}{r}\right), \tag{13.8}$$

and

$$\frac{\partial}{\partial t}(vh) + \frac{\partial}{\partial r}(uvh) = -2uh\left(1+\frac{v}{r}\right). \tag{13.9}$$

In characteristic form, this becomes

$$\begin{bmatrix} h_t \\ u_t \\ v_t \end{bmatrix} + \begin{bmatrix} u & h & 0 \\ 1-S+S\frac{h}{H} & u & 0 \\ 0 & 0 & u \end{bmatrix} \begin{bmatrix} h_r \\ u_r \\ v_r \end{bmatrix} = \begin{bmatrix} -\frac{uh}{r} \\ C^2v\left(2+\frac{v}{r}\right) \\ -u\left(2+\frac{v}{r}\right) \end{bmatrix}. \tag{13.10}$$

The initial conditions are zero velocity in both radial and azimuthal directions, and unit dimensionless height and length at $t = 0$. At $r = 0$ we impose $u = 0$ and a regularity of the angular velocity $\omega = v/r$. The boundary conditions at $r = r_N(t)$ will be discussed latter.

The first two characteristic velocities, $c_\pm$, of this system are like in the two-dimensional case, but the balance on these characteristics is modified by the curvature and Coriolis terms. We obtain

$$dh \pm \frac{1}{\mathcal{G}(h)}du = \left[-\frac{uh}{r} \pm \frac{1}{\mathcal{G}(h)}C^2v(2+\frac{v}{r})\right]dt, \tag{13.11}$$

where

$$\mathcal{G}(h) = \left[\frac{1-S+(Sh)/H}{h}\right]^{1/2}, \tag{13.12}$$

on the characteristics

$$\frac{dr}{dt} = c_\pm = u \pm \left[h(1-S+S\frac{h}{H})\right]^{1/2}. \tag{13.13}$$

This is the axisymmetric (and with possible rotation) counterpart of (12.24)-(12.26).

We reiterate that the azimuthal momentum equation is not influenced by the stratification parameter $S$. The stratification enters into the momentum considerations because the interface $z = h$ encounters different values of $\rho_a$ (actually, different values of the pressure). But in the axisymmetric case $h$ does not vary in the azimuthal direction. Consequently, (13.9) and the third equation in (13.10) are the same as in the non-stratified counterpart discussed in § 8.1.

The third characteristic of the system (13.10) is

$$dv = -u\left(2 + \frac{v}{r}\right) dt \quad \text{on} \quad \frac{dr}{dt} = u. \tag{13.14}$$

This can be rewritten as

$$\frac{dv}{dr} + \frac{1}{r}v = -2 \quad \text{on} \quad \frac{dr}{dt} = u, \tag{13.15}$$

subject to the initial condition $v = 0$ at the position $r = r_{init}$ where the characteristic starts the propagation. This turns out to be identical with the non-stratified case; see (8.30). The solution is

$$\omega = \frac{v}{r} = -1 + \left(\frac{r_{init}}{r}\right)^2. \tag{13.16}$$

The physical interpretation is that a particle (here a ring) of dense fluid that propagates to a larger radius will acquire more and more counter-rotation in the rotating frame of reference. The absolute angular velocity of this particle, in the laboratory frame, becomes smaller than $\Omega$ of the whole system. A differential rotation of the current with respect to the ambient develops.

Boundary conditions for the radial and azimuthal (angular) velocity components at the nose $r = r_N(t)$ are needed. We argue that for small values of $\mathcal{C}^2$ the boundary conditions for $u_N$ are as in the two-dimensional case, and we shall therefore use (12.29) and (12.30). This assumptions is vindicated by the good agreement of the $r_N(t)$ predicted by the present SW formulation with NS computation as discussed later.

Consider the boundary condition for the angular velocity. We recall, see § 8.1.1.1, that the combination $(2\Omega + \hat{z} \cdot \nabla \times \mathbf{v})/h$ (in dimensional form) or $(2 + \hat{z} \cdot \nabla \times \mathbf{v})/h$ (in dimensionless form, scaled with $\Omega$) is defined as the potential vorticity. Using the continuity and azimuthal momentum equations we derived the conservation of potential vorticity for the non-stratified case. Now we recall that the continuity and azimuthal momentum equations for the present $S > 0$ formulation (first and last in (13.10)) are identical with the non-stratified case (see (8.25)). Consequently, the conservation of potential vorticity result is valid also for the present case. We can use

$$\frac{D}{Dt}\left(\frac{\zeta + 2}{h}\right) = 0, \tag{13.17}$$

where

$$\zeta = \frac{1}{r}\frac{\partial}{\partial r}(r^2\omega). \tag{13.18}$$

For the initial conditions $h = 1$ and $\omega = 0$ in the lock, the conservation of potential vorticity can be reformulated as

$$h(r,t) = 1 + \frac{1}{2}\zeta(r,t). \tag{13.19}$$

A combination of the total volume conservation of the dense fluid, (13.19), and (13.18) yields the boundary condition

$$\omega = -1 + \left(\frac{1}{r_N(t)}\right)^2 \quad (r = r_N). \tag{13.20}$$

The same result can be obtained by following the balance of the outer ring of fluid on the aforementioned third characteristic, i.e., use (13.16) with $r_{init} = 1$.

## 13.1.1   Steady-state lens (SL)

For $C > 0$ the system (13.10) admits a non-trivial steady-state solution with $u = 0$ and $r_N = $ constant. This reflects an equilibrium between the pressure and Coriolis-centrifugal forces. The task is to determine $h(r), \omega(r)$, and $r_N$ of the possible steady lens (SL). These flows have important application in oceanography (Csanady 1979, Hedstrom and Armi 1988).

Letting $y = r/r_N$, we can express the radial momentum and potential vorticity equations, (13.8) and (13.19), for $0 \leq y \leq 1$, as

$$A\frac{dh}{dy} = C^2 r_N^2 y[\omega(2+\omega)], \tag{13.21}$$

$$h = 1 + \omega + \frac{1}{2}y\frac{d\omega}{dy}, \tag{13.22}$$

where

$$A = 1 - S + S\frac{h}{H}, \tag{13.23}$$

subject to the boundary conditions (13.20), regularity at $y = 0$, and $h(y = 1) = 0$. Substitution of (13.22) into (13.21) yields a single equation for $\omega$

$$\frac{d^2\omega}{dy^2} + \frac{3}{y}\frac{d\omega}{dy} - 2\frac{C^2}{A}r_N^2\omega(2+\omega) = 0. \tag{13.24}$$

The solution provides $r_N$ of the lens, $\omega(y)$ and $h(y)$. For the non-stratified case, $S = 0$, which is obviously the simplest one because $A = 1$, we recover the classical SL discussed in § 8.1.2. The stratified system is complicated by the additional parameter $S$. In particular, for $S = 1$ we obtain $A = h/H$, and a singularity of (13.24) at $y = 1$ appears due to the conditions $h(1) = 0$. The influence of $S$ is clarified by the following results.

### 13.1.1.1   Analytical approximations for the SL

*For S not close to* 1. The first type of approximate solution can be derived when $\mathcal{C} \ll 1$ and $1 - S \gg \mathcal{C}$ (i.e., $S$ is not very close to 1). We use an expansion of $h(y), \omega(y)$, and $r_N^2$ in powers of $\mathcal{C}$. We find that $h \sim \mathcal{C}$, and hence, to leading order, $A = 1 - S$ is a positive constant. This reduces, to leading order in $\mathcal{C}$, the solution of the present equation (13.24) (with the associated boundary conditions) to the solutions of the non-stratified counterpart (8.48), but with a modified Coriolis coefficient

$$\mathcal{C}_m = \mathcal{C}(1 - S)^{-1/2}, \tag{13.25}$$

instead of the original $\mathcal{C}$. With this change, the results for small values of $\mathcal{C}_m$ (to leading order in this parameter) are

$$r_N = \left(\frac{2}{\mathcal{C}_m}\right)^{1/2}, \tag{13.26}$$

$$\omega = -1 + \mathcal{C}_m\left(1 - \frac{1}{2}y^2\right), \tag{13.27}$$

and

$$h = \mathcal{C}_m(1 - y^2). \tag{13.28}$$

For more details see § 8.1.2.1.

As confirmed by (13.26), for a small $\mathcal{C}_m$ the distance of propagation is significant and a gravity current appears is spite of the hindering Coriolis effects. As expected, the stratification decreases the radial buoyancy force (or pressure gradient) and hence increases the relative importance of the Coriolis effects. The lens is thin, $O(\mathcal{C}_m)$, and practically "feels" the ambient fluid in the proximity of the bottom whose density is $\rho_b$. The resulting lens is like one produced in a homogeneous case with density difference $\rho_c - \rho_b$.

*For $S \approx 1$.* The value of $\mathcal{C}_m$ becomes large and the approximation (13.25)-(13.26) becomes invalid when $S$ approaches 1, i.e., when $\rho_c - \rho_b$ vanishes. This also corresponds to a lens created by an intrusion. A different expansion in powers of $\mathcal{C}$ is needed for this case. This brings us to the second type of approximate solutions, which can be derived when $S = 1$ and $\mathcal{C}^2 H \ll 1$. The density difference between the lens and the ambient is small, and hence, as compared with the previous case, a thicker lens, a smaller radius, and a stronger slope of the interface are needed to counterbalance the Coriolis effects. An order of magnitude consideration indicates that here $h(y), \omega(y)$, and $r_N^2$ can be expanded in powers of $(\mathcal{C}^2 H)^{1/3}$. Substitution of this expansion into the equations (13.21)-(13.24), subject to volume conservation and boundary conditions yields, to leading order,

$$h = (\mathcal{C}^2 H)^{1/3}\left(\frac{3}{2}\right)^{1/3}(1 - y^2)^{1/2}, \tag{13.29}$$

$$\omega = -1 + (C^2 H)^{1/3} \left(\frac{2}{3}\right)^{2/3} \left[1 - (1 - y^2)^{3/2}\right] \frac{1}{y^2}, \quad (13.30)$$

and

$$r_N = \left(\frac{3}{2}\right)^{1/3} (C^2 H)^{-1/6}. \quad (13.31)$$

As $y \to 1$, $h'(y)$ and $\omega''(y) \to \infty$, but $|\omega'(y)|$ is small; a local and relatively small contribution of viscous or turbulent dissipation effects is expected to develop. We are interested in cases with small $C^2 H$ because otherwise the distance of propagation is small, see (13.31), and no real gravity current appears.

When $S = 1$ the dimensionless parameter $C^2 H$ can also be expressed as $(\Omega r_0 / \mathcal{N} h_0)^2 = (f r_0 / 2 \mathcal{N} h_0)^2$, where, again, $\mathcal{N}$ is the usual buoyancy frequency of the ambient, and $f = 2\Omega$; see (13.6). The foregoing approximations yield the following compact results for the radius of propagation and aspect ratio of the lens in the $S = 1$ case

$$\frac{r_N^*}{V^{* \, 1/3}} = \left(\frac{3}{\pi} \frac{\mathcal{N}}{f}\right)^{1/3}, \quad (13.32)$$

$$\frac{h^*(0)}{r_N^*} = \frac{1}{2} \frac{f}{\mathcal{N}}, \quad (13.33)$$

where $V^*$ is the volume of the lens (the upper asterisk denotes dimensional variables). It is remarkable that the shape of the lens is determined only by the volume and $f/\mathcal{N}$; the details of the initial aspect ratio $h_0/r_0$ do not influence the results (to leading order).

The case $S = 1$ corresponds, again, to an intrusion in a stratified fluid at a neutral buoyancy level. In other words, the $S = 1$ results can be applied to the following situation. Suppose that $z = 0$ is now the horizontal midplane in the rotating container. A double-lens, symmetric with respect to this plane, is formed. At this plane the densities of the ambient and lens fluid are equal, which corresponds to $\epsilon_b = \epsilon$, or $S = 1$, in our results. The aspect ratio of thickness to diameter of this double-lens is given by (13.33). Denoting this aspect ratio by $\alpha$, we obtain the prediction $\alpha \mathcal{N}/f = 0.5$. Such (double) lenses have been created in laboratory experiments by injection from a point source of non-rotating fluid at the neutral level of a linearly stratified rotating fluid, for example by Griffiths and Linden (1981) and Hedstrom and Armi (1988). The measured values of $\alpha \mathcal{N}/f$ were about 0.4, and decayed with time after the injection was stopped. The observed lenses were stable and axisymmetric for several hundreds revolutions of the system. This is a good indication that our simplified theory captures the essential effects. In § 8.1.4.1 we explained in some detail why a lens produced by injection attains the shape of a lens produced by lock-release. Briefly, in the present case for a small value of $C^2 H$ (as assumed in the approximation (13.29)-(13.33)) the constant-volume lens spreads out significantly. Therefore the angular velocity of the lens is reduced to almost $-1$; see (13.30). Since the injected fluid has $\omega = -1$ set by the initial condition, the Coriolis-centrifugal accelerations that determine the $dh/dr$ inclination of the interface are similar with the lock-released case. The fixed-volume results for small $\mathcal{C}$ are a good approximation for a slowly-growing lens of non-rotating fluid, for both stratified and non-stratified ambients. (Hedstrom and Armi 1988 also presents data and comparisons for azimuthal speed and decay of the lens, but this is outside our scope.)

### 13.1.1.2 Numerical solution of the SL

In general, the determination of the SL system must be performed by numerical methods. We use a finite difference discretization on a 100 interval grid. Iterations are necessary for the non-linear term (and value of $r_N$) in (13.24). The initial guess is guided by the foregoing approximate results. For each iteration the discretized form of (13.24) produces a simple system for $\omega$ at the grid points. Successive substitution and corrections of $r_N$ lead to the proper non-linear combination of $\omega(y), h(y)$, and $r_N$ which satisfies the equations (13.24), (13.22) and the boundary condition $h(1) = 0$. (The total volume conservation provides an additional check of convergence).

The $S = 1$ case requires special attention. The straightforward finite difference approach fails in the corner region where $y$ approaches 1, because of the singularity which shows up as $A = h/H$ tends to 0 and the slope of $h$ becomes very large. However, $\omega$ remains finite at $y = 1$ ($r = r_N$) according to the boundary condition there. We assume the behavior $h = a_0\xi^\gamma$ for $\xi = (1 - y) \to 0$, where $a_0$ and $\gamma$ are constant. Substitution into (13.21) and subject to (13.20) we obtain for $y$ close to 1 the approximation

$$h = [2HB(1 - y)]^{1/2} \quad \text{where} \quad B = \mathcal{C}^2 r_N^2 \left(1 - \frac{1}{r_N^4}\right). \tag{13.34}$$

This expression is used to evaluate $h$ in the last grid interval when $S = 1$, and thus we avoid the implementation of the finite difference approximations near the singular $y = 1$.

Typical results of the SL shape and internal angular velocity are presented in Fig. 13.2. As the stratification (value of $S$) increases, the lens becomes thicker and shorter and the retrograde angular velocity in the interior decreases. The $S = 1$ case, despite the singularity, does not display any qualitative dissimilarity with the other cases. The agreement between the analytical approximate results and the numerical solution of the SL is good. The approximations were developed for small values of $\mathcal{C}$ and $\mathcal{C}^2H$; in the lower frames of Fig. 13.2 these parameter are not so small, $\mathcal{C} = 0.4$ and $\mathcal{C}^2H = 0.48$, yet fair agreement is obtained for $h$ in all cases (this also implies agreement for $r_N$) and for $\omega$ for $S \leq 0.5$ (in the upper frames $\mathcal{C} = 0.3$ and $\mathcal{C}^2H = 0.18$ and the agreement is very good for all the predicted variables in the full range of $S$).

The lens in the stratified ambient is prone to decay mechanisms of the type mentioned in § 8.1.4. This topic is outside the scope of this text.

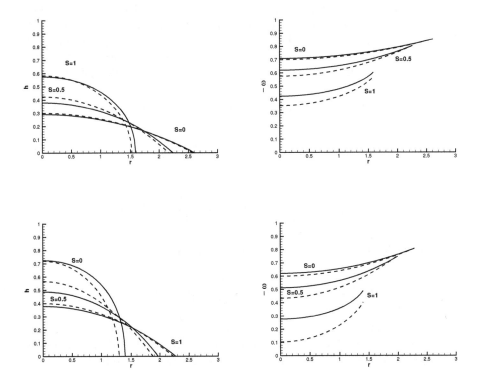

**FIGURE 13.2:** SL profiles $h$ and $\omega$ as functions of $r$ and various $S$, for $C = 0.3$, $H = 2$ (upper frames), and $C = 0.4$, $H = 3$ (lower frames). Numerical solution (solid lines) and approximate solution (dashed lines). (Ungarish and Huppert 2004, with permission.)

### 13.1.2 The energy of the SL

The energy budgets of the time-dependent current will be discussed in Chapter 20. Here we consider briefly the case of the SL. This is a special problem: we look only at the initial and final states, in which the kinetic energy is provided by the angular velocity only.

Since we use a one-layer model, we can discuss the energy of the dense fluid only. In general, the energy is time-dependent. However, here we consider only the initial and steady-state situations. Therefore there is only spatial dependency of the variables $h$ and $\omega$, and the radial velocity $u$ is zero.

The kinetic energy is calculated straightforwardly. The calculation of the potential energy in the reduced-gravity field requires some consideration. We note that the vertical displacement of the dense fluid particles is resisted by the hydrostatic pressure of the embedding ambient fluid. The resulting buoyancy force per unit volume is $[\rho_c - \rho_a(z)]g$ (dimensional), and the corresponding work needed to move a unit volume from the bottom to some $z$, in the present linear $\rho_a(z)$, is $g[(\rho_c - \rho_b)z + (1/2)(\rho_b - \rho_o) z^2/H]$ (dimensional). The integral of this work over the volume of the current gives the potential energy (for more details see Chapter 20). Here we *scale the energy with* $\rho_o g' r_0^2 h_0^2$ (per radian).

The resulting kinetic and potential energies (per radian) can be expressed as

$$K_c = \frac{1}{2}C^2 r_N^4 \int_0^1 [1 + \omega(y)]^2 h(y) y^3 dy. \tag{13.35}$$

$$P_c = r_N^2 \int_0^1 \left[ \frac{1}{2}(1 - S)h^2(y) + \frac{1}{6}\frac{S}{H}h^3(y) \right] y dy. \tag{13.36}$$

In the initial state $h = 1$, $r_N = 1$, and $\omega = 0$, and hence the total energy is

$$E_c = P_c + K_c = \frac{1}{4}(1 - S) + \frac{1}{12}\frac{S}{H} + \frac{1}{8}C^2 \quad (t = 0). \tag{13.37}$$

The stratification reduces the potential energy in the initial system. The energy of the SL is obtained by substituting the appropriate $h(y)$ and $\omega(y)$ into (13.36)-(13.35). Using the approximations (13.28)-(13.31), we find

(1) For $S < 1$

$$E_c = C_m[\frac{1}{6}(1 - S) + C_m\frac{1}{24}\frac{S}{H}] + \frac{23}{240}C_m^3(1 - S) \quad (t = t_1), \tag{13.38}$$

where $C_m$ is given by (13.25).

(2) For $S = 1$

$$E_c = \frac{1}{20}\left(\frac{3}{2}\right)^{2/3}\frac{1}{H}(C^2 H)^{2/3} + 0.120\frac{1}{H}(C^2 H)^{4/3} \quad (t = t_1); \tag{13.39}$$

where $t_1$ denotes the time when a SL is present.

The last term in (13.37)-(13.39) represents the kinetic energy. For the cases considered here (small $C_m$ for $S < 1$ and small $C^2 H$ for $S = 1$) the energy of the SL is significantly smaller than in the initial state. The potential energy is the dominant term in both the SL and initial states. The energy decay is likely to occur along the lines discussed later in Chapter 20. In particular, a significant amount of energy must be spent on the pressure-force work during the expansion to the final $r_N$. The details still require elaboration.

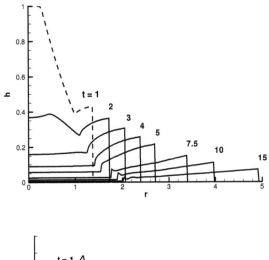

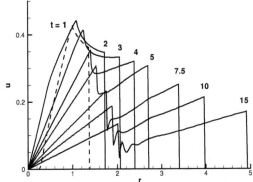

**FIGURE 13.3**:   SW $h$ and $u$ as functions of $r$ at various times for $H = 3$, $S = 0.72$.

## 13.2   SW and NS finite-difference results

   The typical behavior of the axisymmetric (non-rotating) current in a linearly stratified ambient, as predicted by the SW solution, is shown in Fig. 13.3. Both the height and the speed display a quite complex dependency on both time and radius. There is an initial "dam-break" phase during which a backward-moving rarefaction wave sets into motion the fluid in the lock. At $t = 1$ the interface is still quite high near the center, and there is still a core with $u = 0$. At $t = 2$ all the fluid is already moving and the height of the interface is everywhere below 0.4. Next, most of the fluid is concentrated in a quite prominent external ring, followed by a thin horizontal tail. The velocity in this tail is quite large (the small oscillation about the sharp changes are a spurious numerical effect). Eventually, the current becomes a thin layer with a slight upward inclination, and the velocity tends to a linear profile. Unlike the 2D case, we cannot define a clear-cut slumping phase of motion with constant speed. Although the stratification in this example is fairly strong ($S = 0.72$) the qualitative behavior of $h$ and $u$ is similar to that in the non-stratified counterpart; see Fig. 6.1. The major difference is in the speed of propagation. However, the $S = 0$ current attains a self-similar phase at large $t$, while in the stratified case a self-similar motion appears only for $S = 1$.

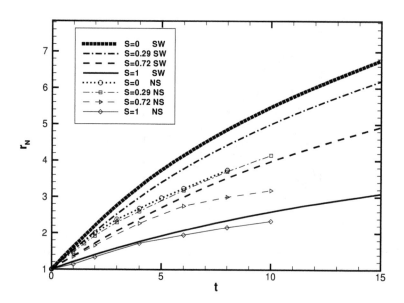

**FIGURE 13.4:**    The propagation of the nose as a function of time for $H = 3$ and various values of $S$. SW results (curves) and NS results (curves with symbols) for various values of $S$. (Ungarish and Huppert 2008, with permission.)

The typical propagation of the nose is presented in Fig. 13.4. As in the 2D case, a stronger stratification (larger $S$) reduces the speed of propagation. The stratified case $S = 0.29$ is super-critical with respect to $u_W$, while $S = 0.72$ and $S = 1$ are sub-critical. (To be more specific, we mention that the SW values of $(u_N, u_W)$ are $(0.54, 0.30)$ for $S = 0.29$ and $(0.38, 0.47)$ for $S = 0.72$ shortly after release.) The super- and sub-criticality of 2D currents with similar initial conditions has been confirmed experimentally by Maxworthy, Leilich, Simpson, and Meiburg (2002). However, there is a significant difference between the 2D and the axisymmetric cases: in the former there is a stage of propagation with constant speed for several lock-lengths, while in the latter the speed decays with $t$ from the beginning. Consequently, the super-critical axisymmetric current will become sub-critical after a relatively short propagation.

The effect of stratification on the propagation of a typical rotating axisymmetric gravity current, as predicted by the SW formulation, is illustrated in Fig. 13.5. The predictions for the interface are illustrated in 13.6. The Coriolis effects hinder, and eventually stop, the radial propagation. The Coriolis-influenced interface develops a downward inclination of the frontal region, and the height of the nose decays to zero in a relatively short time. When $S$ increases both the speed of radial propagation and the maximum radius of spread are significantly reduced. The interpretation of this trend is as follows.

The increase of $S$ decreases the effective reduced gravity that drives the nose, $g'_e$; see (12.53)-(12.54). The approximation is based on the observation that the nose reacts to the density difference at about half height. Thus, the initial radial propagation is expected to decrease with $S$, as in the case of a 2D current, simply because the driving density difference decreases with $S$. On the other hand, the Coriolis-centrifugal forces are not influenced by the axial stratification. This suggests the introduction of an effective Coriolis dimensionless parameter, see (13.5), as follows

$$\mathcal{C}_e = \frac{\Omega r_0}{(g'_e h_0)^{1/2}} = \mathcal{C} \left[ 1 - S(1 - \frac{h_N}{2H}) \right]^{-1/2}. \tag{13.40}$$

During the initial propagation the typical value of $h_N$ is 0.5, then $h_N$ decreases, and hence the effective $\mathcal{C}_e$ is larger than the formal $\mathcal{C}$ for $S > 0$ all the time. Moreover, eventually the Coriolis effects reduce $h_N$ to zero and the propagation stops. This further enhances the effect of stratification because the nose is brought down to encounter levels of larger and larger density. We expect that at this stage of slow radial propagation the dynamic behavior switches to the equilibrium mode, i.e., the SL balances become relevant, in particular the maximum $r_N$ predicted by (13.26) and (13.31); indeed, for $h_N = 0$ we obtain $\mathcal{C}_e = \mathcal{C}_m$; see (13.25). Note that $\mathcal{C}_e$ is associated with the dynamic propagation, and $\mathcal{C}_m$ with the equilibrium lens.

The analysis of the SW results for different values of $\mathcal{C}$ and $H$ leads to the following conclusions: the maximum radius of propagation attained by the current exceeds the radius of the SL; the excess varies from about 30% for $S = 0$ to about 20% for $S = 1$. The time at which the maximum propagation is attained is given (approximately) by $1.7/\mathcal{C}$ for $S = 0$ and decreases (slightly) as $S$ approaches 1. The interval $\mathcal{C}t \approx \pi/2$ from release to the maximal prop-

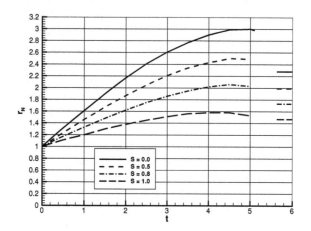

**FIGURE 13.5**: SW results for the propagation of axisymmetric currents rotating with $\mathcal{C} = 0.4$ and $H = 3$ for various $S$. The horizontal lines on the right show the SL asymptote. (Ungarish and Huppert 2004, with permission.)

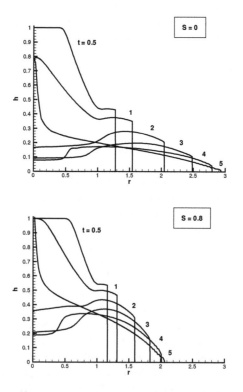

**FIGURE 13.6**: SW predicted profiles of $h$ as a function of $r$ at various times for the rotating gravity currents with $\mathcal{C} = 0.4$ and $H = 2$, for $S = 0$ and $S = 0.8$. (Ungarish and Huppert 2004, with permission.)

agation corresponds to about (1/4) revolution of the system. The fact that the stratification has little influence on this time interval is an interesting outcome, that can be explained by the fact that as $S$ increases both the maximum radius and the velocity of propagation decrease, so that the (mean) ratio of distance over velocity remains unchanged. The prediction that the maximum SW radius is larger than that of the SL indicates that some contraction and oscillations are expected after $t = 1.7/\mathcal{C}$, for any $S$. In the stratified case the classical oscillations of the bulk of dense fluid about the steady-lens form are expected to combine with oscillations of the isopycnals in the ambient. The Ekman layers develop in about one revolution of the system, and hence are expected to not have an influence during the propagation to maximum radius that is attained in a similar time interval. Eventually, these layers and other dissipative mechanisms will smooth out the discontinuity of $\omega$ between the lens and the ambient.

The interaction between the nose and the internal gravity waves is expected to develop as in the two-dimensional case, but complicated by the effects of curvature and Coriolis (inertial waves). Typically, the velocity of axisymmetric currents decays faster than that of 2D currents, and the rotation enhances this trend. We therefore speculate that the major stage of inertia-dominated propagation will be close to its end before the internal waves become influential. This is consistent with the results of NS simulations, see below, but a detailed understanding of this topic requires a great deal of additional investigation.

Now let us consider some NS numerical solutions. The main configurations are the axisymmetric counterpart of the 2D computation for Run 5 and Run 19 which have been considered in detail in the 2D geometry in § 12.3. In the second case (Run 19) rotation will also be added. The results were obtained with a finite-difference code of the type discussed in Appendix B. The numerical tank is of radius $r_w = 8$, $Re = 1.0 \cdot 10^3$, $h_0/r_0 = 0.25$, $H = 3$ and the typical grid is of $320 \times 200$ intervals.

Fig. 13.4 indicates that the NS and SW predictions of $r_N(t)$ agree well only for a short distance of propagation. This behavior seems to reflect a quite general deficiency of axisymmetric NS results: the simulated current tends to disperse after propagation to about $r_N = 2.5$. (The reason for these numerical instabilities is unclear. To overcome this difficulty it may be necessary to use full three-dimensional simulations, e.g., of the type employed by Cantero, Balachandar, and Garcia 2007.) However, the stratification improves the agreements. In any case, the NS results confirm the predicted influence of $S$ on the motion.

Typical density-field results of the NS simulations are shown in Figs. 13.7 and 13.8. We see that the shape of the current is very complex in the region of the nose. Stronger stratification (Fig. 13.8) reduces the height of the head as compared to the case of weaker stratification (Fig. 13.7). The isopycnals of the ambient fluid are considerably displaced above the head of the current almost from the beginning of the process. (At advanced times the density contours also show significant perturbations near the axis; but the volume in this region, $r < 0.1$ approximately, is very small and hence this effect is unimportant.)

We recall that Run 19 uses the values $S = 0.72$, $H = 3$, $\epsilon = 0.0804$, and lock aspect ratio $h_0/r_0 = 0.25$. For this case, simulations for currents rotating with $\mathcal{C} = 0.6$ and 0.8 were also performed. The predictions of $r_N$ as functions of $t$ are shown in Fig. 13.9. The difference between the non-rotating and rotating systems is evident: the Coriolis effects drastically limit the propagation. It is clear that as $\mathcal{C}$ increases the maximum radius of propagation decreases and is attained in a shorter time.

To be specific, the SW predicted maximum $r_N$ of 1.80 and 1.60 for $\mathcal{C} = 0.6$ and 0.8, respectively, attained at $t = 3.0$ and 2.1 (in both cases at $t\mathcal{C} \approx 1.6$). The corresponding NS results attain quasi-maximum radii of 1.8 and 1.6 at $t = 3.0$ and 2.5. By quasi-maximum we mean that the typical head of the current vanishes (this is inferred from the shape and behavior of the interface between the dense fluid and the ambient); the rim of the current still advances very slowly, but its motion seems to be dominated by viscous and diffusion effects. The fact that the NS (and experimental) currents lack a sharp maximum radius of propagation, and actually display a slow spread of the rim after the end of the inertia-Coriolis propagation, has been observed and reported also for non-stratified circumstances (Verzicco et al. 1997, Hallworth et al. 2001), as discussed in § 8.1.5. It could be expected that viscous effects are important when the radial motion of the edge becomes slow and the height there is small, and this explains the reason and the trend of the discrepancy with the inviscid SW results for $t > 1.6/\mathcal{C}$, approximately. Otherwise, the agreement with the SW model is good concerning the radius and time of propagation of the major motion and the influence of the dimensionless parameters.

The present configuration is sub-critical from the beginning of the motion; see Fig. 12.15. Yet we observe that the stratification waves have no significant effect on the motion in the time intervals (or distances of propagation) considered here. Indeed, in the 2D case, see Fig. 12.6, the interaction developed at $t = 12$, after a propagation of about five lock lengths, while in the present rotating axisymmetric cases the maximum radius is reached at $t \approx 3$ and the propagation is about one lock length. In the axisymmetric non-rotating case some hindering of the nose shows up in the NS computation at $t = 12$, but this is not a clear-cut wave effect like in the 2D counterpart. The mean thickness of the axisymmetric current decreases like $1/r_N^2$, and at $t = 10$ is about one tenth of its initial value, while the area of contact with the bottom is about ten times larger than that of the lock. This evidently enhances the relative contribution of viscous friction, and it is difficult to distinguish between this effect and wave hindering of the nose at these times. In any case, the wave-head interaction in the axisymmetric current does not develop sooner than in the 2D counterpart. The dramatic effect of rotation and the complex shape of the interface are illustrated in Fig. 13.10. Contour lines of the density function of value $\phi = 0.72$, obtained from the NS simulations, are plotted at

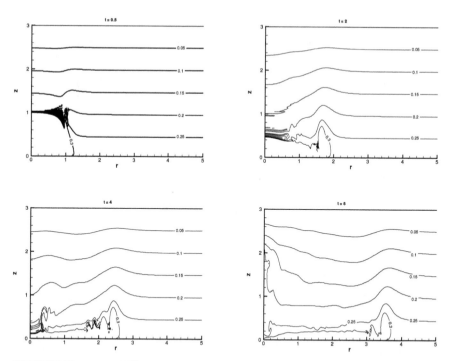

**FIGURE 13.7:** NS density contours at various times for $H = 3$, $S = 0.29$. (Ungarish and Huppert 2004, with permission.)

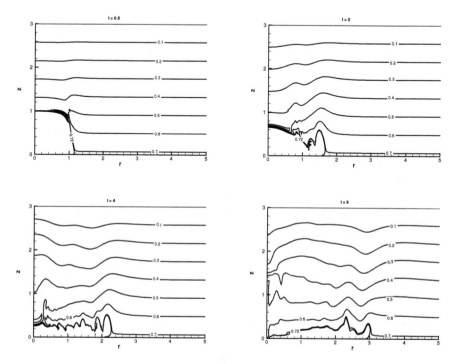

**FIGURE 13.8:** As in previous caption, for $S = 0.72$.

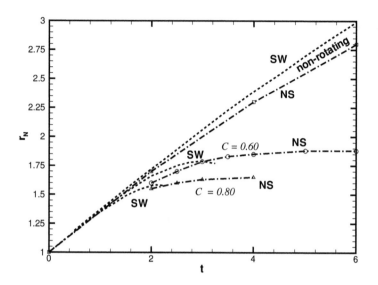

**FIGURE 13.9:** The propagation of axisymmetric gravity currents in a stratified ambient $S = 0.72$ and $H = 3$. NS (symbols and dash-dotted lines) and SW (dashed lines) predictions for $\mathcal{C} = 0.8, 0.6,$ and $0$ (non-rotating). (Ungarish and Huppert 2004, with permission.)

various times for the non-rotating case and for the rotating with $\mathcal{C} = 0.6$ case. At $t = 2$ the difference between the cases is small, but afterward the Coriolis effects dominate the second case. The non-rotating current spreads out for a long time, while the height of the head is reduced gradually. In contrast, the bulk of the rotating current spreads out very little and during a relatively short time only; the bulk of dense fluid even thickens at $t = 4$ (this indicates a reverse motion in the center). Similar features have been reported for a non-stratified ambient; the stratification enhances the differences between the non-rotating and rotating cases in the sense that the maximum radius of propagation decreases with $S$.

Corresponding SW predictions are shown in Fig. 13.11. There is fair agreement in the global behavior, in particular concerning the effect of the rotation on the behavior of the current. The negative radial velocity in the rotating current at $t = 3$ has no counterpart in the non-rotating situation.

The predicted behavior of the angular velocity in the rotating current is displayed in Fig. 13.12. The NS simulations show that the current has a distinct signature of negative $\omega$ (relative to the rotating system) during its propagation. There is, again, fair agreement with the SW approximations. The discrepancies can be attributed to the differences in the local height of

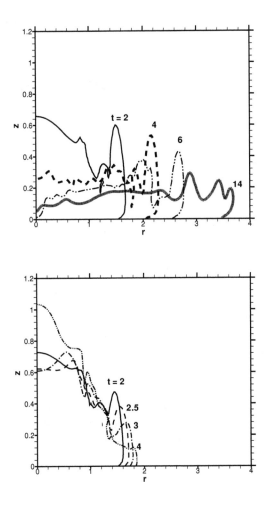

**FIGURE 13.10:** NS predictions: The interface of axisymmetric gravity currents for various times, in non-rotating (upper frame) and rotating with $C = 0.6$ (lower frame). In both cases $S = 0.72$ and $H = 3$. (Ungarish and Huppert 2004, with permission.)

the interface, deviations from the one-dimensional motion, and the friction on the boundary and interface. Again, similar discrepancies have been detected in the non-stratified cases. The SW results predict that the expansion is completed at $t \approx 1.6/C = 2.7$. Indeed, at $t = 2$ and 3 we observe the typical negative angular momentum, but at $t = 4$ we find a significant increase of $\omega$ in the dense fluid. This change can be attributed to the reverse (contraction) radial motion of the current. This reverse motion in the dense fluid is also

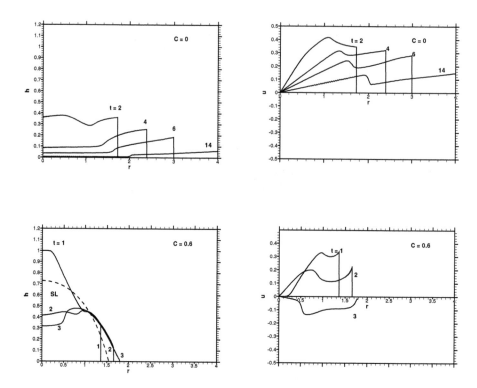

**FIGURE 13.11:**  SW predictions. The interface and radial velocity of axisymmetric gravity currents for various times, in non-rotating (upper frames)
and rotating with $\mathcal{C} = 0.6$ (lower frames). In both cases $S = 0.72$ and $H = 3$.
The SL interface profile is shown by the dashed line. (Ungarish and Huppert
2004, with permission.)

clearly confirmed by the ascent of the interface near the center at $t = 4$; see
Fig 13.10.

Actually, the motion of the interface at the center provides a convenient
detector of the expansion-contraction oscillations that appear in the bulk of
the dense fluid. This is illustrated in Fig. 13.13. In addition to the $S = 0.72$
case discussed above, we also show the lesser stratified $S = 0.43$ and the non-
stratified $S = 0$ counterparts (in all cases $H = 3$ and $\mathcal{C} = 0.6$). In all cases the
height of the interface at the center first decreases and reaches a minimum
at $t \approx 2.7$; this corresponds to the maximum expansion which is expected to
occur, according to the SW estimate, at $t \approx 1.6/\mathcal{C} = 2.7$. Afterward, up and
down oscillations appear, and the period of this motion depends on the value
of $S$. In the non-stratified case the inertial period $\pi/\mathcal{C} = 5.2$ (in dimensional
units, $\pi/\Omega$) is expected to be relevant. The experiments of Hallworth et al.

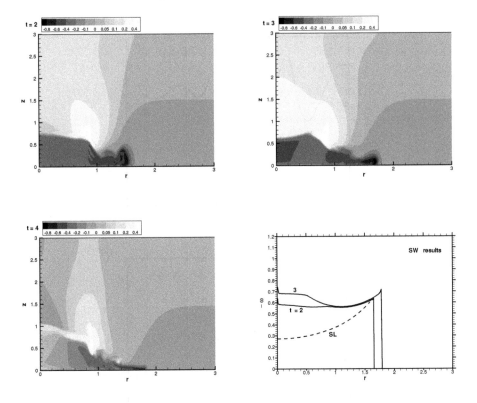

**FIGURE 13.12**: The angular velocity for $\mathcal{C} = 0.6, H = 3, S = 0.72$ configuration. NS prediction contours, and SW prediction (including SL) profiles
for various times. (Ungarish and Huppert 2004, with permission.)

(2001) for a non-stratified ambient, discussed in § 8.1.5, revealed oscillation
with the period of $3.0/\mathcal{C}$. On the other hand, the period of the internal
gravity waves is $2\pi(h_0/r_0)(H/S)^{1/2}$ (in dimensional units, $2\pi/\mathcal{N}$). Thus, the
dimensionless period of these waves is 3.2 for $S = 0.72$ and 4.1 for $S = 0.43$,
which is shorter than the inertial period. We see that the oscillations in the
stratified cases indeed display shorter periods than in the $S = 0$ case, but due
to possible interactions between the waves and viscous damping, a detailed
quantitative description is not available. However, we note that the time
interval between the first minimum and the first peak is (approximately) 1.6,
2.1, and 2.4 for $S = 0.72, 0.43$, and 0, respectively; the first two correspond
to the half-period of the internal waves, the last one to the half-period of
the inertial modes (approximatively). Moreover, the initial amplitude of the
internal waves is larger, because the isopycnal of the interface tends to return
to the initial position of equilibrium $h = 1$ (and the initial displacement is
about 0.55 in the present cases), while the inertial oscillations tend to be

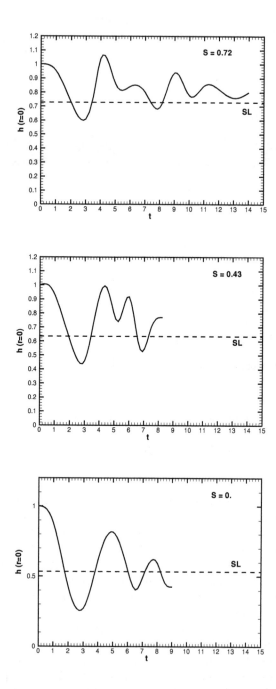

**FIGURE 13.13:** The oscillations as reflected by the height of the interface at the center as a function of $t$ for $\mathcal{C} = 0.6, H = 3$, for three values of $S$: $0.72, 0.43$, and $0$ (non-stratified). NS results (solid line), and also shown the SL results (dashed lines). (Ungarish and Huppert 2004, with permission.)

driven by the displacement from the SL equilibrium (which is about 0.15 in the present cases).

To summarize, the SW approximation for the axisymmetric rotating current in a stratified ambient is consistent with the NS simulations for the initial period of propagation (until Coriolis or viscous forces become dominant). The discrepancies between the SW and NS results are similar to these obtained for the situation of a homogeneous ambient in corresponding cases. Overall, the incorporation of the stratification in the one-layer SW model does not reduce the intrinsic accuracy of this type of analysis.

There are indications that the internal gravity waves are less important in the axisymmetric geometry than in the 2D counterpart because the current decays with the square of the radius and is already slow when the interaction occurs.

In the rotating frame, as stratification increases, the relative importance of the Coriolis effects increases. It is however remarkable that for any value of $S$ the maximum radius of propagation is attained in about 0.3 revolutions of the system.

## 13.3 The validity of the inviscid approximation

The importance of viscous friction in the motion of the current increases with time and distance of propagation. This tendency is enhanced by stratification as shown below. We use the SW results for $u$ and $h$ to estimate the time-dependent ratio of global inertial, $F_I$, to viscous, $F_V$, forces. The inertia per unit volume is well represented by $\rho_c u u_r$, and the viscous force per unit area is expected to be proportional to $\rho_c \nu u/h$. Using dimensionless variables, we obtain

$$\frac{F_I}{F_V} = Re \frac{h_0}{r_0} \frac{\int_0^{x_N(t)} u u_r h r dr}{\int_0^{x_N(t)} (u/h) r dr} = Re \frac{h_0}{r_0} \theta(t). \tag{13.41}$$

The function $\theta(t)$ is expected to be of the order of unity at the beginning of the propagation and decay to quite small values. This function can be calculated from the SW results for $u(r,t)$ and $h(r,t)$; see Fig. 13.14.

Since similarity solutions are not available (except for $S = 1$, see § 16.3.2), for the analytical estimate of $\theta$ we use box-model considerations. Assume that the current is a cylinder box of height $h_N(t)$ and radius $r_N(t)$, and accordingly $u = u_N r/r_N$. With this approximation in the integrals of (13.41), we obtain $\theta \approx u_N h_N^2 r_N^{-1}$. The value of $u_N$ is estimated from (12.29) and (12.30) as: (a) $[(1 - S)h_N]^{1/2}$ for $S$ not close to 1; and (b) $h_N/H^{1/2}$ for $S \approx 1$. In this estimate we consider $Fr$ to be a constant because the current is expected to

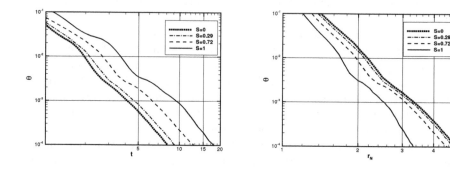

**FIGURE 13.14**: SW results. The coefficient $\theta$, defined in (13.41), for $H = 3$ and various values of $S$: as a function of $t$ (left frame) and of $r_N$ (right frame). (Ungarish and Huppert 2008, with permission.)

be thin when viscous effects are relevant. Finally, we use volume conservation to obtain $h_N = 1/r_N^2(t)$. This yields

$$\theta = \begin{cases} (1 - S)^{1/2} r_N^{-6} & (S \text{ not close to } 1) \\ H^{-1/2} r_N^{-7} & (S \approx 1). \end{cases} \tag{13.42}$$

Using these results, we can estimate the importance of the viscous terms, and also the limit of validity of the inviscid assumption, for a real gravity current with given $Re = Uh_0/\nu$ and $h_0/r_0$. The inviscid theory is expected to be relevant for, roughly, $F_I/F_V > 1$. The contribution of the viscous terms increases very fast with the radius of propagation, and this effect is significantly enhanced when $S$ is close to 1.

# Chapter 14

## The steady-state current

In Chapter 3 we discussed the classical results of Benjamin (1968) concerning the propagation of a steady gravity current into a homogeneous ambient. Here we consider the counterpart problem when the propagation is into a linearly-stratified ambient, following the presentation of Ungarish (2006).

The prototype geometry of Benjamin's solution is a long horizontal channel of height $H$ filled with stationary ambient fluid of constant density $\rho_o$. Into this fluid, at the bottom of the channel, a layer (current) of denser fluid of density $\rho_c$ and thickness $h$ propagates with uniform velocity $U$. The driving force is the reduced gravity

$$g' = \epsilon g, \quad \text{where} \quad \epsilon = \frac{\rho_c - \rho_o}{\rho_o}. \tag{14.1}$$

Another important dimensionless parameter is the fractional depth

$$a = h/H. \tag{14.2}$$

The natural scaling velocity is $(g'h)^{1/2}$. The pertinent scaled velocity of propagation, referred to as the Froude number of the current, is

$$\tilde{U} = Fr = \frac{U}{(g'h)^{1/2}}. \tag{14.3}$$

For further reference, we repeat here Benjamin's main results (see (3.8) and (3.10), denoted by the subscript $B$) for the non-stratified case. First, the scaled speed, $Fr$ is a function of $a$ only,

$$Fr_B(a) = \left[ \frac{(2-a)(1-a)}{1+a} \right]^{1/2}. \tag{14.4}$$

Second, the head loss is given by

$$\Delta_B = h \frac{a(1-2a)}{2(1-a^2)}. \tag{14.5}$$

This is the drop of pressure (scaled with $\rho_o g'$) below the ideal Bernoulli result, along each streamline in the ambient, from the far upstream to the downstream domains. Consequently, a steady-state is physically possible only for

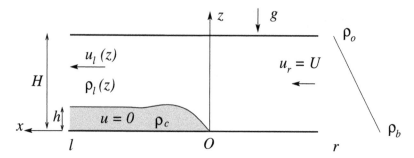

**FIGURE 14.1:** Sketch of the configuration. The lower boundary of the ambient is $z = \chi(x)$. $r$ and $l$ denote the upstream (right) and the downstream (left) positions.

a current which occupies a fractional depth $a \le a_{\max} = 0.5$. A larger $a$ would require an energy increase by some external source.

The stratified configuration is sketched in Fig. 14.1. The density of the unperturbed ambient, on the right-hand side, increases linearly from $\rho_o$ at the top to $\rho_b$ at the bottom; the density of the current is, again, $\rho_c$ (we consider $\rho_c \ge \rho_b$ cases). A new dimensional parameter enters the formulation, the relative magnitude of the stratification

$$S = \frac{\rho_b - \rho_o}{\rho_c - \rho_o}, \tag{14.6}$$

with $0 \le S \le 1$. Now $Fr$ is a function of both $a$ and $S$. We expect that the classical case is recovered in the limit of a very mild stratification, $S \to 0$. We assume a Boussinesq ($\epsilon \ll 1$) almost inviscid ($Re = Uh/\nu \gg 1$) and a shallow configuration in the sense that the typical horizontal length is large compared with $h$.

The complete formulation of the steady-state inviscid Boussinesq problem of interest renders a complicated physical and mathematical system: we are confronted by a two-fluid flow with a sharp interface of unknown shape, speed and density fields depending on $x$ and $z$, mixed boundary conditions, and infinite $x$ domain. Solving analytically this problem is a formidable task. Some indirect approximate approach is necessary for progress.

The idea used below is to explore the similarity between the flow configuration used for the $S = 0$ case in Chapter 3 and an available result of Long's model (Long 1953, Long 1955, Baines 1995) concerning the flow of a stratified fluid in a channel with an upper horizontal solid wall and a prescribed bottom topography. We use a frame of reference attached to the gravity current. As in the non-stratified case, the far upstream velocity of the ambient is constant over the height of the channel, $0 \le z \le H$. This flow climbs the topography (current) at a relatively slow pace (compared with the horizontal propagation). The streamlines become horizontal again in the far downstream region of the thinner channel, $h \leq z \leq H$. For an observer moving with the current (the bottom topography) the flow is steady. Total force, stability, and dissipation considerations are then applied to determine the velocity of a physically valid steady flow - in other words, the acceptable values of $\tilde{U} = Fr$ for a given geometry and stratification.

In the non-stratified case the downstream flow of the ambient fluid regains a uniform velocity, $UH/(H - h)$, and a constant reduced pressure. This facilitates the calculations and simplifies the results. However, when $S > 0$ the linear density stratification in the upstream region must be compressed in a non-trivial manner into a thinner channel. The downstream flow field of the ambient fluid develops a quite complex $z$-dependent structure velocity, density, and pressure. The first task is to specify this flow field.

## 14.1 Steady-state flow pattern

Consider the solution of a two-dimensional (2D) stratified flow field over a rigid bottom topography in a channel with an upper horizontal rigid lid at $z = H$. (The geometry is like that of Fig. 14.1, but the gray domain is a part of the solid bottom.) The horizontal and vertical coordinates are $\{x, z\}$, and the corresponding velocity components are $\{u, w\}$. Gravity acts in the $-z$ direction. To be specific, the obstacle (or topography) encountered by the unperturbed stratified fluid is defined by the bottom elevation function, $z = \chi(x)$. The value of $\chi$ is 0 (i.e., no obstacle) for non-positive $x$. For $x > 0$, the wall is elevated to $\chi(x) > 0$, and far downstream at the left, a parallel geometry is achieved again with $\chi(x) = h = \text{const} > 0$.

The far upstream flow (at the right, $x \to -\infty$), where the bottom is flat, $z = \chi(x) = 0$, consists of parallel horizontal streamlines with constant velocity $U$ and a prescribed stable linearly changing density. Using the subscript $r$ (right) to denote this region, and employing the hydrostatic balance, we write

$$u_r(z) = U; \quad w_r(z) = 0; \quad \rho_r(z) = \rho_b - (\rho_b - \rho_o)\frac{z}{H};$$

$$p_r(z) = -g \int_0^z \rho_r(z')dz' \quad (0 \leq z \leq H), \quad (14.7)$$

where $p$ is the pressure.

Under the assumption of a two-dimensional, steady, Boussinesq, inviscid and hydrostatic flow, the analysis of Long (1953) can be applied to reduce the set of governing Navier-Stokes equations to a single partial-differential equation for the displacement of the streamline, $\delta(x, z)$, with proper boundary conditions, in particular, $\delta = 0$ in the upstream right region, and $\delta = \chi(x)$ at the bottom. The exact solution for the present configuration is reported in

Baines (1995), § 5.7.1. The flow field is conveniently expressed with the aid of the perturbation (about the upstream flow) stream-function, $-\partial\psi/\partial z = u'$, $\partial\psi/\partial x = w$, where $u = U + u'$. The result reads

$$\psi(x, z) = U \cdot \chi(x)\frac{\sin[\beta(1 - z/H)]}{\sin[\beta(1 - \chi(x)/H)]} = U\delta(x, z) \tag{14.8}$$

and

$$\rho(x, z) = \rho_b + (\rho_b - \rho_o)\left[-\frac{z}{H} + \frac{\chi(x)}{H}\frac{\sin[\beta(1 - z/H)]}{\sin[\beta(1 - \chi(x)/H)]}\right] \tag{14.9}$$

where

$$\beta = \frac{[(\rho_b/\rho_o - 1)gH]^{1/2}}{U} = \frac{(Sg'H)^{1/2}}{U} = \frac{\mathcal{N}H}{U} \tag{14.10}$$

and $\mathcal{N} = [(\rho_b/\rho_o - 1)g/H]^{1/2} = (Sg'/H)^{1/2}$ is the buoyancy frequency.

We combine this result with the presumed steady-state bottom gravity current. We replace the solid bottom obstacle with a stationary fluid of density $\rho_c$ in the domain $(x \geq 0, 0 \leq z \leq \chi(x))$. The $\{x, z\}$ system is now a frame of reference attached to the gravity current, and the origin $O$ is the front stagnation point. We assume that, like in the non-stratified case, under certain conditions the structure of the parallel horizontal far up- and down-stream flow regions of the ambient, given by (14.8)-(14.9), is preserved. In this system the right upstream flow, where $\chi(x) = 0$, is unchanged and given by (14.7). In the left (subscript $l$) region, where $\chi(x) = h$, the parallel horizontal flow satisfies

$$u_l(z) = \begin{cases} 0, & (0 \leq z < h) \\ U\left\{1 + \frac{a}{1-a}\frac{\gamma}{\sin\gamma}\cos\left[\frac{\gamma}{1-a}(1 - z/H)\right]\right\} & (h \leq z \leq H) \end{cases} \tag{14.11}$$

$$\rho_l(z) = \begin{cases} \rho_c, & (0 \leq z < h) \\ \rho_r(z) + (\rho_b - \rho_o)\frac{a}{\sin\gamma}\sin\left[\frac{\gamma}{1-a}(1 - z/H)\right] & (h \leq z \leq H) \end{cases} \tag{14.12}$$

$$\delta_l(z) = \frac{h}{\sin\gamma}\sin\left[\frac{\gamma}{1-a}(1 - z/H)\right] \quad (h \leq z \leq H) \tag{14.13}$$

and, since $w_l = 0$, the hydrostatic balance yields

$$p_l(z) = p_O - g\int_0^z \rho_l(z')dz' \tag{14.14}$$

where

$$\gamma = (1 - a)\beta = (1 - a)\sqrt{\frac{S}{a}\frac{1}{\tilde{U}}} \tag{14.15}$$

and $p_O$ is the pressure at the stagnation point $O$.

In the two-fluid flow field a steady-state can be maintained only for certain values of $U$, for which the upstream and downstream properties are balanced. The system (14.7)-(14.14) satisfies volume continuity balances because the replacement of the solid obstacle with a stationary fluid of the same geometry does not affect the volume fluxes. On the other hand, the stresses in the lower fluid which replaces the solid obstacle become a part of the problem, and a new $x$-momentum balance must be satisfied. We consider the momentum balance in a fixed rectangular control volume whose lower and upper boundaries are the zero-stress planes $z = 0, H$ and the vertical boundaries are in the parallel up- and down-stream regions. The assumptions of steady-state and vanishing viscous stress (on the boundaries and in the horizontal flow domains) impose the flow-force balance

$$\int_0^H (\rho_l u_l^2 + p_l)dz = \int_0^H (\rho_r u_r^2 + p_r)dz. \tag{14.16}$$

At the stagnation point $O$ the pressure is $p_O = \frac{1}{2}\rho_b U^2$ (recall that $p_r(z = 0) = 0$). Using (14.7), (14.14), and (14.12), we obtain manageable expressions for the pressure terms in (14.16). The evaluation of the integral of the momentum flux terms in (14.16) is simplified by the Boussinesq assumption (i.e., $\rho_{l,r}(z)u_{l,r}^2(z) \approx \rho_0 u_{l,r}^2(z)$; $u_{l,r}(z)$ are defined by (14.7) and (14.11)). After some algebra we can express the flow-force balance (14.16) as

$$\frac{\tilde{U}^2}{2}\frac{1}{1-a}\left[1 + a - 2a^2 + a^2\left(\gamma^2 + (\gamma\cot\gamma)^2 + \gamma\cot\gamma\right)\right] =$$
$$1 - \frac{1}{2}a + S\left[-1 + a - \frac{1}{3}a^2 + (1-a)^2\frac{1 - \gamma\cot\gamma}{\gamma^2}\right]. \tag{14.17}$$

The RHS of (14.17) can be regarded as the buoyancy pressure driving, and the LHS as the dynamic reaction. The term multiplied by $S$ is the contribution of the stratification to the (integrated) pressure difference. The last term in the brackets on the RHS represents the effect of the displacement of the isopycnals.

Finally, we use (14.15) to eliminate $\tilde{U}$ from (14.17) and obtain the flow-force balance in the form

$$f(\gamma) =$$
$$1 - a + a(2-a)\gamma\cot\gamma + (a\gamma\cot\gamma)^2 + \gamma^2\frac{a}{1-a}\left[(2-a)(1 - \frac{1}{S}) - \frac{1}{3}a^2\right] = 0. \tag{14.18}$$

The root(s) of this equation, for given $a$ and $S$, provides the desired solution $Fr(a, S) = \tilde{U} = (1-a)(S/a)^{1/2}/\gamma$ (see (14.15)). Obviously, only real-valued positive roots of $f(\gamma)$ are of interest in the physical context of the gravity current problem. We note that the stratification enters the relevant equation both explicitly as $S$ and implicitly as $(S/a)$ via $\gamma$. This seems to reproduce a physical property of the system.

## 14.2   Results

### 14.2.1   Small $S$ (small $\gamma$)

Consider the weak stratification, $S \to 0$, case. We expect that the dimensionless velocity $\tilde{U} = Fr$ is of the order of unity. Note, however, that straightforward substitution of $S = \gamma = 0$ into (14.17) and (14.18) yields indeterminate $0/0$ terms. By an expansion of (14.17) in powers of $\gamma$ we obtain, to leading order in $S$, the compact result

$$\tilde{U} = Fr(a, S) = \left[ \frac{(2 - a)(1 - a)}{1 + a} \left( 1 - \frac{2}{3}S \right) \right]^{1/2}$$

$$= Fr_B(a) \left( 1 - \frac{2}{3}S \right)^{1/2}, \quad (14.19)$$

where $Fr_B(a)$ is the classical formula of Benjamin; see (14.4). The resulting behavior is simple and intuitive: the weak stratification causes a mild reduction of $Fr$, which is effectively decoupled from the depth-ratio dependency.

The behavior of a "Froude number" of the current, which is actually a dimensionless velocity, depends of course on the scaling. This may become a confusing issue in the present stratified case where three densities and two lengths are present. We therefore recall our framework. It is convenient to consider the densities at the top of the ambient, $\rho_o$, and of the dense fluid, $\rho_c$, as given, with $\rho_c > \rho_o$. In most experimental settings the top (open surface) is freshwater. The density in the ambient increases linearly with the depth. The bottom density, $\rho_b$, is between $\rho_o$ and $\rho_c$, on a scale expressed by $0 < S < 1$. A weak stratification means that $\rho_b$ is closer to $\rho_o$ than to $\rho_c$. The scaling velocity in the present $Fr$ is independent on $\rho_b$ (or $S$). Thus, according to (14.19), for a given configuration, when a larger $\rho_b$ is produced, the physical velocity of the current decreases. This is indeed simple and intuitive.

### 14.2.2   Large $\gamma$. Validity-stability and criticality

For non-small $S$ and small $a$ the analysis is more complicated. It is necessary to find numerically the solution of (14.18). Now two difficulties, with no counterpart in the non-stratified case, appear.

First, an inspection of the equation indicates that for non-small $S$ several simple positive roots may be obtained. We arrange the roots in increasing order, $\gamma_i$, $(i = 1, 2, ...)$. Second, we note that some roots may be larger than $\pi/2$. This introduces the possibility of negative values of $u_l$, as follows. Considering the contribution of the trigonometric term in (14.11), we find that (1) for $\gamma \le \pi/2$, all values of $u_l(z)$ are positive and larger than $U$; but (2) for certain values of $\gamma > \pi/2$ the velocity profile $u_l(z)$ above the current may attain negative values. This negative $u_l$ situation invalidates the assumption that all the streamlines originate in the right unperturbed "upstream" domain, and is unacceptable in the physical context of real gravity currents which propagate into the unperturbed ambient. Moreover, using (14.12), we find that $-d\rho_l(z)/dz$ has the same sign as $u_l(z)$. This means that the local stratification is unstable when $u_l$ is negative.

To quantify this tendency to negative velocity and instability in the downstream domain, it is convenient to introduce the invalidity (or instability) coefficient

$$\theta = \begin{cases} 0 & (0 < \gamma \leq \frac{\pi}{2} \\ \dfrac{a}{1-a}\gamma|\cot\gamma| & (\frac{\pi}{2} < \gamma < \pi) \\ \dfrac{a}{1-a}\dfrac{\gamma}{|\sin\gamma|} & (\pi < \gamma). \end{cases} \tag{14.20}$$

This is a measure of the most severe relative negative contribution of the perturbation flow to the resulting $u_l$. In other words, for $\gamma > \pi/2$, we obtain $\min(u_l) = U(1 - \theta) < U$, and a tendency to unstable density variation, in at least one horizontal level. If $\theta$ is small there is no danger of spurious behavior, but for $\theta > 1$ the results are physically unacceptable. These unacceptable flow fields are also subject to shear instability, while, on the other hand, the criterion $\theta < 1$ is also sufficient for satisfying the $Ri > 1/4$ stability requirement in the ambient, where $Ri = [N/u'_l(z)]^2$ is the Richardson number of the local flow. This shear instability is beyond the scope of this book, and the interested reader is referred to Long (1955) and Baines (1995).

In the following discussion we are concerned only with "valid" solutions, i.e., with roots of $f(\gamma) = 0$ which satisfy $\theta < 1$. Moreover, we shall monitor the values of $\theta$ in cases of non-unique valid roots; this is as a possible criterion for discrimination between the results (see below). The valid results are in the range $0 < \gamma \leq \max[\pi, (1-a)/a]$. A more detailed sketch of the validity domain in the $a - \gamma$ plane is displayed in Fig. 14.2. For $a < 0.18$ a periodic validity pattern appears, and the number of these repetitions increases as $a$ decreases. It turns out that the number of valid roots of (14.18) follows roughly this pattern. Decreasing $a$ from 0.5 (see below § 14.2.4), the first $\gamma_2$ (valid) solution was reported for $a = 0.1784$ and $S = 0.875$; for larger values of $a$ and $0 < S < 1$, only $\gamma_1$ solutions were found.

Typical valid results of $Fr$, obtained numerically from (14.18), are presented in Figs. 14.3 and 14.4. Fig. 14.3 illustrates the very good accuracy of the approximation (14.19): for a given $a$, $Fr$ starts with Benjamin's value at very small $S$, then decreases with $S$; the slope is very close to -1/3, independent of $a$. The deviation from this simple behavior appears for larger values of $S/a$ ($> 5$, say). Here a second-root branch may also appear.

The ratio of the speed of the current to that of the fastest internal gravity wave in the upstream section given by (12.5) can be expressed as

$$\frac{U}{u_W} = \frac{\pi(1-a)}{\gamma} = \pi\sqrt{\frac{a}{S}}Fr. \tag{14.21}$$

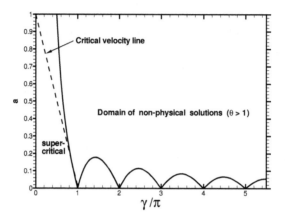

**FIGURE 14.2:**  Domains of validity and criticality in the $\gamma - a$ plane. Points above and to the right of the full-line curve ($\theta = 1$) would produce negative and unstable downstream flows. Points on the left of the dashed critical velocity line correspond to super-critical currents (on the right, to sub-critical). (Ungarish 2006, with permission.)

The critical $Fr$ number, for which this ratio is 1, is

$$Fr_{cr} = \frac{1}{\pi}\sqrt{\frac{S}{a}}. \qquad (14.22)$$

The steady-state current under consideration is super-critical when $Fr > Fr_{cr}$, i.e., $\gamma < \pi(1-a)$. The super-critical domain is also shown in Fig. 14.2. This type of current is more likely to occur for configurations with small $S$ and non-small $a$. The first valid root of $f(\gamma)$ belongs typically to the super-critical regime. The additional roots, when present, yield, without exception, sub-critical propagation.

For lock-release problems with small $a$ and non-small $S$, the sub-critical currents seem more physically acceptable (rather, expected) than the super-critical solution. The reasons are as follows. Since the flow develops from zero velocity conditions, it obviously encounters first the sub-critical quasi-steady balance; if this is sufficiently stable, there is apparently no mechanism to further accelerate the propagation.

### 14.2.3   The effective $g'$ and $Fr$

In the derivation of $Fr$ we scaled the velocity with $g'$ based on the density difference of the current to the top of the ambient, $\rho_c - \rho_o$. This scaling is convenient for homogeneous configurations and for presentation of laboratory results.

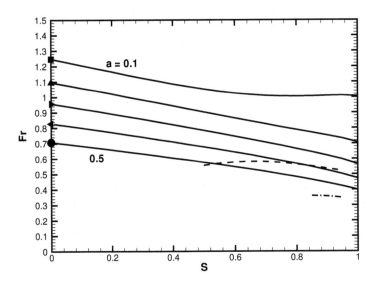

**FIGURE 14.3:** $Fr$ as a function of $S$ for $a = 0.1, 0.2, 0.3, 0.4, 0.5$. The symbols are Benjamin's solution for the corresponding $a$. The dash and dash-dot lines are the second and third roots for $a = 0.1$. (Ungarish 2006, with permission.)

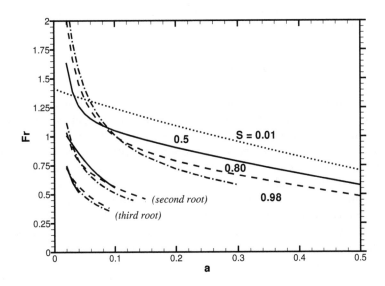

**FIGURE 14.4:** $Fr$ as a function of $a$ for various $S$ ($S = 0.01$ dotted line, $S = 0.5$ full line, $S = 0.8$ dashed line, $S = 0.98$ dash-dotted line). (Ungarish 2006, with permission.)

We argued in § 12.4 that the effective driving force of the current is well correlated to the difference between the density of the dense fluid, $\rho_c$, and the density of the upstream ambient at the mid-height of the head, $\rho_r(z = \frac{1}{2}h)$. We therefore use again the concept of effective (subscript $e$) reduced gravity

$$g'_e = \frac{\rho_c - \rho_r(\frac{1}{2}h)}{\rho_o} g = g'A^2 \qquad (14.23)$$

where

$$A = \left[1 - S\left(1 - \frac{1}{2}a\right)\right]^{1/2}, \qquad (14.24)$$

and (14.7), (14.1), and (14.6) were used. We note in passing that the $\rho_o$ was used at the denominator of (14.23) for simplicity; in the framework of the Boussinesq approximation, a negligible $O(\epsilon)$ change of $g'_e$ appears if $\rho_b$ or $\rho_c$ is used instead. The scaling velocity is now $(g'_e h)^{1/2}$.

The effective Froude number of the steady-state current is therefore defined as

$$Fr_e = \frac{U}{(g'_e h)^{1/2}} = \frac{Fr}{A}; \qquad (14.25)$$

see (14.3). The switch from the previous $Fr$ results to the effective Froude is straightforward. Benjamin's scaling is fully recovered in the limit $S \to 0$ for which $A = 1$.

The rescaled $Fr_e$ as a function of $S$ for various values of $a$ is displayed in Fig. 14.5 (to be compared with the counterpart Fig. 14.3 for the originally defined $Fr$). The conclusions are as follows.

Consider the weak stratification case, $S \leq 0.5$. The leading effect is contributed, again, by Benjamin's $Fr_B$. Indeed, the combination of (14.19) and (14.25) yields, to leading order in $S$,

$$Fr_e = Fr_B(a)\left[1 + S(\frac{1}{3} - \frac{1}{2}a)\right]^{1/2}. \qquad (14.26)$$

For $0.1 \leq a \leq 0.5$, the effective Froude number varies very little with $S$. This is because now the reference velocity takes into account the density stratification effect. In this case, our intuitive expectation of $g'_e$ is confirmed.

However, Fig. 14.5 shows that the counter-intuitive increase of $Fr$ *for the first root* with $S$ when $S/a$ is not small becomes more pronounced in the behavior of the rescaled $Fr_e$. In other words, in this case our intuitive understanding of $g'_e$ is not sufficient for the interpretation of the flow; the real speed is significantly larger than expected. On the other hand, we see that the second and third roots values of $Fr_e$ are of the order of 1. This indicates that our "intuition," based mostly on lock-release experiments, is in accord with the higher roots of the steady-state solution, not with the fastest theoretical flow. Indeed, the dam-break theoretical and experimental results considered in § 12.4 display values of $u_N/A$ which are fairly independent of $S$.

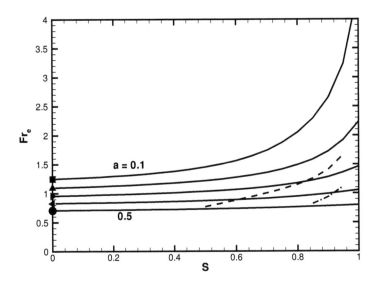

**FIGURE 14.5:** $Fr_e = Fr/A$ as a function of $S$ for $a = 0.1, 0.2, 0.3, 0.4, 0.5$. The symbols are Benjamin's solution for the corresponding $a$. The dash and dash-dot lines are the second and third roots for $a = 0.1$. (Ungarish 2006, with permission.)

Consider the strong stratification $S \to 1$ flow for small $a$, when the second (and third, etc.) root appear (the dashes and dashed-dotted lines in Fig. 14.5). Here the advantage of the rescaling is observed: $Fr_e$ is of the order of 1 (while the original $Fr$ tends to small values). This turns out to fulfill our expectations. Indeed, the case $S \to 1$, i.e., $\rho_b \to \rho_c$, corresponds to an intrusion at the neutral buoyancy level. In this case we can write $g' = \mathcal{N}^2 H$, where, again, $\mathcal{N}$ is the buoyancy frequency. Our estimate of the velocity of propagation, based on physical intuition and supported by experimental and numerical evidence (Wu 1969; Amen and Maxworthy 1980; Ungarish 2005b), is as follows. Since the density difference at $z = 0$ is zero, in this case the buoyancy driving effect is clearly provided by the gradient, $g'/H$, multiplied by the (half) thickness of the current. This produces the effective reduced gravity $g'_e = \frac{1}{2}\mathcal{N}^2 h = \frac{1}{2}ag'$ (exactly as given by (14.23)). Consequently, the relevant expected speed of propagation is $(g'_e h)^{1/2} = (1/\sqrt{2})\mathcal{N}h$. The fact that $Fr_e$ is of the order 1 indicates that there are stable solutions of the generalized Benjamin problem which agree well with our intuitive understanding of the flow field in this limit. We shall return to this issue in § 14.3.

## 14.2.4   Energy dissipation

Consider a streamline at height $z$ in the left-hand-side parallel domain of ambient fluid. This streamline carries fluid of density $\rho_l(z)$ and is elevated $\delta_l(z)$ from the initial right-hand-side position. In ideal conditions the pressure on this streamline satisfies Bernoulli's equation

$$p_l^i(z) + \frac{1}{2}\rho_l(z)u_l^2(z) + \rho_l(z)\delta_l(z)g = p_r(z - \delta_l(z)) + \frac{1}{2}\rho_l(z)U^2 \quad (h \le z \le H) \tag{14.27}$$

where the upperscript $i$ denotes ideal energy-conserving flow. For the non-ideal flow field of the gravity current system, the head loss on the streamline is given by

$$\Delta(z) = \left[p_l^i(z) - p_l(z)\right]/(\rho_0 g') \quad (h \le z \le H). \tag{14.28}$$

In the non-stratified analysis the head loss is independent of $z$. For the present more complex configuration it is convenient to define the average head loss

$$\overline{\Delta} = \frac{1}{H - h} \int_h^H \Delta(z)dz. \tag{14.29}$$

Substituting (14.11)-(14.14) and (14.27) into (14.28)-(14.29) and using the fact that the density is preserved on the streamline we obtain, after some algebra,

$$\frac{\overline{\Delta}}{h} = 1 - S\left[1 - \frac{1}{2}a + \frac{1}{2}\frac{(1-a)^2}{a}\frac{1}{\gamma^2} + \frac{1}{2}a\frac{1}{\sin^2\gamma} + (1-a)\frac{\cot\gamma}{\gamma}\right]. \tag{14.30}$$

This is the ratio of the head loss to the thickness of the current, a convenient dimensionless variable. The results are meaningful for values of $\gamma$ which are valid solutions of $f(\gamma) = 0$.

Consider the behavior for small $S$. We use an expansion of (14.30) in powers of $\gamma$, combined with (14.15) and (14.19). This yields, to leading order in $S$, the compact result

$$\overline{\Delta} = h\frac{a(1 - 2a)}{2(1 - a^2)}\left(1 - \frac{2}{3}S\right) = \Delta_B\left(1 - \frac{2}{3}S\right), \tag{14.31}$$

where $\Delta_B$ is the classical result (14.5) for the non-stratified ambient. As expected, Benjamin's dissipation results, which were discussed in Chapter 3 and § 5.5, are recovered for small values of $S$. The stratification reduces the head loss. In any case, it is evident that, when $S$ is small, $\overline{\Delta} = 0$ only for $a = 1/2$, and $\overline{\Delta}$ becomes negative for larger $a$. This extends the well known outcome of Benjamin's analysis to a slightly stratified ambient: currents of more than half-depth ($a > 0.5$) are non-physical because they require an increase of energy by an external source. As discussed in § 5.5 the head loss is connected with the formal energy dissipation of the system, and a negative $\Delta$

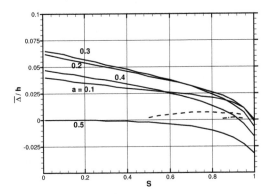

**FIGURE 14.6:** Head loss $\overline{\Delta}/h$ as a function of $S$ for various $a$. The dash and dash-dot lines correspond to the second and third roots for $a = 0.1$. (Ungarish 2006, with permission.)

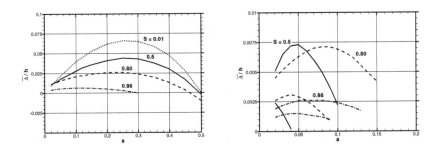

**FIGURE 14.7:** $\overline{\Delta}/h$ as a function of $a$ for various $S$. Left frame: first root; right frame: second and third roots. (Ungarish 2006, with permission.)

implies "negative energy dissipation," which is unacceptable in normal flow fields.

In general, the numerical calculation of the dissipation via (14.30) in the domain of interest shows that for valid solutions of (14.18): (1) For a given $a$, $\overline{\Delta}/h$ decreases as $S$ increases. (2) For $a > 0.5$ only non-physical negative $\overline{\Delta}$ were obtained. Let $a_{\max}(S)$ be the largest fractional depth for which physical non-negative dissipation occurs at a given $S$. Decreasing $a$ from 1, we find that the largest $a_{\max}$ is 0.5 for small $S$; then $a_{\max}$ decreases with $S$ to about 0.3 at $S = 1$.

The energy restriction can be generalized: For both homogeneous and linearly-stratified ambients, a steady flow is impossible in practice if the current occupies more than a certain height portion of the channel, $a_{\max}$.

Typical head-loss results are displayed in Figs. 14.6 and 14.7. In the displayed cases, $0.05 \leq a \leq 0.5$, $0 < S < 1$, the value of $\overline{\Delta}/h$ is quite small, typically 2–5%. The second and third root solutions display very small head loss, $\overline{\Delta}/h < 0.1\%$. However, we must keep in mind that $\Delta/h$ is not a reliable measure of the rate of dissipation of energy in the inviscid two-fluid system, $\dot{\mathcal{D}}$. As indicated by (3.12), $\dot{\mathcal{D}}$ behaves like $\rho_o g' U H \Delta$, and when scaled with $(1/2)\rho_o(g')^{3/2}h^{5/2}$ this yields $2(Fr/a)(\Delta/h)$.

In the non-stratified case $\dot{\mathcal{D}}$ can be quite easily calculated by integrating the in- and out-fluxes $(\rho_a u^2/2 + p)u$ of the ambient fluid in the moving frame. In the stratified case this becomes a cumbersome task. First, the energy flux term is more complex, $[\rho_o u^2/2 + p + \rho_o(\mathcal{N}^2/2)(\delta^2 - z^2)]u$, where $p$ is the pressure reduced with $+\rho_b z$. (The justification follows from the more general energy balances presented later in § 20.2.2.) Second, $p$, $\delta$, and $u$ on the left are not simple functions of $z$. This topic still requires elaboration. The simpler criterion $\overline{\Delta} \geq 0$ is, in any case, a relevant physical condition.

## 14.3    Comparisons and conclusions

Direct experimental data for $Fr(a, S)$ are not available. Actually, even for the homogeneous ambient the direct experimental verification of Benjamin's theory is a complex task. First, it is not easy to produce and control a steady-state gravity current which moves like a slug and whose stagnation point is at the bottom. Second, it is not clear how to define the pertinent height $h$ of a real current, because a thick mixed-fluid interface, with Kelvin-Helmholtz vortices, usually develops in practical circumstances above the head of the current. In the stratified ambient the situation is complicated by the fact that the theory predicts multiple steady-state solutions.

In a simple rectangular lock-release problem the initial (slumping) motion is with (almost) constant velocity in both laboratory experiments and Navier-Stokes simulations. A comparison of the idealized steady-state $Fr$ with this realistic data makes sense. Of particular relevance are the numerical 2D simulations of Birman, Meiburg, and Ungarish (2007) and White and Helfrich (2008) which used slip boundary conditions on the solid boundaries.

The computations of Birman, Meiburg, and Ungarish (2007) were performed for $\epsilon = 0.1$, $Re = 4000$, and various values of $H$ in the range $[1, 5]$ (dimensionless, scaled with $h_0$) and various stratifications. The code used a vorticity-streamfunction formulation (see Appendix B.3). The height $h_N$ of the current was evaluated as the average of first local maximum over a sliding time interval $1.5(h_0/g')^{1/2}$. After an initial adjustment, the speed of propagation $u_N$ and the above-mentioned $h_N$ were fairly constant over $2-3$ lock-lengths of advance. The typical simulated flow field is shown in Fig. 14.8.

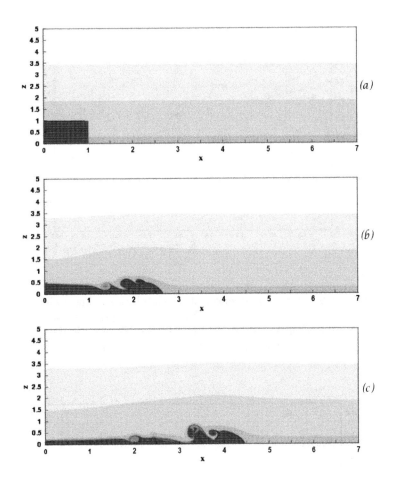

**FIGURE 14.8:** The simulated current with $S = 0.2, H = 5$ at times $t = 0, 2.5$, and 5 (frames $a, b, c$, respectively). The scaling (12.19) is used. (Birman, Meiburg, and Ungarish 2007, with permission.)

Overall, there is consistency between the simulations and theory concerning the propagation. The numerical results are in the domain of validity-stability displayed in Fig. 14.2 and obey the $a_{max}$ restriction inferred from the analytical dissipation considerations.

Comparisons of the resulting numerical $Fr = u_N/(g'h_N)^{1/2}$ with the predictions of the steady-state theory are displayed in Fig. 14.9 and 14.10.

In the geometries with $H = 1$ and 2 the currents do not attain small values of $a = h_N/H$, and hence only the first root of the theoretical $Fr$ solution

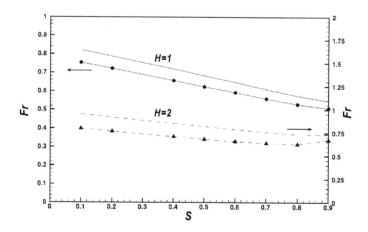

**FIGURE 14.9:** *Fr* as a function of *S*, simulations (lines with symbols), and theory (lines without symbols). Here $H = 1$ (upper lines, left axis, as indicated by the arrow) and $H = 2$ (lower lines, right axis, as indicated by the arrow). (Birman, Meiburg, and Ungarish 2007, with permission.)

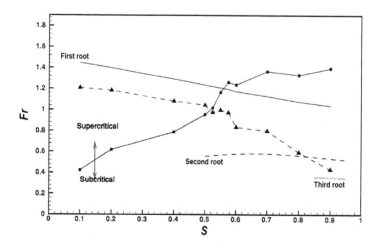

**FIGURE 14.10:** *Fr* as a function of *S*, simulations and theory, for $H = 5$ (dimensionless). The symbols are numerical results. The lines without symbols are the theoretical prediction for first, second, and third roots. Also shown are the super- and sub-critical domains for a current, using the numerical $a = h_N/H$ in (14.22). (Birman, Meiburg, and Ungarish 2007, with permission.)

is relevant. We see in Fig. 14.9 that there is good agreement between the predicted and simulated behavior of $Fr$. The analytical values are larger, an effect already known from the non-stratified counterpart. This can be attributed to friction, mixing, and the uncertainty of the definition of $h_N$ in a real flow. Similar confirmation of the first root behavior is provided by White and Helfrich (2008). (This work also derived and confirmed steady-state first-root solutions for non-linear stratifications.)

For the $H = 5$ case $a = h_N/H \approx 0.1$. The theory indicates that a second and third root of solution appears for sufficiently large $S$. Fig. 14.10 shows that the simulations agree with the first root for $S < 0.55$, approximately. For larger $S$ a departure from the first root toward the second one is observed, and for $S = 0.9$ a departure from the second to the third root appears. The theory suggests that a current released from rest will attain the slowest stable $Fr$ solution. The simulations are consistent with this conjecture, but the transition between the roots is not sharp. The simulations confirm that the second and third roots correspond to sub-critical currents.

Reliable comparisons with experiments are presently not available. The experiments of Maxworthy et al. (2002) were also of the lock-release type, but the height of the current was not properly documented. Consequently, an accurate reduction of the kind displayed for the simulated flows cannot be made. To test the qualitative agreement it is convenient to let $h_N/h_0 = \kappa$, and assume that $\kappa$ is a constant, 0.4 or 0.5 say. With this assumption, (14.19) can be used to predict the speed for currents released from behind a lock of height $h_0$ in a channel of given $H$; actually, $a = \kappa(h_0/H)$. These predictions turn out to be in very good agreement with the experimental results displayed in Fig. 12.10 (the details are left for an exercise; see also Ungarish 2006 Fig. 8).

Finally, we compare the trends of the steady-state propagation with the practical approximation (12.29) conjectured by Ungarish and Huppert (2002) for SW currents in a linearly-stratified ambient. In the present context this reads

$$Fr_{UH}(a, S) = Fr_B(a) \left[ 1 - S(1 - \frac{1}{2}a) \right] = Fr_B(a)\, A, \qquad (14.32)$$

where UH denotes the Ungarish-Huppert conjecture. Let $G$ be the ratio between the "exact" steady-state $Fr$ result and that approximation,

$$G = \frac{Fr}{Fr_{HU}} = \frac{Fr/A}{Fr_B} = \frac{Fr_e}{Fr_B}, \qquad (14.33)$$

where $Fr_e$ is the effective $Fr$ defined in § 14.2.3. This ratio depends on $a$, $S$, and on the number of the root when the exact solution is not unique.

The typical behavior of the ratio $G$ between the practical conjecture and the steady-state solution of $Fr$ is displayed in Fig. 14.11. The conjecture can be considered as "good" when $G$ is close to 1. We see that for small $S/a$ the agreement is indeed very good. In particular, for $S \leq 0.5$ and $a \geq 0.1$ the discrepancy is at most 16%. On the other hand, for $S = 0.8$ the first root solution exceeds the UH prediction by about 25% at $a = 0.3$ and by 65% at $a = 0.10$. However, the second and third roots display a better agreement, in the range of $\pm 20\%$ for $0.02 \leq a \leq 0.12$.

Additional information on the case $S \approx 1$ is presented in Fig. 14.12. Here the values of $G$ and the values of the invalidity-instability coefficient $\theta$, see (14.20), are given for the valid roots corresponding to the shown combinations of $S$ and $a$. For $S = 0.98$ and small $a$ many valid roots are obtained: e.g., 6 and 13 for $a = 0.05$ and 0.02, respectively. The results of the first roots exceed very significantly the HU formula, but for larger $i$ (and $\gamma_i$) a fairly good agreement appears. The value of the invalidity-instability coefficient $\theta$ first decreases with the number of the root, and then increases. In the region of minimum $\theta$ (which can be expected to be the most stable) the values of $G$ are about 1.3. This suggests the following: when several solutions $Fr$ are possible for given $(a, S)$, the one with the smallest $\theta$ is expected to appear in practical cases. The verification and justification of this (or similar criterion) requires further research. Another uncertain point of the steady-state solution is the possible influence of upstream waves on the validity of Long's unperturbed far-field assumption. It is plausible that the initial/boundary conditions play an important role in establishing the multiple-solution regimes indicated by the steady-state theory.

For practical use, the steady-state current is expected to supply a "jump condition" at the nose of a SW current. In the classical Benjamin case the implementation of $Fr_B(a)$ (and $a \leq a_{\max} = 1/2$) as a boundary condition is straightforward. In the stratified case the use of the "exact" $Fr$ solution is awkward: it cannot be expressed by a closed formula, and it is not unique. The use of the approximation (14.32) is mathematically more convenient and physically acceptable. We noticed that the less rigorous HS $Fr(a)$ formula is usually a better choice than $Fr_B(a)$ for SW modeling. Consequently, the combination (12.29)-(12.30) seems to be the best practical choice for a fully linearly-stratified ambient, at least at the present state of knowledge.

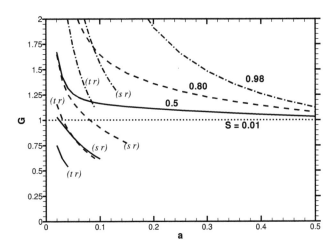

**FIGURE 14.11:** The ratio $G$, see (14.33), as a function of $a$ for various $S$ ($S = 0.01$ dotted line, $S = 0.5$ full line, $S = 0.8$ dashed line, $S = 0.98$ dash-dotted line). The second and third root results are marked by $(sr)$ and $(tr)$, respectively. (Ungarish 2006, with permission.)

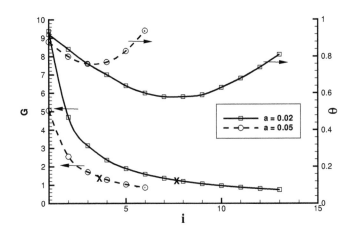

**FIGURE 14.12:** The ratio $G$ and invalidity-instability coefficient $\theta$ as a function of root number $i$ for $S = 0.98$, and two values of $a$: $a = 0.05$ dashed line, $a = 0.02$ full line. The arrows indicate the axis for the line. The most plausible (minimum $\theta$) results are marked by $\times$. (Ungarish 2006, with permission.)

# Chapter 15

## Intrusions in 2D geometry

### 15.1 Introduction

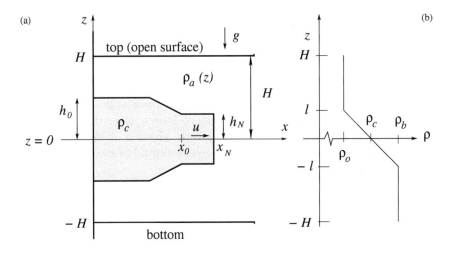

**FIGURE 15.1:** Schematic description of the *fully-symmetric* system (a) the geometry after release from a rectangular lock. (b) density profile in the ambient. In dimensionless form, the horizontal lengths are scaled with $x_0$ and the vertical lengths with $h_0$. The subscripts denote: $N$ - nose (or front); $a$ - ambient; $b$ - bottom; $c$ - current (intrusion); $o$ - top (open surface).

In the stratified ambient fluid the density $\rho_a(z)$ decreases from $\rho_b$ at the bottom to $\rho_o$ at the top (usually open) boundary. Consider a current of given volume of fluid of constant, intermediary density $\rho_c$, which is released from behind a lock into this ambient. The buoyancy forces will push upward in the part of the current embedded in $\rho_a > \rho_c$, and downward in the part of the current embedded in $\rho_a < \rho_c$. As a result, the body of the current will remain suspended between top and bottom, while being squeezed in the horizontal $x$

direction. The main motion is horizontal along a plane where $\rho_a \approx \rho_c$. This is an intrusion. We can sense that many factors are involved in the exact specification of the problem: the stratification function $\rho_a(z)$, the value of $\rho_c$, the geometry of the lock and its position with respect to the top and bottom boundaries. For obtaining the main insights and manageable results we shall introduce some simplifying restrictions. We shall adhere to them unless stated otherwise (§ 15.7).

We shall assume a fully-symmetric configuration. Briefly, this is referred to as "symmetric", and sometimes termed "double-symmetric" intrusion. As we shall see later, various "symmetries" can appear in the flow field of an intrusion. Here the symmetry is in the density field and in the geometry. The typical configuration is sketched in Fig. 15.1. The plane (line) $z = 0$, which is the plane of neutral buoyancy for the intrusion, is in the middle between the top and bottom boundaries, and also in the middle of the lock. The density of the intrusion (current) is the mean between $\rho_b$ and $\rho_o$, and the density of the ambient is antisymmetric with respect to this mean. This we express as

$$\rho_c = \rho_a(z = 0) = \frac{1}{2}(\rho_o + \rho_b). \tag{15.1}$$

In hydrostatic conditions, the bulk of fluid behind the lock will just hang in equilibrium, suspended half above and half below $z = 0$. Moreover, this fluid can be viewed as a "mixed fluid" domain. Indeed, if we trap the original ambient fluid in this lock which is symmetric about $z = 0$, and mix the separated domain, we shall obtain a homogeneous fluid of mean density.

The representative reduced gravity is

$$g' = \epsilon g \tag{15.2}$$

where

$$\epsilon = \frac{\rho_c - \rho_o}{\rho_o} = \frac{1}{2}\left(\frac{\rho_b}{\rho_o} - 1\right). \tag{15.3}$$

This $g'$ is based on the maximum density difference between the current and the ambient. We shall see that the effective buoyancy driving may be significantly smaller than this formal $g'$, depending on the thickness of the density transition layer.

We add the Boussinesq assumption that the density differences in this system are relatively small, i.e., $\epsilon \ll 1$. Now, after release from the lock, the upper and the lower half of the mixed fluid will be pushed forward symmetrically by the pressure difference at the gate. Note that both the upper and the lower half of the mixed fluid are pushed one toward the other by the embedding fluid, and thus actually squeezed to move in the $x$ direction. The resulting flow can be regarded as composed of two gravity currents about the $z = 0$ neutral buoyancy plane. These are like a "top" (or surface) current below, and a "bottom" current above. There is mirror symmetry about $z = 0$.

Also assume that the Reynolds number, $Re$, of the flow is large. We are now at the position to apply the inviscid SW apparatus to the investigation of the intrusion. The underlying idea is that the symmetric intrusion is a superposition of two practically identical gravity currents which flow, like mirror images, on the frictionless impermeable $z = 0$ boundary. Each of these currents contains the volume of half of the intruding fluid.

---

## 15.2 Two-layer stratification

Consider the configuration of Fig. 15.1 in the limit $l \to 0$. The ambient fluid has a sharp, or jump, or two-layer stratification. The contrast is with the continuous stratification case. The intrusion propagates along the interface between the two layers of ambient.

In this special case we can use directly the homogeneous-ambient results. We can solve the upper current by the classical models, in one or two-layer version. The lower current will be the mirror image. The dimensionless parameter is $H$, the ratio of the height (from bottom to top) $2H^*$ of the ambient to the lock, $2h_0$. The relevant reduced gravity is $g'$ given by (15.2). Since the transition is sharp, this indeed is the driving effect encountered by the upper (and lower) current.

There is some experimental and numerical evidence (Sutherland, Kyba, and Flynn 2004, Flynn, Boubarne, and Linden 2008) that the slumping distance of the intrusion is longer than that of the corresponding half-current. The interpretation of this observation is still unclear.

The estimated limit of validity of the inviscid approximation, $x_V$, needs reconsideration. The thickness of the fluid bulk is larger by factor 2, and the propagation is not over a solid boundary. Consequently, a rough rescaling indicates that the ratio of inertial to viscous forces increases by factor 4 as compared with the case of the upper-half gravity current on a solid bottom. In other words, the *crude estimate* is

$$\frac{F_I}{F_V} \approx 4 \frac{(g'h_0)^{1/2}}{\nu} \frac{h_0}{x_0} \theta(t), \tag{15.4}$$

see (2.83), where $\theta(t)$ is calculated for the upper current. In particular, we can use the approximations for $\theta(t)$ which have been developed for a simple current.

## 15.3 Linear transition layer

### 15.3.1 Formulations

The mathematical analysis is performed for the configuration sketched in Fig. 15.1. We follow the presentation of Ungarish (2005b). We use a $\{x, y, z\}$ Cartesian coordinate system with corresponding $\{u, v, w\}$ velocity components. The gravity acts in the $-z$ direction. We assume that the sidewalls of the container are vertical $xz$ planes and the gap between them is large (as compared with the thickness of the intrusion, $h_0$) and hence the inviscid flow does not depend on the coordinate $y$, and $v \equiv 0$. Symmetry of the initial configuration with respect to the horizontal plane $z = 0$ is assumed, as follows. The ambient fluid is in the domain $-H \leq z \leq H$, and is stably stratified; the density of the ambient at $z = 0$ is the mean value over the depth, and the density increase occurs (linearly) either in a part-depth layer $-l \leq z \leq l$, or over the full depth (i.e., $l = H$), from $\rho_o$ to $\rho_b$. The initial position of the intrusive current is in the lock $0 \leq x \leq x_0, -h_0 \leq z \leq h_0$ (assumed first rectangular). The density of the intruding current is equal to that of the ambient at the symmetry plane $z = 0$. Again, the fluid of the intrusion can be regarded as the result of mixing the ambient in the lock, and is also referred to as the "mixed fluid."

It is convenient to use $\rho_o$ as the reference density. We repeat

$$\rho_c = \rho_a(z = 0) = \frac{1}{2}(\rho_o + \rho_b); \tag{15.5}$$

$$\epsilon = \frac{\rho_c - \rho_o}{\rho_o} = \frac{1}{2}\left(\frac{\rho_b}{\rho_o} - 1\right); \quad \text{and} \quad g' = \epsilon g. \tag{15.6}$$

The density in the intruding current and in the ambient can be expressed as

$$\rho_c = \rho_o(1 + \epsilon), \quad \rho_a = \rho_o\left[1 + \epsilon\,\sigma(z)\right], \tag{15.7}$$

where $\sigma(z)$ specifies the form (profile) of the stratification. In general, this is a continuous and typically decreasing (allowing for some piecewise constant regions) function of $z$. We consider the *linear stratification* in a layer $\pm l$ about the midplane,

$$\sigma(z) = \begin{cases} 0 & (z > l) \\ 1 - \frac{z}{l} & (-l \leq z \leq l) \\ 2 & (z < -l) \end{cases} \tag{15.8}$$

where $0 < l \leq H$. Extensions of the analysis to more general forms of $\sigma(z)$ are possible; see Ungarish (2005a). However, in this chapter we shall use the linear density transition. It provides the leading order insights and allows comparisons with available experiments within the simplest mathematical model. The limiting case $l \to 0$ recovers the two-layer sharp stratification ambient configuration considered in § 15.2; this takes us back to the unstratified gravity current. The limiting case $l = H$ represents the important full-depth stratification configuration; this takes us back to the gravity current considered in Chapter 12 with $S = 1$ stratification.

The buoyancy frequency, defined by (12.2), in the linear density transition layer is constant and given by

$$\mathcal{N} = \left(\frac{g'}{l}\right)^{1/2} \quad (|z| \leq l). \tag{15.9}$$

We recall that we consider a Boussinesq system, where the difference between "light" and "heavy" fluids enters only in the context of the reduced gravity forcing. In our case, this means that the current (half intrusion) above $z = 0$ and the counterpart below $z = 0$ will be subjected to the same pressure driving (buoyancy force) in the $x$ direction as long as the shapes are symmetric about $z = 0$. Since the initial conditions satisfy this symmetry, the resulting time-dependent flow field will also be symmetric about $z = 0$. It is therefore sufficient to consider the flow in the domain $z \geq 0$. (The shape of the density interface and the velocity field in the domain $z \leq 0$ are expected to be mirror images.)

For the upper gravity current we use a one-layer approximation. In the ambient fluid domain $u = v = w = 0$ and hence the fluid is in purely hydrostatic balance and maintains the initial density $\rho_a(z)$ given by (15.7) and (15.8). The motion takes place in the intruding layer of fluid only, $0 \leq x \leq x_N(t)$ and $0 \leq z \leq h(x, t)$, where the density is $\rho_c$.

A relationship between the pressure fields and the height $h(x, t)$ can be obtained. In the motionless ambient fluid, the pressure does not depend on $x$, and the hydrostatic balances $\partial p_i/\partial z = -\rho_i g$, where $i = a$ or $c$, and use of (15.7) yield

$$p_a(z,t) = -\rho_o \left[z + \epsilon \int_0^z \sigma(z')dz'\right] g + C, \tag{15.10}$$

$$p_c(x, z, t) = -\rho_o(1 + \epsilon)gz + f(x, t). \tag{15.11}$$

Pressure continuity between the ambient and the intrusion on the interface $z = h(x, t)$ determines the function $f(x, t)$ of (15.11). We obtain

$$p_c(x, z, t) = -\rho_o(1 + \epsilon)gz + \rho_o g' \left[h(x, t) - \int_0^{h(x,t)} \sigma(z)dz\right] + C. \tag{15.12}$$

Consequently, with the aid of Leibniz's rule, we obtain

$$\frac{\partial p_c}{\partial x} = \rho_o g' \left[1 - \sigma(h)\right] \frac{\partial h}{\partial x}. \tag{15.13}$$

This is the fundamental pressure driving equation for the intrusion. The buoyancy driving is related to the shape of the interface, $h(x, t)$. The dependency is on the inclination of the interface, $\partial h/\partial x$, like in the non-stratified case. However, we see that the buoyancy is significantly modified by the stratification via the term $[1 - \sigma(h)]$. This reproduces the fact that the effective buoyancy (reduced gravity) depends on the local density difference between the current and the ambient. To be specific, using (15.8) and (15.9), we obtain

$$g'[1 - \sigma(h)] = \begin{cases} g' & (h \geq l) \\ g'\frac{h}{l} = \mathcal{N}^2 h & (0 \leq h \leq l). \end{cases} \tag{15.14}$$

We keep in mind that at $z = 0$ the difference between $\rho_c$ and $\rho_a$ is zero, and then increases over the layer $l$ to the maximum $\epsilon\rho_a$. Therefore the effective driving is proportional to $h$ in the density transition layer. We conclude that in general the driving force on the intrusion is quite "weak" (that is, compared roughly with the "intuition" of gravity currents in a system with similar density differences). This weakness is more pronounced when the current (intrusion) is thin as compared to $l$; this is expressed by the fact that the effective buoyancy is of the order of $\mathcal{N}^2 h$.

The driving $\partial p_c/\partial x$ is not a function of $z$, which facilitates the derivation of the SW equations.

We note that (15.10)-(15.13), although developed for $z > 0$, are also valid in the $z < 0$ domain. For this extension, it is convenient to define $h(x,t)$ as the thickness of the current measured from $z = 0$ (half-thickness of the intrusion), a positive variable. It turns out that pressure continuity at $z = -h(x,t)$ and $z = h(x,t)$ produces the same driving $\partial p_c/\partial x$. The verification is left for an exercise. This provides the more rigorous confirmation to the argument that the driving forces are symmetric with respect to $z = 0$ for symmetric intrusion configurations.

## 15.3.2  SW equations and boundary conditions

It is convenient to scale the dimensional variables (denoted here by asterisks) as follows

$$\{x^*, z^*, h^*, l^*, H^*, t^*, u^*, p^*\} = \{x_0 x, h_0 z, h_0 h, h_0 l, h_0 H, Tt, Uu, \rho_o U^2 p\}, \tag{15.15}$$

where

$$T = \frac{x_0}{U}, \tag{15.16}$$

$$U = \left[\frac{\rho_c - \rho_a(z=1)}{\rho_o} h_0 g\right]^{1/2} = (h_0 g')^{1/2}\frac{1}{\mathcal{A}} = \begin{cases} (h_0 g')^{1/2} & (l \leq 1) \\ \mathcal{N}h_0 & (l > 1) \end{cases} \tag{15.17}$$

and

$$\mathcal{A} = [1 - \sigma(1)]^{-1/2} = \begin{cases} 1 & (l \leq 1) \\ \sqrt{l} & (l > 1). \end{cases} \tag{15.18}$$

To avoid possible confusion, we specify that $\rho_a$ and $\sigma$ in (15.16)-(15.18) are calculated at the dimensionless $z = 1$.

Here $x_0$ and $h_0$ are the initial length and half-thickness of the intrusion as sketched in Fig. 15.1, $T$ is a typical time period for longitudinal propagation over a typical distance $x_0$, and $U$ is the typical inertial velocity of propagation of the nose.

The definition of the reference velocity $U$ for the intrusion problem involves the insight concerning the dependency of the effective buoyancy on the thickness; see (15.14). The coefficient $\mathcal{A}$ accounts for the fact that the effective driving force is provided by the density difference over the axial dimension of the intrusion. The use of $\mathcal{A}$ allows for a unified formulation to the problems of thin ($l < 1$) and thick ($l > 1$) density transition layers (as compared with the thickness of the intrusion). This is relevant to the practical configurations and experimental results. The details will become evident during the discussion of the solutions of these cases.

We emphasize that hereafter the variables $x, z, u, t, h, H, l, p$ are in dimensionless form unless stated otherwise. The geometry under consideration imposes $H \geq 1$.

The continuity equation is as in the standard 2D SW formulation. The $z$-average of the horizontal momentum equation is obtained as in § 2.2. The difference is that now the pressure-driving term is given by (15.13). (Again, viscous terms are neglected under the assumption that the typical Reynolds number, say $Re = U h_0 / \nu$, is very large.) We obtain a system of equations for $h(x, t)$ and for the averaged longitudinal velocity $u(x, t)$.

In characteristic form the simplified continuity and momentum equations are

$$\begin{bmatrix} h_t \\ u_t \end{bmatrix} + \begin{bmatrix} u & h \\ \mathcal{A}^2[1 - \sigma(h)] & u \end{bmatrix} \begin{bmatrix} h_x \\ u_x \end{bmatrix} = \begin{bmatrix} 0 \\ 0 \end{bmatrix}. \tag{15.19}$$

For the stratification $\sigma$ used here the eigenvalues of the matrix of coefficients are real and the set of eigenvectors is full, and hence the system (15.19) is hyperbolic. The details are left for an exercise. We find that the characteristics are

$$\frac{dx}{dt} = c_\pm = u \pm \mathcal{A}\left[h(1 - \sigma(h))\right]^{1/2}, \tag{15.20}$$

and on which the relationships between the variables are

$$\mathcal{A}\left[\frac{1 - \sigma(h)}{h}\right]^{1/2} dh \pm du = 0. \tag{15.21}$$

The initial conditions for the intrusion at $t = 0$ are: zero velocity, unit length, and prescribed $h(x)$ (the standard rectangular lock, or the horizontal-cylinder lock considered in § 15.5). The boundary conditions at the backwall $x = 0$ is $u = 0$. The additional condition for the velocity $u$ at the nose $x = x_N(t)$ is considered next.

We argued that the upper half of the intrusion can be treated as a gravity current. We can therefore use again the balance between the pressure head and dynamic reaction given by (12.27) (dimensional). In this equation we use the pressure difference provided by the present pressure results (15.10) and (15.12) at $x = x_N$ where $h = h_N$,

$$p_c(z = 0) - p_a(z = 0) = \rho_0 g' \left[ h_N - \int_0^{h_N} \sigma(z) dz \right] \quad \text{(dimensional)}. \quad (15.22)$$

We obtain an expression for the dimensional $u_N$.

After scaling that result according to (15.17), the dimensionless nose velocity can be expressed as

$$u_N = Fr(h_N) h_N^{1/2} \times [1 - \Lambda_N]^{1/2} \mathcal{A}, \quad (15.23)$$

where

$$\Lambda_N = \frac{1}{h_N} \int_0^{h_N} \sigma(z) dz. \quad (15.24)$$

The term in the square brackets of (15.23) is typically smaller than 1, and expresses the explicit slow-down of the head due to the stratification effects.

To close the formulation we shall again use the HS formula

$$Fr = \begin{cases} 0.5 H^{1/3} h_N^{-1/3} & (0.075 \le h_N/H \le 1) \\ 1.19 & (0 < h_N/H < 0.075) \quad \text{(deep intrusion)}. \end{cases} \quad (15.25)$$

It is convenient to use the term "deep" to refer to the case where the ratio $h_N/H$ is so small that $Fr$ is practically a constant. For the correlation (15.25) this ratio is simply $0.075 = 1/13.3$. An intrusion may be deep from the start, or become so eventually as a result of its spreading during the propagation.

An inspection of the closed formulation shows that the form of the stratification, i.e., $\sigma(z)$, enters via three effects: (1) the scaling coefficient $\mathcal{A}$; (2) the pressure gradient in the momentum equation (the coefficient of $h_x$ in the second equation of (15.19)); and (3) the nose velocity driving force (the square brackets term in (15.23)). The second effect also projects on the propagation of the characteristics. For very small $l$ a sharp density transition layer appears, and the case of interfacial intrusion into a two-layer ambient becomes relevant. This $l = 0$ limit is discussed in § 15.2.

In general, the SW system must be solved numerically. The finite-difference methods used in the previous sections can be used. Analytical solutions can be obtained for the initial propagation and for the self-similar developed motion.

To sharpen the insights it is useful to analyze separately several configurations.

## 15.4   Rectangular lock configurations

For release from a rectangular lock (i.e., straight vertical gate) a typical dam-break motion is expected; see § 12.4. The SW prediction is that in the initial slumping stage the intrusion propagates with constant velocity of propagation, $u_N$, and head height, $h_N$. This value of $u_N$ is attained instantaneously in the inviscid model. The slumping distance of the intrusion is one to several lock lengths. This prediction of the SW formulation is supported by experiments and Navier-Stokes simulations.

The analytical solution is obtained by the method of characteristics. We use equations (15.20)-(15.21). Upon the removal of the vertical gate, a rarefaction wave of speed $c_- = -\mathcal{A}[1 - \sigma(1)]^{1/2}$ starts to propagate from the gate $x = 1$ to the backwall $x = 0$, and the front (nose) starts to propagate forward as a discontinuity.

A shrinking domain of stationary fluid of height 1 exists between the wall and the backward-moving rarefaction wave, and a rectangular domain of fluid of constant height $h_N$ and velocity $u_N$ forms behind the nose. A typical characteristic moving forwardly with velocity $c_+$ from the former to the latter domain carries the information $u = 0$ and $h = 1$. Integration of (15.21) along this characteristic gives $u_N$ as a function of $h_N$. On the other hand, $u_N$ must satisfy the nose condition (15.23). The intersection between these relationships yields the slumping-stage results

$$u_N = \Upsilon(1) - \Upsilon(h_N) = Fr(h_N)\, h_N^{1/2} \times [1 - \Lambda_N]^{1/2}\, \mathcal{A}, \qquad (15.26)$$

where

$$\Upsilon(h) = \mathcal{A} \int_0^h \left[\frac{1 - \sigma(h')}{h'}\right]^{1/2} dh'. \qquad (15.27)$$

The needed $\sigma(z)$ and $\Lambda_N$ are given by (15.8) and (15.24). We first solve the non-linear equation (15.27) for $h_N$, then calculate the value of $u_N$. But to this end we need the integrals (15.24) and (15.27). In general, these integrals can be calculated numerically; this allows the use of various stratifications, including these provided by measured data. However, for linear stratifications these integrals can be evaluated analytically. This provides some more explicit and insightful results for typical stratifications.

### 15.4.1   Part-depth transition layer and full-depth lock

Here we consider the cases with fixed $H = 1$ and various $l$. Evidently, $0 < l \leq 1$. This configuration renders a quick glimpse into the influence of the thickness of the transition layer on the speed of propagation after release. The intruding (mixed) fluid occupies initially the full depth of the tank, and the ambient is composed of a lower layer of "heavy" fluid, an upper layer of "light" fluid, and a linear transition layer (see Fig. 15.1).

In this case $\mathcal{A} = 1$, i.e., the reference velocity is $U = (g'h_0)^{1/2}$; see (15.17)-(15.18).

The present SW theory predicts propagation with constant velocity during the initial slumping phase. The theoretical value of the slumping velocity $u_N$ is provided by (15.26)-(15.27). Let us be more specific. Since $0 < l \leq H = 1$ we can calculate by (15.27)

$$\Upsilon(1) = \int_0^l \frac{1}{l^{1/2}} dh + \int_l^1 \frac{1}{h^{1/2}} dh = 2 - l^{1/2}. \tag{15.28}$$

To proceed, we must distinguish between the following possibilities:

1. Thin density transition layer, $l < h_N$. The calculation of $\Lambda_N$ and $\Upsilon(h_N)$, see (15.24) and (15.27), use of (15.28), and substitution into (15.26) yield

$$u_N = 2(1 - h_N^{1/2}) = Fr\, h_N^{1/2}(1 - \frac{1}{2}\frac{l}{h_N})^{1/2}. \tag{15.29}$$

   The largest $u_N$ is expected for $l = 0$, i.e., the two-layer sharply stratified ambient fluid.

2. Thick transition layer, $l > h_N$. Similar calculations give

$$u_N = 2 - l^{1/2} - \frac{h_N}{l^{1/2}} = \frac{Fr}{(2l)^{1/2}} h_N. \tag{15.30}$$

   Here the simplest case is the $l = 1$ limit for which the minimal $u_N$ is expected.

The function $u_N(l)$ is continuous at $l = h_N$ where switch from (15.29) to (15.30) occurs. The right-hand sides of (15.29)-(15.30) predict a notable reduction of velocity with the change from $l = 0$ to $l = 1$. Due to the increasing effect of stratification, the velocity of propagation becomes proportional to $h_N$ (instead of $h_N^{1/2}$) and is further reduced by $\sqrt{2}$ ($Fr$ changes little in this case).

Since $H = 1$, $Fr$ is a function of $h_N$ only. The only free variable in (15.29)-(15.30) is $l$. Consequently, the present SW theory predicts that the velocity of the intrusion, scaled with $U = (g'h_0)^{1/2}$, is a function of the thickness of the density transition layer, scaled with the initial thickness of the intrusion.

The laboratory investigation of Faust and Plate (1984) provides a convenient data set for comparisons with the theoretical results. The experiments were performed with saltwater (clearly in the Boussinesq range) in a tank of 300, 50, and 20 cm length, depth, and width. Intrusions with $Re \gg 1$ produced by various combinations of fluid depth ($h_0$), lock length ($x_0$), thickness of density transition layer ($l$), and total density variation ($\epsilon$) were tested. The major qualitative observation was that, in all experiments, the nose of the intrusion propagated with constant velocity for a relatively long distance (practically, over all the measured interval). The major quantitative results are the values of this velocity of propagation, $u_N^*$, and the fact that $u_N^*/(g'h_0)^{1/2}$ is a function of $l$ (scaled with the (half-) height of the lock, $h_0$) only.

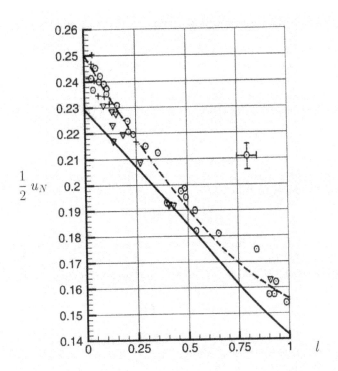

**FIGURE 15.2**: Comparison with Fig. 7 of Faust and Plate (1984): $\frac{1}{2}u_N$ (defined as $F_D$ in that paper) as a function of $l$. The solid line is the SW prediction, the points and dashed line are the experimental measurements and the corresponding curve fit from Faust and Plate (the cross bars show the experimental error). (Ungarish 2005b with permission.)

The SW qualitative predictions are that (a) the speed of propagation is constant for a significant slumping distance, and (b) that this speed, scaled with $(g'h_0)^{1/2}$, depends only on the dimensionless $l$. This is in full agreement with the experimental results. A quantitative comparison is displayed in Fig. 15.2. Both experiment and theory show a monotonic and quite strong decrease of velocity with $l$. The SW results are equal to or slightly below the experimental values (typically by about 5%, the experimental error is about 4%). This gives credence to the SW approach. However, we keep in mind that the one-layer model is expected to be problematic for the full-depth lock. As in the non-stratified case, the speed is well predicted, but the slumping length is significantly underestimated by this model.

## 15.4.2   Fully linearly-stratified tank, part-depth locks

Here we consider the cases with $l = H$ and various $H$. In the scalings (15.16)-(15.18) the coefficient $\mathcal{A} = \sqrt{H}$, and hence the reference velocity and time, can be expressed as

$$U = \left(\frac{g'h_0}{H}\right)^{1/2} = \mathcal{N}h_0; \quad T = \mathcal{N}^{-1}\frac{x_0}{h_0}. \tag{15.31}$$

The dominant waves in the stratified ambient, symmetric about $z = 0$, are a symmetric extension of the leading wave introduced in § 12.1 and their (dimensionless) speed is

$$u_W = \frac{1}{\pi}H. \tag{15.32}$$

(For more details on these linear waves see Baines 1995, Amen and Maxworthy 1980, and Rooij 1999.)

Substitution of $l = H$ and $\mathcal{A} = \sqrt{H}$ renders the equations of motion (15.19) as

$$h_t + (uh)_x = 0; \tag{15.33}$$

$$u_t + \frac{1}{2}(h^2)_x + \frac{1}{2}(u^2)_x = 0; \tag{15.34}$$

the nose velocity condition (15.23) as

$$\frac{dx_N}{dt} = u_N = Fr\frac{1}{\sqrt{2}}h_N, \tag{15.35}$$

and the characteristic balances (15.20)-(15.21) as

$$dh \pm du = 0 \quad \text{on} \quad \frac{dx}{dt} = c_\pm = u \pm h. \tag{15.36}$$

Note: The intrusion (upper half) discussed here is identical with a gravity current of the type discussed in Chapter 12 for $S = 1$. Here the scaling speed $U$ is smaller by factor $\mathcal{A} = \sqrt{H}$. With this rescaling, (12.23)-(12.26) and (12.29) reduce to the present formulation.

The first and very significant phase is, again, propagation with the constant velocity provided by the slumping result. Substitution of $l = H$ into (15.8) and subsequent use of (15.26) yield the slumping velocity

$$u_N = (1 - h_N) = \frac{Fr}{\sqrt{2}}h_N. \tag{15.37}$$

The influence of $H$ enters via $Fr$, see (15.25), which increases with $H$, but not dramatically. Formally, we can rewrite (15.37) as

$$h_N = \frac{\sqrt{2}}{\sqrt{2} + Fr};$$ (15.38a)

$$u_N = \frac{Fr}{\sqrt{2} + Fr};$$ (15.38b)

which can be compared with the one-layer model results for the non-stratified current (2.66). In both cases $u_N$ is of typical magnitude 0.5 and increases with $Fr$ (i.e., with $H$). However, there is a significant difference in the scaling $U$: $(g'h_0)^{1/2}$ for the non-stratified case and $(g'h_0/H)^{1/2} = \mathcal{N}h_0$ in the stratified counterpart. In both cases $g'$ is based on the largest density difference between the current and the ambient. We can summarize that, roughly, under similar circumstances the speed of the intrusion is smaller by a factor $\sqrt{2H}$.

The more precise behavior of $u_N$ as a function of $H$ is shown in Fig. 15.3. The variation is quite small, from 0.28 to 0.46, over the entire range of $H$. This confirms the choice of the scaling quantities. But we must keep in mind that the reference speed $\mathcal{N}h_0$ may decrease when the depth of the ambient increases. This figure also displays the velocity ratio of the intrusion to the dominant wave; see (15.32). This ratio is smaller than 1 for $H = 1$ and decreases as $H$ increases; in other words, the nose velocity is always subcritical.

The SW speed predictions are consistent with the available experiments. A comparison is shown in Fig. 15.4. The big scatter of data and lack of points for deep intrusions suggest that new experiments are needed.

### 15.4.2.1 Comparisons of SW, NS, and experiments

We discuss in some detail results for $H = 2.27$.

The experimental data were taken from Amen and Maxworthy (1980), Run 117 (saltwater, $h_0 = 6.16\,\text{cm}$, $\mathcal{N} = 0.57\,\text{s}^{-1}$). The aspect ratio of the lock is $h_0/x_0 = 0.33$. The NS simulation, see Appendix B, for this configuration used a $260 \times 180$ mesh, the numerical tank was of length $x_w = 5.5$, $\epsilon = 4.64 \cdot 10^{-3}$, and $Re = 2.5 \cdot 10^4$. Distance of propagation as a function of time is displayed in Fig. 15.5. The experimental and SW results are in good agreement. At $t \approx 6$ a strong deceleration of the intrusion is observed in both experiment and NS simulation. This is attributed to the wave-head interaction, discussed later, an effect not resolved by the SW formulation. Indeed, the SW predictions are in very good agreement with the NS simulations and experiments until the time when interaction begins.

Consider the density contours of the NS solution shown in Fig. 15.6, which reveals the complex wavy features of the flow field. The main reason for the pronounced oscillations is the small value of $h_0/x_0 = 0.33$. We estimated that the typical wavelength is $2\pi u_N(h_0/x_0)$ (scaled with $x_0$). Computations

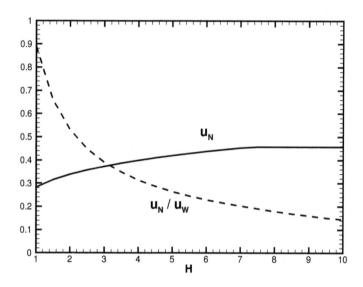

**FIGURE 15.3:** Slumping stage $u_N$ and $u_N/u_W$ as functions of $H$ in intrusion from rectangular lock release. $u_N$ scaled with $\mathcal{N}h_0$. (Ungarish 2005b with permission.)

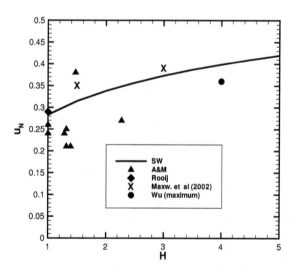

**FIGURE 15.4:** Slumping $u_N$ as a function of $H$ in rectangular lock release: SW prediction, and experimental results of Amen and Maxworthy (1980), Rooij (1999), Maxworthy et al. (2002), and Wu (1969). Wu's result is for a cylindrical lock; the theory predicts that the maximum speed is the same as for the rectangular lock, but no slumping phase exists; see § 15.5. (Ungarish 2005b with permission.)

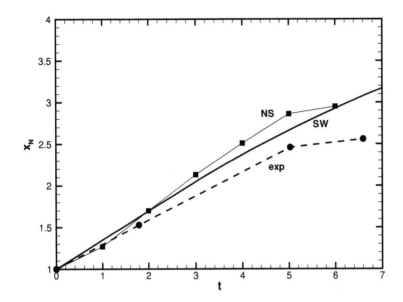

**FIGURE 15.5**: Distance of propagation as a function of time for experiment Run 117 of Amen and Maxworthy (1980) ($H = 2.27$ and $h_0/x_0 = 0.33$). NS, SW, and experimental results are shown. (Ungarish 2005b with permission.)

for a configuration similar to that of Fig. 15.6, but with $h_0/x_0 = 1$ (not displayed here), confirmed that a similar propagation of the intrusion occurs, but with considerably smoother isopycnals (longer wavelengths). The velocity of propagation of the leading waves is $u_W = 0.72$; see (15.32). However, these waves introduce only weak disturbances in front of the intrusion and the length of the tank in the computation was sufficiently large to avoid reflections during the time interval considered.

The velocity field is illustrated by contour lines of $u$ in Fig. 15.7. The return flow in the ambient is quite complex and concentrated in layers (or patches) close to the intrusion. At $t = 6$ the nose of the intrusion is at $x \approx 3$, and we observe that the velocity field in this region is quite weak and smeared. This reproduces the effect of the interaction between the waves and the head. Such interactions occur in sub-critical gravity currents in a linearly-stratified ambient as discussed in § 12.5. The intrusion is a special member of this group: it is always sub-critical, and, for a given $H$, displays the largest ratio of $u_W/u_N$. Consequently, the intrusion is more susceptible to this interaction. Some details are considered below.

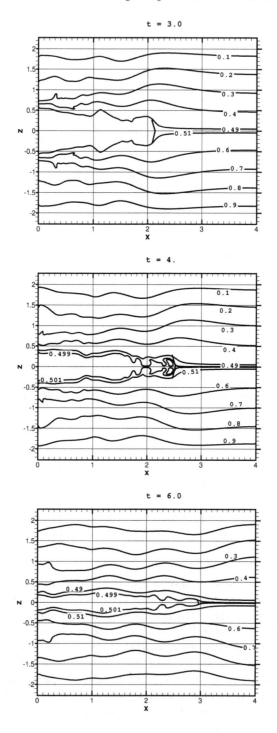

**FIGURE 15.6**: NS density contours at various times, for configuration with $H = 2.27, h_0/x_0 = 0.33$ (simulation of Run 117 of Amen and Maxworthy 1980). The nose is marked by $>$. (Ungarish 2005b with permission.)

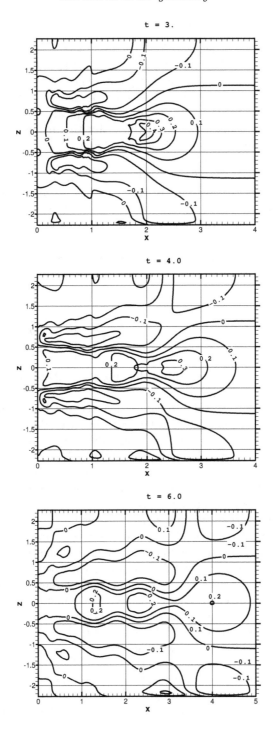

**FIGURE 15.7**: NS $u$ contours at various times; details in previous caption.

### 15.4.2.2 Wave-nose interaction

The experiments of Amen and Maxworthy (1980) are for $1 \leq H \leq 2.5$, $0.29 \leq h_0/x_0 \leq 1.05$, and typical Reynolds number $10^4$. Amen and Maxworthy (1980) recorded the "upper tangent point" where a sharp deceleration of the experimental intrusion starts. This point is denoted here $(t_2, x_2)$. Ungarish (2005b) showed that up to this point the typical propagation was with constant speed, and suggested that at this point the first wave-nose interaction occurs (Figs. 5(c) and (e) of Amen and Maxworthy 1980 indeed display a typical wavy kink at this position).

The theoretical estimate of the position $x_2$ is like in § 12.5. Briefly, the intrusion propagates the distance required for the formation of two waves between the position of the gate $x = 1$, and the front $x = x_N$, plus the additional propagation of the wave crest with relative velocity $u_W - u_N$ over the $(1/4)$ wavelength (to the position where maximum interaction is attained). The result is given by the first line of (12.60). Recall that we estimated $\lambda = 2\pi u_N (h_0/x_0)$. We obtain

$$x_2 = 1 + \lambda \left( 2 + \frac{0.25}{u_W - u_N} u_N \right) = 1 + 2\pi \frac{h_0}{x_0} u_N \left[ 2 + \frac{0.25}{\dfrac{H}{\pi u_N} - 1} \right]. \quad (15.39)$$

(The scaling $U$ now is different from that used in (12.60) therefore the last term in (15.39) looks different.)

The intrusion is always sub-critical, and hence the last term is positive. For a fixed $H$ the resulting $x_2$ is a linear function of the lock aspect ratio $h_0/x_0$, and the slope of the line decreases slightly when $H$ increases. Since the return flow in the ambient will reduce the absolute velocity of the wave, (15.39) underestimates $x_2$. Fig. 15.8 shows the theoretical estimate and experimental values of $x_2 - 1$. The agreement is not always good. We must keep in mind that the start of wave-nose interaction is not a clear-cut experimental feature, and that weak interactions may remain unrecorded.

When $H$ is close to 1 the difference between $u_W$ and $u_N$ is small, see Fig. 15.3, and therefore the interaction is delayed; this is consistent with the observations. Moreover, in this case the return flow in the ambient is significant, and therefore the real $x_2$ is significantly larger than our one-layer model estimate, as expected. These consideration explain why the interval of constant velocity is so large for the experiments for the $H = 1$ configuration. On the other hand, according to (15.39) for large $H$ the interaction is expected practically as soon as the second wave appears.

The effect of the interaction between the wave and the head can be observed in the NS simulation of the experiments of Amen and Maxworthy, see Figs. 15.6-15.7, for the case with $H = 2.7$, $h_0/x_0 = 0.33$. We see that at $t = 6$ the main body of the intrusion becomes separated from the leading blob. The SW approximation, developed for a density-driven flow, is not expected to

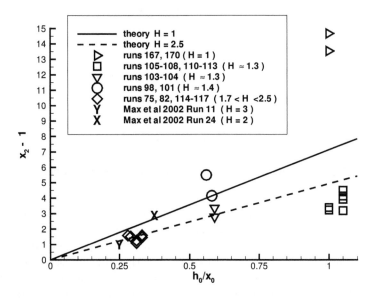

**FIGURE 15.8:** Distance of propagation where sharp deceleration begins. The lines are the theoretical estimate (15.39) of first interaction between waves and nose. The symbols show experiments of Amen and Maxworthy (1980); the $X$ and $Y$ show experiments of Maxworthy et al. (2002) with bottom gravity currents. (Ungarish 2005b with permission.)

be valid in these circumstances of transition to a wave-dominated flow. It is plausible that subsequently the leading portion of the intrusion will behave like the "isolated propagating flow" phenomena investigated for descending thermals by Manasseh et al. (1998) (where other important references are given). This transition and the evolvement into the wave-dominated pattern require more investigation.

## 15.5 Cylindrical lock in a fully linearly-stratified tank

Consider the propagation of an intrusion which is released from a horizontal-cylinder lock as sketched in Fig. 15.9. The geometry of the lock is defined, in dimensional form, by

$$h^* = [1 - (x^*/h_0)^2]^{1/2} \quad (0 \le x^* \le h_0). \tag{15.40}$$

This type of an initial rounded shape is expected to be a better approximation

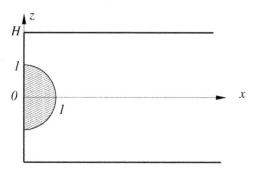

**FIGURE 15.9:**  Schematic description of the horizontal cylinder lock.

to intrusions generated in nature than the classical rectangular lock. The situation in the laboratory seems to be just the opposite; it is indeed much more difficult to assemble and operate a reservoir with cylindrical boundaries than one with straight horizontal and vertical boundaries. In any case, there are some interesting differences between the resulting flows which merit attention.

We consider the propagation of an intrusion into a fully linearly-stratified ambient ($l = H$). Wu (1969) carried out a detailed experimental study for this configuration with $H = 4$. Wu's findings were sometimes considered as prototypical of intrusion in a linearly-stratified ambient, but this is not the case. Actually, Wu's configuration is quite special in several respects: the aspect ratio of the lock $h_0/x_0 = 1$, which renders $T = \mathcal{N}^{-1}$; the shape of the lock which changes the slumping behavior and apparently diminishes the head-wave interaction effect; and the fairly large depth $H = 4$ which promotes transition to self-similar motion. The details will become evident below.

Here $h_0 = x_0$, $\mathcal{A} = \sqrt{H}$, and hence the reference velocity and time, see (15.16)-(15.18), can be expressed as

$$U = \left(\frac{g'h_0}{H}\right)^{1/2} = \mathcal{N}h_0; \quad T = \mathcal{N}^{-1}. \tag{15.41}$$

The speed of the waves in the ambient, the SW equations, and boundary conditions for $u$ are as for the rectangular case discussed in §15.4.2. The motions starts, again, from rest. Up to this point the formulation is identical with that of the counterpart rectangular configuration discussed in § 15.4.2. Thus, (15.32)-(15.36) with $h_0/x_0 = 1$ can be applied. The difference is in the initial condition of the interface, which is now

$$h(x) = (1 - x^2)^{1/2}, \quad (0 \le x \le 1, t = 0). \tag{15.42}$$

This difference introduces a sharp contrast with the counterpart rectangular configuration discussed in § 15.4.2 concerning the resulting propagation.

A simple analytical solution by the method of characteristics of the motion after release (like in the rectangular dam-break problem which led to (15.37)) is not feasible. However, some useful approximations for the initial motion can be made. Since $u$ and the (upward) displacement of the interface $\Delta h$ are initially very small, we expand these variables in powers of $t$. Substitution into (15.33)-(15.34) yields the leading terms

$$u = xt; \tag{15.43}$$

$$\Delta h = \frac{1}{2} \frac{2x^2 - 1}{(1 - x^2)^{1/2}} t^2. \tag{15.44}$$

The details are left for an exercise. Note the stagnation point of the interface at $x = h = 1/\sqrt{2}$ where $\Delta h = 0$. The results indicate that immediately after release the intruding cylindrical body of fluid shrinks at the rear and thickens at the front (in remarkable contrast with a rectangular dam-break release). An inspection of the neglected terms (and comparisons with the finite-difference solution, see below) indicates that the approximations (15.43)-(15.44) are valid for $t \ll 1$ and $x < 1$. The (quite weak) singularity of $\Delta h$ at $x = 1$ is relaxed by the development of a front of increasing height $h_N$ that propagates forwardly to $x_N > 1$. The details cannot be obtained from this approximation, but continuity with (15.43) and characteristic balances indicate that $h_N \approx u_N \approx t$.

The usual finite-difference code for the SW equations, see Appendix A.2, handles the initial conditions of the cylindrical lock with no special difficulties (the numerical starting condition must be some small positive $h_N$, say 0.01, but the sensitivity to this condition in the subsequent solution, say at $t > 0.05$, is very low).

For the configuration of Wu (1969), the SW profiles of the interface and velocity of the intrusion at various times are shown in Fig. 15.10. The behavior of $h$ and $u$ in the rectangular lock counterpart case is shown in Fig. 15.11. The SW predicted velocity of propagation $u_N$ for Wu's configuration and the rectangular counterpart are shown in Fig. 15.12.

The main conclusion is that the initial behavior in the cylindrical lock case is different from that of the rectangular lock counterpart. The reason for this effect is not the kinematic volume difference, but rather the internal dynamics and force distribution. In the hydrostatic state the mixed fluid in the cylindrical lock is subjected to a significant pressure gradient in the $x$ direction. Therefore, the removal of the lock induces accelerations in the whole body of fluid, and, from the start, a linear with $x$ velocity profile develops, and the interface rises quickly in the domain $x > 1/\sqrt{2}$ and descends in the domain $x < 1/\sqrt{2}$. The overall initial trend is to straighten the interface to a horizontal position $h \approx 0.7$ at $t \approx 1$, while the nose develops the height and velocity of almost the slumping rectangular counterpart. However, the moving nose and the trailing region at $x > 1$ cannot maintain a constant-speed motion (as in the rectangular counterpart) because the forward-propagating characteristics that enter this region are bound to carry time-dependent conditions.

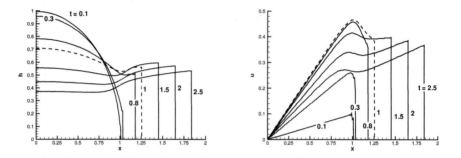

**FIGURE 15.10:** SW predictions for cylindrical lock, $H = 4$ (Wu's configuration): $h$ and $u$ profiles as a function of $x$ at various $t$. (Ungarish 2005b with permission.)

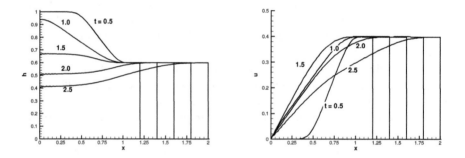

**FIGURE 15.11:** Rectangular lock counterpart configuration of Fig. 15.10: SW predictions of the $h$ and $u$ profiles as a function of $x$ at various $t$. (Ungarish 2005b with permission.)

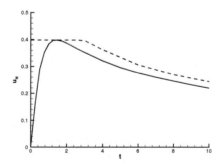

**FIGURE 15.12:** Speed of propagation SW results: cylindrical lock (solid line) and rectangular lock counterpart (dashed line). Here $H = 4$. (Ungarish 2005b with permission.)

The SW velocity of propagation in Wu's configuration, see Fig. 15.12, has a clear two-stage pattern, of acceleration followed by deceleration, with a maximum at $t \approx 2$. The rectangular counterpart displays slumping with constant $u_N$ and the deceleration starts later, at $t \approx 3.2$. The early transition to the deceleration phase renders the cylinder lock configuration more prone to the self-similarity behaviour described in § 15.6 than the rectangular lock counterpart.

It is remarkable that, according to the SW results, the maximum $u_N$ of the cylindrical lock release is equal to the constant slumping velocity of the rectangular lock counterpart. This result can be explained via the balances on the characteristics $c_+$ which propagate from the initial stationary fluid in the lock. The special characteristic from $x = 0, t = 0$ in the cylindrical lock carries to the front the same information as all the characteristics released at $t = 0$ in the rectangular lock. Therefore the agreement in $u_N$ is for a special point only.

We deduced that the SW propagation of an intrusion from a rectangular lock is always sub-critical; see Fig. 15.3. We now combine this result with the previous outcome concerning the maximum $u_N$ attained in release from a cylindrical lock. We reach the conclusion that the propagation from a cylindrical lock release is also always sub-critical. This prediction of the SW theory is consistent with the observations of Wu.

Finally, we compare the SW predictions of $x_N$ as a function of $t$ with the results of Wu (1969). The experimental data was obtained in a saline stratified tank with $H = 4$. The length, depth $(2H^*)$, and width of the ambient fluid in the tank were $90, 48$, and $9$ in, respectively; $\mathcal{N}$ was in the range $0.25 - 1.7\,\mathrm{s}^{-1}$. A quite complex arrangement of two circular cylindrical pieces, moving on circular tracks, was used for creating the horizontal-cylinder reservoir or lock (of $2h_0 = 12$ in). The ambient fluid was trapped in this lock, mixed, colored, and then released.

The curve-fitted data concerning the propagation of these intrusions as a function of time produced the often-quoted formula

$$\frac{x_N^*}{x_0} = \begin{cases} 1 + (0.29 \pm 0.04)\,(\mathcal{N}t^*)^{1.08\pm0.05} & (0 \le \mathcal{N}t^* \le 2.5) \quad \text{(I.C.S.)} \\ (1.03 \pm 0.05)\,(\mathcal{N}t^*)^{0.55\pm0.02} & (3 \le \mathcal{N}t^* \le 25) \quad \text{(P.C.S.)}, \end{cases}$$
$$(15.45)$$

where I.C.P. stands for "initial collapse stage" and P.C.S. stands for "principal collapse stage." (For later times, $\mathcal{N}t^* > 25$, Wu defined a "final collapse stage," which is dominated by viscous effects.)

The experimental outcome (15.45) has been considered as a prototype of the behaviour of an inviscid intrusion. However, there is actually no reason why this result should be applicable to a configuration with a non-cylindrical lock and $H \ne 4$. The deviation bounds of the coefficients indicate a scatter (or error) of about $\pm10\%$. The accuracy of the fit in the first stage is remarkably low, with deviations of up to 14%; this seems to be the result of the oversimplified fit rather than of experimental errors. The agreement between

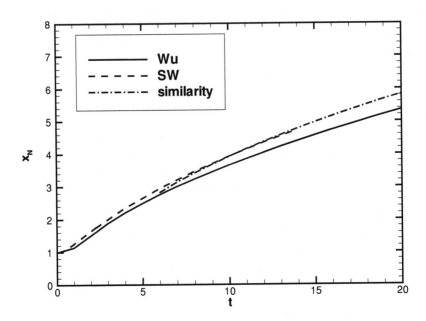

**FIGURE 15.13**:   Results for Wu's configuration (cylindrical lock, $H = 4$), of $x_N$ as a function of time: Wu's experimental curve-fit, SW solution with initial conditions, and SW similarity result for large $t$ with $\gamma = -1.7$. (Ungarish 2005b with permission.)

the SW solution and the experimental fit (15.45) is very good for a significant time of propagation; see Fig. 15.13.

## 15.6  Similarity solution

We consider a full-depth stratified ambient, $l = H$; see Fig. 15.1. We deal formally with the upper half of the intrusion, but we emphasize that the results are actually for a gravity current of the type discussed in Chapter 12 for $S = 1$; for $0 < S < 1$ no similarity solutions are available.

The formulation is like in § 15.4.2. In particular, the reference velocity and time are given by (15.31), i.e.,

$$U = \mathcal{N}h_0; \quad T = \frac{x_0}{U}, \tag{15.46}$$

and the governing equations are (15.33)-(15.35). Now we are interested in solutions for large values of $t$. When the intrusion is deep it is justified to assume that $Fr = $ const $(= 1.19$ in our case; see (15.25)).

We apply the transformation of § 2.3 to the set of equations (15.33)-(15.35). The details are left for an exercise. Let

$$y = x/x_N(t) \quad (0 \le y \le 1) \tag{15.47}$$

and redefine the variables as $h(y,t)$ and $u(y,t)$. We obtain the continuity and momentum equations in the following form

$$\frac{\partial h}{\partial t} + \frac{1}{x_N}(u - y\dot{x}_N)\frac{\partial h}{\partial y} + \frac{1}{x_N}h\frac{\partial u}{\partial y} = 0; \tag{15.48}$$

$$\frac{\partial u}{\partial t} + \frac{1}{x_N}(u - y\dot{x}_N)\frac{\partial u}{\partial y} + \frac{1}{x_N}h\frac{\partial h}{\partial y} = 0; \tag{15.49}$$

subject to the nose condition

$$\dot{x}_N = \frac{Fr}{\sqrt{2}}h(y = 1, t). \tag{15.50}$$

We seek a self-similar solution for the system (15.48)-(15.50) along the lines detailed in § 2.6. In particular, we start with the assumption

$$x_N(t) = Kt^\beta; \quad h(y,t) = \varphi(t)\mathcal{H}(y); \quad u(y,t) = \dot{x}_N\mathcal{U}(y); \tag{15.51}$$

where $K$ and $\beta$ are some positive constants. The obvious boundary conditions are $\mathcal{U}(1) = 1$ and conservation of volume $\mathcal{V}$. The other details are left for an exercise. The result is

$$x_N(t) = Kt^{1/2}; \quad h = (b^2 + y^2)^{1/2}\dot{x}_N(t); \quad u = \dot{x}_N(t)y; \tag{15.52}$$

where

$$b^2 = \frac{2}{Fr^2} - 1. \tag{15.53}$$

The validity of the solution can be verified by substitution into (15.48)-(15.50).

Conservation of the volume, $\mathcal{V}$, determines the value

$$K = (2\mathcal{V})^{1/2}\left[\int_0^1 (b^2 + y^2)^{1/2}dy\right]^{-1/2}. \tag{15.54}$$

In the present scaling the values of $\mathcal{V}$ are 1 for the rectangular lock and $\pi/4$ for the cylindrical lock; see § 15.5. This is the volume of the current or half-intrusion. For $Fr = 1.19$ we obtain $K = 1.537$ and $1.362$, respectively.

We note that for Benjamin's $Fr = \sqrt{2}$ we obtain $b = 0$ and hence the shape of the self-similar $h$ is linear with $y$ in this case.

We recall that a current in a homogeneous ambient propagates with $x_N \sim t^{2/3}$ and $h \sim (C^2 + y^2)\dot{x}_N^2(t))$, where $C$ is a constant; see (2.80). The change of the spread from $t^{2/3}$ to $t^{1/2}$ is also a manifest of the fact that the driving effect for the intrusion in a linearly-stratified ambient is significantly weaker than for the classical gravity current in a homogeneous ambient.

The resulting flow has some "virtual origin" because the physical $u = 0$ conditions and geometry of the lock at $t = 0$ cannot be imposed. In the results (15.52) the time coordinate $t$ can be replaced with the shifted $t + \gamma$ to achieve matching with the real initial slumping phase behavior. This parameter turns out to be negative and of order unity. Evidently, the applicability of the similarity solution to a real intrusion is only after the decay of the slumping stage, i.e., $t > 3$, at least. For practical applications the similarity solution is a quite weak predictive tool: it is restricted to deep intrusions at large times after release, and it is quite plausible that viscous, mixing, and wave effects become influential in these circumstances. Moreover, the value of $\gamma$ depends on the initial conditions and hence some additional assumptions, calculations, or experiments must be involved in the use of (15.52).

The propagation of a SW deep intrusion, $H = 7$, is illustrated in Fig. 15.14. The trends to self-similar profiles of $h$ and $u$ for $t > 5$ are seen.

An experimental test-case for the similarity result is provided by Wu (1969). In this study the intrusions were quite deep and the transition to self-similar behavior was less prone to wave-nose interactions than in experiments with rectangular locks. The expectation is that Wu's "principal collapse stage" overlaps with the theoretical self-similar stage. The comparison is displayed in Fig. 15.13. The details need some clarification. The similarity result for a cylinder lock is $x_N(t) = 1.36(t + \gamma)^{1/2}$. Matching with the finite-difference SW computation yields $\gamma = -1.7$. Wu's curve fit procedure ignored the "virtual origin" constant $\gamma$ and therefore obtained $x_N(t) = (1.03 \pm 0.05)t^{0.55 \pm 0.02}$; see (15.45). It still makes sense to compare the exponents, because, for $t > 10$

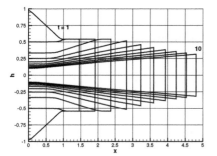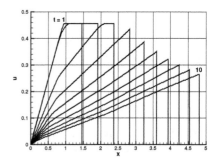

**FIGURE 15.14:** SW predictions for release from rest of a 2D rectangular deep intrusion ($H \geq 7$). The initial volume is a rectangle ($h = 1$). Profiles of $h$ and $u$ as functions of $x$ for $t = 1$ (1) 10.

(say) the slope of $x_N$ vs. $t$ on a log-log plot (as used by Wu) will be little influenced by $\gamma$. The experimentally-derived power of $t$ is by 7–14% larger than the theoretical $1/2$. A plausible explanation for this discrepancy is as follows: the $t^{1/2}$ propagation corresponds to a constant value of $Fr$, i.e., a very deep intrusion. The values of $h_N/H$ in the experiment changed from about 0.15 to 0.10, and therefore $Fr$ still increases during the propagation, see (15.25), and a more rapid than $t^{1/2}$ spread is expected. The similarity solution for a homogeneous ambient predicts spread with $t$ at the power of $2/3$, and Wu's power result is clearly much closer to the prediction $1/2$ obtained for the stratified ambient.

### 15.6.1 The validity of the inviscid approximation

The importance of the viscous forces during the self-similar propagation can be estimated analytically. We start with the ratio of inertial to viscous forces, using dimensionless variables

$$\frac{F_I}{F_V} = Re\frac{h_0}{x_0}\frac{\int_0^{x_N(t)} uu_xhdx}{\int_0^{x_N(t)} (u/h)dx} = Re\frac{h_0}{x_0}\theta(t), \tag{15.55}$$

where $Re = Uh_0/\nu$. On account of the self-similar solution (15.52) we write

$$\theta = \frac{1}{8}K^2Ct^{-2} = \frac{1}{8}K^6Cx_N^{-4}, \tag{15.56}$$

where

$$C = \left[\int_0^1 (b^2 + y^2)^{1/2}ydy\right]\left[\int_0^1 (b^2 + y^2)^{-1/2}ydy\right]^{-1}. \tag{15.57}$$

The value of $C$ is 0.86 for $Fr = 1.19$. The box-model estimate (12.62) for $S = 1$ is in fair agreement with the present more accurate result (the additional $H^{-1/2}$ term is due to the different scaling of $U$ in § 12.7).

We also recall that these estimates were performed for a gravity current on a solid bottom; the $F_I/F_V$ ratio for an intrusion is expected to be larger, by a factor of roughly four, but this is quite insignificant in view of the $x_N^{-4}$ changes.

The corresponding viscous motion is considered in Chapter 19.

## 15.7 Non-symmetric intrusions

Various asymmetric features may be introduced into the flow of the intrusion by the geometry of the lock, profile of stratification, and the density of the current. The rigorous analytical treatment of these cases is cumbersome and incomplete, and beyond the objectives of this book. There are, however, some fairly simple approximations that give a more practical glimpse into the underlying mechanisms. These are concerned with the initial propagation of a fixed volume of fluid released from a rectangular lock. The remarkable *observed* property of the flows discussed in this context is that they display a significant phase of propagation with constant speed. Under proper scaling, we can apply to this flow some of the "slumping" behavior results that were derived for simpler geometries.

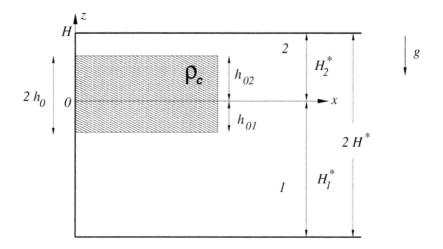

**FIGURE 15.15:**   Schematic description of the *non-symmetric* intrusion system. $\beta = h_{01}/(2h_0)$ and $\gamma = H_1^*/(2H^*)$. The $x$ axis is the interface between $\rho_1$ and $\rho_2$ in the sharp stratification case, and the neutral buoyancy line in the linearly-stratified case.

The system is sketched in Fig. 15.15. The thickness of the intruding current in the lock is $2h_0$ and the thickness of the ambient is $2H^*$, both dimensional. The line $z = 0$ of neutral buoyancy is at height $H_1^*$ above the bottom and $H_2^*$ below the top. Suppose that the stratification is sharp, with $\rho_1$ in the lower layer and a smaller $\rho_2$ in the upper layer. The intrusion is of an in-between density $\rho_c$.

The neutral buoyancy line (plane) is at $z = 0$, the $x$ axis. The sharp stratification interface coincides with this line. Now suppose that this line also divides the intrusion into two separated currents, denoted by the index $i = 1, 2$ for the lower and upper parts, respectively. The "effective" properties will be denoted with subscript $e$. The asymmetry of the given problem is expressed with the aid of three parameters, $\alpha, \beta$, and $\gamma$, each in the range

$(0, 1)$, defined by the following relationships:

$$\rho_c = \rho_2 + \alpha(\rho_1 - \rho_2); \tag{15.58}$$

$$h_{01} = \beta \cdot 2h_0; \quad h_{02} = (1 - \beta) \cdot 2h_0; \tag{15.59}$$

$$H_1^* = \gamma \cdot 2H^*; \quad H_2^* = (1 - \gamma) \cdot 2H^*. \tag{15.60}$$

We see that $\alpha, \beta, \gamma$ express the bias from the upper layer toward the lower layer. When the value of the parameter is $1/2$, the property is at the mean value. The fully symmetric intrusion is obtained when $\alpha = \beta = \gamma = 1/2$.

We use $\rho_2$ as the reference density, and define

$$\epsilon = \frac{1}{2} \frac{\rho_1 - \rho_2}{\rho_2}; \tag{15.61a}$$

$$\epsilon_1 = \frac{\rho_1 - \rho_c}{\rho_2} = (1 - \alpha) \, 2\epsilon; \quad \epsilon_2 = \frac{\rho_c - \rho_2}{\rho_2} = \alpha \, 2\epsilon; \tag{15.61b}$$

$$\epsilon_e = (\epsilon_1 h_{01} + \epsilon_2 h_{02})/(2h_0) = 2\epsilon \left[ (1 - \alpha)\beta + \alpha(1 - \beta) \right]. \tag{15.61c}$$

The corresponding reduced gravities follow as $\epsilon \, g$ and $\epsilon_j g$, where $j = 1, 2, e$. The height ratios of the ambient to current are given by

$$H_1 = \frac{H_1^*}{h_{01}} = \frac{\gamma}{\beta} \left( \frac{H^*}{h_0} \right); \quad H_2 = \frac{H_2^*}{h_{02}} = \frac{1 - \gamma}{1 - \beta} \left( \frac{H^*}{h_0} \right). \tag{15.62}$$

The formal reference speed is given by

$$U_0 = (\epsilon \, gh_0)^{1/2} \tag{15.63}$$

and the effective reference speed (dimensional) is given by

$$U_e = (\epsilon_e gh_0)^{1/2} = \sqrt{2} \left[ (1 - \alpha)\beta + \alpha(1 - \beta) \right]^{1/2} U_0. \tag{15.64}$$

Now we can start our calculations. Under the assumption that the currents are separated, we can calculate the slumping speeds $u_{Ni}$ and heights $h_{Ni}$, $(i = 1, 2)$, in dimensionless form. The regular SW theory provides the necessary results as functions of dimensionless height ratios $H_i = H_i^*/h_{0i}$; see (15.62). For simplicity, we can use the one-layer results (2.66). The dimensional speeds $u_{Ni}^*$ and nose heights $h_{Ni}^*$ are then obtained by multiplication with $(\epsilon_i gh_{0i})^{1/2}$ and $h_{0i}$, respectively.

We argue that the intrusion propagates along the guiding line $z = 0$ as a combination of the above-mentioned two currents. In general, $u_{N1}^*$ differs from $u_{N2}^*$. This mismatch can be adjusted by flow at the nose as follows. Suppose that the moving nose of the upper current is faster than that of the lower one. Then the nose of the upper current will attain direct contact with the lower layer of ambient, and slightly sink; this will reduce the speed of propagation and allow the slower lower current to catch up and merge. In other words, fluid of density $\rho_c$ is added from the faster current to the slower current across the $z = 0$ line. The appearance is of a single merged body of density $\rho_c$ which propagates with speed $U_c^*$. This speed is thus given by the averaged volume fluxes of the upper and lower current, i.e.,

$$U_c^* = \frac{u_{N1}^* h_{N1}^* + u_{N2}^* h_{N2}^*}{h_{N1}^* + h_{N2}^*}. \tag{15.65}$$

The deflection of the faster current towards the slower one also creates a slight distortion of the density division line in the ambient. Volume continuity in the container over a horizontal line ($z = 0$ say) indicates that the density interface is displaced in the opposite direction from the $z$-transfer in the current. In other words, the interface moves into the domain of the faster current. This displacement reduces the pressure head and the speed of the faster current and contributes further to the tendency of the flow field to diminish the speed discrepancy between the lower and upper current.

There is evidence that this simplified prediction works well in two cases: (1) equilibrium intrusions (defined below); and (2) full depth lock release, $H^* = h_0$, i.e., $H_1 = H_2 = 1$.

### 15.7.1 Equilibrium intrusion

The intrusion is in "equilibrium" when the weight of the fluid in the lock is balanced by the buoyancy force, i.e., when the sum of $g(\rho_i - \rho_c)h_{0i}$ vanishes. This is of course the case when we trap the initial ambient fluid and mix it in the lock container. The condition can be expressed as

$$\epsilon_1 h_{01} = \epsilon_2 h_{02}, \tag{15.66}$$

which requires

$$\alpha = \beta. \tag{15.67}$$

Note that in this case the horizontal driving forces, and the typical speeds $(\epsilon_i g h_{0i})^{1/2}$, are equal for the upper and lower currents. We see that actually a great deal of "symmetry" exists in the equilibrium intrusion, at least at the initial stage. The remaining asymmetry is due to the fact that the upper and lower currents may encounter different heights of ambient, $H_1 \neq H_2$ (unless $\gamma = \beta$).

The speed estimate (15.65) can now be expressed in a more explicit form. It is left for an exercise to show that, when $\alpha = \beta$,

$$\frac{U_c^*}{U_0} = 2\left[\alpha(1-\alpha)\right]^{1/2}\frac{u_{N1}h_{N1}\alpha + u_{N2}h_{N2}(1-\alpha)}{h_{N1}\alpha + h_{N2}(1-\alpha)} \tag{15.68}$$

and

$$\frac{U_c^*}{U_e} = \frac{u_{N1}h_{N1}\alpha + u_{N2}h_{N2}(1-\alpha)}{h_{N1}\alpha + h_{N2}(1-\alpha)}. \tag{15.69}$$

We keep in mind that $u_{Ni}$ and $h_{Ni}$ vary in quite small ranges, the former about 0.7 and the latter about 0.5. Consequently, the expression (15.69) turns out to be fairly constant and close to unity for a wide range of $\alpha = \beta$ and $\gamma$. This is a good indication that a proper scaling of the variables and reduction of the flow process were used.

This is illustrated in Fig. 15.16 for configurations in which the ratio of the ambient to lock heights is $H^*/h_0 = 2$. Let the value of $\alpha \, (= \beta)$ vary. We consider two cases of asymmetry of the ambient stratification, expressed by $\gamma$. In the first case (solid line) $\gamma = 3/4$ (i.e., $H_1^*/H_2^* = 3$). For $\alpha = 1/2$ the upper lock is of the thickness of the upper ambient layer, $h_{01} = H_1^*$. Hence only $\alpha > 0.5$ is relevant. As $\alpha$ increases the density of the intrusion increases. Since $\beta = \alpha$, the lock must be submerged more and more into the lower layer of ambient to attain equilibrium. In the second case (dashed line) $\gamma = 1/2$. Again, as $\alpha$ increases the lock must be submerged more and more into the lower layer of ambient. The resulting speed reflects this trade-off of buoyancy and geometry. Overall, the scaled speed changes by less than 50% in the displayed graphs, in spite of the fact that very significant changes of $\epsilon_i$, $h_{0i}$, and $H_i$ occur when $\alpha$ varies.

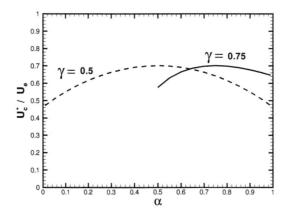

**FIGURE 15.16:** Scaled speed of propagation, $U_c^*/U_e$, of an equilibrium intrusion as a function of $\alpha$. Configurations with $H^*/h_0 = 2$ and: $\gamma = 1/2$ (dashed line), and $\gamma = 3/4$ (solid line). The values $u_N$ and $h_N$ of the currents were calculated with (2.66) and Benjamin's $Fr$.

The real support to this simplistic theory is provided by comparisons with experiments and numerical computations. Such tests were reported, for example, by Flynn, Boubarne, and Linden (2008) (note that the effective speed in that paper is defined as $\sqrt{2}$ times that used here). As mentioned above, the experimental and numerical data show that the speed of propagation is constant over several lock-lengths; the agreements with (15.69) are, roughly, in the range of the experimental errors.

The details of the transition from the constant-speed slumping to the other phases require further investigation.

Note that the fully-symmetric intrusion, discussed in § 15.2, is a particular case of equilibrium intrusion with $\alpha = \beta = \gamma = 1/2$.

A non-equilibrium intrusion will, in general, tend to sink (or float) as a whole upon release, toward an equilibrium position about an upper (lower) level. Vertical oscillations and interface waves may appear. These effects are beyond our scope. There is, however, an interesting exception: the full-depth lock configuration. In this case the bottom and top boundaries prevent vertical motion of the fluid in the reservoir. The estimates for the horizontal motion remain relevant. This is considered next.

### 15.7.2   Full-depth lock

The geometrical constraints impose

$$\beta = \gamma, \tag{15.70}$$

and hence $H_i^*/h_{0i} = 1$ for both $i = 1, 2$ layers. Both the upper and lower currents are expected to display the full-depth lock features. We therefore have (in dimensionless form) $u_{N1} = u_{N2} = u_N \approx 0.43$ and $h_{N1} = h_{N2} = h_N$ (the value is unimportant in our context). Again, in a formally non-symmetric problem we actually find several symmetries: $\beta = \gamma$, $u_{N1} = u_{N2}$, and $h_{N1} = h_{N2}$.

The speed approximation is now drastically simplified. Substitution of the previously mentioned values into (15.65), use of the scaling values, and some reductions yield

$$\frac{U_c^*}{U_0} = 2\,u_N\,[(1-\alpha)^{1/2}\beta^{3/2} + \alpha^{1/2}(1-\beta)^{3/2}], \tag{15.71}$$

and

$$\frac{U_c^*}{U_e} = \sqrt{2}\,u_N\,\frac{(1-\alpha)^{1/2}\beta^{3/2} + \alpha^{1/2}(1-\beta)^{3/2}}{[(1-\alpha)\beta + (1-\beta)\alpha]^{1/2}}. \tag{15.72}$$

Here $u_N \approx 0.43$, the value of the dimensionless speed for the full-depth lock current. Typical results are shown in Fig. 15.17.

Let the full-lock configuration have also a symmetric stratification, $\beta = \gamma = 1/2$ (i.e., $H_1^* = H_2^*$). Substitution into (15.71)-(15.72) gives

$$\frac{U_c^*}{U_0} = \frac{U_c^*}{U_e} = \frac{u_N}{\sqrt{2}}[(1-\alpha)^{1/2} + \alpha^{1/2}]. \tag{15.73}$$

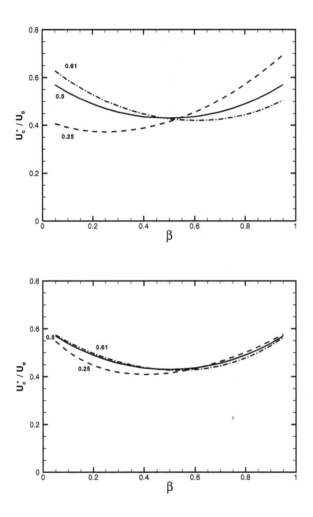

**FIGURE 15.17:** Scaled speed of propagation of a full-depth lock intrusion, as a function of the asymmetry of the interface $\beta$, for three values of $\alpha : 0.25$ (dashed line), 0.5 solid line, and 0.61 (dash-dot line). Shown values of $U_c^*/U_0$ in the upper frame and $U_c^*/U_e$ in the lower frame.

This result turns out to be a very flat function for $0.1 < \alpha < 0.9$, with a maximum of $u_N = 0.43$ at $\alpha = 1/2$.

We see, again, that the scaled speeds vary in a small range for a wide range of the parameters $\alpha$ and $\beta$. Cheong, Kuenen, and Linden (2006) reported experimental and numerical speed values which agree very well with the lines displayed in Fig. 15.17 and with the result (15.73) (in that paper, the speed is scaled with $2U_0$, $\hat{h}_E$ and $h$ correspond to our $\alpha$ and $\beta$, respectively). That work highlighted the propagation when stratification is symmetric, $\beta = 1/2$: then $U^*/U_0$ is (almost) constant for the whole range of $\alpha$ (except near the endpoints 0 and 1, which are non-physical in any case). In other words, the speed of this intrusion is independent of the value of $\rho_c$ (in the valid range $(\rho_2, \rho_1)$). This remarkable result has actually a quite simple interpretation. Equation (15.61c) predicts that the effective driving density difference $\epsilon_e = \epsilon = (\rho_1 - \rho_2)/(2\rho_2)$ when $\beta = 1/2$, for any $\alpha$. When the interface is in the middle of the container, the excess and deficit of buoyancy produced by a given $\rho_c$ act over the same thickness in both the upper and lower layers, and therefore the value of $\rho_c$ cancels out in the resultant driving force.

### 15.7.3    Continuous linear stratification

Consider the configuration of Fig. 15.15 with a continuous linear stratification of the ambient, from $\rho_b$ at the bottom to $\rho_o$ at the top. The density asymmetry is again given by the parameter $\alpha$,

$$\rho_c = \rho_o + \alpha(\rho_b - \rho_o). \tag{15.74}$$

The parameters $\beta$ and $\gamma$ are defined as before.

Again, for $\alpha = \beta = \gamma = 1/2$ the fully-symmetric case is recovered.

Essentially, the arguments used for the sharp stratification can be also applied in the present case. Surprisingly, the linear stratification turns out to be the simpler case, because it contains fewer degrees of freedom (or more intrinsic symmetries). The density of the ambient at $z = 0$ is given by

$$\rho_a(z = 0) = \rho_o + \gamma(\rho_b - \rho_o). \tag{15.75}$$

Since this is the level of neutral buoyancy, the intersection with the constant $\rho_c$ requires

$$\gamma = \alpha. \tag{15.76}$$

Recall that in the two-layer stratification, $\alpha$ and $\gamma$ can be imposed independently even for an equilibrium intrusion.

We use again the averaged flux result (15.65). Under the assumption that the currents are separated, we can calculate the slumping speeds $u_{Ni}$ and heights $h_{Ni}$, in dimensionless form. The regular one-layer SW theory provides the necessary results as functions of dimensionless height ratios $H_i = H_i^*/h_{0i}$; see (15.62). We use the slumping speed results (15.38). The dimensional speeds $u_{Ni}^*$ and nose heights $h_{Ni}^*$ are then obtained by multiplication with $\mathcal{N}h_{0i}$ and $h_{0i}$, respectively.

The scaling speed of the intrusion is $\mathcal{N}h_0$.

Consider the *equilibrium intrusion*. Since the buoyancy force about the neutral-buoyancy line is antisymmetric, the equilibrium is obtained when

$h_{01} = h_{02}$, i.e., $\beta = 1/2$. With this additional symmetry, (15.65) is reduced to

$$\frac{U_c^*}{\mathcal{N} h_0} = \frac{u_{N1} h_{N1} + u_{N2} h_{N2}}{h_{N1} + h_{N2}}. \tag{15.77}$$

We keep in mind that $u_{Ni}$ and $h_{Ni}$ vary in quite small ranges, the former about 0.4 and the latter about 0.5. The prediction is that the propagation of the non-symmetric equilibrium intrusion is quite close to that of a symmetric intrusion with the same $H^*/h_0$ ratio.

Consider the *full-depth lock intrusion* (not necessarily in equilibrium). Now $h_{0i} = H_i^*$. Consequently, $\alpha = \beta = \gamma$ and $H_1 = H_2 = 1$. The upper and lower currents encounter a full-depth linearly-stratified ambient. Therefore, $u_{N1} = u_{N2} = u_N \approx 0.28$ (see Fig. 15.3) and $h_{N1} = h_{N2} = h_N$ (value unimportant). Substitution into (15.65) yields

$$\frac{U_c^*}{\mathcal{N} h_0} = 2 u_N [\alpha^2 + (1-\alpha)^2] \approx 0.56 [\alpha^2 + (1-\alpha)^2]. \tag{15.78}$$

As expected, the speed of an intrusion released in a linearly-stratified ambient from behind a full-depth lock is symmetric with respect to $\alpha = 1/2$. For $\alpha = 1/2$ we recover the fully-symmetric speed result 0.28; see Fig. 15.3 for $H = 1$. When $\alpha = 1$ (or 0), we obtain a full-depth gravity current with $S = 1$. The speed of this bottom (top) current is twice larger than that of a symmetric intrusion in the same channel. This is simply because the same driving force per unit depth above (below) the neutral buoyancy line is present. For a boundary current the total driving force accumulates over the full height of the channel ($2h_0\mathcal{N}$ say), while for a symmetric intrusion only for half this distance ($h_0\mathcal{N}$ say).

These results are in very good agreement with the experimental and Navier-Stokes simulations presented by Bolster, Hang, and Linden (2008) (note that the speed is scaled with $2h_0\mathcal{N}$ in that paper). However, the slumping propagation of the intrusion in a linearly-stratified ambient is usually limited by the interaction with the internal gravity waves. How this is affected by the asymmetric conditions is a topic that still requires clarifications.

### 15.7.4 Summary

Overall, the agreement of the simple theoretical slumping speed result (15.65) for the non-symmetric intrusions with experimental and NS simulation data is quite remarkable. This is because the equilibrium and full-depth lock intrusions contain some intrinsic (but not obvious) symmetries that restrict the differences between the upper and lower flows. Again, the SW theory seems to capture well the underlying mechanisms of the motion.

However, we must keep in mind that the quantitative prediction formulas presented here lack rigor. They are based on a superposition, or some implicit curve-fit (interpolation) manipulation, of SW solutions derived for simple problems. Other approximation formulas for the initial speed in a full-depth-lock geometry are also available, with similar justifications and performances; see Cheong, Kuenen, and Linden (2006) and Bolster, Hang, and Linden (2008). More accurate analytical considerations (e.g., Holyer and Huppert 1980, Flynn and Linden 2006, and Zemach 2007) yield cumbersome steady-state results which cannot be easily matched with realistic initial conditions, or employ adjustable constants. In general, the flow field of the non-symmetric intrusion is complicated by waves that may affect the small $z$-perturbation assumptions, flatten the head, and even bring the current to a full stop (in full-depth lock geometries, at least; see Sutherland, Kyba, and Flynn 2004). These effects still await a comprehensive synthesis.

# Chapter 16

# Intrusions in axisymmetric geometry

## 16.1 Introduction

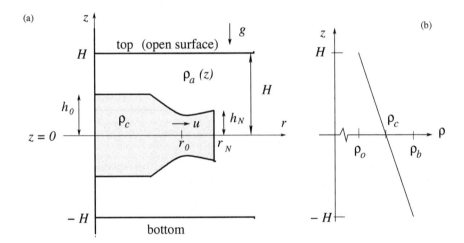

**FIGURE 16.1**: Schematic description of the system (a) the geometry; (b) density profile in the ambient. In dimensionless form, the horizontal lengths are scaled with $r_0$ and the vertical lengths with $h_0$. The subscripts denote: $N$ - nose (or front); $a$ - ambient; $b$ - bottom; $c$ - current (intrusion); $o$ - open surface.

In this chapter we consider axisymmetric counterparts of the 2D intrusions into a stratified ambient discussed in Chapter 15. By axisymmetric we mean that the propagation is in the radial direction in a cylindrical geometry, over a full circle or in a wedge. The propagation starts from rest in a region of radial dimension $r_0$ about the axis, and the velocity has no lateral (azimuthal) component. We assume again that the intrusion is symmetric (or double symmetric) with respect to the horizontal plane $z = 0$ of neutral buoyancy.

Consequently

$$\rho_c = \rho_a(z = 0) = \frac{1}{2}(\rho_o + \rho_b); \qquad (16.1)$$

$$g' = \epsilon g; \quad \epsilon = \frac{\rho_c - \rho_o}{\rho_o} = \frac{1}{2}\left(\frac{\rho_b}{\rho_c} - 1\right); \qquad (16.2)$$

where the subscript $a$ denotes the ambient and $c, b, o$ refer to values of the current (or intrusion, or "mixed fluid"), bottom, and top (or open) open surface. It is convenient to use $\rho_o$ as the reference density.

We add again the Boussinesq and inviscid assumption ($\epsilon \ll 1$, $Re \gg 1$). The resulting flow can be regarded as composed of two gravity currents about the $z = 0$ neutral buoyancy plane. These are like a "top" (or surface) current below, and a "bottom" current above. There is mirror symmetry about $z = 0$. We are now at the position to apply the inviscid SW apparatus to the investigation of the intrusion. The underlying idea is that the symmetric intrusion is a superposition of two practically identical gravity currents which flow on the frictionless impenetrable $z = 0$ boundary. Each of these currents contains the volume of half of the intruding fluid.

## 16.2 Two-layer stratification

This is a straightforward extension of § 15.2. In this special case we can use directly the homogeneous-ambient results for the axisymmetric geometry. We can solve the upper current by the classical models, in one or two-layer version. The lower current will be the mirror image.

## 16.3 Fully linearly-stratified tank, part-depth locks

The typical configuration which will be investigated in detail is sketched in Fig. 16.1. We assume that the density of the ambient fluid varies linearly over the full depth of the container. (In the notation of the previous chapter, this is the $l = H$ case. Solutions for $0 < l < H$ are feasible, but the details need elaboration.) The case considered here corresponds to a gravity current of the type studied in Chapter 13, with $S = 1$ (and non-rotating). We follow the presentation of Ungarish and Zemach (2007) and Zemach and Ungarish (2007).

The 2D counterpart is considered in § 15.4.2. The extension from the rectangular to the axisymmetric geometry is non-trivial and produces some unexpected results. The intruding fluid diverges as a result of the radial propagation and hence the balances along the characteristics of the SW model contain time-dependent source terms. The predicted velocity field and the shape of the interface gain some complexities which have no counterpart in the 2D case. A self-similar behavior emerges, but we shall see that it displays some peculiar features. The results are also expected to be relevant to practical circumstances such as propagation of pollutants. Indeed, in many environmental intrusions the axisymmetric geometry (either a full cylinder or a part of it as a wedge) seems more relevant than the rectangular one.

## 16.3.1 Formulation and SW approximations

The configuration is sketched in Fig. 16.1. We use a $\{r, \theta, z\}$ cylindrical coordinate system with corresponding $\{u, v, w\}$ velocity components. The gravity acts in the $-z$ direction. We assume that the geometry is fully axisymmetric about the axis $z$, or a wedge with smooth vertical sidewalls $rz$ planes (in the latter case, the gap at $r = r_0$ is large as compared with the thickness of the intrusion, $h_0$). The inviscid flow does not depend on the coordinate $\theta$, and $v \equiv 0$. We refer to this configuration as axisymmetric (and non-rotating).

The initial position of the intrusive fluid is in a lock $0 \leq r \leq r_0$. Unless stated otherwise, the lock is a vertical cylinder $-h_0 \leq z \leq h_0$. The density of the intruding current is equal to that of the ambient at the symmetry plane $z = 0$.

The density in the intruding current and in the ambient can be expressed as

$$\rho_c = \rho_o(1 + \epsilon), \quad \rho_a = \rho_o \left[ 1 + \epsilon \left( 1 - \frac{z}{H} \right) \right]. \tag{16.3}$$

The buoyancy frequency, see (12.2), is constant and given by

$$\mathcal{N} = \left( \frac{g'}{H} \right)^{1/2}, \tag{16.4}$$

where $H$ is the dimensional thickness of the ambient fluid above (and below) the neutral buoyancy level midplane, $z = 0$. The speed of the fastest internal wave is $\mathcal{N}H/\pi$ (dimensional).

As explained in detail in the previous chapter, it is sufficient to consider the flow in the domain $z \geq 0$. The shape of the density interface and the velocity field in the domain $z \leq 0$ are expected to be mirror images. We use again the one-layer SW approximations, subject to the modifications imposed by the geometry. In the ambient fluid domain we assume that $u = v = w = 0$ and hence the fluid is in purely hydrostatic balance and maintains the initial density $\rho_a(z)$. The motion is assumed to take place in the intruding layer of fluid only, $0 \leq r \leq r_N(t)$ and $0 \leq z \leq h(r, t)$, where the density is $\rho_c$. We argue that the predominant vertical momentum balance in the intruding fluid is hydrostatic and that viscous effects in the horizontal momentum balance are negligibly small. Hence the motion is governed by the balance between pressure and inertial forces in the $r$ horizontal direction.

A relationship between the pressure fields and the height $h(r,t)$ can be obtained. In the motionless ambient fluid, the pressure does not depend on $r$. The hydrostatic balances are $\partial p_i / \partial z = -\rho_i g$, where $i = a$ or $c$. Integration for the unperturbed ambient gives

$$p_a(z,t) = -\rho_o \left[ 1 + \epsilon \left( 1 - \frac{1}{2} \frac{z}{H} \right) \right] gz + C. \tag{16.5}$$

A similar integration for the current and pressure continuity between the ambient and the intrusion at the interface $z = h(r,t)$ yield

$$p_c(r,z,t) = -\rho_o(1+\epsilon)gz + \rho_o g' \frac{1}{2} \frac{h^2(r,t)}{H} + C, \tag{16.6}$$

and consequently

$$\frac{\partial p_c}{\partial r} = \rho_o \frac{g'}{H} h \frac{\partial h}{\partial r} = \rho_o N^2 h \frac{\partial h}{\partial r}. \tag{16.7}$$

We note in passing that (16.5)-(16.7), although developed for $z > 0$, are also valid in the $z < 0$ domain. Pressure continuity at $z = -h(r,t)$ and $z = h(r,t)$ is equivalent.

The SW equations of motion are next obtained by $z$ averaging the equations of continuity and radial momentum in the mixed-fluid domain, subject to the inviscid, Boussinesq, and thin-layer simplifications; the pressure term is eliminated by (16.7). The results are presented below in dimensionless form.

The dimensional variables (denoted here by asterisks) are scaled as follows

$$\{r^*, z^*, h^*, H^*, t^*, u^*, p^*\} = \{r_0 r, h_0 z, h_0 h, h_0 H, Tt, Uu, \rho_o U^2 p\}, \tag{16.8}$$

where

$$U = \left[ \frac{\rho_c - \rho_a(z=1)}{\rho_o} h_0 g \right]^{1/2} = \left( \frac{h_0 g'}{H} \right)^{1/2} = N h_0; \tag{16.9}$$

$$T = \frac{r_0}{U} = N^{-1} \frac{r_0}{h_0} \tag{16.10}$$

(to avoid possible confusion, we specify that $\rho_a$ in (16.9) is calculated at the dimensionless $z = 1$). Here, again, $r_0$ and $h_0$ are the initial length and half-thickness of the intrusion, $U$ is the typical inertial velocity of propagation of the nose, and $T$ is a typical time period for longitudinal propagation over a typical distance $r_0$. The reference velocity accounts for the fact that the effective driving force is provided by the density difference over the axial dimension of the intrusion. The typical Reynolds number is $Re = Uh_0/\nu$ where $\nu$ is the kinematic viscosity.

We emphasize that hereafter the variables $r, z, u, t, h, H, p$ are in *dimensionless form* unless stated otherwise. The (dimensionless) speed of the dominant wave in the ambient fluid is as in the rectangular counterpart, given by (15.32).

The SW approximation produces a system of equations for the height (or half-thickness) $h(r, t)$ and averaged longitudinal velocity $u(r, t)$. Using these dependent variables, the continuity and momentum equations can be expressed as

$$\begin{bmatrix} h_t \\ u_t \end{bmatrix} + \begin{bmatrix} u & h \\ h & u \end{bmatrix} \begin{bmatrix} h_r \\ u_r \end{bmatrix} = \begin{bmatrix} -\dfrac{uh}{r} \\ 0 \end{bmatrix}. \tag{16.11}$$

The eigenvalues of the matrix of coefficients are real and the set of eigenvectors is full, and hence the system (16.11) is hyperbolic. The characteristic balances are

$$dh \pm du = -\frac{uh}{r} dt \tag{16.12}$$

$$\text{on} \quad \frac{dr}{dt} = c_\pm = u \pm h. \tag{16.13}$$

The boundary conditions are (1) the obvious $u = 0$ at the center; and, (2) at the nose $r = r_N(t)$

$$\frac{dr_N}{dt} = u_N = Fr \frac{1}{\sqrt{2}} h_N, \tag{16.14}$$

where the HS $Fr$ formula is used; see (15.25). The derivation is discussed in § 15.3.2 for the 2D case. The extension is supported by the fact that comparisons of SW propagation results with NS simulations, discussed later, show good agreements also in the axisymmetric geometry.

The "standard" initial conditions in the lock are simply: $r_N = 1$, $u = 0$ and $h = 1$ at $t = 0$. The $h = 1$ can be changed to reproduce release from a different initial shape, e.g., an ellipsoid $h = \sqrt{1 - r^2}$; this is the counterpart of the geometry discussed in § 15.5.

In general, the solution of this hyperbolic-type initial-value problem is performed by numerical methods. The source term on the right-hand side of (16.12) defies simple analytical solution during the initial dam-break stage. However, we know that the SW propagation is sub-critical. To reach this conclusion we compare the speed of propagation of the axisymmetric intrusion with that of the 2D rectangular counterpart. The balance along a forward-moving characteristic, see (16.12), indicates that the initial speed (at $t = 0^+$) is the same in both cases, but afterward the axisymmetric intrusion moves more slowly. This is a result of the negative source term in (16.12) which expresses the divergence of the radial flow in the cylindrical geometry (this term is obviously zero in the rectangular case). We showed in § 15.4.2 that the propagation of the 2D intrusion is always sub-critical (i.e., $u_N < u_W$ given by (15.32)). We conclude that the speed of propagation of the axisymmetric intrusion is also always sub-critical.

A useful analytical result can be obtained under the assumption that a self-similar behavior develops for large $t$. This is our next topic.

## 16.3.2   Similarity solution

The axisymmetric intrusion displays some special self-similar propagation features which merit attention. We shall now discuss this topic in some detail. Formally, we deal with the upper half of the intrusion, but we emphasize that the results are actually for a gravity current of the type discussed in Chapter 13 for $S = 1$ (and non-rotating); for $0 < S < 1$ no similarity solutions are available.

We apply the transformation of § 2.3 to the set of equations (16.11) and (16.14). The details are left for an exercise. Let

$$y = r/r_N(t)   (0 \le y \le 1) \tag{16.15}$$

and redefine the variables as $h(y,t)$ and $u(y,t)$. We obtain the continuity and momentum equations in the following form

$$\frac{\partial h}{\partial t} + \frac{1}{r_N}(u - y\dot{r}_N)\frac{\partial h}{\partial y} + \frac{1}{r_N}h\frac{\partial u}{\partial y} = -\frac{uh}{yr_N}; \tag{16.16}$$

$$\frac{\partial u}{\partial t} + \frac{1}{r_N}(u - y\dot{r}_N)\frac{\partial u}{\partial y} + \frac{1}{r_N}h\frac{\partial h}{\partial y} = 0; \tag{16.17}$$

subject to the nose condition

$$\dot{r}_N = \frac{Fr}{\sqrt{2}}h(y = 1, t). \tag{16.18}$$

We seek a self-similarity solution to the system of SW governing equations and boundary conditions, for a deep intrusion and large values of $t$. If the intrusion is deep it is justified to assume that $Fr = \text{const}$ (= 1.19 in the present formulation). We follow the procedure of § 6.5. Assume

$$r_N(t) = Kt^\beta;   h(y,t) = \varphi(t)\mathcal{H}(y);   u(y,t) = \dot{r}_N U(y); \tag{16.19}$$

where $K$ and $\beta$ are some positive constants. The obvious boundary conditions are $\mathcal{U}(1) = 1$ and conservation of volume $\mathcal{V}$ (per radian).

Using the boundary condition (16.18) of $u$ at $y = 1$, we obtain

$$\varphi(t) = \beta K t^{\beta-1} = \dot{r}_N;   \mathcal{H}(1) = \frac{\sqrt{2}}{Fr}. \tag{16.20}$$

The integral volume continuity of the current, upon use of (16.19) and (16.20), gives

$$\mathcal{V} = \int_0^{r_N} h(r,t)r\,dr = \int_0^1 h(y,t)r_N^2 y\,dy = r_N^2\varphi(t)\int_0^1 \mathcal{H}(y)y\,dy =$$

$$= \beta K^3 t^{3\beta-1}\int_0^1 \mathcal{H}(y)y\,dy. \tag{16.21}$$

This equation imposes

$$\beta = \frac{1}{3}. \tag{16.22}$$

The continuity equation (16.16) and the momentum equation (16.17) are next used. The substitution of the similarity variables and $\beta = 1/3$ decouples the continuity equation, and the solution is $\mathcal{U}(y) = y$. To determine $\mathcal{H}(y)$ we employ the momentum equation. After some algebra (16.17) is reduced to

$$(\beta - 1)y + \frac{1}{2}\beta(\mathcal{H}^2)' = 0. \tag{16.23}$$

The solution of this equation for $\beta = 1/3$ and subject to the boundary condition $\mathcal{H}(1) = \sqrt{2}/Fr$ produces an unexpected feature: the resulting $\mathcal{H}$ is real-valued only for $y \geq y_1 > 0$.

We summarize the similarity solution as

$$r_N(t) = Kt^{1/3}; \quad h(y,t) = \dot{r}_N \sqrt{2}(y^2 - y_1^2)^{1/2}; \quad u(y,t) = \dot{r}_N \, y; \tag{16.24}$$

$$y_1 = \sqrt{1 - \frac{1}{Fr^2}}, \tag{16.25}$$

for $y \geq y_1$. To make this result physically acceptable, we postulate that $h = \mathcal{H} = 0$ for $y < y_1$. The "virtual origin" difficulty remains: the use of $t + \gamma$ instead of $t$ does not affect the validity of the self-similar results.

The constant $K$ is determined by volume continuity considerations. Combining (16.21) with the results concerning the height of the interface and value of $\beta$, we write

$$\mathcal{V} = \frac{1}{3}K^3 \int_{y_1}^{1} \sqrt{2}(y^2 - y_1^2)^{1/2} y \, dy = \frac{\sqrt{2}}{9Fr^3}K^3, \tag{16.26}$$

and hence

$$K = 3Fr \left(\frac{\mathcal{V}}{3\sqrt{2}}\right)^{1/3}. \tag{16.27}$$

Again, $\mathcal{V}$ is the constant volume of half of the intrusion (per unit azimuthal angle). For the present value of $Fr = 1.19$ we obtain: (a) for the "standard" initial cylinder of mixed fluid $\mathcal{V} = 1/2$, and hence $K = 1.750$; (b) for an initial ellipsoid of mixed fluid $\mathcal{V} = 1/3$, and hence $K = 1.529$.

Note that $y_1 > 0$ for $Fr > 1$; for the present $Fr = 1.19$, we obtain $y_1 = 0.54$. Moreover, we note that $h(y_1) = 0$. In other words, in the self-similar stage the intrusion is a ring of mixed fluid embedded in the original ambient fluid. The inner radius, $r_1(t) = y_1 r_N(t)$, is slightly larger than half the outer radius in the present case. The thickness of the intrusion decreases continuously to zero as $y \to y_1^+$. The typical behavior of the shape of the resulting self-similar intrusion is illustrated in Fig. 16.2. A self-similar current which leaves behind a bare (or dry) domain in which the ambient is in direct contact with the plane

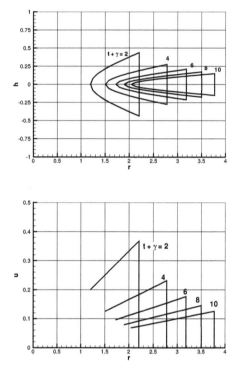

**FIGURE 16.2:** Self-similar shape of interface and velocity for an intrusion of volume $\mathcal{V} = 0.5$ at various times $t + \gamma$. The "negative" $h$ is the mirror image of the calculated $h$. (Ungarish and Zemach 2007 with permission.)

of propagation $z = 0$ has been encountered in non-Boussinesq systems in which the density ratio is larger than 2; see § 10.4. Here this behavior appears in a Boussinesq system due to the stratification and divergent geometry.

In a non-stratified ambient fluid the similarity behavior of an axisymmetric intrusion/current, discussed in § 6.5, is very different. There is no bare region restriction $y < y_1$, and the propagation is with $t^{1/2}$. As expected, the intrusion in the stratified ambient spreads out more slowly, with $t^{1/3}$, and the dominance of the inertial terms decays faster. However, the striking novel feature of the present self-similar case is the appearance of a clear inner domain, while the whole volume of the intrusion is confined to an annular domain with a fixed ratio between the inner and outer radii; see Fig. 16.2. A self-similar current is expected to "forget" the initial conditions. It is still relevant to ask if and how this special shape can develop from simple initial conditions. The answers will emerge from the solutions of release with physical initial boundary conditions, to which we proceed now.

### 16.3.3 SW results and comparisons with NS

The solution of the initial-value SW problem (again, with $u = 0$ and $h = h(r)$ in $0 \leq r \leq r_N = 1$ at $t = 0$) can be obtained by the usual finite-difference scheme; see Appendix A.2. The typical grid has 300 intervals in $y \in [0, 1]$ and the typical time step is $1 \cdot 10^{-3}$.

The "standard" problem is the release of a cylinder of mixed fluid, i.e., $h = 1$ for $0 \leq r < 1, t = 0$. Pertinent results are presented in Fig. 16.3. Here the initial depth ratio is $H = 9$. This provides, from the beginning, a sufficiently deep nose and hence a constant $Fr = 1.19$; see (15.25). (It is left for an exercise to verify that identical results are obtained for $H = 7$ and $H = 20$.) During the initial phase of propagation (see $t = 1$) the shape resembles the

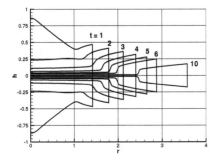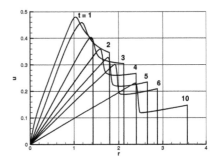

**FIGURE 16.3:** SW predictions for release of a deep intrusion $(H \geq 7)$ from rest. The initial volume is a cylinder $(h = 1)$. Profiles of thickness and $u$ as functions of $r$ for various $t$. (Ungarish and Zemach 2007 with permission.)

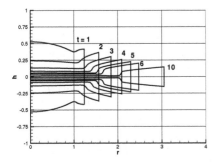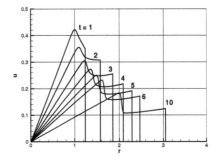

**FIGURE 16.4:** As in the previous caption, but the initial volume is an ellipsoid $(h = \sqrt{1 - r^2})$.

behavior of the dam-break problem: the height of the interface of the fluid propagating from the lock (the "head") is about half the initial height, while in the lock domain (the "tail") the interface becomes negatively-inclined due to a backwardly propagating rarefaction wave. However, the radial velocity in the lock region increases with $r$ and reaches a maximum at $r \approx 1$. This is a result of the cylindrical-geometry constraint on the characteristic balances. Consequently, fluid is quite quickly drained from the "tail" into the $r > 1$ domain. At $t = 3$, the "tail" contains less than 10% of the initial volume. It is clearly seen that the mixed fluid, released from rest in the standard vertical cylinder lock, has the tendency to evolve into the ring-shape of the self-similar analytical prediction displayed in Fig. 16.2. The sharp difference to the corresponding rectangular current, of both $h$ and $u$ profiles, is illustrated by Fig. 15.14.

Further support to these insights is provided by the SW solution of a deep intrusion which is released as an initial ellipsoid volume of mixed fluid, i.e., $h = \sqrt{1 - r^2}$ for $0 \leq r < 1, t = 0$. Fig. 16.4 shows that the essential propagation is like in the previous "standard" problem. The initial spread is slightly delayed by the fact that the height of the nose must develop from zero, but the tendency to the similarity shape at $t > 3$ is evident.

An additional test of the tendency to self-similar behavior is as follows. The value of $\gamma$ can be estimated by matching, at a moderate $t = t_m$, the numerically computed $r_N$ with the similarity form $K(t_m + \gamma)^{1/3}$, where $K$ is known as specified above. For $t_m = 5$ the result is $\gamma = -1.53$ for the cylinder and $-1.63$ for the ellipsoid. Using this outcome, a comparison is performed of the similarity predictions and numerical SW results for the radius and velocity of propagation over a wide range of $t > t_m$; see Fig. 16.5. The observed agreement is good. However, this comparison becomes problematic for really large values of $t$ because of the accumulation of numerical errors in the initial-value problem solution. A sharper description of the large-time behavior requires a specially designed code, as discussed later in § 16.3.3.1.

### 16.3.3.1 The tail and long-time variables stretching

The behavior of the thin disk-tail which is left inside the major self-similar ring turns out to be also quite interesting. We see in Fig. 16.3 and 16.4 that for $t \geq 2$ the interface of the tail is horizontal and the corresponding velocity increases linearly with $r$ (as first approximations). This feature can be explained analytically as follows. Consider the equations of motion (16.11) subject to the boundary condition $u = 0$ at $r = 0$. For $h(r, t)$ and $u(r, t)/r$ we seek approximations of the form $f_0(t) + f_1(r, t)$ under the assumption that $|f_1|/|f_0| \ll 1$ for large $t$. Substitution of the corresponding expansions into (16.11) yields, after some algebra, the leading terms

$$h(r, t) = \frac{C_1}{(t + C_2)^2}; \quad u(r, t) = \frac{1}{t + C_2}r; \quad (16.28)$$

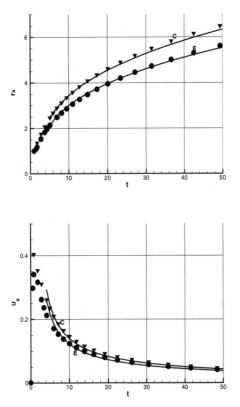

**FIGURE 16.5**: Comparison of similarity (line) and initial-value results (symbols) for a cylinder (C) and ellipsoid (E) initial shape. (Ungarish and Zemach 2007 with permission.)

where $C_1, C_2$ are constants which represent some (as yet unspecified) initial conditions. We note that (16.28) is actually an exact solution of (16.11). Formally, this can also be referred to as a self-similar inner-region behavior. An order of magnitude consideration shows that $C_1, C_2$ are positive and of order unity.

The behavior (16.28) of the thin tail provides additional insight into the development of the major similarity shape from realistic initial conditions. For large $t$ this tail is expected to extend from the axis to the position $r_1 \approx y_1 K t^{1/3}$ where the similarity profile (16.24) starts. First, combining this $r_1$ with $h \sim t^{-2}$, we infer that the volume of fluid in the tail, which is (approximately) $0.5 r_1^2 h$, decays like $t^{-4/3}$ for large $t$. Second, substituting this $r_1$ into (16.28), we obtain $u(r_1, t) = y_1 K t^{-2/3}$ for large $t$ in the tail. On the other hand, the similarity solution (16.24) predicts that $u$ varies from $(1/3) y_1 K t^{-2/3}$ (at $r_1$) to $(1/3) K t^{-2/3}$ (at $r_N$). This means that the value of $u$ is expected to reach the maximum in the tail (recall that $y_1 = 0.54 > 1/3$). Third, we notice that in the region about $r_1$ the value of $c_- = u - h$ is positive. This implies that information from the head is not propagated into the tail, and hence the behavior (16.28) for $y \leq y_1$ remains unaffected from the connection with the self-similar region (16.24). These predictions are consistent with the $u$ profiles displayed in Figs. 16.3 and 16.4. Indeed, at a given time, $u$ increases linearly with $r$ until about the position where the thin tail ends. There $u$ reaches the maximum. Then a sharp decrease of $u$ with $r$ occurs, followed by, again, a linear increase with $r$. These features are observed for both the cylinder and ellipsoid initial volume.

Additional details of the dependencies predicted by (16.28) were verified and confirmed by comparisons with the SW solutions. The fit to the numerical SW results yields: $C_1 = 1.19, C_2 = 0.194$ for the cylinder; and $C_1 = 0.815, C_2 = 1.946$ for the ellipsoid. Practically, the tail domain is expected to be of little importance to the flow field after the formation of the main self-similar ring (say, $t > 4$). In a real flow, the decay of thickness, volume, and energy in the tail is augmented by mixing and friction. Consequently, the present inviscid immiscible SW description of the tail may lose physical relevance in a quite early stage.

We can estimate the ratio of inertial to viscous forces by using a formula of the type of (6.54). We find that in the tail domain

$$\frac{F_I}{F_V} \approx Re \frac{h_0}{r_0} \frac{\int_0^{r_1} uu_r hr dr}{\int_0^{r_1} (u/h) r dr} \approx Re \left(\frac{h_0}{r_0}\right) t^{-5} = Re \left(\frac{h_0}{r_0}\right) K^{15} r_N^{-15}, \quad (16.29)$$

where $Re = U h_0 / \nu$ and $K$ is the coefficient of the similarity propagation (16.24) (this constant is included here because it contributes a factor of about $10^3$ to the above ratio). We used $h$ and $u$ given by (16.28). The details are left for an exercise.

It is interesting that a tail of this type also appears in the propagation of an axisymmetric current in a non-stratified ambient, as discussed in § 6.3 and seen in Fig. 6.1 (for $t \geq 3$). There is, however, an essential difference. In the non-stratified case the tail appears as a transient feature during the development of the self-similar solution; eventually, the radial domain of the tail, $r < r_j$, shrinks to zero and the self-similar profiles take control over the full radial domain from the axis to $r_N(t)$. In the stratified case the tail is a part of the long-time solution in the expanding radial domain $r < r_1 = y_1 r_N(t)$.

To verify these insights it is necessary to solve (numerically) the SW equations with the lock-release initial conditions for long times. The straightforward solution is unsuitable for this task because the accumulations of numerical errors may become dominant in the regions where $h \propto t^{-2}$. This difficulty can be removed by using a transformation or stretching of $r, t$ and of the dependent variables. The new variables are designed to remain of order unity during propagation to large $t$ and $r_N$. Such transformations and solutions were presented by Grundy and Rottman (1985a) and Zemach and Ungarish (2007). To illustrate the procedure, consider the transformation

$$y = r/r_N(t); \quad \tau = \beta \ln(t); \tag{16.30}$$

$$r_N(t) = Kt^\beta R(\tau); \tag{16.31}$$

$$h(r,t) = \varphi(t)\mathcal{H}(y,\tau); \quad u(r,t) = \beta Kt^{\beta-1}\mathcal{U}(y,\tau). \tag{16.32}$$

The values of $\beta, K$, and $\varphi$ are taken from the appropriate similarity solution; for example, for the intrusion discussed in this chapter, $\beta = 1/3$, $\varphi = (K/3)t^{-2/3}$, and $K$ is given by (16.27). The initial conditions are prescribed at $t = 1$ ($\tau = 0$).

The stretched variables $y$ and $\tau$ change in a fairly restricted range while $t$ varies from 1 to large values (say $10^3$). A similar behavior is expected for $R, \mathcal{H}$, and $\mathcal{U}$. Substitution of (16.30)-(16.32) into the SW equations of motion and boundary conditions yields relatively simple equations in terms of the new variables. These equations can be solved numerically by the same methods that were used for the original equations. The details are presented in Grundy and Rottman (1985a) and Zemach and Ungarish (2007). These studies provide accurate SW results for $t \approx 10^3$ and beyond. The inference that for large $t$ a tail of the form (16.28) remains in the stratified solution, but disappears in the non-stratified case, has been confirmed. The results also corroborate the stability of the self-similar solution for various initial conditions.

### 16.3.3.2 Comparison to NS simulation

Laboratory experiments for this type of intrusions are not available. Consider a "numerical experiment" for $H = 2.27$. The configuration is a realistic counterpart of the two-dimensional case discussed in § 15.4.2.1. The relevant 2D experiment was performed by Amen and Maxworthy (1980), Run 117 (saltwater, $h_0 = 6.16$ cm, $\mathcal{N} = 0.57\,\mathrm{s}^{-1}$, $h_0/x_0 = 0.33$, where $x_0$ is the length of the lock.).

The SW results are presented in Fig. 16.6. We observe that for $t \geq 3$ the motion of this intrusion resembles the similarity behavior, cf. Fig. 16.2. In particular, the volume is concentrated in an annular domain (with a thin tail whose thickness decays with time). The tail region develops a horizontal interface for $t > 2$, where next $h \propto t^{-2}$. The velocity in the intrusion, $u$, displays, again, two regions of linear increase with $r$, with a quite sharp drop of value in a transition region about the inner radius of the annulus. We infer that the main qualitative features of the SW similarity solution, derived for a deep intrusion, are valid also for release at moderate depths ($H = 2.27$ here).

In the NS simulations the (axisymmetric) tank had the outer radius $r_w = 4$ and height $2H = 4.54$. This $r, z$ domain was covered by a $260 \times 180$ unstretched mesh. In the simulation $\epsilon = 4.65 \cdot 10^{-3}$ and $Re = 2.3 \cdot 10^4$. Rigid boundary conditions were applied at the bottom and outer cylindrical wall, and no-stress

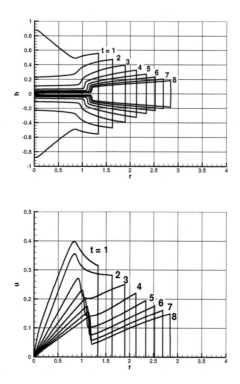

**FIGURE 16.6:** SW predictions for $H = 2.27$. Profiles of $h$ and $u$ as functions of $r$ for various $t$. (Ungarish and Zemach 2007 with permission.)

and regularity conditions at the axis $r = 0$. It was assumed that the top is a free-slip flat surface. It was estimated that the value of $r_w$ is sufficiently large for preventing interference between the intrusion and the reflected wave during the time intervals of interest here ($t < 7$), and that the grid is sufficiently fine for resolving the expected features. Initially, the current in the lock is covered by 65 radial and 79 axial grid intervals. At the maximal propagation considered here, there still are about 20 axial grid intervals in the main nose region of the intrusion. The rich and smooth details of the computed flow fields presented below, and in particular the clear-cut differences between the results in the axisymmetric and 2D geometries, indicate that the numerical resolution was satisfactory. Density contours of the NS solution are shown in Fig. 16.7. The nose of the intrusion is indicated by the large > (this is the foremost position of the domain with scaled density $\phi = 0.5$; the contours 0.51 and 0.49 embed this domain). The NS simulations, performed for the full domain $-H \leq z \leq H$, confirm the SW assumption of symmetry about $z = 0$. Some small deviations appear near the boundaries because in the NS

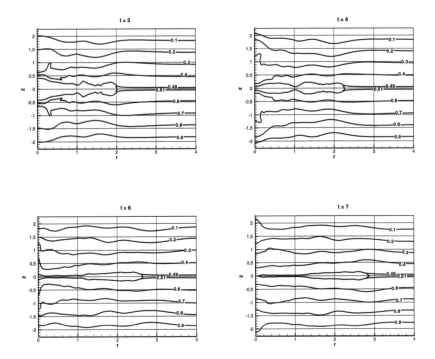

**FIGURE 16.7**: NS density contours at various $t$, for $H = 2.27, h_0/r_0 = 0.33$. The nose is marked by $>$. (Ungarish and Zemach 2007 with permission.)

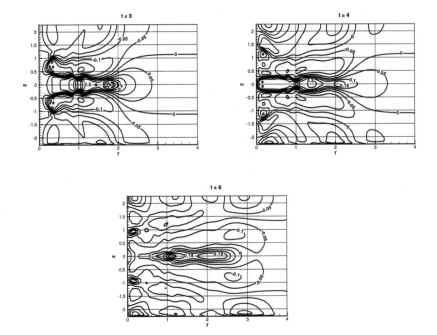

**FIGURE 16.8**: NS $u$ contours; details in previous caption.

simulations free slip and no-slip conditions are applied at the top and bottom, respectively, to reproduce a real experimental tank.

We observe that at $t = 6$ in the central domain, $r < 1.1$ (approximately), the intrusion is very thin, and most of the mixed-fluid volume is concentrated at larger radii. This shape is more pronounced at $t = 7$. This behavior confirms the SW predictions concerning both the (a) peculiar similarity shape (see Fig. 16.2), and (b) the thin tail in the inner region. When the NS-simulated layer of intruding fluid becomes very thin, viscous effects are expected to enter into the balance and hinder the motion in this region. Actually, the SW approximation is also expected to lose validity at the matching point $y = y_1+$ because $\partial h/\partial y \to \infty$ there; see (16.24). We mention in passing that a similar thin residual layer is also expected to develop in a laboratory experiment.

The isopycnals in the ambient above and below the intrusion display a wavy structure. The typical wavelength (scaled with $r_0$) is expected to be $2\pi(h_0/r_0)u_N \approx 0.7$ during the initial stages of motion. This is consistent with the NS results. However, at the present state of knowledge, it is difficult to predict the amplitude of these waves. The order of magnitude is implied by the displacement of the neutral buoyancy line by the intruding nose, which is expected to be $\pm h_N \approx \pm 0.4$ during the initial stages of motion. Since both $u_N$ and $h_N$ decay with time, and the bulk of the intrusion is affected by the backward moving "dam break" rarefaction wave, the resulting amplitude of the displaced isopycnals is quite irregular. How to quantify the effect of these displacements into a SW model error is a topic for further research. A propagation of the density perturbation ahead of the nose is also observed. However, the density perturbations induced by the leading wave are small (during the time considered here, at least). This supports the SW model assumption that the ambient in front of the intrusion maintains the initial $t = 0$ structure, in spite of the fact that the propagation is sub-critical.

It is interesting to note that in the present axisymmetric case there is almost no interaction between the waves and the head of the current. This is in sharp contrast with the 2D Cartesian counterpart, where a strong interaction appeared from $t \approx 5$ in both the numerical and experimental data. The possible explanation is the very different geometry of the head. In the axisymmetric case the head carries a significant portion of the total volume, while in the 2D case a significant volume is contained in a thick and long tail (see Fig. 15.6).

The behavior of the velocity field is illustrated by contour lines of $u$ in Fig. 16.8. The NS results display the difference between the tail and front of the mixed-fluid region, which has been indicated by the SW profiles. The radial velocity is large near $r = 1$ and near the nose, as predicted by the SW approximation. As expected, a return flow develops in the ambient between the boundaries and the head of the intrusion. This flow is strongly $z$-dependent and therefore it would be difficult to incorporate it in a simple two-layer SW model.

The radius of propagation as a function of time is displayed in Fig. 16.9. The agreement between the NS and SW predictions is very good. The small oscillation of the NS result can be attributed to the influence of the internal waves, but this turns out to be a small effect.

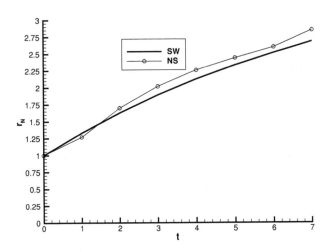

**FIGURE 16.9**: Radius of propagation as a function of $t$ for axisymmetric $H = 2.27$ configuration, NS and SW results. (Ungarish and Zemach 2007 with permission.)

### 16.3.4 The validity of the inviscid approximation

The importance of the viscous forces during the self-similar propagation can be estimated using

$$\frac{F_I}{F_V} = Re\frac{h_0}{r_0}\frac{\displaystyle\int_{r_1}^{r_N(t)} uu_r hr\,dr}{\displaystyle\int_{r_1}^{r_N(t)} (u/h)r\,dr} = Re\frac{h_0}{r_0}\theta(t). \tag{16.33}$$

Here $Re = Uh_0/\nu$ and the variables are dimensionless. We substitute the similarity solution (16.24) into this equation. It is left for an exercise to show that $\theta(t) = (2/27)K^9Cr_N^{-7}$ where $C \approx 0.3$ is the ratio of two integrals involving $(y^2 - y_1^2)^{1/2}$. To be specific, we obtain $\theta = 3.2r_N^{-7}$ for the standard cylinder volume $\mathcal{V} = 1/2$ and $Fr = 1.19$. This we compare with the result (13.42) for $S \approx 1$ obtained from box-model consideration. An additional $H^{1/2}$

coefficient is present in (13.42), because the definition of $U$ in $Re$ is different: there $U = (g'h_0)^{1/2}$, and here $U = \mathcal{N}h_0$. With this correction in mind, there still is a difference between the coefficients 3.2 and 1. The similarity solution ring profile is typically thicker than estimated by the box model, and hence the viscous effects are less pronounced. The major conclusion remains that the importance of viscous effects increases like $r_N^7$. We also recall that these estimates were performed for a gravity current on a solid bottom; the $F_I/F_V$ ratio for an intrusion is expected to be larger, by a factor of roughly four, but this is quite insignificant in view of the $r_N^7$ changes.

## 16.3.5   Summary

The SW theory predicts that, after an initial propagation to about 2.5 times the initial radius, the intrusion attains a self-similar behavior. In this stage the spread is with $t^{1/3}$ and the bulk of mixed fluid is in a ring of fixed ratio of the inner to outer radii (of about one half). Inside the inner radius of the ring there is a thin residual layer (disk) of mixed fluid, whose thickness decreases like $t^{-2}$. The speed of propagation of the nose (front) of the intrusion is sub-critical in all stages of propagation.

The NS simulations confirm the SW predictions. These results incorporate the effect of the internal gravity waves, which is ignored in the SW approximation. It turns out that the wave-head interaction in the axisymmetric geometry is quite insignificant, in contrast to the 2D rectangular counterpart problem.

The ratio of the inertial to viscous terms in the horizontal momentum balance decays like $r_N^{-7}$ (in the non-stratified counterpart, the result is $r_N^{-6}$). This poses quite a strong limitation on the applicability of the inviscid theory. The corresponding viscous motion is considered in § 19.2.

# Chapter 17

## Box models for 2D geometry

Box models have been successfully used in the investigation of some problems of homogeneous ambients, as discussed in Chapter 4. Here we extend these models to the current in a linearly-stratified ambient. Our main concern will be with the $S = 1$ intrusion-like configuration with changing volume $\mathcal{V} = qt^\alpha$, which will be considered in § 17.2; this has some simple analytical solution and is of practical importance.

We assume that the current has a simple shape, a well defined "box." In the present 2D case the box is a rectangle of length $x_N(t)$ and height $h_N(t)$. The main objective is to determine the behavior of these two variables, which means that we need two equations. We use the Boussinesq approximation.

The first governing equation of the box model is provided by the obvious volume continuity requirement

$$x_N(t)h_N(t) = \mathcal{V}. \tag{17.1}$$

The second equation must involve some dynamic considerations. For a simplified flow field the global buoyancy, inertial, and viscous effects for the volume are calculated. Then a governing momentum integral balance is applied from which an equation for $u_N(t)$ can be derived. The integration of this equation provides $x_N(t)$.

The stratification is of the type defined in § 12.2 and sketched in Fig. 12.1. The densities of the current and ambient are given by

$$\rho_c = \rho_o(1 + \epsilon), \quad \rho_a = \rho_o\left[1 + \epsilon S(1 - \frac{z}{H})\right], \quad (0 \le z \le H) \tag{17.2}$$

where $\epsilon = \rho_c/\rho_o - 1$, and $0 \le S \le 1$. We also recall

$$g' = \epsilon g; \quad \text{and} \quad \mathcal{N} = \left(\frac{Sg'}{H}\right)^{1/2}. \tag{17.3}$$

The current propagates on the plane $z = 0$.

Attention: in § 17.1 we use dimensionless variables, and in § 17.2 we use dimensional variables.

## 17.1 Fixed volume and inertial-buoyancy balance

*In this section we use dimensionless variables.* The scaling is given by (12.19)-(12.20), which we repeat here for convenience

$$\{x^*, z^*, h^*, H^*, t^*, u^*\} = \{x_0 x, h_0 z, h_0 h, h_0 H, Tt, Uu\}, \qquad (17.4)$$

with

$$U = (h_0 g')^{1/2} = \mathcal{N} \sqrt{\frac{H}{S}} h_0 \quad \text{and} \quad T = \frac{x_0}{U}. \qquad (17.5)$$

This will facilitate the comparison of the box-model results with the previous SW derivations in Chapter 12.

In the particular case of inertial-buoyancy dominated motion we can skip the integral balance and go directly to the boundary condition (12.29)

$$u_N = \frac{dx_N}{dt} = Fr(a) \left( 1 - S + \frac{S}{2} \frac{h_N}{H} \right)^{1/2} [h_N(t)]^{1/2}, \qquad (17.6)$$

where $a = h_N(t)/H$. The justification is provided by the expectation that the box-model and SW propagation are governed by the same mechanism, so we can start with the best available result which is already in a sufficiently simple form for use in the box model.

Combining (17.1) and (17.6) we obtain one equation for $x_N(t)$ (or for $h_N(t)$). Formally, for the case of a fixed volume $\mathcal{V}$, the solution is

$$t - t_I = \frac{1}{\mathcal{V}^{1/2}} \int_{x_I}^{x_N(t)} [Fr(a(x))]^{-1} \left( 1 - S + \frac{S}{2H} \frac{\mathcal{V}}{x} \right)^{-1/2} x^{1/2} dx + C, \quad (17.7)$$

where $a = \mathcal{V}/(Hx)$. The initial condition is $x_N = x_I$ at $t = t_I$. The numerical quadrature is straightforward.

For $S = 1$ some compact analytical results can be obtained, as follows.

Consider the case of a deep current with $S = 1$ and *constant* $Fr$. The result of (17.7) can be expressed as

$$x_N = \left[ \sqrt{2} Fr \mathcal{V} H^{-1/2} (t - t_I) + x_I^2 \right]^{1/2} = \left( \sqrt{2} Fr \mathcal{V} \right)^{1/2} H^{-1/4} (t + \gamma)^{1/2}. \qquad (17.8)$$

Here $\gamma$ is a constant which combines the values of the initial $t_I$ and $x_I$. This result is remarkably close to the prediction of the similarity solution (15.52). The propagation is with $t^{1/2}$, and the coefficient is proportional to $\mathcal{V}^{1/2}$ in both solutions (see (15.54)). The present result contains $H^{-1/4}$ in this coefficient because the present time scale differs from that used in Chapter 15. Note that the box-model approximation $h(x,t) = h_N(t)$ implies a $x$-independent axial velocity and hence $u(x,t)$ is a linear function of $x$. This is, again, in agreement with the self-similar solution. Moreover, as in the self-similar result, there is a time-shift constant $\gamma$ which must be determined by some initial conditions that may depend on the previous stage of propagation. This renders the box-model result also a quite weak prediction tool. We summarize that the box-model result (17.8) provides a plausible approximation for large times. This $S = 1$ result is relevant to an intrusion, and reproduces early models which have been suggested to approximate this type of flow (see Kao 1976). However, if an analytical solution of the SW equations is available, there is actually no need for this model.

For $S = 1$ and the non-constant branch $Fr = (1/2)a^{-1/3}$ of the HS formula, the analytical solution of (17.7) with initial condition $x_I = 1$ at $t_I = 0$ is

$$x_N(t) = \left[ \frac{5H^{1/3}}{6\sqrt{2}} \mathcal{V}^{2/3} H^{-1/2} t + 1 \right]^{3/5} \qquad (0.075 \leq h_N/H \leq 1). \qquad (17.9)$$

The speed of propagation decays with time and hence the initial (slumping) stage of motion (with constant or increasing speed) is not properly described by this approximation. However, this result indicates that, after the slumping phase, a non-deep intrusion propagates with $t$ at the power $3/5$ which is larger than the power $1/2$ of a deep intrusion. We recall that Wu (1969) reported the experimental value of $0.55 \pm 0.02$ which fits well into this range.

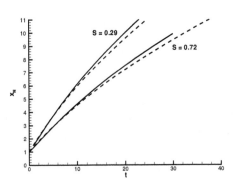

**FIGURE 17.1:** Propagation as a function of time, box model (dashed line) and SW (solid line) predictions for $H = 3$ and two values of $S$. (Ungarish and Huppert 2004 with permission.)

A comparison for typical configurations between the box model and SW predictions of the propagation is presented in Fig. 17.1. The agreement is initially good, but eventually the box-model results lag more and more behind the SW. The box model captures well the influence of the stratification represented by the parameter $S$.

The fair agreements at both $S = 1$ and $0 < S < 1$ indicate that the present model is a reliable extension of the similar box-model approximations for homogeneous currents, and can be used, with due care, as a first and quick estimate of the flow of a current in a linearly-stratified ambient.

## 17.2    $S = 1$, inflow volume change $\mathcal{V} = qt^\alpha$

*In this section we use dimensional variables.* This provides a more direct insight into the balances and results which appear.

We consider a gravity current with $S = 1$, i.e., the density of the intruding fluid is equal to that of the ambient at the plane on which propagation occurs, $\rho_c = \rho_a(z = 0)$. This also corresponds to a symmetric intrusion. This current (or half-intrusion) is of changing volume in time according to $qt^\alpha$ where $q$ ($> 0$) and $\alpha$ ($\geq 0$) are given constants. The $\alpha = 0$ case gives the simple fixed-volume current, and the $\alpha = 1$ case corresponds to a current supplied by a constant-rate influx (say, by an open tap). The $\alpha > 1$ cases may represent a leak or eruption which worsens with time. This problem is an extension of the non-stratified cases considered in § 4.2.

The volume continuity requirement of the box model is

$$x_N(t)h_N(t) = \mathcal{V} = qt^\alpha. \tag{17.10}$$

In the particular case $\alpha = 0$ we can replace $q = \mathcal{V}$, where $\mathcal{V}$ is the volume (per unit width) and the dimensions are cm$^2$. In general, the dimensions of $q$ are cm$^2$ s$^{-\alpha}$. For definiteness, let the source be at $x = 0$. We assume that the inflow velocity is sufficiently small to prevent formation of a jet. This can be expressed as $u_i/(\mathcal{N}h_i) < 1$, where the subscript $i$ denotes the inlet position. This condition implies that characteristics from the interior can reach the inlet.

The following simplified behavior is assumed:

$$x_N = Kt^\beta; \quad h_N = \frac{\mathcal{V}}{x_N} = \frac{q}{K}t^{\alpha-\beta}; \tag{17.11a}$$

$$u_N = \beta K t^{\beta-1}; \quad u(x,t) = u_N\frac{x}{x_N} = \beta t^{-1}x; \tag{17.11b}$$

where $\beta$ and $K$ are constants to be determined by the solution. For later use we also rewrite the first dependency in (17.11a) as

$$t = \left(\frac{x_N}{K}\right)^{\frac{1}{\beta}}. \tag{17.12}$$

The relevant forces can now be estimated as we did for the homogeneous case in Chapter 4. The major difference from the homogeneous case is due to the effective driving buoyancy. Now the effective reduced gravity changes from zero to $\mathcal{N}^2 h_N$ over the thickness of the current (see second line of (15.14)). In the spirit of the momentum-integral approximation, the straightforward change of formulation is to replace $g'$ in the non-stratified buoyancy with $(1/2)\mathcal{N}^2 h_N$.

The relevant force results (per unit width or the current) are as follows. The driving "buoyancy" force is

$$F_B = \frac{1}{4}\rho_c \mathcal{N}^2 h_N^3 \tag{17.13}$$

(in the non-stratified case, the value is $(1/2)g' h_N^2$). The inertial forces are

$$F_I = \rho_c h_N \int_0^{x_N} u u_x \, dx = \frac{1}{2}\rho_c u_N^2 h_N, \tag{17.14}$$

and the viscous forces are

$$F_V = \rho_c \nu \int_0^{x_N} \frac{u}{h} \, dx = \frac{1}{2}\rho_c \nu \frac{u_N}{h_N} x_N. \tag{17.15}$$

The evaluation of the last two forces is not affected by the stratification. The stratification will enter into the results via the effect on the value of $u_N$.

The ratio of inertial to viscous forces can be expressed as

$$\frac{F_I}{F_V} = \frac{1}{\nu} u_N h_N^2 \frac{1}{x_N} = \frac{1}{\nu}\beta \left(\frac{q}{K}\right)^2 t^{2\alpha - 2\beta - 1}. \tag{17.16}$$

The thickness ratio $h_N/x_N$ behaves like $t^{\alpha - 2\beta}$; see (17.11a). The assumption of a thin current restricts the validity of the analysis to

$$\alpha - 2\beta < 0. \tag{17.17}$$

We shall see that this is always satisfied by the current (intrusion) under consideration.

Our objective is to determine the values of $\beta$ and $K$.

## 17.2.1 Inertial-buoyancy balance

We start again with the result that the propagation of the inertial current is governed by the jump conditions at the nose. The speed of propagation of the $S = 1$ current, in dimensional form, is

$$u_N = \frac{Fr}{\sqrt{2}}\mathcal{N} h_N. \tag{17.18}$$

(See (15.35), where the speed is scaled with $\mathcal{N} h_0$ and $h_N$ with $h_0$; or use (17.6) for $S = 1$ with the appropriate scaling.)

Combining this result with (17.11) we write

$$u_N = \beta K t^{\beta-1} = \frac{Fr}{\sqrt{2}} \mathcal{N}\left(\frac{q}{K}\right) t^{\alpha-\beta}. \tag{17.19}$$

Assuming that $Fr$ is a *constant*, we equate the powers of $t$ and the coefficients in the last equation to obtain

$$\beta = \beta_I = \frac{\alpha+1}{2}; \tag{17.20}$$

$$K = K_I = \left(\frac{Fr}{\sqrt{2}\beta_I}\right)^{1/2} [\mathcal{N}q]^{1/2}. \tag{17.21}$$

Here the subscript $I$ indicates that the flow is dominated by inertial (or inviscid) effects.

We summarize: the propagation of the 2D inviscid box-model gravity current (intrusion) is (in dimensional form)

$$x_N(t) = \left(\frac{\sqrt{2}Fr}{\alpha+1}\right)^{1/2} [\mathcal{N}q]^{1/2} t^{(\alpha+1)/2}. \tag{17.22}$$

The ease and simplicity of this result is of course pleasing, but, again, we must keep in mind not to confuse it with accuracy.

The propagation in a non-stratified ambient is with $t$ at the power $(\alpha+2)/3$. Therefore, the propagation in the non-stratified ambient is faster for $\alpha < 1$ and slower for $\alpha > 1$.

Substitution of (17.20) into (17.16) and use of (17.12) yield

$$\frac{F_I}{F_V} = \frac{1}{\nu}\beta_I\left(\frac{q}{K_I}\right)^2 t^{\alpha-2} = \frac{1}{\nu}\beta_I\left(\frac{q}{K_I}\right)^2 \left(\frac{x_N}{K_I}\right)^{2(\alpha-2)/(\alpha+1)}. \tag{17.23}$$

For $\alpha = 0$, these results reduce satisfactorily to the self-similar solution derived in § 15.6. In particular, we see that $F_I/F_V$ decays like $x_N^{-4}$.

Finally, we verify the thickness ratio requirement (17.17). We find that this equation is satisfied for any $\alpha$. We conclude that the present 2D inviscid box model is consistent with the thin-layer assumption for all practical values of $\alpha$.

For $\alpha > 0$ flows in a stratified ambient no analytical SW solutions have been developed, and hence the present box-model approximation is the only tool of this type. The price is a loss of accuracy which is difficult to estimate *a priori*. Numerical solutions of the SW equations are expected to provide more reliable results.

## 17.2.2 Viscous-buoyancy balance

When the inertial terms are negligibly small, the buoyancy is counteracted by viscous drag. For $F_V = F_B$ equations (17.13) and (17.15) yield

$$u_N = \frac{1}{2} \frac{\mathcal{N}^2}{\nu} \frac{h_N^4}{x_N}. \tag{17.24}$$

In the context of the present text, this is an interesting and novel result. We shall see later a more rigorous derivation (Chapter 19). The difference with the speed of propagation of the inertial current, $Fr\mathcal{N}h_N/\sqrt{2}$, is very significant.

Now we combine the assumed behavior (17.11) with (17.24) to obtain

$$\beta K t^{\beta-1} = \frac{\mathcal{N}^2}{2\nu} \frac{q^4}{K^5} t^{4\alpha-5\beta}. \tag{17.25}$$

The solution is

$$\beta = \beta_V = \frac{4\alpha+1}{6}; \tag{17.26}$$

$$K = K_V = \left(\frac{1}{2\beta_V}\right)^{1/6} \left[\frac{\mathcal{N}^2}{\nu} q^4\right]^{1/6}. \tag{17.27}$$

Here the subscript $V$ indicates that the flow is dominated by viscous effects.

We summarize: in dimensional form, the box-model propagation of a viscous current is

$$x_N(t) = \left(\frac{3}{4\alpha+1}\right)^{1/6} \left[\frac{\mathcal{N}^2}{\nu} q^4\right]^{1/6} t^{(4\alpha+1)/6}. \tag{17.28}$$

Again, the ease and simplicity of the results illustrate the appeal of the box models.

Substitution of (17.26) into (17.16) and use of (17.12) yield

$$\frac{F_I}{F_V} = \frac{1}{\nu} \beta_V \left(\frac{q}{K_V}\right)^2 t^{2(\alpha-2)/3} = \frac{1}{\nu} \beta_V \left(\frac{q}{K_V}\right)^2 \left(\frac{x_N}{K_V}\right)^{4(\alpha-2)/(4\alpha+1)}. \tag{17.29}$$

The previous result for $\beta_V$ satisfies the thickness ratio requirement (17.17) for any $\alpha$.

The viscous box-model propagation results are in good agreement with the similarity solution which will be derived later in Chapter 19. However, we shall see that the more rigorous similarity solution is quite simple and can be easily implemented for any practical value of $\alpha$. This reduces the motivation for using the box-model results in these cases.

### 17.2.3   Critical $\alpha$

We showed in § 4.2.3 that changes of regime from inertial to viscous, or from viscous to inertial, may occur. The type of change depends on the value of $\alpha$, or, to be more specific, on the sign of $\alpha - \alpha_c$. For the critical $\alpha = \alpha_c$ no change of regime occurs during the propagation.

The critical $\alpha$ effect appears also for the stratified case discussed here. An inspection of (17.23) and (17.29) shows that $\alpha_c = 2$. If the filling is with this value of $\alpha$, the ratio of inertial to viscous forces does not change with time. For $\alpha < \alpha_c$ the ratio of inertial to viscous forces decreases with time, and for $\alpha > \alpha_c$ this ratio increases with time. For $\alpha = \alpha_c$ the propagation is with $t$ at the power $\beta_c = 3/2$ for both inertial and viscous dominance. The experimental confirmation is still lacking. The derivation of the Julian number counterpart, see § 4.2.3, for the stratified current is left for an exercise.

# Chapter 18

## Box models for axisymmetric geometry

We assume that the current has a simple shape, a well defined "box." Here we are concerned with extensions of Chapter 7 to linearly-stratified ambients. In the present axisymmetric case the box is a cylinder of radius $r_N(t)$ and height $h_N(t)$. The main objective is to determine the behavior of these two variables, which means that we need two equations. We use the Boussinesq approximation.

The first governing equation is provided by the obvious volume continuity requirement

$$\frac{1}{2}r_N^2(t)h_N(t) = \mathcal{V}. \tag{18.1}$$

The second equation must involve some dynamic considerations. For a simplified flow field the global buoyancy, inertial, and viscous effects for the volume are calculated. Then a governing momentum integral balance is applied from which an equation for $u_N(t)$ can be derived. The integration of this equation provides $r_N(t)$.

The stratification is of the type defined in § 12.2 and sketched in Fig. 13.1. The densities of the current and ambient are given by

$$\rho_c = \rho_o(1+\epsilon), \quad \rho_a = \rho_o\left[1 + \epsilon S(1 - \frac{z}{H})\right], \quad (0 \le z \le H) \tag{18.2}$$

where $\epsilon = \rho_c/\rho_o - 1$, and $0 \le S \le 1$. We also recall

$$g' = \epsilon g; \quad \text{and} \quad \mathcal{N} = \left(\frac{Sg'}{H}\right)^{1/2}. \tag{18.3}$$

The current propagates on the plane $z = 0$.

Attention: in § 17.1 we use dimensionless variables, and in § 17.2 we use dimensional variables.

## 18.1   Fixed volume and inertial-buoyancy balance

*In this section we use dimensionless variables.* The scaling is given by (13.1)-(13.2), which we repeat here

$$\{r^*, z^*, h^*, H^*, t^*, u^*\} = \{r_0 r, h_0 z, h_0 h, h_0 H, T t, U u\}, \tag{18.4}$$

where

$$U = (h_0 g')^{1/2} = \mathcal{N}\sqrt{\frac{H}{S}} h_0 \quad \text{and} \quad T = \frac{r_0}{U}. \tag{18.5}$$

This will facilitate the comparison of the box-model approach with the previous SW derivation in Chapter 13.

In the particular case of inertial-buoyancy dominated motion we can skip the integral balance and go directly to the boundary condition (12.29)

$$u_N = \frac{dr_N}{dt} = Fr(a)\left(1 - S + \frac{S}{2}\frac{h_N}{H}\right)^{1/2}[h_N(t)]^{1/2}, \tag{18.6}$$

where $a = h_N(t)/H$. The justification is provided by the expectation that the box-model and SW propagation are governed by the same mechanism, so we can start with the best available result which is already in a sufficiently simple form for use in the box model.

Combining (18.1) and (18.6) we obtain one equation for $r_N(t)$ (or for $h_N(t)$). Formally, for the case of a fixed volume $\mathcal{V}$, the solution is

$$t - t_I = \frac{1}{(2\mathcal{V})^{1/2}}\int_{r_I}^{r_N(t)}[Fr(a(r))]^{-1}\left(1 - S + \frac{S}{H}\frac{\mathcal{V}}{r^2}\right)^{-1/2} r\, dr, \tag{18.7}$$

where $a = 2\mathcal{V}/(Hr^2)$. The initial condition is $r_N = r_I$ at $t = t_I$. The numerical quadrature is straightforward.

For $S = 1$ some compact analytical results can be obtained as follows.

Consider the case of a deep current with $S = 1$ and *constant Fr*. The result (18.7) can be expressed as

$$r_N = \left[3\sqrt{2}Fr\mathcal{V}H^{-1/2}(t - t_I) + r_I^3\right]^{1/3} = \left(3\sqrt{2}Fr\mathcal{V}\right)^{1/3}H^{-1/6}(t + \gamma)^{1/3}. \tag{18.8}$$

Here $\gamma$ is a constant which combines the values of the initial $t_I$ and $r_I$. This box-model result is fairly close to the prediction of the similarity solution (16.24). The propagation is with $t^{1/3}$, and the coefficient is proportional to $\mathcal{V}^{1/3}$ in both solutions (see (16.27)). The present result contains $H^{-1/6}$ in this coefficient because the present time scale differs from that used in Chapter 16. However, we recall that the self-similar solution predicts the concentration of the intrusion in a ring, in sharp contrast with the box assumed in the present approximation.

For $S = 1$ and the non-constant branch $Fr = (1/2)a^{-1/3}$ of the HS formula, the analytical solutions of (18.7) with initial condition $r_I = 1$ at $t_I = 0$ is

$$r_N(t) = \left[ \frac{7}{3} 2^{-5/6} H^{1/3} \mathcal{V}^{2/3} H^{-1/2} t + 1 \right]^{3/7} \quad (0.075 \leq h_N/H \leq 1). \quad (18.9)$$

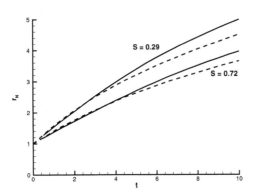

**FIGURE 18.1:** Propagation as a function of time, box model (dashed line) and SW (solid line) predictions for $H = 3$, and two values of $S$. (Ungarish and Huppert 2004 with permission.)

A comparison for typical configurations between the box model and SW predictions of the propagation is presented in Fig. 18.1. The agreement is initially good, but eventually the box-model results lag more and more behind the SW; this trend is more pronounced in the axisymmetric geometry than in the 2D counterpart. Again, the box model captures well the influence of the stratification represented by the parameter $S$. We therefore conclude that the present model is a reliable extension of the similar box-model approximations for homogeneous currents, and can be used, with due care, as a first and quick estimate of the flow.

## 18.2  $S = 1$, inflow volume change $\mathcal{V} = qt^\alpha$

*In this section we use dimensional variables.* This provides a more direct insight into the balances and results which appear. The 2D case is discussed in § 17.2.

We consider a gravity current with $S = 1$, i.e., the density of the intruding fluid is equal to that of the ambient at the plane on which propagation occurs, $\rho_c = \rho_a(z = 0)$. This also corresponds to a symmetric intrusion. This current (or half-intrusion) is of changing volume in time according to $qt^\alpha$ where $q \, (> 0)$ and $\alpha \, (\geq 0)$ are given constants. The $\alpha = 0$ case gives the simple fixed-volume current, and the $\alpha = 1$ case corresponds to a current supplied by a constant-rate influx (say, by an open tap). The $\alpha > 1$ cases may represent a leak or eruption which worsens with time. We assume that the source is at the center $r = 0$.

The volume continuity requirement of the box model is

$$\frac{1}{2}r_N^2(t)h_N(t) = \mathcal{V} = qt^\alpha. \tag{18.10}$$

In the particular case $\alpha = 0$ we can replace $q = \mathcal{V}$, where $\mathcal{V}$ is the volume (per radian) and the dimensions are $\text{cm}^3$. In general, the dimensions of $q$ are $\text{cm}^3 \, \text{s}^{-\alpha}$.

The following simplified behavior is assumed:

$$r_N = Kt^\beta; \quad h_N = \frac{2\mathcal{V}}{r_N^2} = \frac{2q}{K^2}t^{\alpha-2\beta}; \tag{18.11a}$$

$$u_N = \beta K t^{\beta-1}; \quad u(r,t) = u_N\frac{r}{r_N} = \beta t^{-1}r; \tag{18.11b}$$

where $\beta$ and $K$ are constants to be determined by the solution. For later use we also rewrite the first dependency in (18.11a) as

$$t = \left(\frac{r_N}{K}\right)^{\frac{1}{\beta}}. \tag{18.12}$$

The relevant forces can now be estimated. As explained for the 2D case in § 17.2, the reduced gravity which acts on the intrusion increases from $0$ to $\mathcal{N}^2 h_N$ as the distance from the plane of motion $z = 0$ increases to $h_N$. Consequently, the reduced gravity $g'$ in the driving buoyancy force of the homogeneous ambient case is changed with the representative average $(1/2)\mathcal{N}^2 h_N$. This approximation facilitates a straightforward extension of the non-stratified counterpart results derived in § 7.2.

The buoyancy force, represented by the pressure force on the periphery "dam" which bounds the current in the box (per radian), is approximated by

$$F_B = \frac{1}{4}\rho_c \mathcal{N}^2 r_N h_N^3 \tag{18.13}$$

(in the non-stratified case, the value is $(1/2)g'r_N h_N^2$).

The inertial forces are

$$F_I = \rho_c h_N \int_0^{r_N} u u_r r \, dr = \frac{1}{3}\rho_c u_N^2 r_N h_N, \tag{18.14}$$

and the viscous forces are

$$F_V = \rho_c \nu \int_0^{r_N} \frac{u}{h} r\, dr = \frac{1}{3} \rho_c \nu \frac{u_N}{h_N} r_N^2.$$ (18.15)

The evaluation of the last two forces is not affected by the stratification. The stratification will enter into the results via the effect on the value of $u_N$.

The ratio of inertial to viscous forces can be expressed as

$$\frac{F_I}{F_V} = \frac{1}{\nu} u_N h_N^2 \frac{1}{r_N} = \frac{1}{\nu} \beta \left(\frac{2q}{K^2}\right)^2 t^{2\alpha-4\beta-1}.$$ (18.16)

The thickness ratio $h_N/r_N$ behaves like $t^{\alpha-3\beta}$; see (18.11a). The assumption of a thin current restricts the validity of the analysis to

$$\alpha - 3\beta < 0.$$ (18.17)

We shall see that this is always satisfied by the current (intrusion) under consideration.

Our objective is to determine the values of $\beta$ and $K$.

## 18.2.1   Inertial-buoyancy balance

We start again with the result that the propagation of the inertial current is governed by the jump conditions at the nose. The speed of propagation of the $S = 1$ current, in dimensional form, is

$$u_N = \frac{Fr}{\sqrt{2}} \mathcal{N} h_N.$$ (18.18)

(We can use (18.6) for $S = 1$, recast in dimensional form.) We therefore write

$$u_N = \beta K t^{\beta-1} = \frac{Fr}{\sqrt{2}} \mathcal{N} \frac{2q}{K^2} t^{\alpha-2\beta}.$$ (18.19)

Assuming that $Fr$ is a *constant*, we equate the powers of $t$ and the coefficients in the last equation to obtain

$$\beta = \beta_I = \frac{\alpha+1}{3};$$ (18.20)

$$K = K_I = \left(\frac{\sqrt{2}Fr}{\beta_I}\right)^{1/3} [\mathcal{N}q]^{1/3}.$$ (18.21)

Here the subscript $I$ indicates that the flow is dominated by inertial (or inviscid) effects. Comparing with the 2D case, we observe that the radially-divergent geometry entered into the behavior of $h_N$ in the equation (18.19) for $u_N$. As a result, for a given $\alpha$, the propagation of the axisymmetric current is with a smaller power $\beta$ than the 2D counterpart.

We summarize: the propagation of an inviscid axisymmetric box-model gravity current is (in dimensional form)

$$r_N(t) = \left(\frac{3\sqrt{2}Fr}{\alpha + 1}\right)^{1/3} [\mathcal{N}q]^{1/3} t^{(\alpha+1)/3}. \tag{18.22}$$

Substitution of (18.20) into (18.16) and use of (18.12) yields

$$\frac{F_I}{F_V} = \frac{1}{\nu}\beta_I \left(\frac{2q}{K_I^2}\right)^2 t^{(2\alpha-7)/3} = \frac{1}{\nu}\beta_I \left(\frac{2q}{K_I^2}\right)^2 \left(\frac{r_N}{K_I}\right)^{(2\alpha-7)/(\alpha+1)}. \tag{18.23}$$

For $\alpha = 0$ these time-dependency results reduce well to the SW self-similar solution. In particular, we see that $F_I/F_V$ decays like $r_N^{-7}$.

When $\alpha > 0$, the inflow conditions are inconsistent with the assumption that $u = u_N r/r_N$ at $r = 0$. We assume that this discrepancy is resolved in a small domain $r/r_N \ll 1$ which has a negligible effect on the global balances (18.13)-(18.15).

Substituting the previous result for $\beta_I$ into the thickness ratio requirement (18.17), we conclude that the present axisymmetric inviscid box model is valid for any $\alpha$.

## 18.2.2   Viscous-buoyancy balance

When the inertial terms are negligibly small, the buoyancy is counteracted by viscous drag. For $F_V = F_B$ equations (18.13) and (18.15) yield

$$u_N = \frac{3}{4}\frac{\mathcal{N}}{\nu}\frac{h_N^4}{r_N}. \tag{18.24}$$

In the context of the present text, this is an interesting and novel result. We shall see later a more rigorous derivation (§ 19.2). The difference with the speed of propagation of the inertial current, $Fr\mathcal{N}h_N/\sqrt{2}$, is very significant.

Now we substitute the assumed behavior (18.11) into (18.24) to obtain

$$\beta K t^{\beta-1} = 12\frac{\mathcal{N}^2}{\nu}\frac{q^4}{K^9}t^{4\alpha-9\beta}. \tag{18.25}$$

The solution is

$$\beta = \beta_V = \frac{4\alpha + 1}{10}; \tag{18.26}$$

$$K = K_V = \left(\frac{12}{\beta_V}\right)^{1/10} \left[\frac{\mathcal{N}^2}{\nu}q^4\right]^{1/10}. \tag{18.27}$$

Here the subscript $V$ indicates that the flow is dominated by viscous effects.

We summarize: in dimensional form, the box-model propagation of the axisymmetric viscous current is

$$r_N(t) = \left(\frac{120}{4\alpha + 1}\right)^{1/10} \left[\frac{\mathcal{N}^2}{\nu} q^4\right]^{1/10} t^{(4\alpha+1)/10}. \qquad (18.28)$$

Substitution of (18.26) into (18.16) and use of (18.12) yield

$$\frac{F_I}{F_V} = \frac{1}{\nu}\beta_V \left(\frac{2q}{K_V^2}\right)^2 t^{(2\alpha-7)/5} = \frac{1}{\nu}\beta_V \left(\frac{2q}{K_V^2}\right)^2 \left(\frac{r_N}{K_V}\right)^{2(2\alpha-7)/(4\alpha+1)}. \qquad (18.29)$$

The previous result for $\beta_V$ satisfies the thickness ratio requirement (18.17) for any relevant $\alpha$.

The viscous box-model propagation results are in good agreement with the similarity solution which will be derived later in § 19.2. However, we shall see that the more rigorous similarity solution is quite simple and can be easily implemented for any practical value of $\alpha$. This reduces the motivation for using the box-model results. When $\alpha > 0$, the inflow conditions are inconsistent with the assumption that $u = u_N r/r_N$ at $r = 0$. We assume that this discrepancy is resolved in a small domain $r/r_N \ll 1$ which has a negligible effect on the global balances. In the viscous case this difficulty is less severe than in the inertial case. The viscous diffusion smooths out large gradients which may develop about the source, while in the inviscid system the source conditions can be propagated along characteristics deep into the current.

## 18.2.3 Critical $\alpha$

As for the 2D current with inflow, a critical value of $\alpha$, $\alpha_c = 7/2$, for the behavior of the solutions exists. At this point $\beta_I = \beta_V = \beta_c = 3/2$, and for larger values of $\alpha$ the rate of spread of the viscous current is larger than that of the inertial current.

The implications of the $\alpha_c$ were discussed in § 4.2.3. We recall that in the non-stratified case both the 2D and axisymmetric currents have the same value of critical $\beta_c = 5/4$. A similar feature is observed for the stratified current, but now this critical value is $\beta_c = 3/2$.

## 18.2.4 Some experimental support

To our knowledge, no experimental confirmation of the critical behavior has been reported for the stratified axisymmetric current.

For the constant flux $\alpha = 1$ some comparisons with experimental data were presented, e.g., Didden and Maxworthy (1982), Zatsepin and Shapiro (1982), Lemckert and Imberger (1993), and Kotsovinos (2000), but the results are not sharp. The first two investigations considered a viscous intrusion, and recorded propagation is with $t^{1/2}$, in agreement with (18.28) for $\alpha = 1$. This will be discussed further in § 19.2.2.

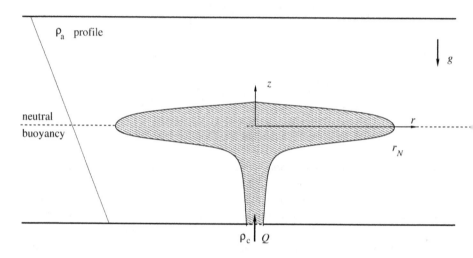

**FIGURE 18.2:**   Sketch of plume-to-intrusion experiment.

The third and fourth studies considered a system of the type sketched in Fig. 18.2. A source of constant volume flux releases a plume of density $\rho_c$ at the bottom of a linearly stratified saline ambient ($\rho_a$ at the bottom is larger than $\rho_c$). This plume first moves upward and then spreads out at the level of neutral buoyancy in the linearly-stratified saline. Lemckert and Imberger (1993) performed the experiments in large-scale reservoirs ($r_N \sim 100\,\text{m}, h_N \sim 10\,\text{m}$); Kotsovinos (2000) used tanks of typical dimension 2 m. The conversion of the vertical plume into a horizontally-spreading intrusion involves mixing with the ambient fluid at various levels, and introduces uncertainties about the values of $q, \rho_c$, and even $\alpha$. Kotsovinos (2000) covered a fairly wide range of parameters, but the data are presented in scalings which do not allow a simple comparison with the results given above. On the other hand, Lemckert and Imberger (1993) concluded that their data are in the inertial-buoyancy regime, and are well represented by the rule

$$r_N = C[\mathcal{N}Q]^{1/3} t^{2/3}, \tag{18.30}$$

where $Q = 4\pi q$ is the rate of volume increase for the whole body of intruding fluid. This propagation is in qualitative agreement with (18.22) for $\alpha = 1$. The fit to experiments produced the value $C = 0.40$. The theoretical value for the corresponding coefficient in (18.22), for $Fr = 1.19$ and $\alpha = 1$, is 0.59. The slower propagation can be attributed to buoyancy reduction due to mixing.

# Chapter 19

## Lubrication theory for viscous currents with $S = 1$

Here we consider the prototype thin viscous current which propagates over a horizontal solid boundary in a linearly-stratified ambient such that $\rho_c = \rho_b = \rho_a(z = 0)$. This corresponds to the $S = 1$ case and is also relevant to an intrusion. The non-stratified counterpart is considered in Chapter 11. The system satisfies the Boussinesq assumption.

We assume that in the current the viscous forces, $F_V$, are dominant over the inertial forces, $F_I$. This means that the properly defined Reynolds number is small. This also implies that viscous diffusion smooths out velocity gradients in a very short time interval as compared to the typical time of propagation. For a current of fixed volume released from a rectangular lock we can use the results derived in § 15.6.1 and § 16.3.4 to determine when the viscous dominance appears. That analysis makes it evident that the typical viscous current is a very thin layer of fluid. The box-model approximations considered in Chapters 17 and 18 also provide fairly easy estimates of the position where $F_I/F_V$ becomes small.

## 19.1  2D geometry

We use a one-layer approximation. The objective is to derive and solve the governing equations for the thickness of the layer, $h(x, t)$, and the $z$-averaged speed of motion, $\bar{u}(x, t)$. This also provides the formula for the propagation, $x_N(t)$.

We assume a motionless hydrostatic ambient fluid of density $\rho_a(z)$. The variables of the ambient are denoted by the subscript $a$. For definiteness, the current propagates over a solid bottom, $z = 0$, which is also the neutral-density plane, i.e., $\rho_c = \rho_a(z = 0)$. Usually the variables in this layer are without a subscript, and when emphasis is necessary we use the subscript $c$. For example, $\nu$ means the kinematic viscosity of the dense fluid current layer.

Consider the current. We apply the considerations discussed in the context of the non-stratified ambient in § 11.1. The fact that we deal with a thin layer of fluid validates the same lubrication-theory simplification. In particular,

(a) the pressure obeys the hydrostatic balance in the $z$ direction; and (b) the pressure is continuous on the interface $z = h(x,t)$. Consequently, the fundamental relationship (12.18) (here for $S = 1$) between the driving pressure gradient and the slope of the interface developed for the inviscid one-layer model is valid also for the viscous case. Again, the longitudinal pressure gradient in the current can be written as

$$\frac{\partial p_c}{\partial x} = \rho_o \mathcal{N}^2 h \frac{\partial h(x,t)}{\partial x}. \tag{19.1}$$

Note that the same result follows from (15.13) and the second line of (15.14).

In the $x$-momentum equation of the current the leading balance is between the pressure gradient and viscous shear

$$0 = -\frac{1}{\rho_c}\frac{\partial p_c}{\partial x} + \nu \frac{\partial^2 u}{\partial z^2}, \tag{19.2}$$

which, using (19.1), is expressed as

$$0 = -\mathcal{N}^2 h \frac{\partial h}{\partial x} + \nu \frac{\partial^2 u}{\partial z^2}. \tag{19.3}$$

The momentum equation (19.3) is readily integrated to yield

$$u(x,z,t) = \frac{\mathcal{N}^2}{\nu} h \frac{\partial h}{\partial x}\left[\frac{1}{2}z^2 + Az + B\right] \quad (0 \le z \le h(x,t)), \tag{19.4}$$

where $A$ and $B$ are functions of $x,t$ to be determined by boundary conditions.

In the prototype problem, sketched in Fig. 11.1, the base of the current is a no-slip wall at $z = 0$ and hence $u = 0$ at this position. At the top $z = h$ the current is matched to a thick layer of ambient fluid. As discussed in § 11.1, we apply to (19.4) the conditions

$$u(x, z = 0, t) = 0; \quad \frac{\partial u}{\partial z}(x, z = h, t) = 0; \tag{19.5}$$

to obtain

$$u(x,z,t) = \frac{\mathcal{N}^2}{\nu} h \frac{\partial h}{\partial x}\left(\frac{1}{2}z^2 - hz\right). \tag{19.6}$$

Finally, the $z$-averaged velocity in the viscous current is

$$\bar{u}(x,t) = \frac{1}{h}\int_0^h u\,dz = -\frac{1}{3}\frac{\mathcal{N}^2}{\nu}h^3\frac{\partial h}{\partial x}. \tag{19.7}$$

The fundamental features of the viscous current are as in the non-stratified counterpart. The speed of propagation decreases with $\nu^{-1}$, and forward propagation requires a negative $\partial h/\partial x$. The negative slope of the interface produces a higher pressure in the back which pushes forward the dense fluid.

This forcing is opposed by the viscous friction. While in the non-stratified case the speed behaves like $h^3$, see (11.9), now the behavior is like $h^4$. This is because the driving force is now proportional to $h$. This indicates a slower propagation and faster decrease of speed than in the non-stratified case. The speed is fully determined by the local behavior of the interface, so the next major objective is to calculate $h(x, t)$.

Following the methodology of § 11.1, we attempt to determine the behavior of $h(x, t)$ by the use of the continuity equation (2.24), in which $\bar{u}$ is defined by (19.7). This combination of continuity and momentum balances yields

$$\frac{\partial h}{\partial t} - \frac{1}{3}\frac{N^2}{\nu}\frac{\partial}{\partial x}h^4\frac{\partial h}{\partial x} = 0. \tag{19.8}$$

The governing equations of the viscous current in a linear stratification with $S = 1$ are (19.7) and (19.8). The boundary conditions will be discussed later. Contrary to the inviscid SW case: (a) Here the equations for $h$ and $u$ can be easily decoupled. Actually, only a single PDE must be solved; this is equation (19.8) for the unknown $h(x, t)$. (b) The dominant PDE (19.8) is parabolic (the SW system is hyperbolic).

To both generalize the results and simplify the manipulation, we perform two preliminary steps:

1. We assume that the volume of the current is given by

$$\mathcal{V} = qt^\alpha, \tag{19.9}$$

where $q$ ($> 0$) and $\alpha$ ($\geq 0$) are constants, with the particular case

$$q = \mathcal{V} \quad \text{for} \quad \alpha = 0. \tag{19.10}$$

The source is at $x = 0$. This covers a quite general variety of currents. In addition to the fixed volume current, we have these produced by constant discharge, $\alpha = 1$, by an open tap, say; and also currents produced by a worsening leakage, $\alpha > 1$.

2. We switch to dimensionless variables. The horizontal and vertical lengths are scaled with $x_0, h_0$, respectively, and volume (per unit width) is scaled with $x_0 h_0$. The velocity and time are scaled with

$$U = \frac{1}{3}N^2\frac{1}{\nu}h_0^4\frac{1}{x_0}; \quad T = \frac{x_0}{U}. \tag{19.11}$$

Note that the coefficient $q$ is scaled with $x_0 h_0 T^{-\alpha}$, a quite unusual quantity. We also keep in mind that the proper choice of $x_0$ and $h_0$ may require some analysis of the particular problem. For example, for a fixed-volume current the intuitive choice for $x_0$ and $h_0$ is the dimensions of the lock. However, the initial motion may well be in the inviscid regime. Therefore a more relevant choice will be the length and mean thickness of this current when transition to the viscous regime occurs.

The following analysis uses *dimensionless variables* unless stated otherwise. The system for solution is the dimensionless form of the equations (19.7) and (19.8), subject to the volume continuity (19.9). This is expressed as follows: The momentum equation is

$$\bar{u}(x,t) = -h^3 \frac{\partial h}{\partial x};$$

(19.12)

and the continuity equation reads

$$\frac{\partial h}{\partial t} - \frac{\partial}{\partial x} h^4 \frac{\partial h}{\partial x} = 0;$$

(19.13)

subject to

$$\int_0^{x_N(t)} h(x,t)dx = V = qt^\alpha.$$

(19.14)

Clearly, we must concentrate on the solution of the PDE (19.13) in this system for $h$, then the velocity follows. This is facilitated, again, by the transformation introduced in § 2.3. Letting $y = x/x_N(t)$ and using (2.46)-(2.47) we obtain the following equation for $h(y,t)$

$$\frac{\partial h}{\partial t} - y \frac{\dot{x}_N}{x_N} \frac{\partial h}{\partial y} - \frac{1}{x_N^2} \frac{\partial}{\partial y} h^4 \frac{\partial h}{\partial y} = 0 \quad (0 \le y \le 1);$$

(19.15)

subject to

$$x_N(t) \int_0^1 h(y,t)dy = V = qt^\alpha.$$

(19.16)

In general, the solution can be obtained by numerical methods. The behavior of $h$ poses a difficulty at the nose, when $y$ approaches 1. Since the equation is parabolic a nose jump condition of the type encountered for the SW inviscid hyperbolic formulation is not expected. The propagation of the viscous fluid is supported at any $y$ by a negative slope of $h$; see (19.12). We thus expect that $h$ decreases continuously with $y$, to $h = 0$ at $y = 1$. However, to keep this point moving we need to assume that $h^3(\partial h/\partial y)$ is finite when $h \to 0$ as $y \to 1$. This is consistent with the formulation. To work out the details, an analytical solution of (19.15) about $y = 1$ must be considered. The pertinent results are obtained below as a part of the similarity solution. A numerical finite-difference solution of (19.15) can be performed for the $y$ domain $[0, 1 - \Delta]$, with the boundary condition at $y = 1 - \Delta$ given by the analytical results, where $\Delta$ is reasonably small (say 0.005). The volume in this subdomain is negligibly small, as shown below. However, there is little incentive for calculating the numerical solution of the PDE for $h(y,t)$, because, as for the non-stratified counterpart, a convenient analytical result is available. The initial conditions for this equation may also be difficult to specify, because in many cases of interest the current propagates first in an inertial-dominated regime where (19.15) is not valid.

The analytical solution of (19.15)-(19.16) is obtained by self-similar assumptions. We seeks a solution of the form

$$x_N = Kt^\beta; \quad h = \frac{\nu}{x_N}\mathcal{H}(y) = \frac{q}{K}t^{\alpha-\beta}\mathcal{H}(y); \tag{19.17}$$

where $K$ and $\beta$ are some positive constants. The objective is to determine $\beta$, $K$, and $\mathcal{H}(y)$. Substitution into (19.15) and arrangement yields

$$(\alpha - \beta)\mathcal{H} - \beta y\mathcal{H}' - t^{4\alpha-6\beta+1}\frac{q^4}{K^6}(\mathcal{H}^4\mathcal{H}')' = 0. \tag{19.18}$$

Similarity requires that the power of $t$ is zero. This determines

$$\beta = \frac{4\alpha + 1}{6}. \tag{19.19}$$

Further simplification is achieved upon the substitution

$$\mathcal{H}(y) = \frac{K^{3/2}}{q}\lambda(y), \tag{19.20}$$

which reduces (19.18) to

$$(\lambda^4\lambda')' + \beta y\lambda' - (\alpha - \beta)\lambda = 0. \tag{19.21}$$

The boundary condition for this equation is $\lambda(1) = 0$. The pressure gradient drives the current against the viscous forces, and this requires a decreasing $h$ with $x$. This implies that the self-similar profiles $\mathcal{H}$ (and $\lambda$) are positive monotonically decreasing functions of $y$ for $0 < y < 1$. This behavior must end with $h = 0$ at $x_N$, i.e., $\mathcal{H}(1) = \lambda(1) = 0$.

Equation (19.21) has a singularity at $y = 1$. Using a Frobenius series expansion, $\lambda = \xi^\gamma(a_0 + a_1\xi + \cdots)$, where $\xi = (1 - y)$, we obtain the leading term

$$\lambda(y) = \left[\frac{2}{3}(4\alpha + 1)\right]^{1/4}(1 - y)^{1/4}. \tag{19.22}$$

We can now calculate good approximations for the values of $\lambda$ and $\lambda'$ at $y = 1-\Delta$ for some small $\Delta$ (0.005 say). These values can be used as boundary conditions for the numerical integration of (19.21) in the domain $[0, 1 - \Delta]$. This method determines uniquely the solution $\lambda(y)$ for relevant values of $\alpha$. The typical thickness profiles are shown in Fig. 19.1.

Next, we calculate $K$ using the global volume balance (19.16), into which we substitute (19.17) and (19.20). After some algebra this yields

$$\int_0^1 \mathcal{H}(y)dy = \frac{K^{3/2}}{q}\int_0^1 \lambda(y)dy = 1, \tag{19.23}$$

and hence

$$K = q^{2/3}\left[\int_0^1 \lambda(y)dy\right]^{-2/3} = q^{2/3}\tilde{\eta}_N(\alpha). \tag{19.24}$$

Since in general $\lambda(y)$ is calculated numerically, so is also $\tilde{\eta}_N(\alpha)$; see Fig. 19.2. The values of $\tilde{\eta}_N$ are of the order of unity, which vindicates the scaling used in the derivation.

The calculation of the speed $\bar{u}(y,t)$ is left for an exercise.

There are significant differences between the inviscid and the viscous cases. In the inviscid case $\mathcal{H}$ is an increasing function of $y$ whose maximum is at the nose $y = 1$, while in the viscous case $\mathcal{H}$ decreases with $y$ to $\mathcal{H}(1) = 0$ at the nose; see Figs. 15.14 and 19.1. This is, again, a manifest of the essential difference: the inviscid current is controlled by the pressure-jump conditions at the nose, while the viscous current is pushed from behind against the local shear. For $\alpha = 0$, in the inviscid case the speed of propagation decreases like $x_N^{-1}$, while in the viscous case like $x_N^{-5}$.

Even for the simple $\alpha = 0$ case the transition process from the inviscid to the viscous solution is not clear. We can patch the two solutions as suggested for the non-stratified case. For the more general $\alpha > 0$ case the situation is more difficult because we do not have simple similarity solutions for the inviscid regime. In this book, the inviscid-inertial current with volume changing like $qt^\alpha$ is covered only by the imprecise box-model approximation. These differences can be related to the hyperbolic properties of the inviscid system, which support formation and propagation of steep changes of the dependent variables. On the other hand, the viscosity tends to smooth out gradients in a shorter time interval than the time of propagation. Therefore the viscous current is intrinsically simpler than the inviscid-inertial current: it is governed by one parabolic PDE for $h$.

### 19.1.1 Summary

The behavior of the viscous 2D current of volume $\mathcal{V} = qt^\alpha$ is predicted by the similarity solution of the lubrication formulation as (in dimensionless form)

$$x_N(t) = Kt^{(4\alpha+1)/6}; \quad h(y,t) = K^{1/2}t^{(2\alpha-1)/6}\lambda(y); \qquad (19.25)$$

where $K = q^{2/3}\tilde{\eta}_N(\alpha)$. $\lambda(y)$ and $\tilde{\eta}_N(\alpha)$ are displayed in Figs. 19.1 and 19.2. The scaling is given by (19.11).

The transformation of the foregoing dimensionless results to dimensional counterparts requires some work, but is straightforward. We must keep in mind the unusual scaling of $q$ with $x_0h_0T^{-\alpha}$. The result is

$$x_N(t) = \frac{\tilde{\eta}_N(\alpha)}{3^{1/6}}\left[\frac{\mathcal{N}^2}{\nu}q^4\right]^{1/6}t^{(4\alpha+1)/6}, \qquad (19.26)$$

where the variables are dimensional (except for $\alpha$ and $\tilde{\eta}_N$ which are numerical coefficients by definition).

A comparison of this self-similarity result with the box-model approximation (17.28) shows a remarkable agreement. This comparison gives some credence to the box-model approximation, but, on the other hand, the present

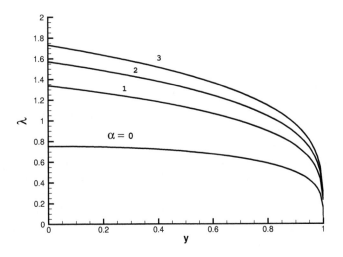

**FIGURE 19.1:** Similarity profiles of interface for 2D viscous current in $S = 1$ stratification. $\lambda(y)$ is given for various values of $\alpha$.

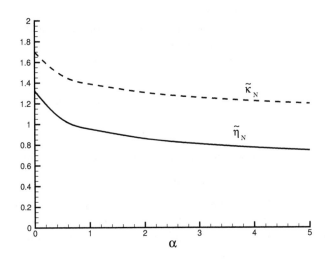

**FIGURE 19.2:** Coefficients for the self-similar viscous current as functions of $\alpha$. (a) $\tilde{\eta}_N$ for the 2D case; and (b) $\tilde{\kappa}_N$ for the axisymmetric case (dash-dot line).

solution is more accurate and sufficiently simple for making it the preferred modeling tool.

---

## 19.2 Axisymmetric geometry

Consider an axisymmetric configuration. The objective is to formulate and solve the governing equations for the thickness $h(r,t)$ and $z$-averaged speed $\bar{u}(r,t)$ in the viscous current. This also provides the behavior of $r_N(t)$.

We again consider a one-layer system, and use the same type of simplifications as for the 2D geometry; see § 19.1. The horizontal coordinate $x$ of the 2D case is replaced with the cylindrical radius $r$. The momentum considerations used in the 2D case can be extended to the cylindrical propagation. To be specific, (19.1)-(19.7) are applicable when $x$ is replaced by $r$. Here the volumes and volume fluxes are per radian.

We start the analysis with the *dimensional form* of the averaged radial velocity

$$\bar{u}(r,t) = \frac{1}{h}\int_0^h u\,dz = -\frac{1}{3}\frac{\mathcal{N}^2}{\nu}h^3\frac{\partial h}{\partial r}, \qquad (19.27)$$

and the corresponding continuity equation derived from (6.6), which is

$$\frac{\partial h}{\partial t} - \frac{1}{3}\frac{\mathcal{N}^2}{\nu}\frac{1}{r}\frac{\partial}{\partial r}rh^4\frac{\partial h}{\partial r} = 0. \qquad (19.28)$$

Again, the equations for speed and thickness can be decoupled. The major task is to solve a single parabolic PDE for the unknown $h(r,t)$.

To both generalize the results and simplify the manipulation, we perform two preliminary steps:

1. We assume that the volume of the current (per radian) is given by

$$\mathcal{V} = qt^\alpha, \qquad (19.29)$$

   where $q\ (>0)$ and $\alpha\ (\geq 0)$ are constants, with the particular case

$$q = \mathcal{V} \quad \text{for} \quad \alpha = 0. \qquad (19.30)$$

   The source is, for simplicity, at $r = 0$. This introduces a non-physical behavior of $u$ near the source for the $\alpha > 0$ case, but we assume that this is a local effect which has minor influence on the main results.

2. We switch to dimensionless variables. The horizontal and vertical lengths are scaled with $r_0, h_0$, respectively, and volume (per radian) is scaled with $r_0^2 h_0$. The velocity and time are scaled with

$$U = \frac{1}{3}\mathcal{N}^2\frac{1}{\nu}h_0^4\frac{1}{r_0}; \quad T = \frac{r_0}{U}. \qquad (19.31)$$

Note that the coefficient $q$ is scaled with $r_0^2 h_0 T^{-\alpha}$, a quite unusual quantity. We also keep in mind that the proper choice of $r_0$ and $h_0$ may require some analysis of the particular problem. For example, for a fixed-volume current the intuitive choice for $r_0$ and $h_0$ is the dimensions of the lock. However, the initial motion may well be in the inviscid regime. Therefore a more relevant choice will be the radius and mean thickness of this current when transition to the viscous regime occurs.

The following analysis uses *dimensionless variables* unless stated otherwise. The momentum equation is

$$\bar{u}(r,t) = -h^3 \frac{\partial h}{\partial r}; \tag{19.32}$$

and the continuity equation (19.28) reads

$$\frac{\partial h}{\partial t} - \frac{1}{r}\frac{\partial}{\partial r} r h^4 \frac{\partial h}{\partial r} = 0, \tag{19.33}$$

subject to

$$\int_0^{r_N(t)} h(r,t) r \, dr = \mathcal{V} = q t^\alpha. \tag{19.34}$$

Clearly, we must concentrate on the solution of the PDE equation for $h$, then the velocity follows. This is facilitated, again, by the transformation introduced in § 2.3 and § 6.2. Letting $y = r/r_N(t)$ and using (2.46)-(2.47) we obtain the following equation for $h(y,t)$

$$\frac{\partial h}{\partial t} - y\frac{\dot{r}_N}{r_N}\frac{\partial h}{\partial y} - \frac{1}{r_N^2}\frac{1}{y}\frac{\partial}{\partial y} y h^4 \frac{\partial h}{\partial y} = 0 \quad (0 \leq y \leq 1); \tag{19.35}$$

subject to

$$r_N^2(t) \int_0^1 h(y,t) y \, dy = \mathcal{V} = q t^\alpha. \tag{19.36}$$

We seek a similarity solution of the form

$$r_N = K t^\beta; \quad h = \frac{\mathcal{V}}{r_N^2}\mathcal{H}(y) = \frac{q}{K^2}t^{\alpha-2\beta}\mathcal{H}(y); \tag{19.37}$$

where $K$ and $\beta$ are some positive constants. Substitution into (19.35) and arrangement yields

$$(\alpha - 2\beta)y\mathcal{H} - \beta y^2 \mathcal{H}' - t^{4\alpha-10\beta+1}\frac{q^4}{K^{10}}\left(y\mathcal{H}^4\mathcal{H}'\right)' = 0. \tag{19.38}$$

Similarity requires that the power of $t$ is zero. This determines

$$\beta = \frac{4\alpha + 1}{10}. \tag{19.39}$$

Further simplification is achieved upon the substitution

$$\mathcal{H}(y) = \frac{K^{5/2}}{q}\lambda(y),$$ (19.40)

which reduces (19.38) to

$$(y\lambda^4\lambda')' + \beta y^2\lambda' - (\alpha - 2\beta)y\lambda = 0.$$ (19.41)

The boundary condition for this equation is $\lambda(1) = 0$. The justification is as in the 2D case. The pressure gradient drives the current against the viscous forces, and this requires a decreasing $h$ with $r$. This behavior must end with $h = 0$ at $r_N$, i.e., $\mathcal{H}(1) = \lambda(1) = 0$.

Equation (19.41) has a singularity at $y = 1$. Using a Frobenius series expansion $\lambda = \xi^\gamma(a_0 + a_1\xi + \cdots)$, where $\xi = (1 - y)$, we obtain the leading term

$$\lambda(y) = \left[\frac{2}{5}(4\alpha + 1)\right]^{1/4}(1 - y)^{1/4}.$$ (19.42)

In general, we use this approximation to calculate numerically $\lambda(y)$ as explained for the 2D case. Results for typical values of $\alpha$ are displayed in Fig. 19.3. For $\alpha > 0$ a non-physical increase of thickness appears near the center $y = 0$. This is because the finite imposed influx $\alpha q t^{\alpha-1}$ must be matched by $ruh$ with $r \to 0$ (theoretically). This introduces unrealistic large values of $u$ and $h$ near the center. The accepted remedy is to simply ignore the small region about the center ($y < 0.05$ say). The volume in this domain is small, and in any case a real source has a finite radius.

Next, we calculate $K$ using the global volume balance (19.36) in which we substitute (19.37) and (19.40). After some algebra this yields

$$\int_0^1 \mathcal{H}(y)y\,dy = \frac{K^{5/2}}{q}\int_0^1 \lambda(y)y\,dy = 1,$$ (19.43)

and hence

$$K = q^{2/5}\left[\int_0^1 \lambda(y)y\,dy\right]^{-2/5} = q^{2/5}\tilde{\kappa}_N(\alpha).$$ (19.44)

The value of $\tilde{\kappa}_N(\alpha)$ is calculated numerically; see Fig. 19.2.

The calculation of the speed is left for an exercise.

For the $\alpha = 0$ case we can compare to the SW inviscid current (intrusion) discussed in § 16.3. In the self-similar inviscid stage the propagation is with $t^{1/3}$; see (16.24). Here the propagation is significantly slower, with $t^{1/10}$. In the $\alpha = 0$ case the boundary condition for $u$ at the center $y = 0$ is simply satisfied. For $\alpha > 0$ the behavior of $u$ for very small $y$ poses a difficulty, as discussed above.

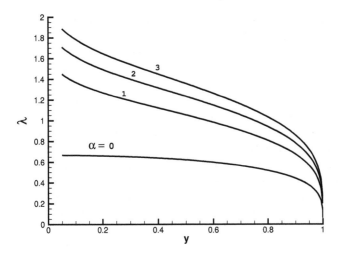

**FIGURE 19.3:** Similarity profiles of interface for the axisymmetric viscous current. $\lambda(y)$ is given for various values of $\alpha$.

## 19.2.1  Summary

The behavior of the viscous axisymmetric current of volume $\mathcal{V} = qt^\alpha$ (per radian) is predicted by the similarity solution of the lubrication formulation as (in dimensionless form)

$$r_N(t) = Kt^{(4\alpha+1)/10}; \quad h(y,t) = K^{1/2}t^{(\alpha-1)/5}\lambda(y); \qquad (19.45)$$

where $K = q^{2/5}\tilde{\kappa}_N(\alpha)$. The profile $\lambda(y)$ and the values $\tilde{\kappa}_N(\alpha)$ are displayed in Figs. 19.3 and 19.2.

The transformation of the foregoing dimensionless results to dimensional counterparts requires some work, but is straightforward. We must keep in mind the unusual scaling of $q$ with $r_0^2 h_0 T^{-\alpha}$. It is left for an exercise to show that the dimensional result is

$$r_N(t) = \frac{\tilde{\kappa}_N(\alpha)}{3^{1/10}} \left[\frac{\mathcal{N}^2}{\nu}q^4\right]^{1/10} t^{(4\alpha+1)/10}, \qquad (19.46)$$

where variables are dimensional (except for $\alpha$ and $\tilde{\kappa}_N$ which are numerical coefficients by definition). We also recall that for the fixed volume current, $\alpha = 0$, the value of $q$ is that of the volume (per radian).

It is interesting that for $\alpha = 1$ both the stratified $S = 1$ and the non-stratified $S = 0$ solutions predict propagation with $t^{1/2}$, cf. (11.63).

### 19.2.2   Some comparisons

We compare the prediction (19.46) with the box-model approximation results (18.28) (both are in dimensional form). The agreement is remarkable. This, again, gives some credence to the box-model approximation, but, on the other hand, the present solution is more accurate and sufficiently simple for making it the preferred modeling tool.

Zatsepin and Shapiro (1982) reported a similarity solution as above for $\alpha = 1$, but applied it to an intrusion created by slow injection from a tube in the center of a tank with linearly-stratified saline. The associated experiments confirmed propagation with $t^{1/2}$, and a fair collapse to the theoretical parameter dependencies, in particular concerning the scaling of $r_N$ with $(\mathcal{N}^2 q^4/\nu)^{1/10}$ (but the experimental value of the numerical coefficient is unclear). However, we must keep in mind that the boundary conditions of the intrusion are not fully compatible with these of the gravity current, as discussed in § 11.1.5. Some of the experiments of Didden and Maxworthy (1982) with influx in the corner of a rectangle, as discussed in § 11.2.4, were performed with linearly-stratified ambients. The conditions reproduced closely the $S = 1$ viscous $\alpha = 1$ case. The plot of the measured $r_N$ vs. $t$ on log-log axes is a straight line with slope $1/2$ in agreement with the theoretical prediction (19.46). The details of the coefficient were not investigated.

# Chapter 20

## Energy

## 20.1 Introduction

In this chapter we consider the exchange of energy for an inviscid gravity current (or intrusion) which is released from a lock and then propagates over a horizontal boundary (or plane of neutral buoyancy). In § 5.5 we discussed some essentials of this problem in the framework of a two-layer SW model for the unstratified ambient. Here we generalize the topic by also considering effects due to stratification in the ambient and results of Navier-Stokes finite-difference simulations. (On the other hand, here we use a one-layer shallow-water model.) The energy "dissipation" problem for a steady-state current, or about the nose jump, has been considered in Chapters 3 and 14.

We follow the presentations of Ungarish and Huppert (2006,2008).

## 20.2 2D geometry

The system under consideration is sketched in Fig. 12.1 (page 279): a deep layer of ambient fluid, of density $\rho_a(z)$, lies above a horizontal surface at $z = 0$. Gravity acts in the negative $z$ direction. In the rectangular case the system is bounded by vertical, smooth, impermeable $xz$ planes and the current propagates in the $x$ direction. At time $t = 0$ a given volume of homogeneous fluid of density $\rho_c \geq \rho_a(0) \equiv \rho_b$ and kinematic viscosity $\nu$, initially at rest in a rectangular box of height $h_0$ and length $x_0$, is instantaneously released into the ambient fluid. A 2D current commences to spread. We assume that the Reynolds number of the flow is large.

We use an $\{x, y, z\}$ Cartesian coordinate system with corresponding $\{u, v, w\}$ velocity components, assuming that the flow does not depend on the coordinate $y$ and that $v \equiv 0$. We use *dimensional variables* until specified otherwise.

Initially, the height of the propagating current is $h_0$, its length is $x_0$, and its density $\rho_c$. The height of the ambient fluid is $H$ (dimensional). The ambient fluid is stably stratified over the full depth. The density at the top (usually, an open boundary) is $\rho_o$, and increases linearly with depth by the increment

$\widetilde{\Delta\rho}$ to the value $\rho_b$ at the bottom. This corresponds to the flow discussed in Chapter 12. The $S = 1$ case also describes a symmetric intrusion which propagates about the level of neutral buoyancy.

We use $\rho_o$ as the reference density and to introduce, again, the reduced density differences and ratios between them

$$\epsilon = \frac{\rho_c - \rho_o}{\rho_o}, \quad \epsilon_b = \frac{\rho_b - \rho_o}{\rho_o} = \frac{\widetilde{\Delta\rho}}{\rho_o} \tag{20.1}$$

and

$$S = \epsilon_b/\epsilon, \tag{20.2}$$

from which it follows that

$$\rho_c = \rho_o(1+\epsilon), \quad \rho_a = \rho_b - \frac{\widetilde{\Delta\rho}}{H}z = \rho_o\left[1 + \epsilon S(1 - \frac{z}{H})\right]. \tag{20.3}$$

Obviously, $S$ represents the magnitude of the stratification in the ambient fluid, and we consider only $0 \le S \le 1$. The homogeneous ambient is recovered by setting $S = 0$. We also define the reference reduced gravity,

$$g' = \epsilon g, \tag{20.4}$$

where $g$ is the gravitational acceleration. We recall that the buoyancy frequency is defined by

$$\mathcal{N}^2 = g\frac{\widetilde{\Delta\rho}}{\rho_o H} = g'\frac{S}{H}. \tag{20.5}$$

## 20.2.1  SW formulation

We use the one-layer model, which omits the motion in the ambient. In this respect, the analysis in this chapter is different from that of § 5.5, where we used a two-layer model. However, we showed there that the SW energy results for the ambient layer are unreliable: the current transfers energy to the ambient via work, but a significant part of this energy is "lost" by the SW-approximated ambient as "formal dissipation." Moreover, we reiterate that in § 5.5 the ambient is unstratified.

We assume that in the ambient fluid domain $u = v = w = 0$ and hence the fluid is in purely hydrostatic balance and maintains the initial density $\rho_a(z)$ given by (20.3). Motion is assumed to take place in the lower layer only, $0 \le x \le x_N(t)$ and $0 \le z \le h(x,t)$.

The equations of motion and their solutions are as discussed in Chapter 12. The SW energy calculations are a by-product of the solution $h(x,t), u(x,t)$, and $x_N(t)$.

### 20.2.1.1 Energy and work

We now consider the energy in the dense fluid domain. In the SW framework the significant speed of motion is represented by the $z$-independent average horizontal velocity, and hence the kinetic energy of the current (denoted by subscript $c$) is simply

$$K_c(t) = \frac{1}{2}\rho_c \int_0^{x_N(t)} u^2(x,t)h(x,t)dx. \tag{20.6}$$

The vertical displacement of the dense fluid particles is resisted by the hydrostatic pressure of the embedding ambient fluid. The resulting buoyancy force per unit volume is $[\rho_c - \rho_a(z)]g$ and the corresponding work needed to move a unit volume from the bottom to some $z$, in the present linear $\rho_a(z)$, is $g[(\rho_c - \rho_b)z + (1/2)\widetilde{\Delta\rho}\, z^2/H]$. The potential energy of the current is therefore

$$P_c(t) = g\int_0^{x_N(t)} dx \int_0^{h(x,t)} [(\rho_c - \rho_b)z + \frac{1}{2}\frac{\widetilde{\Delta\rho}}{H}z^2]dz =$$
$$g\int_0^{x_N(t)} \left[\frac{1}{2}(\rho_c - \rho_b)h^2 + \frac{1}{6}\frac{\widetilde{\Delta\rho}}{H}h^3\right] dx. \tag{20.7}$$

Initially, at $t = 0$, $h = h_0, x_N = x_0$, and $u = 0$. Consequently, $K_c(0) = 0$ and, using (20.1)-(20.4), we obtain

$$P_c(0) = \left(\frac{1}{2}\rho_o x_0 h_0^2 g'\right)\left[1 - S + \frac{1}{3}\frac{S}{H/h_0}\right]. \tag{20.8}$$

This indicates that energy in the problem under investigation is conveniently scaled with $(1/2)\rho_o x_0 h_0^2 g'$. For a particular system it is convenient to refer the energy to $P_c(0)$. These scalings are used later in this section.

The nose of the current pushes against the pressure force $0.5\Delta p\, h_N$, approximately, where $\Delta p = p_c(z = 0) - p_a(z = 0)$ at $x = x_N$. This pressure head at the front is also given by (12.28). Using this result, we estimate that during the propagation $\Delta x$ the current does the work

$$\Delta W_N \approx 0.5\rho_o g'\left[1 - S + \frac{1}{2}S\frac{h_N}{H}\right]h_N^2\Delta x. \tag{20.9}$$

This work is the major energy "sink" for the current. This energy is transferred to the ambient fluid, but the one-layer SW model does not attempt to give details on what happens next to this energy.

### 20.2.2 Navier-Stokes considerations

We consider, again, a two-dimensional rectangular domain, with gravity acting in the negative $z$ direction. The velocity is $\mathbf{v} = u\hat{x} + w\hat{z}$. We employ the following dimensional balance equations:

1. Conservation of volume

$$\nabla \cdot \mathbf{v} = 0. \tag{20.10}$$

2. Momentum balance

$$\rho \frac{D\mathbf{v}}{Dt} = -\nabla p - (\rho - \rho_b)g\hat{z} + \mu \nabla^2 \mathbf{v}, \tag{20.11}$$

where $p$ is the pressure reduced with $\rho_b g z$ and $\mu$ is the dynamic viscosity coefficient, assumed constant and equal for both fluids.

3. Density transport

$$\frac{D\rho}{Dt} = 0. \tag{20.12}$$

For energy considerations, we multiply the horizontal and vertical components of (20.11) by $u$ and $w$, respectively, and add the results. After some algebra and use of (20.10), we obtain

$$\rho \frac{D}{Dt}\left[\frac{1}{2}(u^2 + w^2)\right] = -\nabla \cdot p\mathbf{v} - (\rho - \rho_b)gw + \Phi_v, \tag{20.13}$$

where $\Phi_v$ is the standard viscous dissipation function for the inviscid NS formulation, contributed by $\mu\mathbf{v} \cdot \nabla^2 \mathbf{v}$ (a more explicit form is given in Appendix C).

Consider the integral of (20.13) over the volume $\Omega$ of the two-fluid system. Here Reynolds' transport theorem is useful (see Appendix C). For simplicity, we consider a closed rectangular domain $0 \le x \le x_w, 0 \le z \le H$. This domain is denoted $\Omega$, and we observe that it is the union of two sub-domains: $\Omega_c$ (of the current) and $\Omega_a$ (of the ambient fluid).

Within the Boussinesq approximation, the slightly varying $\rho$ on the LHS of (20.13) can be replaced by the constant $\rho_o$. The integrated LHS term thus yields the rate of change of the total kinetic energy, defined straightforwardly by

$$K_i(t) = \rho_o \int_{\Omega_i} \frac{1}{2}(u^2 + w^2)dV, \tag{20.14}$$

where $i$ is $a$ for the ambient, $c$ for the current, and no subscript for the whole system. On the RHS of (20.13), the integrated contribution of the pressure term vanishes on account of the boundary conditions. The effect of the viscous dissipation term on the time-dependent energy behavior of the system enters via

$$\mathcal{D}_v(t) = \int_0^t dt \int_\Omega \Phi_v dV. \tag{20.15}$$

We note in passing that this loss of energy is very different from the "formal" dissipation effect considered in § 5.5.

The integral of the second term on the RHS of (20.13), which represents the work of the buoyancy force, is associated with the potential energy of the system. In general, the potential energy of a stratified system is a problematic property, because the change of density and $z$-position of the fluid particles is controlled by several processes like advection, mixing, and diffusion; the reader is referred to Winters et al. (1995) for a detailed discussion. However, in the present simple (non-diffusive, laminar and closed) system, the reversible energy exchange due to buoyancy can be manipulated into an informative form as follows. To this end, we introduce the vertical upward displacement $\eta(x, z, t)$ of a particle of density $\rho$ in the ambient fluid, from its initial position in the linear density profile given by (20.3). The density of the particle is conserved, which means that $\rho(x, z, t) = \rho_a[z - \eta(x, z, t)]$, where $\rho_a$ is the initial linear stratification. Solving for the displacement, we obtain

$$\eta(x, z, t) = z + [\rho(x, z, t) - \rho_b]\frac{H}{\widetilde{\Delta\rho}}, \tag{20.16}$$

and hence

$$\rho - \rho_b = \frac{\widetilde{\Delta\rho}}{H}(\eta - z). \tag{20.17}$$

Consequently, we can write the overall contribution of the buoyancy work as

$$\int_\Omega g(\rho - \rho_b)w dV =$$

$$g\int_{\Omega_c}(\rho_c - \rho_b + \frac{\widetilde{\Delta\rho}}{H}z - \frac{\widetilde{\Delta\rho}}{H}z)w dV + g\int_{\Omega_a}\frac{\widetilde{\Delta\rho}}{H}(\eta - z)w dV =$$

$$\frac{D}{Dt}\left[\int_{\Omega_c}g[(\rho_c - \rho_b)z + \frac{1}{2}\frac{\widetilde{\Delta\rho}}{H}z^2]dV\right] + \frac{D}{Dt}\left[\rho_o\mathcal{N}^2\int_{\Omega_a}\frac{1}{2}\eta^2 dV\right]$$

$$- \rho_o\mathcal{N}^2\int_\Omega zw dV =$$

$$\frac{D}{Dt}[P_c(t) + P_a(t)] - \rho_o\mathcal{N}^2\int_\Sigma \frac{1}{2}z^2\mathbf{v}\cdot\hat{n}dA. \tag{20.18}$$

In this derivation we used (20.5) and (20.12) and the relationships $w = Dz/Dt$ and $2zw = \nabla\cdot[z^2\mathbf{v}]$. The details are left for an exercise. The last term vanishes because the boundary $\Sigma$ of the present closed system is impermeable. The relevant outcome is that the work of the buoyancy term can be identified with the rate of change of the potential energy of the current and ambient. The expression for the $P_c$ term has been determined before, see (20.7), as the work performed to elevate the particles of dense fluid from the bottom to present position in a force field $[\rho_c - \rho_a(z)]g$ (per unit volume). The form of $P_a$ reproduces a similar effect. For a particle of the ambient fluid, the resisting force (per unit volume) at $\zeta$ displacement from the initial height is $\rho_o\mathcal{N}^2\zeta$, approximately; the accumulated work needed to achieve the displacement $\eta$ is therefore $\rho_o\mathcal{N}^2\eta^2/2$. A summarizing inspection of the resulting terms inside the control volume and of the boundary conditions is now in place. We see that the energy flux in the inviscid non-stratified ambient is concerned with $\rho|\mathbf{v}|^2/2 + p$; in the stratified Boussinesq case the $(\rho_o \mathcal{N}^2/2)(\eta^2 - z^2)$ must be added. Here $p$ is the hydrodynamic pressure reduced with $+\rho_b g z$ (and for simplicity we can assume the value 0 at the bottom of the unperturbed fluid). In the unperturbed ambient $\eta = 0$ and $p - (\rho_o \mathcal{N}^2/2)z^2 = 0$.

Hereafter, *dimensionless variables* are used unless stated otherwise. The scaling is provided by (12.19)-(12.20); the energy is scaled with $(1/2)\rho_o x_0 h_0^2 g'$ and $\eta$ with $h_0$.

The energy terms for the current are similar to these derived for the SW case, see (20.6) and (20.7), except for the fact that in the SW approximation the contribution of $w$ to the kinetic energy has been discarded. According to (20.18) the scaled form of the potential energy of the ambient is

$$P_a(t) = \frac{S}{H} \int_{\Omega_a} \eta^2 dV. \qquad (20.19)$$

This indicates that for weak stratification (small $S$) most of the energy transferred to the ambient is of kinetic type, and hence more prone to viscous dissipation. This trend is consistent with the observation of Necker et al. (2005) that higher levels of kinetic energy are associated with larger values of viscous dissipation.

We are concerned with the behavior of the mechanic energies; by total energy we refer to the sum of kinetic and potential components. The total energy of the two-fluid system is expected to decay due to irreversible viscous dissipation.

Suppose that the numerical solution of the NS formulation is obtained numerically, for example by a finite-difference code of the type discussed in Appendix B. The code results provide the values of $u$, $w$, and density function $\phi$ at grid points $(x_i, z_j)$. Recall that

$$\phi(x, z, t) = \frac{\rho(x, z, t) - \rho_o}{\rho_c - \rho_o} = \frac{1}{\epsilon} \left[ \frac{\rho(x, z, t)}{\rho_o} - 1 \right]. \qquad (20.20)$$

The values of $\eta_{i,j}$ at the grid points $(x_i, z_j)$ can then be calculated with the aid of (20.16) as

$$\eta_{i,j} = z_j + H \left( \frac{\phi_{i,j}}{S} - 1 \right). \qquad (20.21)$$

Using the values of $u$, $w$, $\phi$, and $\eta$ at the grid points, the potential and kinetic energies of the current and of the ambient can be calculated by numerical integration (summation) over the appropriate grid-volume elements.

## 20.2.3   Results

We consider the influence of the stratification, $S$, on the energy behavior of the gravity current released from behind a lock in a configuration with $H$

=3 (dimensionless). This geometry is expected to be typical of currents in a non-shallow ambient, and is also compatible with experimental Runs 5 and 19 of Maxworthy, Leilich, Simpson, and Meiburg (2002).

The results are presented in dimensionless form subject to the scaling (12.19)-(12.20); the energy and work are scaled with $(1/2)\rho_o x_0 h_0^2 g'$.

The SW governing equations with the appropriate boundary conditions were solved by a standard finite-difference method. The resulting discrete values of $h$ and $u$ were used to calculate the energy integrals (20.6) and (20.7) by the trapezoidal rule.

The predicted propagation of the nose is presented in Fig. 20.1. As the stratification parameter $S$ increases, the speed of propagation is reduced. Cases $S = 0$ and $S = 0.29$ are super-critical with respect to $u_W$, and cases $S = 0.72$ and $S = 1$ are sub-critical. Note that the configurations for $S = 0.29$ and $S = 0.72$ discussed here correspond to the experimental Runs 5 and 19 respectively of Maxworthy, Leilich, Simpson, and Meiburg (2002). Figures 20.2 and 20.3 display the SW energy behavior of the current as functions of time and of distance for various values of the stratification parameter $S$. The energies of each system are referred to the initial potential energy $P_c(0)$; see (20.8). In all cases, the potential energy decays monotonically, while the kinetic energy has an initially increasing and then decreasing profile with a maximum of about 0.3 at $t \approx 2$ to 3. A similar behavior has been observed by Necker et al. (2005) in the context of homogeneous and particle-driven gravity currents. The kinetic energy of the current develops from zero during the dam-break phase of the motion. The rarefaction wave which travels backward into the bulk of stationary fluid sets this fluid into motion and increases the kinetic energy. Then the current evolves to a rectangle of height $h_N$, approximately; here the kinetic energy reaches the maximum. This happens before the end of the slumping phase of constant-velocity propagation. The ratio of kinetic to potential energy becomes close to 1 at a quite advanced stage of propagation ($t \approx 20$), when the potential energy has decayed to about 10% of its initial value.

The energy of the current $E_c = P_c + K_c$ decreases with propagation due to the pressure work of the nose. According to (20.8) and (20.9) the ratio of work performed during the propagation $\Delta x$ to the initial potential energy is

$$\frac{\Delta W_N}{P_c(0)} \approx \left[ \frac{1 - S + S\dfrac{h_N}{2H}}{1 - S + S\dfrac{1}{3H}} \right] h_N^2 \Delta x. \qquad (20.22)$$

The coefficient in the brackets is of order 1 for both small stratification and strong stratification (as long as $h_N \approx 0.5$). Indeed, we see that for a propagation $\Delta x \approx 2$ from the initial $x_N = 1$, about half of the energy of the current is lost. For $S$ close to 1 this decay is faster than for smaller $S$.

Further analytical insights can be obtained for the self-similar propagation. For the $S = 0$ case we derived in § 5.5.4 the results $E_c \sim t^{-2/3}$ or $E_c \sim x_N^{-1}$.

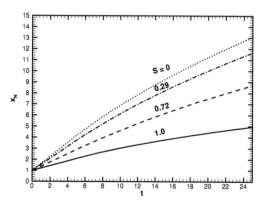

**FIGURE 20.1:** SW propagation of the nose as a function of time for $H = 3$ and various values of $S$. (Ungarish and Huppert 2006, with permission.)

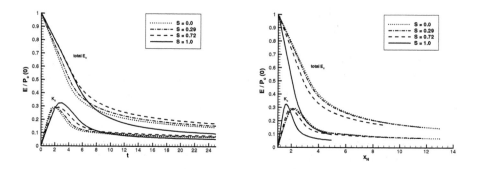

**FIGURE 20.2:** SW kinetic and total energy of the current as functions of time (left frame) and $x_N$ (right frame), for various values of $S$. (Ungarish and Huppert 2006, with permission.)

It is left for an exercise to show that for the $S = 1$ stratification the use of (15.52) produces $E_c \sim t^{-1}$ or $E_c \sim x_N^{-2}$ for the self-similar phase.

The corresponding NS simulations were performed on grids of $320 \times 200$ intervals. In these computations $H = 3$ and $h_0/x_0 = 0.25$, $x_w = 8$, $Re = 1.0 \cdot 10^4$ and the values of $S$ were $0, 0.29, 0.72$, and $1$. The values of $\epsilon$ were $0.115, 0.115, 0.0804, 0.115$, respectively. These parameters were chosen so that the second and third cases reproduce the experimental Runs 5 and 19 of Maxworthy, Leilich, Simpson, and Meiburg (2002).

The flow fields were discussed in Chapter 12. Typical results of the NS simulations are shown in Figs. 12.4 and 12.7. We recall that for $S = 0.72$ (typical for sub-critical currents) a strong wave-head interaction occurs. We see in Fig. 12.6 that at $t \approx 12$ the velocity of propagation enters a state of strong deceleration, which practically brings the nose to rest at $t \approx 14$. However, at $t \approx 16$ the original speed is recovered. What happens to the energy of the current in the special stage of interaction with the ambient?

A comparison between SW and NS energy results for the current is shown in Fig. 20.4. We observe that the SW and NS results are fairly close for a significant period of propagation. Eventually, the NS results decay more quickly to zero than the SW predictions. This is a result of viscous effects, which are not included in the SW model. For weak stratifications, shortly after release, the potential and total energy of the current as predicted by the NS simulations are slightly larger than those determined by the SW results. This can be attributed to the initial adjustment motion of the head which is incompatible with the SW assumptions. We conclude that, overall, the NS simulations confirm the energy predictions of the simplified SW model. This strengthens the confidence in the insights derived about the effects of stratification.

Additionally, an interesting outcome is that the wave-nose interaction in the sub-critical current has no significant projection on the energy exchange. This is illustrated by the configuration for $S = 0.72$. The NS simulations show clearly a change of behavior of $x_N(t)$ in the time interval $10-15$; see Fig. 12.6. However, the corresponding graphs of energy as a function of time, produced by the same NS simulations, do not display a change of behavior during this time interval. The interpretation is as follows. The interaction occurs when the kinetic energy of the current is already low (see Fig. 20.3) and the head is relatively shallow. In these circumstances, a relatively small increase of the velocity of the main bulk which follows the head is able to recover the energy lost by the head during the deceleration phase. Afterward, this process is reversed and the head regains its velocity. This is an interesting outcome regarding the predictive ability of the SW model. This model does not detect the interaction, and hence it could be expected that its predictions become invalid after this occurrence. However, we found that the energy balances of a real current are also unaffected by this interaction. Thus, the prediction of energy as a function of time by this model remains valid for a longer period of time than could be anticipated for a sub-critical propagation.

Fig. 20.5 and 20.6 show the time-dependent behavior of the total mechanical energy (kinetic plus potential) of the two-fluid system, and of the energies of the ambient. This information is obtained from the NS computations. Evidently, the total (mechanical) energy is not conserved because of irreversible viscous dissipation. We note that stratification hinders the decay of the total mechanical energy. The interpretation can be inferred from the behavior of the kinetic and potential energies in the ambient. The stratification enhances the ability of the ambient to accumulate potential energy. Thus, as $S$ in-

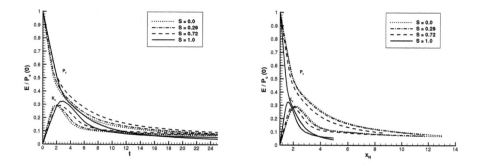

**FIGURE 20.3:** SW results. The kinetic and potential energy of the current as functions of time (left frame) and $x_N$ (right frame), for various values of $S$. (Ungarish and Huppert 2006, with permission.)

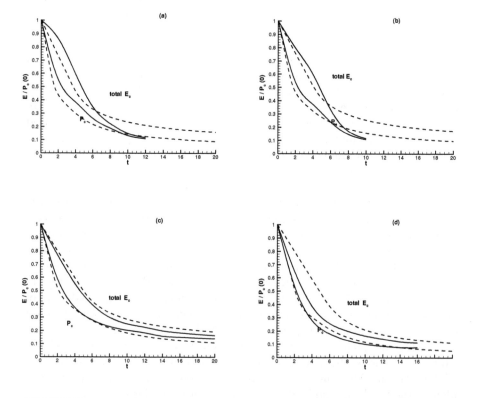

**FIGURE 20.4:** The potential and total (mechanical) energy of the current as functions of time. Comparisons of NS (full lines) and SW (dashed lines) results. The frames are for $S = 0$ (a), 0.29 (b), 0.72 (c), and 1.0 (d). (Ungarish and Huppert 2006, with permission.)

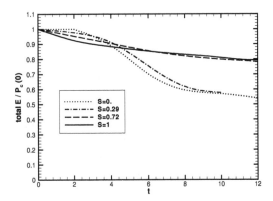

**FIGURE 20.5**: Total mechanical energy of the two-fluid system as a function of time for various $S$, NS results. (Ungarish and Huppert 2006, with permission.)

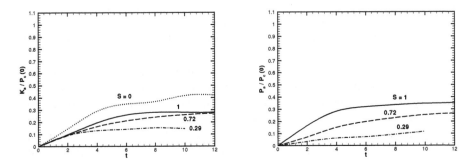

**FIGURE 20.6**: Kinetic and potential energy of the ambient as a function of time for various $S$, NS results. (The potential energy result for $S = 0$ is zero.) (Ungarish and Huppert 2006, with permission.)

creases, the ratio of kinetic to potential energy in the ambient is reduced. For $S = 0.72$ the ratio of potential to kinetic energies in the ambient is typically 1, while for $S = 0.29$ the ratio is typically 0.5. The smaller kinetic energy of the ambient fluid indicates a reduction of the velocity differences in the corresponding domain and hence also a smaller viscous dissipative friction.

Considering the behavior of the total mechanical energy of the system as a function of time, we notice a striking similarity between the super-critical currents $S = 0$ and $S = 0.29$, on one hand, and between the sub-critical currents $S = 0.72$ and $S = 1$, on the other hand.

## 20.3   Axisymmetric geometry

The extension of the foregoing energetic behavior to the axisymmetric current analysis has both practical and academic motivations. In various natural and industrial applications gravity currents and intrusions can propagate in an unrestricted (fully cylindrical) manner, or in an expanding wedge. We recall that there are significant qualitative and quantitative differences between the 2D Cartesian and axisymmetric currents. For example, the 2D currents display a slumping stage of propagation with constant speed, while the speed of an axisymmetric current decays from the beginning; also, the ratio of inertial to viscous effects decays much faster in an axisymmetric geometry. An axisymmetric intrusion develops a peculiar self-similar propagation in which all the dense fluid is in a ring of constant inner to outer radius ratio of about 0.5; this has no counterpart in the 2D geometry. What is the behavior of the energy during this motion? Moreover, the numerical simulation of the axisymmetric current is more problematic than the 2D counterpart. The axisymmetric numerical simulations (with a fairly large number of discretization points) reproduce successfully the initial motion, up to about 2.5 to 3 lock-lengths, but afterward the simulated current is significantly slower than experimental observations. For this problematic range the shallow-water predictions are the only available practical theoretical tool. (Three dimensional simulation are still very expensive.)

The system under consideration is sketched in Fig. 13.1 (page 307): a deep layer of ambient fluid, of density $\rho_a(z)$, lies above a horizontal surface at $z = 0$. Gravity acts in the negative $z$ direction. The system is cylindrical about the $z$ axis, either as a full circle, or as a wedge bounded by vertical, smooth, impermeable surfaces. The current propagates in the radial $r$ direction. At time $t = 0$ a given volume of homogeneous fluid of density $\rho_c \geq \rho_a(0) \equiv \rho_b$ and kinematic viscosity $\nu$, initially at rest in a cylinder of height $h_0$ and radius $r_0$, is instantaneously released into the ambient fluid. An axially-symmetric current commences to spread. We assume that the Reynolds number of the flow is large.

We use a cylindrical coordinate system, with the radial coordinate $r$ along the horizontal bottom and $z$ pointing upward. The radial and vertical velocity components are $u$ and $w$. We assume that the flow does not depend on the azimuthal coordinate and that there is zero velocity in this direction. This is appropriate for the description of flow in a wedge or in a full circular geometry (with no rotation).

The height of the ambient fluid is $H$ (dimensional). The ambient fluid is stably stratified: the density at the top (usually an open boundary) is $\rho_o$, and it increases linearly with depth by the increment $\widetilde{\Delta \rho}$ to the value $\rho_b$ at the bottom. This corresponds to the flow fields discussed in Chapter 13. The $S = 1$ case also describes a symmetric intrusion which propagates about the level of neutral buoyancy. The reference density is $\rho_o$. The definitions of $\epsilon$, $\epsilon_b$, $\widetilde{\Delta\rho}$, $S$, $g'$, $\mathcal{N}$, $\rho_c$, and $\rho_a(z)$ are like in the 2D case and given by (20.1)-(20.5).

## 20.3.1 SW model

We use a one-layer approximation which omits the motion in the ambient. To be specific, we assume that in the ambient fluid domain the velocity is zero and hence the fluid is in purely hydrostatic balance and maintains the initial density $\rho_a(z)$ given by (20.3). Motion is assumed to take place in the lower layer only, $0 \leq r \leq r_N(t)$ and $0 \leq z \leq h(r,t)$.

The equations of motion were presented in Chapter 13. We now consider the energy terms in this model.

### 20.3.1.1 Energy and work

In the SW framework the speed of motion is represented by the $z$-independent average horizontal velocity, and hence the kinetic energy of the current (denoted by subscript $c$) is given by

$$K_c(t) = \frac{1}{2}\rho_c \int_0^{r_N(t)} u^2(r,t)h(r,t)rdr. \tag{20.23}$$

The vertical displacement of the dense fluid particles is resisted by the hydrostatic pressure of the surrounding ambient fluid. The resulting buoyancy acceleration is $[\rho_c - \rho_a(z)]g$ and the corresponding work needed to move a unit volume from the bottom to some $z$, in the present linear density profile $\rho_a(z)$, is $g[(\rho_c - \rho_b)z + (1/2)\widetilde{\Delta\rho}\, z^2/H]$. The potential energy of the current is therefore

$$P_c(t) = g \int_0^{r_N(t)} rdr \int_0^{h(r,t)} [(\rho_c - \rho_b)z + \frac{1}{2}\frac{\widetilde{\Delta\rho}}{H}z^2]dz$$

$$= g \int_0^{r_N(t)} \left[\frac{1}{2}(\rho_c - \rho_b)h^2 + \frac{1}{6}\frac{\widetilde{\Delta\rho}}{H}h^3\right] rdr. \tag{20.24}$$

Initially, at $t = 0$, $h = h_0, r_N = r_0$, and $u = 0$. Consequently, $K_c(0) = 0$ and, using (20.1)-(20.4), we obtain

$$P_c(0) = \left(\frac{1}{4}\rho_o r_0^2 h_0^2 g'\right)\left[1 - S + \frac{1}{3}\frac{S}{H/h_0}\right]. \tag{20.25}$$

This indicates that energy in the problem under investigation is conveniently scaled with $(1/4)\rho_o r_0^2 h_0^2 g'$. For a particular system it is convenient to refer the energy to $P_c(0)$. These scalings are used later in this section.

The nose of the current pushes against the pressure force $0.5\Delta p\, r_N h_N$ (per radian), approximately, where $\Delta p = p_c(z = 0) - p_a(z = 0)$ at $r = r_N$. This

$\Delta p$ pressure head at the front is also given by (12.28). Using this result, we estimate that during the propagation $\Delta r$ the current does the work

$$\Delta W_N \approx 0.5 \rho_0 g' \left[ 1 - S + \frac{1}{2} S \frac{h_N}{H} \right] h_N^2 r_N \Delta r. \qquad (20.26)$$

This work is the major energy "sink" for the current. This energy is transferred to the ambient fluid, but the one-layer SW model does not attempt to give details on what happens next to this energy.

## 20.3.2 Navier-Stokes considerations

We consider, again, an axisymmetric cylindrical domain with velocity $\mathbf{v} = u\hat{r} + w\hat{z}$. The NS balance equations (dimensional) are given by (20.10)-(20.12).

For energy considerations, we perform again the manipulations described in the context of the 2D geometry in § 20.2.2. In the present case the involved volumes are cylindrical. The volume $\Omega$ of the two-fluid system is a closed cylindrical domain $0 \le r \le r_w, 0 \le z \le H$. This domain is denoted $\Omega$, and we observe that it is the union of two sub-domains: $\Omega_c$ (for the current) and $\Omega_a$ (for the ambient fluid). The integral energy equations (20.13) - (20.15) and (20.18) carry over to the axisymmetric geometry.

In this case the vertical upward displacement of a particle of density $\rho(r, z, t)$ in the ambient fluid from its initial position in the linear density profile is $\eta(r, z, t)$. The conservation of density of the particle, combined with (20.3), yields

$$\eta(r, z, t) = z - [\rho_b - \rho(r, z, t)] \frac{H}{\widetilde{\Delta \rho}}. \qquad (20.27)$$

Again, the kinetic energy (dimensional) is given by

$$K_i(t) = \rho_o \int_{\Omega_i} \frac{1}{2} (u^2 + w^2) dV, \qquad (20.28)$$

where $i$ is $a$ for the ambient, $c$ for the current, and no subscript for the whole system.

Hereafter, *dimensionless variables* are used unless stated otherwise. The scaling is provided by (13.1)-(13.2); the energy is scaled with $(1/4) \rho_o r_0^2 h_0^2 g'$ and $\eta$ with $h_0$.

As expected, the energy terms for the current are similar to these derived for the SW case. In the SW approximation the contribution of $w$ to the kinetic energy has been discarded. The scaled form of the potential energy of the ambient is

$$P_a(t) = 2 \frac{S}{H} \int_{\Omega_a} \eta^2 dV. \qquad (20.29)$$

As in the 2D case, for weak stratification (small $S$) most of the energy transfered to the ambient is expected to be of kinetic type, and hence more prone to viscous dissipation.

The numerical solution of the NS axisymmetric formulation was obtained by using a finite-difference method. Some flow-field results were discussed in Chapter 13. Various comparisons indicate that the axisymmetric NS results tend to become incoherent for $r_N > 3$, approximately. The finite-difference results provide the values of $u$, $w$, and $\phi$ at grid points $(r_i, z_j)$. This allows the calculation of the potential and kinetic energies of the current and of the ambient, in a quite similar way as for the 2D case (but now the integration is over cylindrical volumes elements). The values of $\eta_{ij}$ were calculated with the aid of (20.21), which carries over to the present axisymmetric field.

### 20.3.3 Results

We consider the influence of the stratification, $S$, on the the energy behavior of the gravity current released from behind a lock in a configuration with $H$ =3 (dimensionless). This geometry is expected to be typical of currents in a non-shallow ambient. The parameters $H$ and $S$ considered here are also compatible with these used in the previous investigation of the 2D current. The results are presented in dimensionless form subject to the scaling (13.1)-(13.2); the energy is scaled with $(1/4)\rho_o r_0^2 h_0^2 g'$ and $\eta$ with $h_0$.

In general, the SW governing equations with the appropriate initial and boundary conditions must be solved numerically. For large times analytical similarity solutions for the cases $S = 0$ and $S = 1$ are available; these will be used later.

The main SW results were obtained by a finite-difference method. The resulting discrete values of $h$ and $u$ were used to calculate the energy integrals (20.23) and (20.24) by the trapezoidal rule. The predicted propagation of the nose is presented in Fig. 20.7. As expected, as the stratification parameter $S$ increases, the speed of propagation is reduced. Cases $S = 0$ and $S = 0.29$ are super-critical with respect to $u_W$, while $S = 0.72$ and $S = 1$ are sub-critical.

The present inviscid approach is valid as long as $F_I/F_V$ remains large (at least $> 1$). As discussed in § 13.3, for the typical value of $Re(h_0/r_0) = 10^3$, viscous effects may become dominant after a propagation to about $r_N = 3.5$ for small $S$ and $r_N = 2.5$ for $S$ close to 1. Although this is not a very significant propagation, the mean thickness of the current is reduced by a factor of about 10. This implies a very significant decay of the potential energy $P_c$, see (20.24), and hence the main driving energy is more or less exhausted during this propagation.

Figs. 20.8 and 20.9 display the SW energy balances of the current as functions of time and of distance for various values of the stratification parameter $S$. The energies of each system are referred to the initial potential energy $P_c(0)$; see (20.25). In all cases, the potential energy decays monotonically, while the kinetic energy has an initially increasing and then decreasing profile with a maximum of about 0.4 at $t \approx 2$ to 3. These energy trends can be understood in view of the $h$ and $u$ profiles shown in Fig. 13.3. The kinetic energy of the current develops from zero during the "dam-break" phase of the motion.

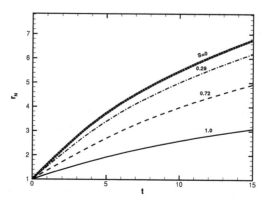

**FIGURE 20.7:** SW radius of the nose as a function of time for $H = 3$ and various values of $S$. (Ungarish and Huppert 2008, with permission.)

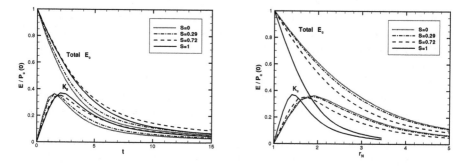

**FIGURE 20.8:** SW results. The kinetic and total energy of the axisymmetric current as functions of time (left frame) and $r_N$ (right frame), for various values of $S$. (Ungarish and Huppert 2008, with permission.)

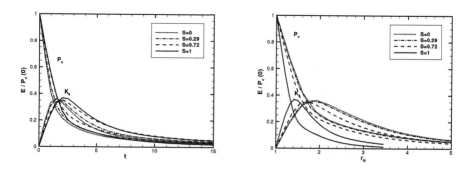

**FIGURE 20.9:** SW results. The kinetic and potential energy of the axisymmetric current as functions of time (left frame) and $r_N$ (right frame), for various values of $S$. (Ungarish and Huppert 2008, with permission.)

The rarefaction wave which travels backward into the bulk of stationary fluid sets this fluid into motion and increases the kinetic energy. Then the current evolves to, roughly, a cylinder of height $h_N \approx 0.4$. This geometry contains less than half of the original potential energy. Here the kinetic energy reaches the maximum, and the kinetic and potential energy are roughly equal. Next, both the kinetic and potential energies of the current decay at about the same rate.

We note that the energy decay as a function of the distance of propagation $r_N$ is quite insensitive to the value of $S$ for $S < 0.7$ (say). The case $S = 1$ (i.e., $\rho_c = \rho_b$, the strongest stratification in the present framework) displays a different behavior with $r_N$ to the other cases: the total energy of the current, $E_c = P_c + K_c$, decays faster, and the maximum of the kinetic energy is attained after a shorter propagation.

The energy of the current $E_c = P_c + K_c$ decreases with propagation due to the pressure work of the nose. According to (20.25) and (20.26) the ratio of work performed during the propagation $\Delta r$ to the initial potential energy is

$$\frac{\Delta W_N}{P_c(0)} \approx 2 \left[ \frac{1 - S + S\dfrac{h_N}{2H}}{1 - S + S\dfrac{1}{3H}} \right] h_N^2 r_N \Delta r. \tag{20.30}$$

This expression is more complex than the 2D counterpart (20.22) because of dependency on $r_N$ and lack of a slumping phase with constant $h_N$. However, the coefficient of $\Delta r$ is evidently of order 1 for both small stratification and strong stratification during propagation from $r_N = 1$ to 2. About half of the energy of the current is indeed lost in this initial stage. For $S$ close to 1 this decay is faster than for smaller $S$.

We can supplement the trends derived from the numerical SW solution with some available analytical results of the SW equations at large $t$. At this stage the initial conditions are "forgotten" except for the prescribed fixed volume $\mathcal{V}$ (per radian), and $Fr$ has attained a constant value because $h_N/H$ is very small. A self-similar behavior appears for $S = 0$ and $S = 1$. The space-similarity coordinate is

$$y = r/r_N(t) = r/At^\beta \quad (0 \le y \le 1), \tag{20.31}$$

where $A$ and $\beta$ are constants. (Here $A$ replaces the usual $K$, in order to avoid confusion with the kinetic energy.)

The self-similar result for the non-stratified $S = 0$ axisymmetric case has been derived in § 6.5. Here we derive the corresponding energetic results. We start here with the solution of the equations and nose boundary condition

$$r_N = At^{1/2}; \quad h = \dot{r}_N^2 \left[ \frac{1}{Fr^2} + \frac{1}{2}(y^2 - 1) \right]; \quad u = \dot{r}_N y; \tag{20.32}$$

where the overdot denotes time derivative and

$$A = 2 \left( \frac{2Fr^2}{4 - Fr^2} \right)^{1/4} \mathcal{V}^{1/4}. \tag{20.33}$$

Since $\dot{r}_N = (1/2)At^{-1/2}$, (20.32) can also be expressed as

$$r_N = At^{1/2}; \quad h = \frac{1}{4}A^4 r_N^{-2} \left[ \frac{1}{Fr^2} + \frac{1}{2}(y^2 - 1) \right]; \quad u = \frac{1}{2}A^2 r_N^{-1} y. \tag{20.34}$$

These results cover the entire domain $0 \le y \le 1$.

We calculate the energy by using the definitions (20.23) and (20.24). In dimensionless form we obtain

$$K_c = 4r_N^2 \int_0^1 \frac{1}{2} u^2(y,t) h(y,t) y \, dy = 2 \left( \frac{A}{2} \right)^6 \left( \frac{1}{Fr^2} - \frac{1}{6} \right) t^{-1}, \tag{20.35}$$

$$P_c = 4r_N^2 \int_0^1 \frac{1}{2} h^2(y,t) y \, dy = 2 \left( \frac{A}{2} \right)^6 \left( \frac{2}{Fr^4} - \frac{1}{Fr^2} + \frac{1}{6} \right) t^{-1}, \tag{20.36}$$

and

$$E_c = K_c + P_c = \left( \frac{1}{2Fr} \right)^4 A^6 t^{-1}. \tag{20.37}$$

(The integrals in (20.35) and (20.36) are multiplied by 4 because of the energy scaling.) The conclusion is that the energy of the axisymmetric current in a non-stratified ambient decays like $t^{-1}$ or $r_N^{-2}$. In the self-similar stage $K_c > P_c$. In the present case $\mathcal{V} = 1/2$. This analytical result is in good agreement with the numerical solution of the SW equations.

For the $S = 1$ case the similarity solution, derived in § 16.3.2, is

$$r_N(t) = At^{1/3}; \quad h(y,t) = \dot{r}_N \sqrt{2H}(y^2 - y_1^2)^{1/2}; \quad u(y,t) = \dot{r}_N y; \tag{20.38}$$

where

$$y_1 = \left( 1 - \frac{1}{Fr^2} \right)^{1/2}, \tag{20.39}$$

and

$$A = 3Fr \left[ \frac{\mathcal{V}}{3\sqrt{2H}} \right]^{1/3}. \tag{20.40}$$

(Note that here the scalings of speed and time differ from these used in § 16.3.2; this introduces an additional $H^{1/2}$ factor in the formulas.)

For the present value of $Fr = 1.19$ we obtained $y_1 = 0.54$. The entire volume of the dense fluid is in the ring $y_1 r_N(t) \le r \le r_N(t)$. We showed that an intrusion released from behind a lock indeed converges to this peculiar solution. (In addition, there is a thin disk-tail left behind, whose volume is small compared with that in the ring and decays like $t^{-4/3}$. This we discard in the energy estimates.)

We calculate the energy by using the definitions (20.23) and (20.24). Now the integration is for the $y$ domain $[y_1, 1]$. In dimensionless form we obtain

$$K_c = \frac{4}{3^4}\sqrt{2H}\left(\frac{1}{2Fr^3} - \frac{1}{5Fr^5}\right)A^5 t^{-4/3},\qquad(20.41)$$

$$P_c = \frac{4}{3^4 \cdot 5}\sqrt{2H}\frac{1}{Fr^5}A^5 t^{-4/3},\qquad(20.42)$$

and

$$E_c = K_c + P_c = \frac{2}{3^4}\frac{1}{Fr^3}\sqrt{2H}A^5 t^{-4/3}.\qquad(20.43)$$

The conclusion is that the energy of the axisymmetric current in a $S = 1$ stratified ambient, at large times, decays like $t^{-4/3}$ or $r_N^{-4}$. In the self-similar stage $K_c > P_c$. In the present case $\mathcal{V} = 1/2$. This analytical result is in good agreement with the numerical solution of the SW equations. The decay of energy in the $S = 1$ stratified case is significantly faster than for the non-stratified $S = 0$ counterpart. We recall that this case corresponds to an intrusion which propagates about the plane of neutral buoyancy.

The difference between the axisymmetric self-similar behavior for $S = 1$ and the 2D counterpart is sharp. In the latter case $x_N \sim t^{1/2}$ and hence the energies of the current decay like $t^{-1}$ or $x_N^{-2}$.

The corresponding NS computations were performed on grids of $400 \times 220$ intervals. In these simulations $H = 3$ and $h_0/r_0 = 0.25$, $r_w = 8$, $Re = 1.0 \cdot 10^4$ and the values of $S$ were $0, 0.29, 0.72$, and $1$. The values of $\epsilon$ were $0.115, 0.115, 0.0804, 0.115$, respectively. These parameters were chosen so that the second and third cases reproduce axisymmetric counterparts of the experimental 2D Runs 5 and 19 of Maxworthy, Leilich, Simpson, and Meiburg (2002). Similar parameters were used in the previous simulations of the 2D case. The simulated axisymmetric flow fields are discussed in § 13.2.

A comparison between NS and SW energy results is shown in Fig. 20.10. We observe that the SW and NS results are fairly close during the initial propagation. Eventually, the NS results decay more quickly to zero than the SW predictions. The agreement between the SW and NS potential energies of the current is better than that for the total energy of the current. This can be attributed to the same effects as the discrepancy observed for $r_N(t)$, because a slower propagation implies an even more significantly reduced kinetic energy.

Fig. 20.11 and 20.12 display the time-dependent behavior of the total mechanical energy (kinetic plus potential) of the two-fluid system and of the energies of the ambient. This information is obtained from the NS computations. Evidently, the total (mechanical) energy is not conserved because of irreversible viscous dissipation. We note that stratification hinders the decay of the total mechanical energy. The interpretation can be inferred from the behavior of the kinetic and potential energies in the ambient. The stratification enhances the ability of the ambient to accumulate potential energy. Thus, as $S$ increases, the ratio of kinetic to potential energy in the ambient is

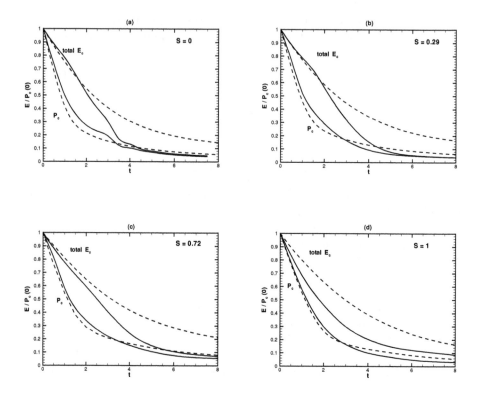

**FIGURE 20.10:** The potential and total (mechanical) energy of the axisymmetric current as functions of time. Comparisons of NS (full lines) and SW (dashed lines) results. The frames are for $S = 0$ (unstratified), 0.29, 0.72, and 1.0. (Ungarish and Huppert 2008, with permission.)

reduced. This reduces the velocity differences in the ambient fluid and hence the viscous dissipative friction is less influential.

To summarize: The one-layer shallow-water analysis, which neglects motion in the ambient, seems to capture well the energy exchange of the gravity current in the inertia-dominated stage of propagation. In particular, this approach is able to elucidate by comparatively simple means the effects of stratification on the energy balances of the gravity current. The analysis indicates that for sufficiently deep currents ($H \geq 2$ say), the motion in the ambient has little influence on the energetic behavior of the dense fluid. In other words, the ambient reacts to, rather than interacts with, the motion of the dense fluid. It is somewhat surprising that the one-layer shallow-water model is relevant in spite of the fact that the ambient gains a significant part of the energy of the current. This can be attributed to two effects: (1) The energy transfer is a rather one-sided process. The current performs

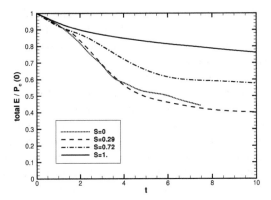

**FIGURE 20.11**: NS results for the total mechanical energy of the two-fluid axisymmetric system as a function of time for various $S$. (Ungarish and Huppert 2008, with permission.)

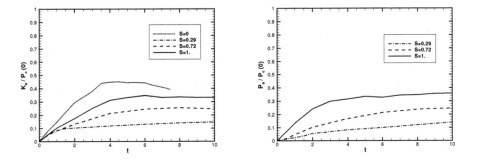

**FIGURE 20.12**: NS results for the kinetic (left) and potential (right) energy of the ambient as a function of time for various $S$. (The potential energy result for $S = 0$ is zero.) (Ungarish and Huppert 2008, with permission.)

work on the ambient (mostly at the propagating nose), and thus loses energy. From the point of view of the current, it is unimportant how this energy is further propagated. (2) The volume of the ambient is large, and the initial disturbances there are zero. Consequently, on a space and time average, the feedback to the current from the perturbations in the ambient is rather small for a significant distance of propagation.

The propagation of the 2D sub-critical current is prone to interactions of the head with the waves, during which the velocity of the head decreases significantly and then increases. However, the NS simulations (which detect this effect well) indicate that no significant counterpart effect occurs in the energy exchange behavior. The SW model provides valid energy results as a function of time during this problematic stage, in spite of the fact that it does not incorporate the interaction effect.

In general, the trends of the energy budgets in the axisymmetric system are similar to these in the two-dimensional counterpart. However, the transfers are more rapid in the diverging geometry: when $r_N \approx 2$ the kinetic energy attains a maximum, and then both potential and kinetic energy decay: (a) like $r_N^{-2}$ for a weak stratification; and (b) like $r_N^{-4}$ for a strong stratification. The latter case is relevant to an intrusion at the neutral buoyancy level.

As stratification increases, the capability of the ambient to store potential energy by the wavy displacement of the isopycnals increases. This reduces the portion of kinetic energy in the ambient and the viscous friction dissipation.

We observed that the agreements between the predictions of the NS simulations and SW model are better for 2D cases than for the axisymmetric counterpart. We speculate that this is due to a deficiency of the numerical finite-difference method used in the simulation. This intriguing issue should be resolved by experiments and more sophisticated simulations. In this context we note that the three-dimensional high resolution simulations of Cantero, Balachandar, and Garcia (2007) provided good agreements with experiments for up to $r_N \approx 6$ in a full-depth lock release. Our expectation is that the more accurate data will show better agreements with the SW predictions than the available axisymmetric NS computations, and will sharpen the understanding of the process under discussion. Also puzzling is the lack of a clear-cut difference between super- and sub-critical axisymmetric currents. Perhaps this is because in the axisymmetric case the speed of the nose decays from the beginning of the motion, and hence (a) in any case the sub-critical state prevails after a short propagation, and (b) when the interaction of the waves with the nose occurs (after a propagation of several lock-lengths), the speed of the current is already so small that viscous effects dominate in our calculations.

The clarification of these and related issues are topics of current and future research, where the readers of this text are expected to make their contributions. With this open accord, which connects past to future, we conclude this book.

# Appendix A

## SW equations: characteristics and finite-difference schemes

### A.1 Characteristics

The method of characteristics is a very useful tool in the analysis of the SW equations. The characteristics are special lines (trajectories) in the real space on which the hyperbolic PDEs can be replaced by ODEs. We shall present here the basic technical details in the context of the present discussion. (For the mathematical derivation and further details see Anderson, Tannehill, and Pletcher 1984, Morton and Mayers 1994, and Mattheij, Rienstra, and Boonkkamp 2005.)

Consider the system

$$\begin{bmatrix} h_t \\ u_t \end{bmatrix} + \begin{bmatrix} \mathcal{U} & \mathcal{H} \\ A & B \end{bmatrix} \begin{bmatrix} h_x \\ u_x \end{bmatrix} = \begin{bmatrix} S_1 \\ S_2 \end{bmatrix}, \tag{A.1}$$

where $h$ and $u$ are functions of $x$ and $t$, and $A, B, \mathcal{H}, \mathcal{U}, S_1$, and $S_2$ are functions of $h, u, x$, and $t$. The coordinate $x$ stands for a general coordinate in the direction of propagation, i.e., $r$ in the cylindrical case, or the transformed $y$. To be specific, in our context the first line expresses the continuity equation, and the second line the horizontal momentum equation. Addition of dependent variables, such as azimuthal/angular velocity and particle concentration (volume fraction) and corresponding equations increases the dimension of the system. In general, let $n$ be the number of dependent variables ($n = 2$ in the illustrated case). The hyperbolicity justifies the expectation that the $n \times n$ matrix which multiplies the vector of $x$-derivatives has real distinct eigenvalues $\lambda_j$ and a full set of corresponding eigenvectors, ($j = 1, 2, \cdots, n$). Let the eigenvector corresponding to $\lambda_j$ be of the form $(\nu_1^j, \nu_2^j, \cdots, \nu_n^j)$.

Calculate the eigenvalues of the matrix of coefficients of (A.1) by

$$\begin{vmatrix} \mathcal{U} - \lambda & \mathcal{H} \\ A & B - \lambda \end{vmatrix} = 0. \tag{A.2}$$

We obtain

$$\lambda_1 = \frac{1}{2}\left[\mathcal{U} + B + \sqrt{(\mathcal{U} + B)^2 + 4(A\mathcal{H} - B\mathcal{U})}\right], \tag{A.3}$$

$$\lambda_2 = \frac{1}{2}\left[\mathcal{U} + B - \sqrt{(\mathcal{U} + B)^2 + 4(A\mathcal{H} - B\mathcal{U})}\right]. \tag{A.4}$$

Calculate the left-hand-side eigenvectors by

$$[\nu_1^j, \nu_2^j]\begin{bmatrix} \mathcal{U} & \mathcal{H} \\ A & B \end{bmatrix} = \lambda_j[\nu_1^j, \nu_2^j], \tag{A.5}$$

for $j = 1, 2$. Suppose that the system admits $\nu_1^1 \neq 0$. We set $\nu_1^1 = 1$. We obtain

$$\nu_2^j = (\lambda_j - \mathcal{U})\frac{1}{A} \quad \text{and} \quad \nu_2^j = \frac{\mathcal{H}}{B - \lambda_j}, \tag{A.6}$$

which are different forms of the same result. (In general, the value 1 can be assigned to one component of the eigenvector; but we must be careful not to assign this value to a component that must be 0.)

Finally consider

$$[\nu_1^j, \nu_2^j]\begin{bmatrix} dh \\ du \end{bmatrix} = [\nu_1^j, \nu_2^j]\begin{bmatrix} S_1 \\ S_2 \end{bmatrix} dt \tag{A.7}$$

$$\text{on} \quad \frac{dx}{dt} = \lambda_j \quad (j = 1, 2). \tag{A.8}$$

This yields the characteristic equations

$$\nu_1^1 dh + \nu_2^1 du = (\nu_1^1 S_1 + \nu_2^1 S_2)dt \quad \text{on} \quad \frac{dx}{dt} = \lambda_1; \tag{A.9}$$

$$\nu_1^2 dh + \nu_2^2 du = (\nu_1^2 S_1 + \nu_2^2 S_2)dt \quad \text{on} \quad \frac{dx}{dt} = \lambda_2. \tag{A.10}$$

Explicit integrals of the characteristic equations can be obtained in some restricted cases, for example when the system is homogeneous ($S_1 = S_2 = 0$) and the $\nu^j$ components are constants or simple functions.

This technique can be extended to more variables which depend on $x, t$. Consider for example the addition of angular velocity $\omega(x, t)$ to the system. The additional variable introduces a new column and a new row in the matrix of coefficients. This also produces the additional $\lambda_3$ (i.e., another characteristic trajectory), an additional eigenvector, and additional component of eigenvector $\nu_3^j$ ($j = 1, 2, 3$). The results will be of the form

$$\nu_1^j dh + \nu_2^j du + \nu_3^j d\omega = (\nu_1^j S_1 + \nu_2^j S_2 + \nu_3^j S_3)dt \quad \text{on} \quad \frac{dx}{dt} = \lambda_j \quad (j = 1, 2, 3). \tag{A.11}$$

## A.2    Numerical solution of the SW equations

The domain of solution is $0 \le x \le x_N(t)$. It is convenient to use $y = x/x_N$ and apply the transformation of § 2.3. The dependent variables are now functions of $y, t$ with $0 \le y \le 1$ and $t \ge 0$. The typical system, in dimensionless conservation form, is written as

$$\frac{\partial \mathbf{V}}{\partial t} = \frac{1}{x_N}\left[y\dot{x}_N\frac{\partial}{\partial y}\mathbf{V} - \frac{\partial}{\partial y}\mathbf{W}\right] - \mathbf{s}. \tag{A.12}$$

(In the axisymmetric case $y = r/r_N$ and $x_N$ in these equations is replaced with $r_N$.) For $\mathbf{V}$ we have initial values; we also assume that the boundary values for $\mathbf{V}$ at $y = 0, 1$ are known or can be easily calculated. The components of $\mathbf{W}$ and $\mathbf{s}$ are functions of the components of $\mathbf{V}$.

We introduce the grid

$$t^n = n\delta t, \quad (n = 0, 1, \dots); \tag{A.13}$$

$$y_j = j\delta y, \quad (j = 0, 1, \dots, J), \tag{A.14}$$

where $\delta y = 1/J$. Let $v_j^n$ denote the discretized value of the variable $v(y, t)$ at position $y_j$ and time $t^n$.

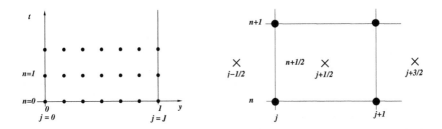

**FIGURE A.1:**    Schematic description of the numerical grid (left) and enlarged cell (right) for the SW equations. The $\times$ shows the (virtual) intermediate points.

To discretize (A.12) we can adapt the Lax-Wendroff two-step scheme, following Press, Teukolski, Vetterling, and Flannery (1992) and Bonnecaze, Huppert, and Lister (1993). Consider the representative

$$\frac{\partial v}{\partial t} = -\frac{\partial}{\partial y}F(v), \tag{A.15}$$

and the schematic grid of Fig. A.1. Suppose that the values of $v_j^n$ are known (at all grid points $y_j$). We wish to obtain the corresponding values after one time interval, at $t^{n+1}$. This is achieved in two steps. First, intermediate values at half-mesh space intervals and half time interval advance are defined by the the finite-difference approximation of (A.15) as follows

$$v_{j+1/2}^{n+1/2} = \frac{1}{2}(v_{j+1}^n + v_j^n) - \frac{1}{2}\frac{\delta t}{\delta y}(F_{j+1}^n - F_j^n) \quad (j = 0, 1, \dots, J-1). \quad \text{(A.16)}$$

Note that the boundary conditions at $j = 0$ and $J$ enter into this calculation.

With the intermediate values of $v$ we calculate $F_{j+1/2}^{n+1/2}$ at the internal points $j + 1/2 = 1/2, 3/2, \dots, J - 1/2$. Next we advance the solution to the new time by a discretization of (A.15) which is formally centered about the midpoint

$$v_j^{n+1} = v_j^n - \frac{\delta t}{\delta y}(F_{j+1/2}^{n+1/2} - F_{j-1/2}^{n+1/2}) \quad (j = 1, \dots, J-1). \quad \text{(A.17)}$$

The value of $v^{n+1}$ at $j = 0$ and $J$ is updated via the boundary conditions. The intermediate values of $v$ can be discarded. If a source term $-s(y, v)$ is present on the RHS of (A.15), the contributions $-s\,\delta t/2$ and $-s\,\delta t$, at the appropriate points, should be added on the RHS of (A.16) and (A.17), respectively.

The truncation error of the two-step Lax-Wendroff method is $O(\delta y^2, \delta t^2)$ and the amplitude damping is relatively small (in particular when a properly defined Courant number is close to 1). The speed of shocks is usually well reproduced. However, some spurious oscillations appear; this effect can be reduced with the aid of an artificial dissipation (viscosity) term.

The extension of the two-step algorithm (A.15)-(A.17) to the components of (A.12) is straightforward. Linear interpolation is used to calculate values at mid-interval positions $j + 1/2$. The speed of the current $\dot{x}_N = u_N$ is provided by the nose condition, and $x_N$ is advanced by $u_N \delta t$.

Consider the boundary conditions. For some variables these are known by explicit formulas or values; for example, $u(y = 0, t) = 0$ at the backwall or at the center of the lock. For other variables some finite-difference approximations must be used; for example, for the height $h$ at $y = 0$ and 1. Two convenient methods can be used.

First, we can integrate on a characteristic which propagates from the interior to the appropriate boundary. For example, for the 2D one-layer SW model the characteristic $dx/dt = u - h^{1/2}$ can be used to calculate $h(x = 0)$ with the aid of the (discretized) relationship $h^{-1/2}dh - du = 0$; see (2.38).

Second, we can use the equations of motion at the boundary with one-sided discretization of the spatial derivatives. We recall that the values at the inner points $j = 1, \dots, J - 1$ are obtained from the original scheme. For example, at $y = 0$ the 2D continuity equation, with $u(y = 0) = 0$, yields

$$\frac{dh}{dt} = -h\frac{1}{x_N}\frac{\partial u}{\partial y} \quad (y = 0) \quad \text{(A.18)}$$

which can be used to calculate $h_{j=0}^{n+1}$. Some care may be required to smooth out local oscillations which tend to appear near the endpoints.

Consider the artificial dissipation term. This is usually added to the RHS of $x$-momentum equation written for $uh$. A practical form is

$$\frac{\partial}{\partial x}\left[b\delta x(|u|+c_+)h\frac{\partial u}{\partial x}\right] = \frac{1}{x_N}\frac{\partial}{\partial y}\left[b\delta y(|u|+c_+)h\frac{\partial u}{\partial y}\right], \qquad (A.19)$$

where $b$ is a constant (typically 0.1) and $c_+$ is the speed of the relevant forward-moving characteristic. This term can be incorporated as a component of $\mathbf{W}$ in (A.12). The rigor is unimportant because this is an artificial addition. The time step is limited by the CFL condition which, very roughly, requires $\delta t/\delta y < 1$. On a typical personal computer or work stations the computation time for a $J = 100$ grid and $t \approx 10$ is insignificant. The optimal time step can be easily determined by a trial-end-error process.

Another simple scheme which can be used along similar lines is the explicit MacCormack method (see Anderson, Tannehill, and Pletcher 1984). We sketch the procedure for (A.15). The first step is a predictor represented by

$$v_j^{n+*} = v_j^n - \frac{\delta t}{\delta y}(F_{j+1}^n - F_j^n), \qquad (A.20)$$

in which the space difference of $\partial F/\partial x$ is forward. The second corrector step is

$$v_j^{n+1} = \frac{1}{2}\left[v_j^n + v_j^{n+*} - \frac{\delta t}{\delta y}(F_j^{n+*} - F_{j-1}^{n+*})\right], \qquad (A.21)$$

in which the space difference is backward. The predictor values can be discarded. The errors and stability properties in linear problems are similar with these of the Lax-Wendroff scheme, but some (usually small) differences show up in non-linear problems, and an improved resolution of discontinuities may be achieved with MacCormack's method upon a proper choice of the order of forward/backward differencing in the predictor/corrector steps.

The calculation of the boundary conditions and artificial diffusions were discussed above.

The numerical solution of hyperbolic systems is a well-developed topic. For further information on the above-mentioned schemes and other relevant methods the reader is referred to the available literature, for example Anderson, Tannehill, and Pletcher (1984), Morton and Mayers (1994), and Mattheij, Rienstra, and Boonkkamp (2005).

In any case, some trial-and-error and fine tuning of the scheme (concerning the points number $J$, ratio $\delta t/\delta y$, artificial diffusion coefficient $b$, boundary conditions approximation) may be necessary for obtaining satisfactory results in a particular problem. The analytical solutions for slumping and similarity provide useful guiding lines and test cases.

# Appendix B

## Navier-Stokes numerical simulations

Here we illustrate some principles and considerations concerning the simulation of gravity currents and intrusion. We follow the presentations of Hallworth, Huppert, and Ungarish (2001) and Ungarish and Huppert (2004). We focus attention on the more complex axially-symmetric rotating configuration; the two-dimensional case follows by a straightforward modification.

### B.1 Formulation

Let $z$ be the vertical coordinate, with the gravity $g$ pointing in the negative $z$ direction. For generality, we use a cylindrical $\{r, \theta, z\}$ coordinate system which rotates with constant $\Omega$ about the axis $z$. The velocity components are $\{u, v, w\}$ . The geometries under consideration are

(1) Axisymmetric with rotation. All three velocity components are of interest.

(2) Axisymmetric non-rotating. In this case we set $\Omega = 0, v \equiv 0$. Only the $u$ and $w$ velocity components are of interest.

(3) Cartesian non-rotating 2D. This case is obtained by setting $\Omega, v$, and the curvature terms equal to zero. The $r$ coordinate is replaced by $x$, and the regularity boundary condition at $r = 0$ is replaced by the no-slip condition $w = 0$ at $x = 0$.

Let $\rho_c$ be the density of the current. The density of the unperturbed ambient varies from $\rho_o$ at the top (usually, an open interface) to $\rho_b (\leq \rho_c)$ at the bottom. In the non-stratified case, $\rho_o = \rho_b$. In the stratified case $\rho_o < \rho_b \leq \rho_c$.

We define again

$$\epsilon = \frac{\rho_c - \rho_o}{\rho_o}; \quad S = \frac{\rho_b - \rho_o}{\rho_c - \rho_o}; \quad g' = \epsilon g. \tag{B.1}$$

Here $\epsilon > 0$ and $0 \leq S \leq 1$.

We introduce the density function

$$\phi(\mathbf{r}, t) = [\rho(\mathbf{r}, t) - \rho_o]/\epsilon, \tag{B.2}$$

which represents the concentration of the dense component on the $[0, 1]$ scale. We expect $0 \leq \phi \leq S$ in the ambient and $\phi = 1$ in the current.

The density in the system is now given by

$$\rho(\mathbf{r}, t) = \rho_o[1 + \epsilon\, \phi(\mathbf{r}, t)]. \tag{B.3}$$

The lengths are scaled with $L = r_0$, speed with $U = (g'r_0)^{1/2}$, time with $r_0/U$, and pressure with $\rho_o U^2$. The reduced pressure is obtained by addition of $\rho_o gz - (1/2)\Omega^2 r^2$ to the thermodynamic pressure.

We employ the following dimensionless balance equations:

1. Continuity of volume

$$\nabla \cdot \mathbf{v} = 0; \tag{B.4}$$

2. Momentum balance

$$\frac{D\mathbf{v}}{Dt} + 2\mathcal{C}\hat{z} \times \mathbf{v} = \frac{1}{1 + \epsilon\phi}\left[-\nabla p + \phi(\epsilon\, \mathcal{C}^2 r\hat{r} - \mathcal{F}^2\hat{z}) + \frac{1}{Re}\nabla^2\mathbf{v}\right]; \tag{B.5}$$

3. Density transport

$$\frac{\partial\phi}{\partial t} + \nabla \cdot \mathbf{v}\phi = \mathcal{D}\nabla^2\phi. \tag{B.6}$$

The relevant dimensionless parameters, in addition to $\epsilon$, are the Reynolds number,

$$Re = \frac{UL}{\nu} = \frac{(g'r_0)^{1/2}r_0}{\nu}; \tag{B.7}$$

the Coriolis to inertia ratio parameter

$$\mathcal{C} = \frac{\Omega L}{U} = \frac{\Omega r_0}{(g'r_0)^{1/2}}; \tag{B.8}$$

the global Froude number (squared)

$$\mathcal{F}^2 = \frac{g'L}{U^2} = 1 \tag{B.9}$$

and the dimensionless diffusion coefficient $\mathcal{D} = 1/Pe = 1/(\sigma Re)$ where $Pe$ and $\sigma$ are the Peclet and Schmidt numbers.

*The scaling for the code* is slightly different from that used in the analytical investigations and discussion in the main text. In the latter it is convenient (and actually more insightful) to use different length scales for the horizontal and vertical directions. The rescaling is straightforward. The numerical results presented in the main text have been rescaled, unless stated otherwise.

The components of the vector operators in these equations are presented in § C.2.

We are interested in flows with large values of $Re$, moderately small $\mathcal{C}$, small $\epsilon$, and very small $\mathcal{D}$. Actually, the typical physical value of $\mathcal{D}$ is negligibly small (recall that $\sigma \gg 1$ for saline solutions in water), but here a non-vanishing

$\mathcal{D}$ is used as an artificial diffusion (or dissipation) coefficient for numerical smoothing of the large density gradients of the moving interface.

The Ekman number can be defined in terms of the previous parameters by

$$\mathcal{E} = \nu/(\Omega L^2) = (\mathcal{C}Re)^{-1} \tag{B.10}$$

and is a small number.

In the lock-release problem in a bounded tank, three geometric parameters appear: the height of the lock, $h_0/r_0$; the height of the ambient fluid, $H$ (scaled with $r_0$); and the outer radius of the container, $r_w$ (also scaled with $r_0$).

The initial conditions at $t = 0$ are

$$\mathbf{v} = \mathbf{0} \quad (0 \le r \le r_w, \ \ 0 \le z \le H) \tag{B.11}$$

$$\phi = \begin{cases} 1 & (0 \le r \le 1, \ \ 0 \le z \le h_0/r_0) \\ S(1 - z/H) & \text{elsewhere.} \end{cases} \tag{B.12}$$

The boundary conditions for $t \ge 0$ are

$$\mathbf{v} = \mathbf{0} \quad (\text{on the bottom and side walls}); \tag{B.13}$$

$$w = 0, \quad \text{and no shear,} \quad (z = H); \tag{B.14}$$

$$u = 0, \quad \text{and regularity} \quad (r = 0); \tag{B.15}$$

and

$$\hat{n} \cdot \nabla \phi = 0 \quad (\text{ on all boundaries}). \tag{B.16}$$

These conditions contain some simplifications. The initial interfaces (between the ambient and dense fluids and also the free surface) deviate from the horizontal by an amount $0.5\epsilon\,\mathcal{C}^2 r^2$. The free surface may have an additional height perturbation of magnitude $\epsilon$ during the flow. Neglecting these deviations from the horizontal is justified for the small values of $\epsilon$ and $\mathcal{C}^2$ used in the calculations. In addition, we assume that the lock is removed instantaneously and without any perturbation to the fluid.

Useful modifications of the initial and boundary conditions can be easily implemented. The first line of (B.12) describes a gravity current released from a straight lock at the bottom. The shape of the lock can be changed by specifying the domain $z(r)$ in which $S = 1$. To simulate a symmetric intrusion, we can set the value $\phi = 0.5$ in the lock, and place the half lock above and half lock below the $z = 0.5$ plane. And so on.

The boundary conditions (B.13)-(B.13) can be modified for various slip and no-slip conditions on a part, or all, of the boundary of the tank.

The equations are for a non-Boussinesq case. However, the value of $\epsilon$ has been scaled out. It is therefore possible to use very small values of $\epsilon$ without losing accuracy; this will recover the Boussinesq case. We must keep in mind that this formulation assumes zero mutual diffusion between the components.

When diffusion is present, the fluxes of the "dense" and "ambient" components are different from the center-of-mass velocity $\mathbf{v}$. In this case the RHS of the volume continuity and momentum equations contain some additional $O(\epsilon)$ terms, and a more precise definition of $\mathcal{D}$ is needed. For Boussinesq systems the difference caused by mutual diffusion is negligible, but when the current and ambient are gases of high density ratio it makes sense to incorporate the diffusion in the simulation. Such studies were performed by Etienne, Hopfinger, and Saramito (2005), to which the reader is referred for more details.

---

## B.2    A finite-difference code

The solution of the system given in the previous section can be obtained by finite difference methods. The dependent variables are $u, v, w, p,$ and $\phi$. The object is to calculate them by a sequence of time steps on a fixed grid. A practical computer code (applicable to various configurations, see for example Hallworth, Huppert, and Ungarish 2001; Ungarish and Huppert 2002, 2004, 2006; Ungarish 2005a,b; and Ungarish and Zemach 2007) is as follows.

Consider the time advance of a flow field variable denoted by $f$ at time $t$ to the new value denoted $f^+$ at time $t + \delta t$. We use a forward time, finite difference technique. One time step for the momentum equation (B.5) with the Coriolis and pressure terms treated implicitly and other terms treated explicitly yields

$$\mathbf{v}^+ + 2\mathcal{C}\delta t \hat{z} \times \mathbf{v}^+ = -\frac{\delta t}{1 + \epsilon\phi}\nabla p^+ + \delta t \mathbf{X} + \mathbf{v} \equiv \mathbf{B}, \qquad (B.17)$$

where

$$\mathbf{X} = -\mathbf{v} \cdot \nabla\mathbf{v} + \frac{1}{1 + \epsilon\phi}\left[(\epsilon\mathcal{C}^2 r\hat{r} - \mathcal{F}^2\hat{z})\phi + \frac{1}{Re}\nabla^2\mathbf{v}\right]. \qquad (B.18)$$

With some vector algebra manipulations (as discussed in detail by Ungarish 1993, p.303), we obtain from (B.17) an explicit expression for $\mathbf{v}^+$,

$$\mathbf{v}^+ = \frac{1}{1 + 4\mathcal{C}^2\delta t^2}\left[\mathbf{B} + 4\mathcal{C}^2\delta t^2(\hat{z} \cdot \mathbf{B})\hat{z} - 2\mathcal{C}\delta t\hat{z} \times \mathbf{B}\right]. \qquad (B.19)$$

We next apply the divergence operator to both sides of this expression and impose the continuity equation (B.4) on $\mathbf{v}^+$. The result is an elliptic equation for the pressure $p^+$ at $t + \delta t$,

$$\nabla \cdot \frac{1}{1 + \epsilon\phi}\nabla p^+ + 4\delta t^2\mathcal{C}^2\frac{\partial}{\partial z}\frac{1}{1 + \epsilon\phi}\frac{\partial p^+}{\partial z} - \nabla \cdot \mathbf{X} - 4\delta t^2\mathcal{C}^2\frac{\partial}{\partial z}\hat{z} \cdot \mathbf{X}$$

$$- 2\delta t\mathcal{C}\hat{z} \cdot \nabla \times \mathbf{X} - 4\delta t\mathcal{C}^2\frac{\partial w}{\partial z} - 2\mathcal{C}\hat{z} \cdot \nabla \times \mathbf{v} - \frac{\nabla \cdot \mathbf{v}}{\delta t} = 0. \quad (B.20)$$

(Theoretically the last term on the LHS is zero, but to prevent accumulation of numerical errors it is sometimes useful to keep it in the calculations.) The boundary conditions for (B.20) are of Newmann type: values of $\hat{n} \cdot \nabla p^+$ are provided by the velocity conditions. Hence the solution $p^+$ is well defined up to a constant; we can set $p^+ = 0$ at the first grid point. Using the solution $p^+$ we can straightforwardly obtain the velocity field $\mathbf{v}^+$ from (B.19). The $\phi^+$ field can be calculated next using the scalar equation (B.6). This completes, in principle, the time step advance, and a new cycle can be attempted. The accuracy of the time discretization is $O(\delta t^2)$.

The spatial discretization is performed on a staggered grid with $il$ radial intervals and $jl$ axial intervals as sketched in Figure B.1. The variables $p$ and $\phi$ are defined at mid-cell position denoted $(i, j)$; $u$ and $v$ are both defined at the positions $(i \pm \frac{1}{2}, j)$ (to allow straightforward implementation of the Coriolis coupling) and $w$ is defined at $(i, j \pm \frac{1}{2})$.

The boundaries of the tank are at half-cell position: $r_{1/2} = 0, r_{il+\frac{1}{2}} = r_w$; $z_{1/2} = 0, z_{jl+\frac{1}{2}} = H$. This facilitates the implementation of boundary conditions. Both the $r$ and $z$ grid coordinates are stretched by simple mapping functions $r(R)$ and $z(Z)$. The grids $R_i = (i + \frac{1}{2})\delta R$ and $Z_j = (j + \frac{1}{2})\delta Z$ are uniform in the domain $(0 \leq R \leq r_w, \quad 0 \leq Z \leq H)$ with intervals

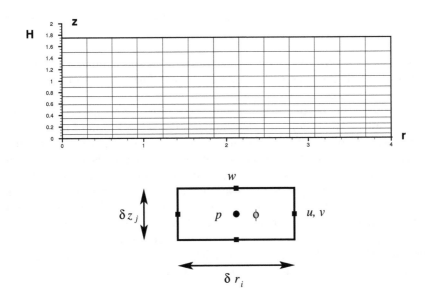

**FIGURE B.1:** The numerical grid, and the details of one cell. The center of the computational cell is denoted by $(i, j)$, corresponding to the position $r_i, z_j$, $(1 \leq i \leq il, \ 1 \leq j \leq jl)$.

$\delta R = r_{\rm w}/il$, and $\delta Z = H/jl$. Linear interpolation is used to obtain values at intermediary grid positions. The truncation error is $O(\delta R^2 + \delta Z^2)$. Dummy cells are added for easy implementation of the boundary conditions. For example, consider the condition at the grid point $r_i, z_{1/2} = 0$. Let the points $i, j = 0$ be the centers of dummy cells at the position $r_i, Z_0$, where $Z_0 = -\delta Z/2$ (a virtual position). To reproduce the no-slip condition for $u$ we set $u_{i,0} = -u_{i,1}$; to reproduce the slip condition we set $u_{i,0} = u_{i,1}$.

An illustration of the finite difference approximation approach is

$$\left(\frac{1}{r}\frac{\partial}{\partial r}r\frac{1}{1+\epsilon\phi}\frac{\partial p}{\partial r}\right)_{r_i,z_j} \approx \frac{1}{r_i}\frac{1}{r_i'\delta R}\left(Y_{i+\frac{1}{2},j} - Y_{i-\frac{1}{2},j}\right), \qquad (\text{B.21})$$

where

$$Y_{i+\frac{1}{2},j} = r_{i+\frac{1}{2}}\frac{1}{1+\epsilon\phi_{i+\frac{1}{2},j}}\frac{1}{r_{i+\frac{1}{2}}'\delta R}(p_{i+1,j} - p_{i,j}) \qquad (\text{B.22})$$

and $r_i'$ is the derivative of $r(R)$ at $R_i$. (To calculate $Y_{i-\frac{1}{2},j}$ we substitute $i-1$ in place of $i$ in (B.22).)

This method of central differences was employed for all terms, except for the advection terms in the $\phi$ transport equation (B.6). For this equation, according to MacCormack's explicit method (Anderson, Tannehill, and Pletcher 1984 § 4-4.3), we use at each time step the predictor-corrector relationships

$$\begin{aligned}
\phi_{i,j}^p &= \phi_{i,j} - \delta t(Adv_f\,\phi)_{i,j}^f + \delta t(Dif\,\phi)_{i,j}, \\
\phi_{i,j}^c &= \phi_{i,j}^p - \delta t(Adv_b\,\phi^p)_{i,j} + \delta t(Dif\,\phi^p)_{i,j}, \\
\phi^+ &= 0.5(\phi_{i,j} + \phi_{i,j}^c),
\end{aligned} \qquad (\text{B.23})$$

where $Adv_f$ and $Adv_b$ denote the advection terms as approximated by forward and backward differencing, and $Dif$ denotes the diffusion terms approximated by central differences.

The combination of the foregoing time and space discretizations is the core of the computer code. For each time step the discretized form of the Poisson equation (B.20) for the discretized variables $p_{i,j}^+$, $1 \le i \le il$, $1 \le j \le jl$, must be solved. This yields, after the implementation of the boundary conditions, a block tri-diagonal linear system. The matrix corresponding directly to (B.20) is non-symmetric, but upon multiplication of each of the discretized equations for point $(i,j)$ by $r_i r_i' z_j'$ a symmetric system can be obtained (in computation tests, no remarkable difference between the performances of these versions was noticed). The linear system was solved by a bi-conjugate gradient iterative algorithm (Press et al. 1992). The iterations in the first time step start with the analytical solution of the hydrostatic approximation

$$p(r, z) = -S(z - \frac{1}{2}\frac{z^2}{H}). \qquad (\text{B.24})$$

Subsequently the $p_{i,j}$ field provides the starting values for $p_{i,j}^+$. The computations use real-8 variables. The typical grid has $il = 200$ constant horizontal intervals and $jl = 250$ stretched axial intervals, with $z_1' \approx 0.7$ and $z_{jl}' \approx 1.4$; this was motivated by a compromise between computational limitations and physical considerations. The typical time step was $\delta t = 10^{-3}$. In several test cases the grid size was changed and the time step was halved, without causing any significant differences in the results. The diffusion coefficient in the density transport equation was usually taken as $\mathcal{D} = 0.3\delta R^2$ and $\mathcal{D} = 0.3\delta Z^2$ for the radial and axial fluxes, respectively, and hence this artificially augmented effect is expected to be of the order of magnitude of the numerical truncation error in the physical advection term.

The bi-conjugate gradient method needs typically 200 iterations per time step to reach the allowed error $e = 10^{-4}$, where $e$ is the Euclidean seminorm of the residues divided by the seminorm of the right-hand side of the system of equations. The typical average relative error, $e/(il \times jl)$, in the solution of $p_{i,j}^+$ is therefore less than $10^{-8}$.

The solution of the elliptic equation for $p_{i,j}^+$ is the major computational effort at each time step.

The choice of the numerical grid parameters is usually motivated by the compromise between physical accuracy considerations and computational limitations. Essentially, the mesh intervals are considerably smaller than the expected typical corresponding geometrical dimensions of the simulated current, e.g., the length of the "head," and the average thickness. In rotating systems the Ekman layer thickness (estimated as $3\,\mathcal{E}^{1/2}$) should also be resolved. Nominally, the magnitude of the spatial errors $\delta R^2 + \delta Z^2$ is typically about 0.1 %, and of the accumulated time stepping errors $N\delta t^2$ (where N is the number of time steps) is also typically about 0.1 %; the value of $\delta R^4 + \delta Z^4$ which typifies the smoothing dissipation terms (see below) is about $10^{-6}$, smaller than $1/Re$. We therefore expect that the numerical results provide an acceptable simulation of an observable gravity-current process, at least during the initial period. Eventually, when the current becomes thin (say, about 10 axial intervals) and the interface very irregular, the numerical errors may become significant and even dominant.

Numerical tests indicated that the numerical stability can be improved by the addition of small $O(\delta r^4 + \delta z^4)$ fourth order derivative dissipation terms in the momentum equations, of the form

$$-\varepsilon_e \left[ (\delta R)^4 \frac{\partial^4 \mathbf{v}}{\partial R^4} + (\delta Z)^4 \frac{\partial^4 \mathbf{v}}{\partial Z^4} \right], \tag{B.25}$$

where $\varepsilon_e$ is a constant of approximately 0.06 (see Anderson et al. 1984 § 8-3.3).

## B.3 Other codes

Various other numerical codes have been also used for the simulation of gravity currents. In general, some finite-differences method is used for time discretization. The combination with spectral and pseudo-spectral space discretization allows for high-resolution simulations. These methods are computationally more efficient than the second-order finite-difference code described in the previous section, but require a more difficult programming and debugging process. The details are complicated and the interested reader should consult the appropriate papers for more information. Some typical examples are as follows. Härtel, Meiburg, and Necker (2000) reported two- and three-dimensional high resolution direct numerical simulation in a Cartesian box using the Boussinesq equations. A non-Boussinesq extension for the 2D case was used by Birman, Martin, and Meiburg (2005). Three- and two-dimensional simulations, including cylindrical geometries, with high resolution were presented by Cantero, Balachandar, and Garcia (2007). Particle settling effects were incorporated in the simulations by Necker, Härtel, Kleiser, and Meiburg (2002) and Necker, Härtel, Kleiser, and Meiburg (2005). Effects of mutual diffusion in non-Boussinesq systems were treated numerically by Etienne, Hopfinger, and Saramito (2005), using a 2D finite element grid in space. A finite-volume front capturing method that allows simulations of non-diffusive fluids of arbitrary density ratio was used for 2D simulations of non-Boussinesq currents by Bonometti, Balachandar, and Magnaudet (2008) and Bonometti and Balachandar (2009).

Some of the numerical 2D simulations use the vorticity-streamfunction formulation. With $u = \partial\psi/\partial z, w = -\partial\psi/\partial x$, where $\psi$ is the stream-function, the continuity equation $\nabla \cdot \mathbf{v} = 0$ is identically satisfied. The introduction of the vorticity $\omega = \partial w/\partial x - \partial u/\partial z$ allows the elimination of the pressure term from the momentum equations. However, it is still needed to solve an elliptic equation, $\nabla^2\psi = -\omega$, instead of (B.20). The density function equation is unchanged. (The resulting equations, for a non-stratified ambient, can be seen in Birman, Martin, and Meiburg 2005.)

# Appendix C

## Some useful formulas

### C.1 Leibniz's Theorem

The differentiation of an integral satisfies

$$\frac{d}{d\eta} \int_{a(\eta)}^{b(\eta)} f(z,\eta)dz = \int_{a(\eta)}^{b(\eta)} \frac{\partial}{\partial \eta} f(z,\eta)dz + f(b,\eta)\frac{db}{d\eta} - f(a,\eta)\frac{da}{d\eta}. \quad \text{(C.1)}$$

### C.2 Vectors and coordinate systems

Notation: $\phi, \psi$ scalars; $\mathbf{A}, \mathbf{B}, \mathbf{C}, \mathbf{v}$ vectors, $\mathbf{r}$ radius vector; $\mathbf{I}$ unit tensor; $\tau$ second order tensor and $\tau^T$ its transpose; $\nabla[\ ] \equiv \text{grad}\ [\ ], \nabla \cdot [\ ] \equiv \text{div}\ [\ ]$. A combination of the form $\mathbf{AB}$ is a dyadic, a special form of $\tau$. Several useful identities involving these variables and operations are summarized below.

$$\mathbf{A} \cdot (\mathbf{B} \times \mathbf{C}) = \mathbf{B} \cdot (\mathbf{C} \times \mathbf{A}) = \mathbf{C} \cdot (\mathbf{A} \times \mathbf{B}) \quad \text{(C.2)}$$

$$\mathbf{A} \times (\mathbf{B} \times \mathbf{C}) = \mathbf{B}(\mathbf{A} \cdot \mathbf{C}) - \mathbf{C}(\mathbf{A} \cdot \mathbf{B}) \quad \text{(C.3)}$$

$$\mathbf{A} \cdot \mathbf{BC} = (\mathbf{A} \cdot \mathbf{B})\mathbf{C} \quad \text{(C.4)}$$

$$\nabla(\psi\phi) = \psi\nabla\phi + \phi\nabla\psi \quad \text{(C.5)}$$

$$\nabla \cdot (\psi\mathbf{A}) = \psi\nabla \cdot \mathbf{A} + \mathbf{A} \cdot \nabla\psi \quad \text{(C.6)}$$

$$\nabla \times (\psi\mathbf{A}) = \psi\nabla \times \mathbf{A} + (\nabla\psi) \times \mathbf{A} \quad \text{(C.7)}$$

$$\nabla^2\psi \equiv \nabla \cdot (\nabla\psi) \quad \text{(C.8)}$$

$$\mathbf{A} \cdot \nabla\psi = (\mathbf{A} \cdot \nabla)\psi \quad \text{(C.9)}$$

$$\mathbf{A} \cdot \nabla\mathbf{B} = (\mathbf{A} \cdot \nabla)\mathbf{B} \quad \text{(C.10)}$$

$$\nabla(\mathbf{A} \cdot \mathbf{B}) = \mathbf{A} \cdot \nabla\mathbf{B} + \mathbf{B} \cdot \nabla\mathbf{A} + \mathbf{A} \times (\nabla \times \mathbf{B}) + \mathbf{B} \times (\nabla \times \mathbf{A}) \quad \text{(C.11)}$$

$$\frac{1}{2}\nabla(\mathbf{A} \cdot \mathbf{A}) = \mathbf{A} \cdot \nabla\mathbf{A} + \mathbf{A} \times (\nabla \times \mathbf{A}) \quad \text{(C.12)}$$

$$\nabla \cdot (\mathbf{A} \times \mathbf{B}) = \mathbf{B} \cdot (\nabla \times \mathbf{A}) - \mathbf{A} \cdot (\nabla \times \mathbf{B}) \quad \text{(C.13)}$$

$$\nabla \times (\mathbf{A} \times \mathbf{B}) = \mathbf{B} \cdot \nabla \mathbf{A} - \mathbf{B}(\nabla \cdot \mathbf{A}) - \mathbf{A} \cdot \nabla \mathbf{B} + \mathbf{A}(\nabla \cdot \mathbf{B}) \quad \text{(C.14)}$$

$$\nabla^2 \mathbf{A} \equiv \nabla \cdot (\nabla \mathbf{A}) = \nabla(\nabla \cdot \mathbf{A}) - \nabla \times (\nabla \times \mathbf{A}) \quad \text{(C.15)}$$

$$\nabla \cdot (\mathbf{AB}) = \mathbf{B}(\nabla \cdot \mathbf{A}) + \mathbf{A} \cdot \nabla \mathbf{B} \quad \text{(C.16)}$$

$$\nabla \psi \mathbf{A} = (\nabla \psi)\mathbf{A} + \psi \nabla \mathbf{A} \quad \text{(C.17)}$$

$$\nabla \cdot (\nabla \mathbf{A})^T = \nabla(\nabla \cdot \mathbf{A}) \quad \text{(C.18)}$$

$$\nabla \cdot (\psi \tau) = (\nabla \psi) \cdot \tau + \psi \nabla \cdot \tau \quad \text{(C.19)}$$

$$\nabla \cdot \mathbf{r} = 3; \quad \nabla \mathbf{r} = \mathbf{I}. \quad \text{(C.20)}$$

The "substantial" derivative is defined by:

$$\frac{D\psi}{Dt} \equiv \frac{\partial \psi}{\partial t} + \mathbf{v} \cdot \nabla \psi; \quad \text{(C.21)}$$

$$\frac{D\mathbf{v}}{Dt} \equiv \frac{\partial \mathbf{v}}{\partial t} + \mathbf{v} \cdot \nabla \mathbf{v} = \frac{\partial \mathbf{v}}{\partial t} + \frac{1}{2}\nabla(\mathbf{v} \cdot \mathbf{v}) - \mathbf{v} \times (\nabla \times \mathbf{v}). \quad \text{(C.22)}$$

The components of $\mathbf{v} \cdot \nabla \mathbf{v}$ in Cartesian and Cylindrical systems are given by (C.31) and (C.41).

The dyadic product: $(A_1, A_2, A_3)$ and $(B_1, B_2, B_3)$ are the components of $\mathbf{A}$ and $\mathbf{B}$, respectively; $\mathbf{AB}$ is the tensor

$$\tau = \begin{pmatrix} A_1 B_1 & A_1 B_2 & A_1 B_3 \\ A_2 B_1 & A_2 B_2 & A_2 B_3 \\ A_3 B_1 & A_3 B_2 & A_3 B_3 \end{pmatrix}; \quad \text{(C.23)}$$

Reynolds' transport theorem for a specific mass of fluid in the control volume $V$, that carries a property $\psi$, is

$$\frac{D}{Dt}\int_V \psi dV = \int_V \left[ \frac{\partial \psi}{\partial t} + \nabla \cdot (\psi \mathbf{v}) \right] dV. \quad \text{(C.24)}$$

## C.2.1   Cartesian coordinates $x, y, z$

Here

$$\nabla = \hat{x}\frac{\partial}{\partial x} + \hat{y}\frac{\partial}{\partial y} + \hat{z}\frac{\partial}{\partial z}. \quad \text{(C.25)}$$

The derivatives of the unit vectors in respect with $x, y, z$ are equal to zero.

Let $\mathbf{v} = u\hat{x} + v\hat{y} + w\hat{z}$, and $\tau$ symmetric (i.e., $\tau_{xy} = \tau_{yx}$, etc.).

$$\nabla \psi = \frac{\partial \psi}{\partial x}\hat{x} + \frac{\partial \psi}{\partial y}\hat{y} + \frac{\partial \psi}{\partial z}\hat{z} \quad \text{(C.26)}$$

$$\nabla^2 \psi = \frac{\partial^2 \psi}{\partial x^2} + \frac{\partial^2 \psi}{\partial y^2} + \frac{\partial^2 \psi}{\partial z^2} \quad \text{(C.27)}$$

$$\nabla \cdot \mathbf{v} = \frac{\partial u}{\partial x} + \frac{\partial v}{\partial y} + \frac{\partial w}{\partial z} \tag{C.28}$$

$$\nabla \times \mathbf{v} = \left(\frac{\partial w}{\partial y} - \frac{\partial v}{\partial z}\right)\hat{x} + \left(\frac{\partial u}{\partial z} - \frac{\partial w}{\partial x}\right)\hat{y} + \left(\frac{\partial v}{\partial x} - \frac{\partial u}{\partial y}\right)\hat{z} \tag{C.29}$$

$$[\nabla^2 \mathbf{v}]_x = \frac{\partial^2 u}{\partial x^2} + \frac{\partial^2 u}{\partial y^2} + \frac{\partial^2 u}{\partial z^2}$$

$$[\nabla^2 \mathbf{v}]_y = \frac{\partial^2 v}{\partial x^2} + \frac{\partial^2 v}{\partial y^2} + \frac{\partial^2 v}{\partial z^2} \tag{C.30}$$

$$[\nabla^2 \mathbf{v}]_z = \frac{\partial^2 w}{\partial x^2} + \frac{\partial^2 w}{\partial y^2} + \frac{\partial^2 w}{\partial z^2}$$

$$[\mathbf{v} \cdot \nabla \mathbf{v}]_x = u\frac{\partial u}{\partial x} + v\frac{\partial u}{\partial y} + w\frac{\partial u}{\partial z}$$

$$[\mathbf{v} \cdot \nabla \mathbf{v}]_y = u\frac{\partial v}{\partial x} + v\frac{\partial v}{\partial y} + w\frac{\partial v}{\partial z} \tag{C.31}$$

$$[\mathbf{v} \cdot \nabla \mathbf{v}]_z = u\frac{\partial w}{\partial x} + v\frac{\partial w}{\partial y} + w\frac{\partial w}{\partial z}$$

$$[\nabla \cdot \tau]_x = \frac{\partial \tau_{xx}}{\partial x} + \frac{\partial \tau_{xy}}{\partial y} + \frac{\partial \tau_{xz}}{\partial z}$$

$$[\nabla \cdot \tau]_y = \frac{\partial \tau_{xy}}{\partial x} + \frac{\partial \tau_{yy}}{\partial y} + \frac{\partial \tau_{yz}}{\partial z} \tag{C.32}$$

$$[\nabla \cdot \tau]_z = \frac{\partial \tau_{xz}}{\partial x} + \frac{\partial \tau_{yz}}{\partial y} + \frac{\partial \tau_{zz}}{\partial z}$$

For an incompressible Newtonian fluid the dissipation function is

$$\Phi_v/\mu = 2\left[\left(\frac{\partial u}{\partial x}\right)^2 + \left(\frac{\partial v}{\partial y}\right)^2 + \left(\frac{\partial w}{\partial z}\right)^2\right]$$
$$+ \left(\frac{\partial v}{\partial x} + \frac{\partial u}{\partial y}\right)^2 + \left(\frac{\partial w}{\partial y} + \frac{\partial v}{\partial z}\right)^2 + \left(\frac{\partial u}{\partial z} + \frac{\partial w}{\partial x}\right)^2. \tag{C.33}$$

## C.2.2 Cylindrical coordinates $r, \theta, z$

Here

$$\nabla = \hat{r}\frac{\partial}{\partial r} + \hat{\theta}\frac{1}{r}\frac{\partial}{\partial \theta} + \hat{z}\frac{\partial}{\partial z}. \tag{C.34}$$

Derivatives of unit vectors:

$$\frac{\partial}{\partial \theta}\hat{r} = \hat{\theta}; \quad \frac{\partial}{\partial \theta}\hat{\theta} = -\hat{r}; \tag{C.35}$$

otherwise, the unit vectors are constant.

Let $\mathbf{v} = u\hat{r} + v\hat{\theta} + w\hat{z}$, and $\tau$ symmetric.

$$\nabla \psi = \frac{\partial \psi}{\partial r}\hat{r} + \frac{1}{r}\frac{\partial \psi}{\partial \theta}\hat{\theta} + \frac{\partial \psi}{\partial z}\hat{z} \tag{C.36}$$

$$\nabla^2 \psi = \frac{1}{r}\frac{\partial}{\partial r}\left(r\frac{\partial \psi}{\partial r}\right) + \frac{1}{r^2}\frac{\partial^2 \psi}{\partial \theta^2} + \frac{\partial^2 \psi}{\partial z^2} \tag{C.37}$$

$$\nabla \cdot \mathbf{v} = \frac{1}{r}\frac{\partial}{\partial r}(ru) + \frac{1}{r}\frac{\partial v}{\partial \theta} + \frac{\partial w}{\partial z} \tag{C.38}$$

$$\nabla \times \mathbf{v} = \left(\frac{1}{r}\frac{\partial w}{\partial \theta} - \frac{\partial v}{\partial z}\right)\hat{r} + \left(\frac{\partial u}{\partial z} - \frac{\partial w}{\partial r}\right)\hat{\theta} + \left(\frac{1}{r}\frac{\partial}{\partial r}rv - \frac{1}{r}\frac{\partial u}{\partial \theta}\right)\hat{z} \tag{C.39}$$

$$[\nabla^2\mathbf{v}]_r = \frac{\partial}{\partial r}\left[\frac{1}{r}\frac{\partial}{\partial r}(ru)\right] + \frac{1}{r^2}\frac{\partial^2 u}{\partial \theta^2} - \frac{2}{r^2}\frac{\partial v}{\partial \theta} + \frac{\partial^2 u}{\partial z^2}$$

$$[\nabla^2\mathbf{v}]_\theta = \frac{\partial}{\partial r}\left[\frac{1}{r}\frac{\partial}{\partial r}(rv)\right] + \frac{1}{r^2}\frac{\partial^2 v}{\partial \theta^2} + \frac{2}{r^2}\frac{\partial u}{\partial \theta} + \frac{\partial^2 v}{\partial z^2} \tag{C.40}$$

$$[\nabla^2\mathbf{v}]_z = \frac{1}{r}\frac{\partial}{\partial r}\left(r\frac{\partial w}{\partial r}\right) + \frac{1}{r^2}\frac{\partial^2 w}{\partial \theta^2} + \frac{\partial^2 w}{\partial z^2}$$

$$[\mathbf{v} \cdot \nabla\mathbf{v}]_r = u\frac{\partial u}{\partial r} + \frac{v}{r}\frac{\partial u}{\partial \theta} - \frac{v^2}{r} + w\frac{\partial u}{\partial z}$$

$$[\mathbf{v} \cdot \nabla\mathbf{v}]_\theta = u\frac{\partial v}{\partial r} + \frac{v}{r}\frac{\partial v}{\partial \theta} + \frac{uv}{r} + w\frac{\partial v}{\partial z} \tag{C.41}$$

$$[\mathbf{v} \cdot \nabla\mathbf{v}]_z = u\frac{\partial w}{\partial r} + \frac{v}{r}\frac{\partial w}{\partial \theta} + w\frac{\partial w}{\partial z}$$

$$[\nabla \cdot \tau]_r = \frac{1}{r}\frac{\partial}{\partial r}(r\tau_{rr}) + \frac{1}{r}\frac{\partial}{\partial \theta}\tau_{r\theta} - \frac{1}{r}\tau_{\theta\theta} + \frac{\partial \tau_{rz}}{\partial z}$$

$$[\nabla \cdot \tau]_\theta = \frac{1}{r}\frac{\partial \tau_{\theta\theta}}{\partial \theta} + \frac{\partial \tau_{r\theta}}{\partial r} + \frac{2}{r}\tau_{r\theta} + \frac{\partial \tau_{\theta z}}{\partial z} \tag{C.42}$$

$$[\nabla \cdot \tau]_z = \frac{1}{r}\frac{\partial}{\partial r}(r\tau_{rz}) + \frac{1}{r}\frac{\partial \tau_{\theta z}}{\partial \theta} + \frac{\partial \tau_{zz}}{\partial z}$$

For an incompressible Newtonian fluid, axisymmetric case, the dissipation function is

$$\Phi_v/\mu = 2\left[\left(\frac{\partial u}{\partial r}\right)^2 + \left(\frac{u}{r}\right)^2 + \left(\frac{\partial w}{\partial z}\right)^2\right]$$

$$+ \left[r\frac{\partial}{\partial r}\left(\frac{v}{r}\right)\right]^2 + \left(\frac{\partial v}{\partial z}\right)^2 + \left(\frac{\partial u}{\partial z} + \frac{\partial w}{\partial r}\right)^2. \tag{C.43}$$

# *Notation guide*

The list below gives the main use of the symbols. In some places *ad hoc* different use is made, with appropriate explicit definition.

| | |
|---|---|
| 2D | two-dimensional Cartesian geometry. |
| $a$ | depth ratio at nose, $h/H$ or $h_N/H$. |
| $c_\pm$ | speed of the main SW characteristics. |
| $C$ | a constant. |
| $\mathcal{C}$ | Coriolis to inertial effects ratio. |
| $E$ | energy. |
| $F$ | force. |
| $Fr$ | (function for) Froude number. |
| $g'$ | reduced gravity, see (1.2a). |
| $h$ | thickness of current; half-thickness of symmetric intrusion. |
| $h_0$ | reference height, usually of the lock. |
| $H$ | height of ambient; dimensionless height ratio of ambient/lock. |
| $i$ | index. |
| $k$ | permeability of porous medium. |
| $K$ | (1) a constant; |
| | (2) kinetic energy. |
| LHS | left-hand side. |
| $L_{Ro}$ | Rossby "radius". |
| $\mathcal{N}$ | buoyancy frequency. |
| NS | Navier-Stokes. |
| ODE | ordinary differential equation. |
| $p$ | pressure. |
| $P$ | potential energy. |
| PDE | partial differential equation. |
| $q$ | (1) $hu$ variable; |
| | (2) constant coefficient of inflow volume change. |
| $r$ | radius. |
| $r_0$ | reference length, usually the radius of the lock. |
| $Re$ | Reynolds number. |
| RHS | right-hand side. |
| $S$ | stratification parameter. |
| SL | steady lens. |
| SW | shallow-water. |
| $T$ | time scale. |
| $u$ | speed of propagation. |

| $U$ | (1) speed scale for major motion (usually); |
|---|---|
| | (2) the typical speed of propagation of the current; |
| | (3) the speed of the steady-state current. |
| $v$ | velocity component. |
| $V$ | (1) speed of shock; |
| | (2) speed scale for $y$ direction. |
| $\mathcal{V}$ | volume of current (per unit width in 2D; per radian in axisymmetric). |
| $w$ | velocity component in $z$ direction. |
| $W$ | (1) work; |
| | (2) velocity in $z$ direction. |
| $\mathcal{W}$ | Coriolis to inertial effects ratio in a channel. |
| $x_0$ | reference length, usually of the lock. |
| $y$ | (1) Cartesian coordinate; |
| | (2) similarity coordinate. |
| $z$ | coordinate in vertical direction. |

### Subscript

| $a$ | denotes the ambient fluid domain. (Not to be confused with $a = h_N/\mathit{l}$ |
|---|---|
| $b$ | (1) bottom (usually); |
| | (2) bore. |
| $c$ | (1) denotes the current (usually); |
| | (2) denotes a "critical" value. |
| $e$ | effective. |
| $h$ | at the interface (the inclined and horizontal portions, except for the ju |
| $i$ | (1) general index, see the local use in the text; |
| | (2) of the inclined portion of the interface. |
| $I$ | (1) inertial (usually); |
| | (2) initial. |
| $N$ | nose (front) of the current |
| $o$ | open surface, top boundary. |
| $p$ | particle. |
| $ro$ | runout (in particle-driven case). |
| $s$ | (1) slumping, at end of slumping stage; |
| | (2) in Chapter 9: sedimentation, sediment. |
| $v$ or $V$ | viscous. |
| w | wall of container. |
| $W$ | wave (internal). |
| $+,-$ | forward and backward (or rapid and slow) characteristics. |
| $0$ | usually denotes a reference value or scaling length. |

Upper- and over-script

| | |
|---|---|
| $*$ | used to emphasize a dimensional variable. |
| $-$ | used to emphasize a $z$-averaged variable. |
| $\wedge$ | unit vector. |

Greek

$\alpha$    (1) a constant, usually the power of $t$ in volume-growth function;
       (2) in Chapter 9: the particle volume fraction.

$\beta$    (1) a constant, usually the power of $t$ in similarity propagation;
       (2) in Chapter 9: the scaled settling speed of particle.

$\tilde{\gamma}$    density ratio, see (9.16).

$\Delta\rho$    $\rho_c - \rho_a$.

$\widetilde{\Delta\rho}$    $\rho_b - \rho_o$ in stratified ambient.

$\epsilon$    density difference, relative (or reduced).

$\varepsilon_p$    density difference of particle relative to the interstitial fluid.

$\zeta$    vorticity ($z$ component).

$\eta$    (1) in § 8.2: thickness of current;
       (2) in Chapter 20: upward displacement from initial height.

$\theta$    (1) azimuthal angle in cylindrical coordinate system;
       (2) function which represents the decay of inertial/viscous ratio.
       (3) in Chapter 14: invalidity (or instability) coefficient.

$\lambda$    (1) eigenvalue, speed of characteristic;
       (2) wavelength;
       (3) coefficient for porous boundary, see (9.109).

$\mu$    viscosity coefficient.

$\nu$    kinematic viscosity coefficient.

$\omega$    angular velocity of the fluid with respect to system.

$\Omega$    (1) angular velocity of the system;
       (2) in Chapter 20: volume.

$\rho$    density.

$\varsigma$    velocity shape factor, see (2.21)

$\phi$    (1) in NS simulation: the density function, see (B.2);
       (2) in Chapter 9: the scaled particle volume fraction.

$\Phi$    pressure at the interface.

# Bibliography

Abbott, M. B. (1961). On the spreading of one fluid over another. Part ii. The wave front. *La Houille Blanche 6*, 827–846.

Acton, J. M., H. E. Huppert, and M. G. Worster (2002). Two-dimensional viscous gravity currents flowing over a deep porous medium. *J. Fluid Mech. 440*, 359–380.

Amen, R. and T. Maxworthy (1980). The gravitational collapse of a mixed region into a linearly stratified fluid. *J. Fluid Mech. 96*, 65–80.

Anderson, D. A., J. C. Tannehill, and R. M. Pletcher (1984). *Computational Fluid Mechanics and Heat Transfer*. Hemisphere, NY.

Baines, P. G. (1995). *Topographic Effects in Stratified Flows*. Cambridge University Press.

Baines, P. G. (2001). Mixing in flows down gentle slopes into stratified environments. *J. Fluid Mech. 443*, 237–270.

Batchelor, G. K. (1981). *An Introduction to Fluid Dynamics*. Cambridge Un. Press, G.B.

Benjamin, T. B. (1968). Gravity currents and related phenomena. *J. Fluid Mech. 31*, 209–248.

Billingham, J. and A. King (2000). *Wave Motion*. Cambridge Un. Press.

Birman, V. K., J. E. Martin, and E. Meiburg (2005). The non-Boussinesq lock exchange problem. Part 2: High-resolution simulations. *J. Fluid Mech. 537*, 125–144.

Birman, V. K., E. Meiburg, and M. Ungarish (2007). On gravity currents in stratified ambients. *Phys. Fluids 19*, (086602) 1–10, [DOI: 10.1063/1.2756553].

Bolster, D., A. Hang, and P. F. Linden (2008). The front speed of intrusions into a continuously stratified medium. *J. Fluid Mech. 594*, 369–377.

Bonnecaze, R. T., M. A. Hallworth, H. E. Huppert, and J. R. Lister (1995). Axisymmetric particle-driven gravity currents. *J. Fluid Mech. 294*, 93–121.

Bonnecaze, R. T., H. E. Huppert, and J. R. Lister (1993). Particle-driven gravity currents. *J. Fluid Mech. 250*, 339–369.

Bonometti, T. and S. Balachandar (2009). Simulation of non-Boussinesq gravity currents of various initial fractional depths: a comparison with a recent shallow-water model. *Submitted.*

Bonometti, T., S. Balachandar, and J. Magnaudet (2008). Wall effects in non-Boussinesq density currents. *J. Fluid Mech. 616*, 445–475.

Britter, R. E. (1979). The spread of a negatively buoyant plume in a calm environment. *Atmos. Environ. 13*, 1241–1247.

Cantero, M., S. Balachandar, and M. H. Garcia (2007). High resolution simulations of cylindrical density currents. *J. Fluid Mech. 590*, 437–469.

Cheong, H.-B., J. J. P. Kuenen, and P. F. Linden (2006). The front speed of intrusive gravity currents. *J. Fluid Mech. 552*, 1–11.

Choboter, P. F. and G. E. Swaters (2000). On the baroclinic instability of axisymmetric rotating gravity currents with bottom slope. *J. Fluid Mech. 408*, 149–177.

Csanady, G. T. (1979). The birth and death of a warm core ring. *J. Geophys. Res. 84*, 777–780.

Dade, W. and H. Huppert (1995a). A box model for non-entraining, suspension driven gravity surges on horizontal surfaces. *Sedimentology 42*, 453–471.

Dade, W. and H. Huppert (1995b). Runout and fine sediment deposits of axisymmetric turbidity curents. *J. Geophys. Res. 100*, 18,597–18,609.

D'Alessio, S., T. Moodie, J. Pascal, and G. Swaters (1996). Gravity currents produced by sudden release of a fixed volume of heavy fluid. *Studies in Applied Mathematics 96*, 359–385.

Davidson, P. A. (2001). *An Introduction to Magnetohydrodynamics.* Cambridge University Press.

Dewar, W. and P. Killworth (1990). On the cylinder collapse problem: mixing and merger of isolated eddies. *J. Phys. Oceanog. 20*, 1563–75.

Didden, N. and T. Maxworthy (1982). The viscous spreading of plane and axisymmetric gravity currents. *J. Fluid Mech. 121*, 27–42.

Einstein, H. A. (1968). Deposition of suspended particles in a gravel bed. *J. Hydraul. Div. ASCE 94*, 1197–1205.

Ermanyuk, E. V. and N. V. Gavrilov (2007). A note on the propagation speed of a weakly dissipative gravity current. *J. Fluid Mech. 574*, 393–403.

Etienne, J., E. J. Hopfinger, and P. Saramito (2005). Numerical simulations of high density ratio lock-exchange flows. *Physics of Fluids 17*(3), 036601–12.

Fanneløp, T. K. and Ø. Jacobsen (1984). Gravity spreading of heavy clouds instantaneously released. *ZAMP 35*, 559–584.

Fanneløp, T. K. and G. D. Waldman (1972). Dynamics of oil slicks. *AIAA J. 10*, 506–510.

Faust, K. M. and E. J. Plate (1984). Experimental investigation of intrusive gravity currents entering stably stratified fluids. *J. Hydraulic Res. 22*, 315–325.

Fay, J. A. (1969). *The spread of oil slicks on a calm sea*. In *Oil on the Sea*, pp. 33–63. editor: D.P. Hoult, Plenum.

Felix, M. (2002). Flow structure of turbidity currents. *Sedimentology 49*, 397–419.

Fernandez, R. L. and J. Imberger (2008). Time-varying underflow into a continuous stratification with bottom slope. *J. Hydraulic Engineering 134 (9)*, 1191–1198.

Flierl, G. R. (1979). A simple model for the structure of warm and cold core rings. *J. Geophys.Res. 84*, 781–785.

Flynn, M. R., T. Boubarne, and P. F. Linden (2008). The dynamics of steady, partial-depth intrusive gravity curents. *Atmosphere-Ocean 46*, 421–432.

Flynn, M. R. and P. F. Linden (2006). Intrusive gravity curents. *J. Fluid Mech. 568*, 193–202.

Gladstone, C., J. C. Philips, and R. S. J. Sparks (1998). Experiments on bidisperse, constant-volume gravity currents: propagation and sediment deposition. *Sedimentology 45*, 833–843.

Gratton, J. and C. Vigo (1994). Self-similar gravity currents with variable inflow revisited: plane currents. *J. Fluid Mech. 258*, 77–104.

Greenspan, H. P. (1968). *The Theory of Rotating Fluids*. Cambridge Un. Press, Cambridge, G.B. (Reprinted by Breukelen Press, Brookline, MA, 1990).

Griffiths, R. W. (1986). Gravity currents in rotating systems. *Annu. Rev. Fluid Mech. 18*, 59–89.

Griffiths, R. W. and E. J. Hopfinger (1983). Gravity currents moving along a lateral boundary in a rotating fluid. *J. Fluid Mech. 134*, 357–399.

Griffiths, R. W. and P. Linden (1981). The stability of vortices in a rotating, stratified fluid. *J. Fluid Mech. 105*, 283–316.

Gröbelbauer, H. P., T. K. Fanneløp, and R. E. Britter (1993). The propagation of intrusion fronts of high density ratios. *J. Fluid Mech. 250*, 669–687.

Grundy, R. E. and J. Rottman (1985a). The approach to self-similarity of the solutions of the shallow-water equations representing gravity current releases. *J. Fluid Mech. 156*, 39–53.

Grundy, R. E. and J. Rottman (1985b). Self-similar solutions of the shallow-water equations representing gravity current with variable inflow. *J. Fluid Mech. 169*, 337–351.

Hacker, J. N. and P. F. Linden (2002). Gravity currents in rotating channels. Part 1. Steady-state theory. *Journal of Fluid Mechanics 457*, 295–324.

Hallworth, M., H. E. Huppert, J. Phillips, and R. Sparks (1996). Entrainment into two-dimensional and axisymmetric turbulent gravity currents. *J. Fluid Mech. 308*, 289–311.

Hallworth, M. A., A. J. Hogg, and H. E. Huppert (1998). Effects of external flow on compositional and particle gravity currents. *J. Flyuid Mech. 359*, 109–142.

Hallworth, M. A., H. E. Huppert, and M. Ungarish (2001). Axisymmetric gravity currents in a rotating system: experimental and numerical investigations. *J. Fluid Mech. 447*, 1–29.

Hallworth, M. A., H. E. Huppert, and M. Ungarish (2003). On inwardly propagating axisymmetric gravity currents. *J. Fluid Mech. 494*, 255–274.

Harris, T., A. J. Hogg, and H. E. Huppert (2002). Polydisperse particle-driven gravity currents. *J. Fluid Mech. 472*, 333–371.

Härtel, C., F. Carlsson, and M. Thunblom (2000). Analysis and direct numerical simulation of the flow at a gravity-current head. Part 2. The lobe-and-cleft instability. *J. Fluid Mech. 418*, 213–229.

Härtel, C., E. Meiburg, and F. Necker (2000). Analysis and direct numerical simulation of the flow at a gravity-current head. Part 1. Flow topology and front speed. *J. Fluid Mech. 418*, 189–212.

Hedstrom, K. and L. Armi (1988). An experimental study of homogeneous lenses in a stratified rotating fluid. *J. Fluid Mech. 191*, 535–556.

Helfrich, K. R., A. C. Kuo, and L. J. Pratt (1999). Nonlinear rossby adjustment in a channel. *J. Fluid Mech. 390*, 187–222.

Helfrich, K. R. and J. C. Mullarney (2005). Gravity currents from a dam-break in a rotating channel. *J. Fluid Mech. 536*, 253–283.

Hogg, A., M. Ungarish, and H. E. Huppert (2000). Particle-driven gravity currents: asymptotic and box-model solutions. *European J. Mech. B-Fluids 19*, 139–165.

Hogg, A. J., M. A. Hallworth, and H. E. Huppert (1999). Reversing buoyancy of particle-driven gravity currents. *Phys. Fluids 11*, 2891–2900.

Hogg, A. J. and D. Pritchard (2004). The effects of hydraulic resistance on dam-break and other shallow inertial flows. *J. Fluid Mech. 501*, 179–212.

Holford, J. M. (1994). *The evolution of a front*. Ph. D. thesis, University of Cambridge, UK.

Holyer, J. Y. and H. E. Huppert (1980). Gravity currents entering a two-layered fluid. *J. Fluid Mech. 100*, 739–767.

Hoult, D. P. (1986). Oil spreading on the sea. *Annu. Rev. Fluid Mech. 4*, 341–368.

Huppert, H. E. (1982). The propagation of two-dimensional and axisymmetric viscous gravity currents over a rigid horizontal surface. *J. Fluid Mech. 121*, 43–58.

Huppert, H. E. (2006). Gravity currents: a personal perspective. *Journal of Fluid Mechanics 554*, 299–322.

Huppert, H. E. and J. E. Simpson (1980). The slumping of gravity currents. *J. Fluid Mech. 99*, 785–799.

Huppert, H. E. and A. W. Woods (1995). On gravity driven flows in a porous medium. *J. Fluid Mech. 292*, 52–69.

Huq, P. (1996). The role of aspect ratio on entrainment rates of instantaneous, axisymmetric finite volume releases of dense fluid. *J. Hazardous Materials 49*, 89–101.

Kao, T. W. (1976). Principal stage of wake collapse in a stratified fluid: Two-dimensional theory. *Phys. Fluids 19*, 1071–1074.

Kármán, T. v. (1940). The engineer grapples with nonlinear problems. *Bull. Am. Math. Soc. 46*, 615–683.

Keller, J. J. and Y.-P. Chyou (1991). On the hydraulic lock-exchange problem. *ZAMP 42*, 874–910.

Killworth, P. (1992). The time-dependent collapse of a rotating fluid cylinder. *J. Phys. Oceanog. 22*, 390–397.

Klemp, J. B., R. Rotunno, and W. C. Skamarock (1994). On the dynamics of gravity currents in a channel. *J. Fluid Mech. 269*, 169–198.

Kneller, B. and C. Buckee (2000). The structure and fluid mechanics of turbidity currents: a review of some recent studies and their geological implications. *Sedimentology 47*, 62–94.

Kotsovinos, N. E. (2000). Axisymmetric submerged intrusion in stratified fluid. *J. Hydr. Engrg. 126*, 446–456.

Kubokawa, A. and K. Hanawa (1984). A theory of semigeostrophic gravity waves and its application to the intrusion of a density current along a coast. Part 2. Intrusion of a density current along a coast of a rotating fluid. *J. Oceanogr. Soc. Japan 40*, 260–270.

Kunsch, J. P. and D. M. Webber (2000). Simple box model for dense-gas dispersion in a straight sloping channel. *J Haz. Mat. 75 (1)*, 29–46.

Lemckert, C. and J. Imberger (1993). Axisymmetric intrusive gravity currents in stratified reservoirs. *J. Hydraulic Engineering 119(6)*, 662–679.

Leppinen, D. and A. Kay (2006). Gravity currents near the density maximum. *APS/DFD meeting 59.*

Lister, J. R. and R. C. Kerr (1989). The propagation of two-dimensional and axisymmetric viscous gravity currents at a fluid interface. *J. Fluid Mech. 203,* 215–249.

Long, R. R. (1953). Some aspects of the flow of stratified fluids. I. A theoretical investigation. *Tellus 5,* 42–58.

Long, R. R. (1955). Some aspects of the flow of stratified fluids. III. Continuous density gradients. *Tellus 7,* 341–357.

Lowe, R. J., J. W. Rottman, and P. F. Linden (2005). The non-Boussinesq lock exchange problem. Part 1: Theory and experiments. *J. Fluid Mech. 537,* 101–124.

Lyle, S., H. E. Huppert, M. Hallworth, M. Bickle, and A. Chadwick (2005). Axisymmetric gravity currents in a porous medium. *Journal of Fluid Mechanics 543,* 293–302.

Manasseh, R., C.-Y. Ching, and H. J. S. Fernando (1998). The transition from density-driven to wave-dominated isolated flows. *J. Fluid Mech. 361,* 253–274.

Martin, J. R. and G. F. Lane-Serff (2005). Rotating gravity currents. Part 1. Energy loss theory. *Journal of Fluid Mechanics 522,* 35–62.

Martin, J. R., D. A. Smeed, and G. F. Lane-Serff (2005). Rotating gravity currents. Part 2. Potential vorticity theory. *Journal of Fluid Mechanics 522,* 63–89.

Mattheij, R. M. M., S. Rienstra, and J. Boonkkamp (2005). *Partial Differential Equations; Modeling, Analysis, Computation.* SIAM.

Maxworthy, T. (1983a). Experiments on solitary internal Kelvin waves. *J. Fluid Mech. 128,* 365–383.

Maxworthy, T. (1983b). Gravity currents with variable inflow. *J. Fluid Mech. 128,* 247–257.

Maxworthy, T., J. Leilich, J. E. Simpson, and E. H. Meiburg (2002). The propagation of gravity currents in a linearly stratified fluid. *J. Fluid Mech. 453,* 371–394.

Morton, K. W. and D. F. Mayers (1994). *Numerical Solutions of Partial Differential Equations.* Cambridge Un. Press.

Necker, F., C. Härtel, L. Kleiser, and E. Meiburg (2002). High-resolution simulations of particle-driven gravity currents. *Int. J. Multiphase Flow 28,* 279–300.

Necker, F., C. Härtel, L. Kleiser, and E. Meiburg (2005). Mixing and disipation in particle-driven gravity currents. *J. Fluid Mech. 545,* 339–372.

Nof, D. (1987). Penetrating outflows and the dam-break problem. *J. Mar. Res. 45*, 557–577.

Nof, D. and L. M. Simon (1987). Laboratory experiments on the merging of nonlinear anticyclonic eddies. *J. Phys. Oceanog. 17*, 343–357.

Patterson, M. D., J. E. Simpson, S. B. Dalziel, and G. J. F. van Heijst (2006). Vortical motion in the head of an axisymmetric gravity current. *Phys. Fluids 18*, 046601(1–7).

Pedlosky, J. (1997). *Geophysical Fluid Dynamics.* Springer.

Press, W. H., S. A. Teukolski, W. T. Vetterling, and B. P. Flannery (1992). *Numerical Recipes in Fortran.* Cambridge University Press.

Rooij, F. d. (1999). *Sedimenting particle-laden flows in confined geometries.* Ph. D. thesis, DAMTP, University of Cambridge.

Ross, A. N., S. B. Dalziel, and P. F. Linden (2006). Axisymmetric gravity currents on a cone. *J. Fluid Mech. 565*, 227–253.

Rottman, J. and J. Simpson (1983). Gravity currents produced by instantaneous release of a heavy fluid in a rectangular channel. *J. Fluid Mech. 135*, 95–110.

Rubino, A., K. Hessner, and P. Brandt (2002). Decay of stable warm-core eddies in a layered frontal model. *J. Phys. Oceanog. 32*, 188–202.

Saunders, P. M. (1973). The instability of a baroclinic vortex. *J. Phys. Ocean. 3*, 61–65.

Shin, J. O., S. B. Dalziel, and P. F. Linden (2004). Gravity currents produced by lock exchange. *J. Fluid Mech. 521*, 1–34.

Simpson, J. and R. E. Britter (1979). The dynamics of the head of a gravity current advancing over a horizontal surface. *J. Fluid Mech. 94*, 477–495.

Simpson, J. E. (1997). *Gravity Currents in the Environment and the Laboratory.* Cambridge University Press.

Slim, A. C. and H. E. Huppert (2004). Self-similar solutions of the axisymmetric shallow-water equations governing converging inviscid gravity currents. *J. Fluid Mech. 506*, 331–355.

Sparks, R. S. J., R. T. Bonnecaze, H. E. Huppert, J. R. Lister, M. A. Hallworth, H. Mader, and J. C. Phillips (1993). Sediment-laden gravity currents with reversing buoyancy. *Earth Planet. Sci. Lett. 114*, 243–257.

Stansby, P. K., A. Chegini, and T. C. D. Barnes (1998). The initial stages of dam-break flow. *J. Fluid Mech. 374*, 407–424.

Stegner, A., P. Bouruet-Aubertot, and T. Pichon (2004). Nonlinear adjustment of density fronts. Part 1. The Rossby scenario and the experimental reality. *J. Fluid Mech. 502*, 335–360.

Stern, M. E., J. A. Whitehead, and B. L. Hua (1982). The intrusion of a density current along the coast of a rotating fluid. *J. Fluid Mech. 123*, 237–265.

Sutherland, B. R., P. J. Kyba, and M. R. Flynn (2004). Intrusive gravity currents in two-layer fluids. *J. Fluid Mech. 514*, 327–353.

Thomas, L. P., B. M. Marino, and P. F. Linden (1998). Gravity currents over porous substrate. *J. Fluid Mech. 366*, 239–258.

Ungarish, M. (1993). *Hydrodynamics of Suspensions: Fundamentals of Centrifugal and Gravity Separation.* Springer.

Ungarish, M. (1996). A note on the effects of bulk density vs. interstitial fluid density in stability considerations of a suspension overlain by a heavy fluid. *Int. J. Multiphase Flow 22*, 621–625.

Ungarish, M. (2005a). Dam-break release of a gravity current in a stratified ambient. *European J. Mech. B/Fluids 24*, 642–658.

Ungarish, M. (2005b). Intrusive gravity currents in a stratified ambient - shallow-water theory and numerical results. *J. Fluid Mech. 535*, 287–323.

Ungarish, M. (2006). On gravity currents in a linearly stratified ambient: a generalization of Benjamin's steady-state propagation results. *J. Fluid Mech. 548*, 49–68.

Ungarish, M. (2007a). Axisymmetric gravity currents at high Reynolds number - on the quality of shallow-water modeling of experimental observations. *Phys. Fluids 19*, 036602/7.

Ungarish, M. (2007b). A shallow water model for high-Reynolds gravity currents for a wide range of density differences and fractional depths. *J. Fluid Mech. 579*, 373–382.

Ungarish, M. (2008). Energy balances and front speed conditions of two-layer models for gravity currents produced by lock release. *Acta Mechanica 201*, 63–81.

Ungarish, M. and H. E. Huppert (1998). The effects of rotation on axisymmetric particle-driven gravity currents. *J. Fluid Mech. 362*, 17–51.

Ungarish, M. and H. E. Huppert (1999). Simple models of Coriolis-influenced axisymmetric particle-driven gravity currents. *Int. J. Multiphase Flow 25*, 715–737.

Ungarish, M. and H. E. Huppert (2000). High Reynolds number gravity currents over a porous boundary: shallow-water solutions and box-model approximations. *J. Fluid Mech. 418*, 1–23.

Ungarish, M. and H. E. Huppert (2002). On gravity currents propagating at the base of a stratified ambient. *J. Fluid Mech. 458*, 283–301.

Ungarish, M. and H. E. Huppert (2004). On gravity currents propagating at the base of a stratified ambient: effects of geometrical constraints and rotation. *J. Fluid Mech. 521*, 69–104.

Ungarish, M. and H. E. Huppert (2006). Energy balances for propagating gravity currents: homogeneous and stratified ambients. *J. Fluid Mech. 565*, 363–380.

Ungarish, M. and H. E. Huppert (2008). Energy balances for axisymmetric gravity currents in homogeneous and linearly stratified ambients. *J. Fluid Mech. 616*, 303–326.

Ungarish, M. and J. Mang (2003). Spin-up from rest of a two-layer fluid about a vertical axis. *J. Fluid Mech. 474*, 117–145.

Ungarish, M. and T. Zemach (2003). On axisymmetric rotating gravity currents: two-layer shallow-water and numerical solutions. *J. Fluid Mech. 481*, 37–66.

Ungarish, M. and T. Zemach (2005). On the slumping of high Reynolds number gravity currents in two-dimensional and axisymmetric configurations. *European J. Mech. B/Fluids 24*, 71–90.

Ungarish, M. and T. Zemach (2007). On axisymmetric intrusive gravity currents in a stratified ambient - shallow-water theory and numerical results. *European J. Mech. B/Fluids 26*, 220–235.

Verzicco, R., F. Lalli, and E. Campana (1997). Dynamics of baroclinic vortices in a rotating, stratified fluid: A numerical study. *Phys. Fluids 9*, 419–432.

White, B. L. and K. R. Helfrich (2008). Gravity currents and internal waves in a stratified fluid. *J. Fluid Mech. 616*, 327–356.

Winters, K. B., P. N. Lombard, J. J. Riley, and E. A. D'Asaro (1995). Available potential energy and mixing in density-stratified fluids. *J. Fluid Mech. 289*, 115–128.

Woods, A. W. and R. Mason (2000). The dynamics of two-layer gravity-driven flows in permeable rock. *J. Fluid Mech. 421*, 83–114.

Wu, J. (1969). Mixed region collapse with internal wave generation in a density-stratified medium. *J. Fluid Mech. 35*, 531–544.

Zatsepin, A. G. and G. I. Shapiro (1982). A study of axisymmetric intrusions in a stratified fluid. *Izvestia Akademii Nauk SSSR, Fizika Atmosfery i Okeana 18(1)*, 101–105.

Zemach, T. (2007). *Simulation of non-homogeneous flow fields subject to rotation and gravity effects*. Ph. D. thesis, Technion, Haifa, Israel.

Zemach, T. and M. Ungarish (2007). On axisymmetric intrusive gravity currents in a deep fully-linearly stratified ambient: the approach to self-similarity solutions of the shallow-water equations. *Proc. Royal Soc. London 463*, 2165–2183.

# Index

9 780367 385682